NUCLEAR POWER PLANT ENGINEERING

James H. Rust

Professor of Nuclear Engineering
Georgia Institute of Technology

S.W. HOLLAND COMPANY
P.O. BOX 250254
ATLANTA, GA 30325

Nuclear Power Plant Engineering

Printed in the United States of America

Library of Congress Catalog Number 79-88918

ISBN 0-934534-00-4

PREFACE

The majority of nuclear engineering departments in the United States place their greatest emphasis in areas which could be considered an outgrowth from physics. These areas include interactions of nuclear radiation with matter, neutron physics, reactor kinetics, etc. In the course of the approximate twenty year history of graduate level nuclear engineering, we have seen thousands of students obtaining M. S. and Ph.D. degrees with specialties in physics oriented areas of nuclear engineering. As a consequence of this effort we now have a reasonable understanding of the physics of nuclear reactors and a multitude of workers in this field. It is now reasonable to expect that the next major effort in nuclear engineering should be in technological applications of nuclear energy. This will require educating students to a high level of understanding of fundamental principles of engineering analysis. Nuclear engineers with a strong capability in both classic engineering and reactor physics will produce the technological improvements necessary to insure a continuation in the growth of the use of nuclear energy.

This book evolved from the author's lecture notes on nuclear power design taught the past twelve years in the graduate nuclear engineering departments of The University of Virginia and the Georgia Institute of Technology. Because graduate level nuclear engineering has not evolved from a well established undergraduate discipline, the undergraduate backgrounds of students taking nuclear power plant design courses range from engineering to chemistry, physics, and mathematics. Naturally this creates great problems in course organization because many students have no prior exposure to traditional engineering subjects such as thermodynamics, strength of materials, fluid flow, and heat transfer.

The purpose of this text is to provide basic insight into some of the aspects of engineering analysis used in the design of nuclear reactor systems. Since nuclear reactor system design spans such a wide range of disciplines, it is not considered practical to treat in detail or at all many of the facets of nuclear reactor system analysis. Hopefully this text will stimulate the reader to continue studies in additional engineering specialties of interest.

The subject matter of this text is considered suitable for a two or three quarter course on nuclear reactor system analysis for seniors or first-year graduate students in nuclear or mechanical engineering. It is assumed that students have some training in reactor physics and interactions of nuclear radiation with matter. With this training students should be able to identify internal heat sources in reactor cores and shields due to fission, gamma attenuation, and neutron slowing down. In addition, it is assumed students have the standard undergraduate training in mathematics through differential equations. For non-engineering students, it is strongly recommended that they take one or more courses in undergraduate engineering thermodynamics before using this text. This will help ease the pain in adjusting to the English engineering system of units and provide some feeling for conservation principles and order of magnitudes of quantities.

The development of the subject matter in this text is as follows: Chapter 1 contains a brief description of the various types of nuclear power plants in use or under serious consideration in the United States. Chapter 2 deals with thermodynamics of nuclear power plants. It is the author's opinion that the application of thermodynamics is the most logical starting point for developing some of the theory used in engineering analysis. Chapter 3 is a rapid review of some of the basic concepts of fluid flow which may be omitted when teaching courses to students with an adequate background in fluid mechanics. Chapter 4 covers fluid flow in nuclear reactor systems which is of great importance in dealing with the subject of heat transfer covered in a later chapter. Chapter 5 discusses heat generation in nuclear reactors. Chapter 6 covers single-phase heat transfer to ordinary fluids and liquid metals, boiling heat transfer, and heat conduction in solid fuels. Chapter 7 deals with core thermal-hydraulic design including the use of hot-channel factors and subchannel analysis. Chapter 8 covers basic principles of stress analysis and applications to nuclear reactor analysis including use of the ASME Boiler Code. The final chapter, Chapter 9, deals with the analysis of fluid transients and their application to nuclear power plants.

A final remark with regard to the homework problems at the end of each chapter is necessary for both students and instructors. Engineering can be defined as solving problems of importance to the welfare of society. As a consequence, a good engineer is a problem solver and this requires using the skills acquired throughout a lifetime, employing sensible and mature judgment, and obtaining the best solution to a problem with the means available. In order to train good engineers it is necessary for them to attack numerous problems which require careful thought and good judgment. Therefore, instructors using this text are urged to make many homework assignments and, if practical, to adjust the difficulty of problems to the level of understanding of the students. It is further recommended that instructors require the same level of formality to the solutions handed in as suggested by Professor W. C. Kays of Stanford University. Each homework problem should be treated as a formal engineering assignment and the following steps incorporated, when applicable, in presenting solutions handed in:

1. Concisely state the objective of the problem.
2. Sketch a diagram of the system being analyzed if possible.
3. Give a qualitative estimate of the expected solution to the problem. This may be a diagram of a temperature distribution, etc.
4. Perform the analysis of the problem using fundamental principles. Concisely state the assumptions involved with any equations used. Provide enough verbal description of the analysis so that each step is clearly understood.
5. Finish the analysis with discussion of the results and specify conclusions from the analysis when possible.

A text of this nature usually contains little of the author's work and in a sense represents his best estimate of the items of importance to the subject being treated. Consequently, it is most difficult to properly acknowledge contributors to this text and the author would like to extend his thanks to the many engineers and scientists whose names appear in the bibliographies at the end of each chapter. Thanks are also due the reactor venders who were most cooperative in providing technical information and descriptive material on nuclear power plants.

James H. Rust

Atlanta, Georgia

CONTENTS

Chapter 1

DESCRIPTION OF NUCLEAR REACTORS

1.1 Introduction

In order to place in perspective design approaches discussed in later chapters, it is necessary to have some understanding of plant layouts and reactor configurations for nuclear power plants. In the United States four types of nuclear power plants are expected to play the major role in nuclear power generation—**pressurized water reactors** (PWRs), **boiling water reactors** (BWRs), **high temperature gas-cooled reactors** (HTGRs), and **liquid metal-cooled fast breeder reactors** (LMFBRs). Although no plans have been made for construction as of this writing, **high temperature gas-cooled fast breeder reactors** (HTGFBRs) may also play an important role in future United States nuclear power programs.

Other types of nuclear power plants have been considered in the United States and the rest of the world. Some examples are **sodium-cooled graphite-moderated reactors** (SGRs), **organic-cooled and moderated reactors** (OMRs), **molten salt breeder reactors** (MSBRs), **heavy water-cooled and moderated reactors** (HWRs), and **light-water breeder reactors** (LWBRs). References [1 - 6] contain descriptive material about these power plant concepts.

This chapter deals with only the first four mentioned nuclear power plants. All of these power plants are highly complex systems and it is impossible for a book of this size to treat any of them in great detail. Consequently, no discussion is given of control systems, emergency cooling systems, containment design, waste disposal systems, makeup and purification systems, etc. For more complete descriptions of nuclear power plants, interested readers are urged to consult the source references or individual power plant Preliminary Safety Analysis Reports (PSARs) or Final Safety Analysis Reports (FSARs).

1.2 Pressurized Water Reactors

Babcock and Wilcox, Combustion Engineering, and Westinghouse Electric are the three vendors supplying pressurized water nuclear steam supply systems in the United States. All three systems are somewhat similar with a notable exception that the Babcock and Wilcox system uses once-through steam generators (OTSGs) for their boilers while the other vendors employ recirculating boiler (vertical U-tube) steam generators.

Figure 1.1 illustrates a simplified diagram of a PWR system for electric power generation. The left-hand portion is the nuclear steam supply system (NSSS) which consists of the reactor vessel, pressurizer, and two or more primary coolant loops with associated reactor coolant pumps and steam generators. The NSSS is totally enclosed within a containment building to prevent radiation leakage to the environment in case of accidents. The secondary system is external to the containment building and consists of the turbine-generator, condenser, feedwater heaters, feed pump, and piping.

Reactor Vessel

Figure 1.2 illustrates an elevation view of a PWR reactor vessel and internals. A cross section view of the reactor vessel and internals is shown in Fig. 1.3. High pressure water at an approximate pressure of 2250 psia enters four inlet nozzles located above the midpoint of the pressure vessel as shown in Fig. 1.2. The water flows downward through the annular space enclosed by the reactor vessel and thermal shield and then upward through the fuel assemblies. Water exits from the fuel assemblies into the upper plenum chamber and then down around the inside of the upper pressure vessel and out the two outlet nozzles as shown in Fig. 1.2.

The core is supported by the core support shield and core barrel which hang from the seal ledge near the closure head. The core barrel provides some gamma attenuation to augment shielding from the thermal shield. This shielding is used to reduce gamma heating in the reactor vessel wall. The reactor vessels for the three venders are quite similar and Table 1.1 gives design parameters for the reactor vessel shown in Fig. 1.2.

Core

A cross section of the reactor core is shown in Fig. 1.3. For the high power reactor being designed the irregular shape of the core has a cross section area equivalent to that of a circle of an approximate diameter of 12 ft. The core consists of 200 or more fuel assemblies with a square cross section approximately 8 1/2 in. on a side.

Figure 1.4 illustrates a typical PWR fuel assembly. Fuel rods are arranged vertically in a 16 x 16 or 17 x 17 square array. Approximately one-third of the fuel assemblies contain control rods which are incorporated into those fuel assemblies by means of control rod guide tubes replacing fuel rods. Figure 1.4 shows a fuel assembly containing

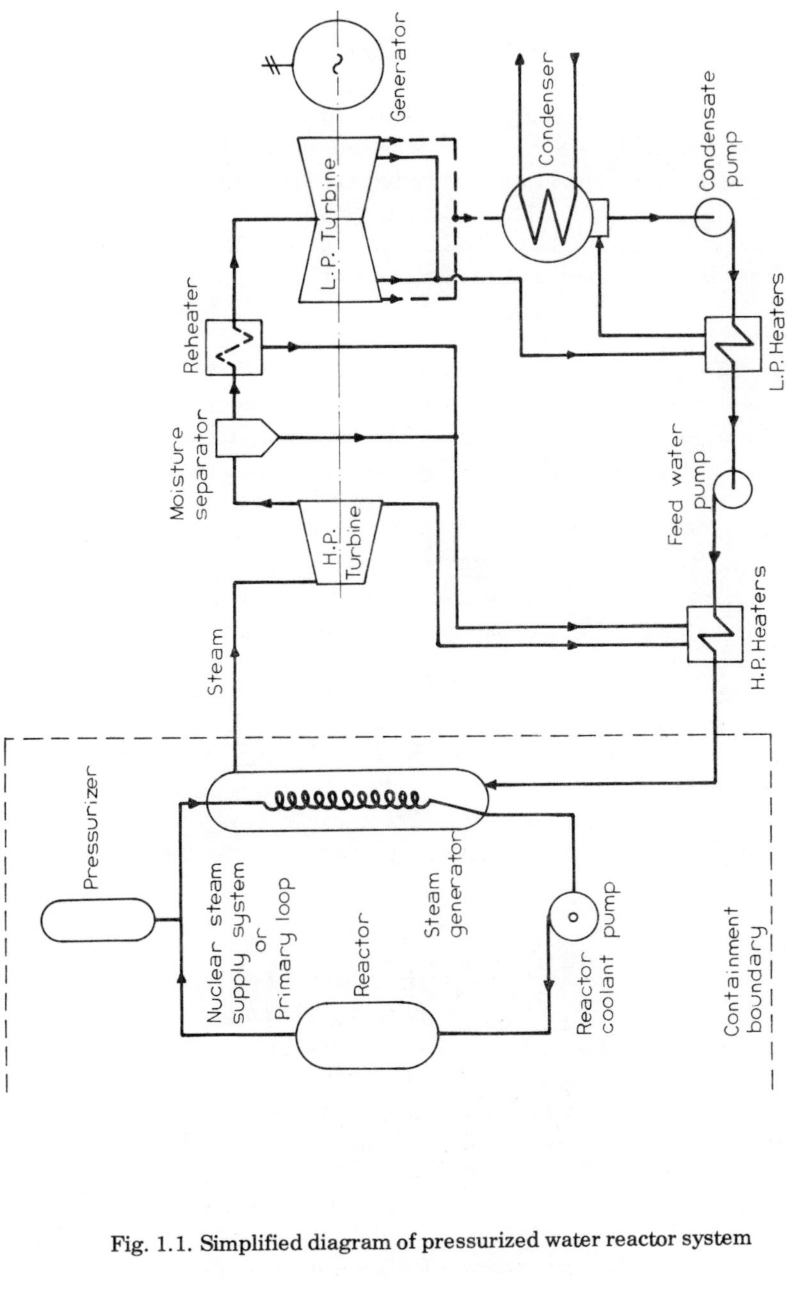

Fig. 1.1. Simplified diagram of pressurized water reactor system

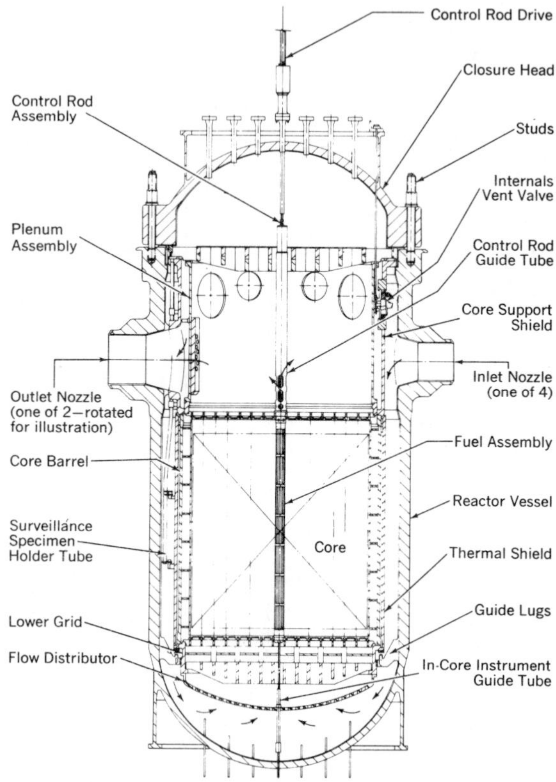

Fig. 1.2. Elevation view of PWR reactor vessel and internals [7]

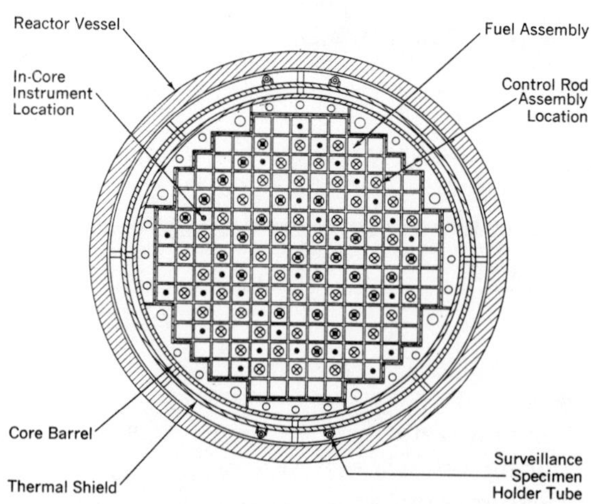

Fig. 1.3. Cross section of PWR reactor vessel [7]

control rods. Other rods within fuel assemblies contain no fuel so as to allow space for burnable poisons and in-core instrumentation.

Table 1.1

Design parameters for PWR reactor vessel [8]

Base Material	Low-Alloy Steel
Cladding Material	Stainless Steel and Inconel
Design Pressure	2500 psia
Design Temperature	670°F
Operating Pressure	2250 psia
Hydrostatic Test Pressure	3125 psia
O. D. Including Nozzles	275 in.
I. D.	182 in.
Overall Height	43 ft.
Vessel Wall Thickness	9.125 in.
Inlet Nozzle, I. D.	28 in.
Outlet Nozzle, I. D.	38 in.
Core Flood Nozzle, I. D.	2.765 in.
Water Volume	4730 ft^3

The fuel consists of UO_2 pellets with enrichment ranging from 2.1 to 3.1 percent U^{235}. The pellets are approximately 0.32 in. in diameter and about 0.6 in. in length. The pellets are contained in a Zircaloy-4 tube with a wall thickness of 0.025 in. Zircaloy is used as a fuel cladding because of its low neutron absorption cross section. A fuel rod is also shown in Fig. 1.4. From the fabrication process the UO_2 pellets have an average radial clearance from the cladding slightly greater than 0.003 in.

Table 1.2 lists geometry and thermal-hydraulic characteristics for a PWR core.

Pressurizer

The pressurizer establishes and maintains reactor coolant system pressure during steady state operation and transients. It also functions as a surge chamber and water reserve to accomodate primary coolant volume changes during operation. The typical pressurizer, shown in Fig. 1.5, is a vertical cylindrical pressure vessel with a bottom surge line inlet connected to the reactor coolant piping. Design data for a

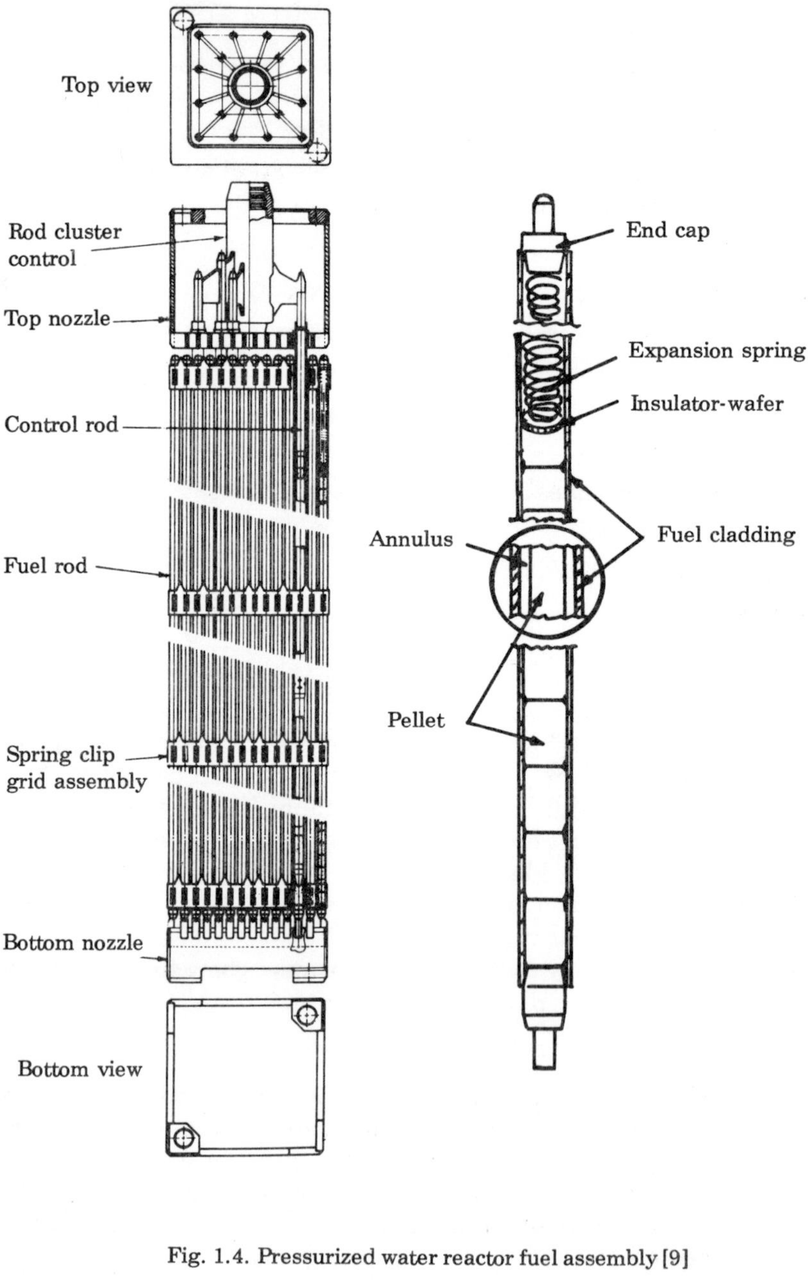

Fig. 1.4. Pressurized water reactor fuel assembly [9]

Table 1.2

PWR core geometry and thermal-hydraulic characteristics [10]

Thermal Output	3800 MW(t)
Active Height	150 in.
Equivalent Active Diameter	142.9 in.
Height/Diameter	1.05
Active Core Volume	1413 ft^3
Average Core Power Density	2690 kW/ft^3
Fuel Weight	103,000 kg
Specific Power	36.9 kW/kgU
Burnup	33,000 MWD/MTU
Conversion Ratio	0.5
Number of Fuel Assemblies	241
Fuel Element Array	16 x 16
Assembly Dimensions	8 in. x 8 in.
Assembly Pitch	8.15 in.
Number of Fuel Rods/Assembly	236
Total Number of Fuel Rods	56,876
Fuel Element Pitch	0.504 in.
Fuel Element O. D.	0.382 in.
Pitch/Diameter	1.33
Clad Thickness	0.025 in.
Fuel Pellet Diameter	0.325 in.
Pellet-Clad Gap	0.0035 in.
Fuel Enrichment	1.9/2.4/2.9
System Pressure	2,250 psia
Coolant Flow	1.59×10^8 lb/hr
Average Linear Heat Rate	5.34 kW/ft
Maximum Linear Heat Rate	12.51 kW/ft
Average Heat Flux	206,000 Btu/hr-ft^2
Maximum Heat Flux	550,000 Btu/hr-ft^2
Minimum DNBR	1.3
Inlet Temperature	565°F
Outlet Temperature	622.4°F
Maximum Fuel Temperature	3420°F

typical pressurizer is listed in Table 1.3.

The pressurizer contains removable electric heaters in its bottom section which is normally submerged in water. The upper section contains a water spray nozzle in a region which is normally filled with steam. When pressure in the primary system decreases, some of the water in the pressurizer flashes into steam to maintain pressure. The electric heaters are then activated until required system pressure is

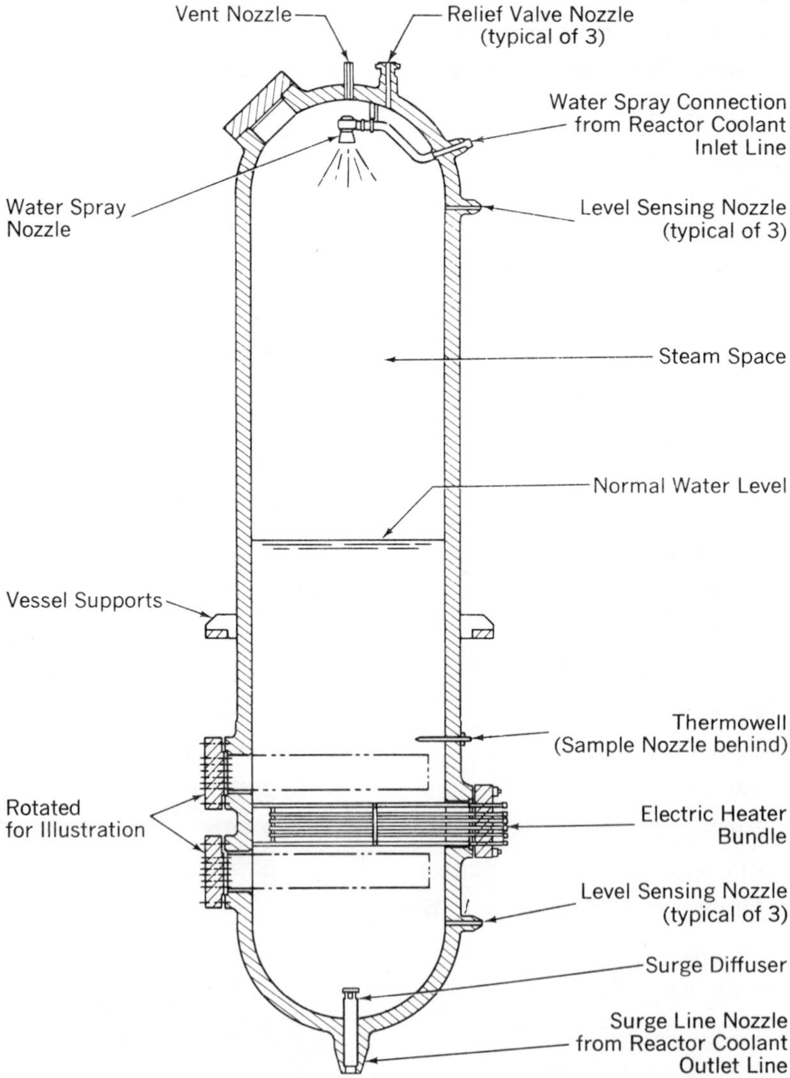

Fig. 1.5. PWR pressurizer [7]

re-established. Conversely when system pressure increases, steam is condensed by the water spray from the reactor coolant inlet line to reduce pressure back to the normal operating value.

Water level is maintained in the pressurizer by varying the amount of water charged into and removed from the primary coolant system. An additional function of the pressurizer is to provide relief from sudden overpressurization. Safety valves automatically open at a preset overpressure. In order to reduce the frequency of operation of the safety valves, an electromagnetically controlled relief valve is also provided.

Table 1.3

PWR pressurizer design parameters [11]

Overall Height	52.75 ft
Overall Diameter	7.75 ft
Normal Operating Water Volume	1080 ft^3
Normal Operating Steam Volume	720 ft^3
Design Temperature	650°F
Operating Temperature	618.5° F
Design Pressure	2500 psia
Operating Pressure	2335 psia
Installed Heater Power	1800 kw
Number of Safety Valves	6
Safety Valve Size	3 in.
Number of Relief Valves	2
Relief Valve Size	2.5 in.
Spray Rate	
Maximum	800 gpm
Continuous	1 gpm
Shell Material	Mn-Mo
Cladding	Stainless Steel
Dry Weight	235,000 lb
Flooded Weight	347,320 lb

Steam Generators

Steam generators transfer heat from the primary coolant to secondary feedwater in order to generate steam for driving the turbine-generator. PWR steam generators are normally shell-and-tube heat exchangers with high pressure primary water passing through the tube side and lower pressure secondary feedwater and steam passing through the shell side. In addition to providing steam, steam generators also function as a barrier between the radioactive primary coolant and the secondary coolant.

PWR nuclear steam supply systems incorporate two or more steam generators in order to reduce equipment size and increase plant safety through redundant systems. Babcock and Wilcox and Combustion Engineering PWR systems use two steam generators while Westinghouse Electric plants incorporate two, three, or four steam generators depending upon plant size.

Both Westinghouse Electric and Combustion Engineering plants use a vertical U-tube steam generator similar to that shown in Fig. 1.6. Hot leg primary coolant enters the steam generator inlet nozzle and flows upward through the tubes to the top of the U-tube bundle and then downward to exit from the outlet nozzle. Feedwater enters the side of the steam generator and flows upward through an economizer region. The feedwater continues upward as it gains energy until it flows past the tube bundle where a water level is maintained above the tubes and below the moisture separators. Wet saturated steam passes through swirl vane moisture separators where the majority of entrained water is removed. Entrained water from the moisture separators flows downward through a downcomer and mixes with incoming feedwater. Steam leaving the swirl vane moisture separators enters steam dryers where the moisture content is reduced to less than 0.25 percent. Design data for a vertical U-tube steam generator is given in Table 1.4.

Babcock & Wilcox employs a once-through steam generator which is shown in Fig. 1.7. Hot leg primary coolant enters the top of the steam generator through a single inlet nozzle and flows downward inside the tubes, discharging through two outlet nozzles. Feedwater enters the side of the steam generator and flows downward along the shell until it reaches the tubesheet. Flow is then upward countercurrent to the primary coolant. As the feedwater flows upward through the shell side it is converted to steam and then superheated 50°F. At the top of the tube bundle the superheated steam flow is outward toward the shell and then downward in the annular region formed by the shell and upper baffle to the outlet nozzles. Table 1.5 lists design data for a once-through steam generator.

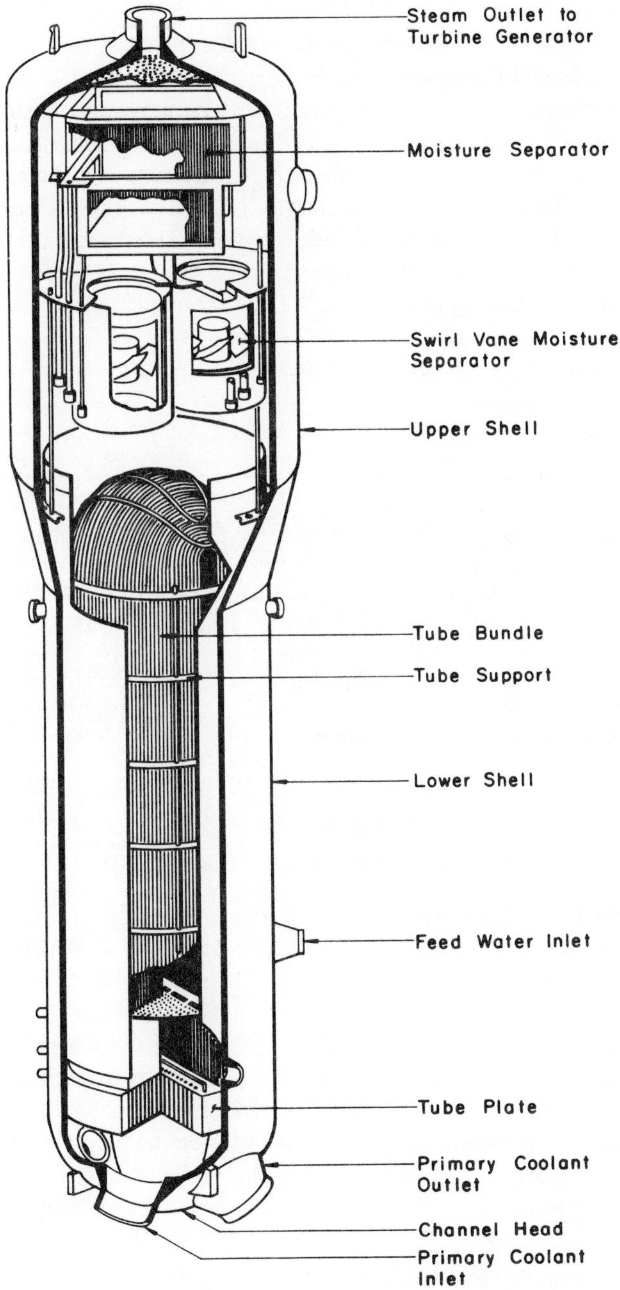

Fig. 1.6. Vertical U-tube steam generator [9]

Table 1.4

Vertical U-tube steam generator design data [10]

Weights and Dimensions	
Dry Weight, lb	1,428,900
Flooded Weight, lb	2,220,000
Normal Operating Weight, lb	1,725,000
Overall Height , ft	68
Upper Shell, O. D. , ft	22.2
Lower Shell, O. D., ft	15.8
Functional Performance	
Primary Side (or tube side)	
Design Pressure, psia	2,500
Design Temperature, °F	650
Operating Pressure, psia	2,250
Inlet/Outlet Temperature, °F	621.2/564.5
Secondary Side (or shell side)	
Design Pressure, psia	1,270
Design Temperature, °F	575
Feedwater Inlet Temperature, °F	450
Steam Outlet Pressure, psia	1,070
Steam Outlet Temperature, °F	552.9
Steam Quality, %	99.75
Steam Flow (per Generator), lb/hr	8.999×10^6

Table 1.5

Once-through steam generator design data [8]

Weights and Dimensions	
Dry weight, lb	1,080,000
Overall Height, ft	75.5
Shell I. D., in.	139
Number of Tubes	16,000
Tube Size, O. D., in.	0.625 x 0.034
Effective Tube Length, ft.	52
Functional Performance	
Steam Flow per Steam Generator, lb/hr	7.757×10^6
Steam Temperature, SH, °F	601.6/50
Steam Pressure, psia	1060
Feedwater Temperature, °F	465
Primary Flow per Steam Generator, lb/hr	72.25×10^6
Primary Temperature (inlet/outlet), °F	625/569
Heat Transferred per Steam Generator, (Btu/hr)	6.174×10^9

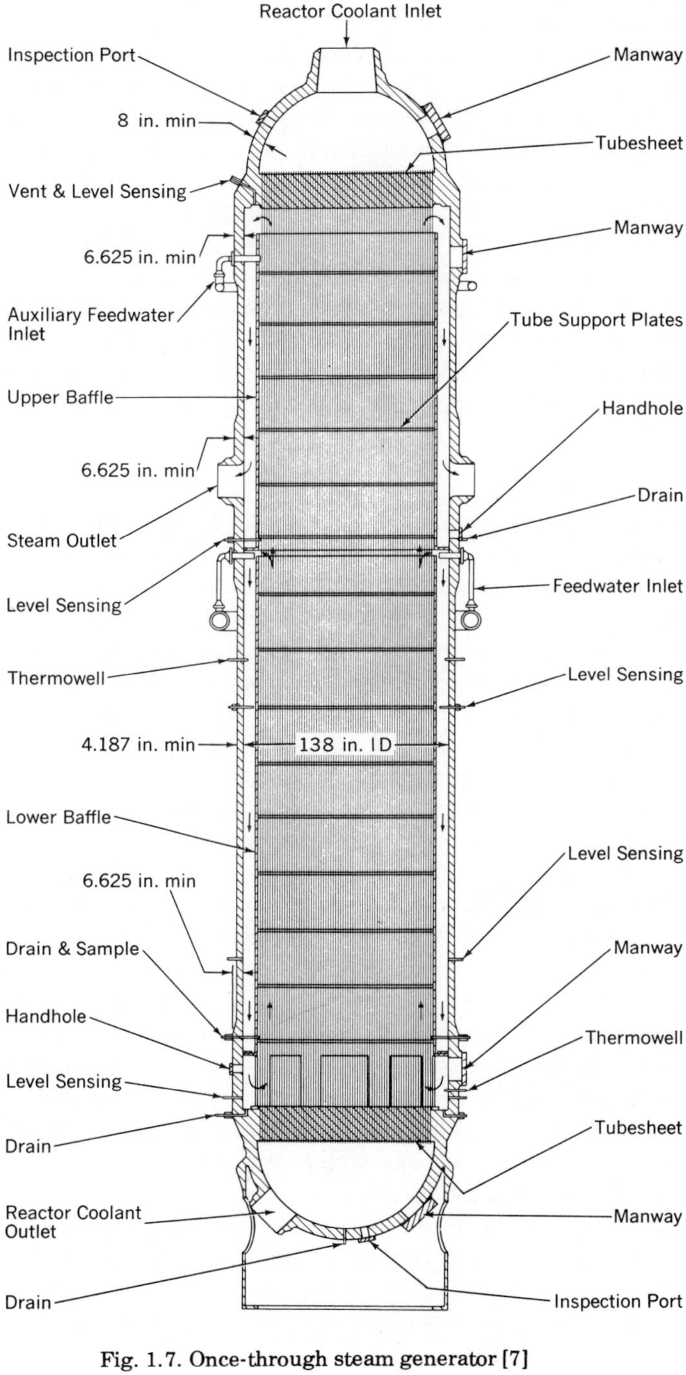

Fig. 1.7. Once-through steam generator [7]

1.3 Boiling Water Reactors

Boiling water reactors operate with a reactor vessel pressure of 1040 psia which is considerably lower than for PWRs. Their nuclear steam supply systems differ from PWRs in that steam is produced within the core and fed directly to the turbine-generator. Consequently, this system uses a direct cycle as opposed to the indirect cycles used with other nuclear power plants. Figure 1.8 illustrates a simplified diagram of a BWR system for electric power generation.

An advantage of BWRs is the direct cycle eliminates the need for steam generators. An additional feature is that most of the energy transfer in the core is by latent heat as opposed to sensible heat in other nuclear power plants. This allows smaller flow rates. However these advantages are offset by radioactive contamination of the steam which requires additional shielding for the piping, turbine, and feedwater heaters.

Reactor Vessel

Figure 1.9 shows the reactor vessel and internals for a BWR/6, the sixth evolutionary step in BWR design. A schematic diagram of the coolant recirculation system is illustrated in Fig. 1.10. Table 1.6 gives design parameters for the boiling water reactor vessel.

The nuclear fuel assemblies are arranged inside a core shroud contained within the reactor vessel. Subcooled waters enter the bottom of the core and flows upward through vertical fuel assemblies. A two-phase mixture of about 14 percent quality exits from the fuel assemblies and enters an upper plenum region with a dome-shaped top. Attached to the dome is an array of standpipes with a three-stage steam separator at the top of each standpipe.

An internal steam separator is shown in Fig. 1.11. This steam separator is an axial-flow type which has no moving parts. The two-phase mixture rising from the standpipe impinges on the turning vanes which induces a vortex motion to the flow. Centrifugal force separates water from the mixture in each of the three stages. The separated water exits from the bottom of each stage to mix with water in a pool surrounding the standpipes. The wet steam from the last stage enters a plenum region below the steam dryers. This steam enters a steam dryer consisting of chevron-type drying vanes. The steam dryers remove more moisture so that steam with a quality of at least 99.6 percent leaves the reactor vessel.

The water in the pool surrounding the standpipes mixes with feedwater by means of the feedwater sparger. This water flows to the downcomer annulus where flow is forced to the lower plenum region below the fuel assemblies by means of 20 jet pumps mounted internally within the reactor vessel.

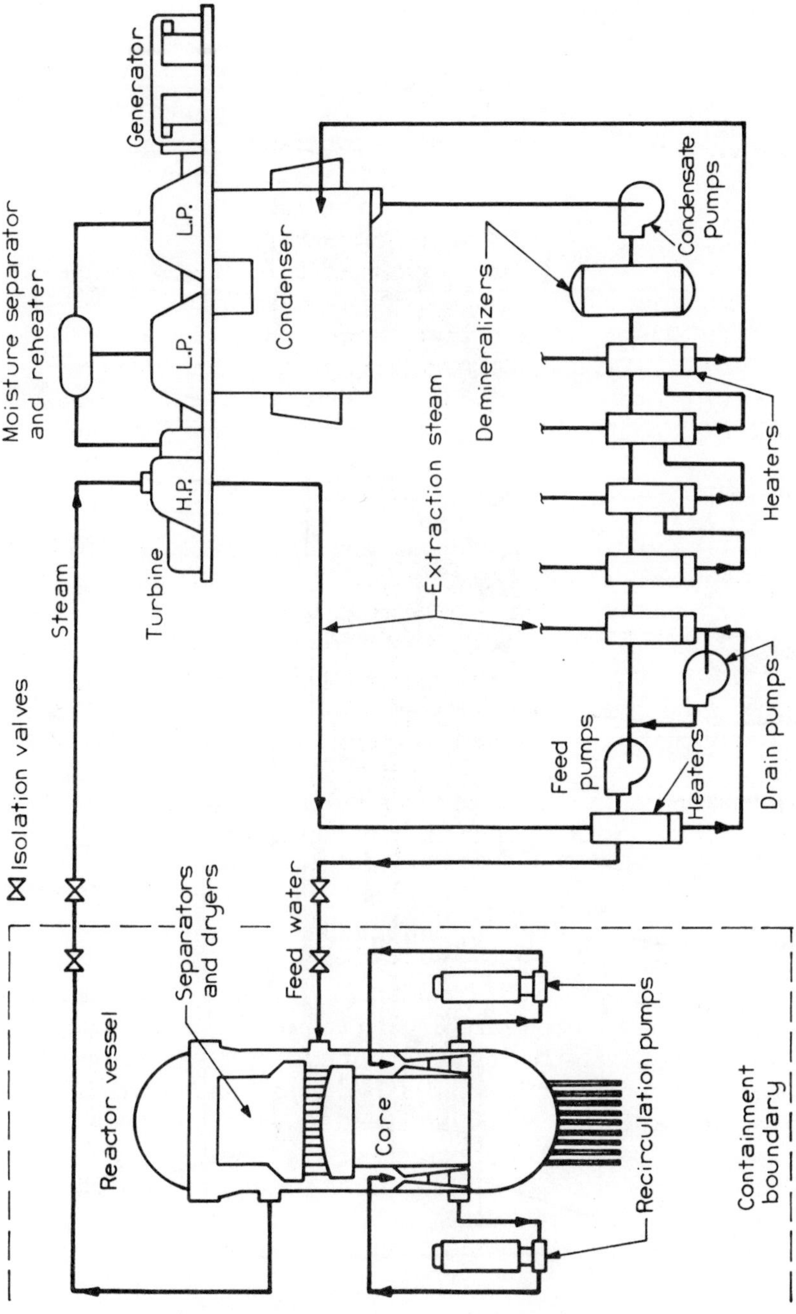

Fig. 1.8. Direct-cycle reactor system (BWR).

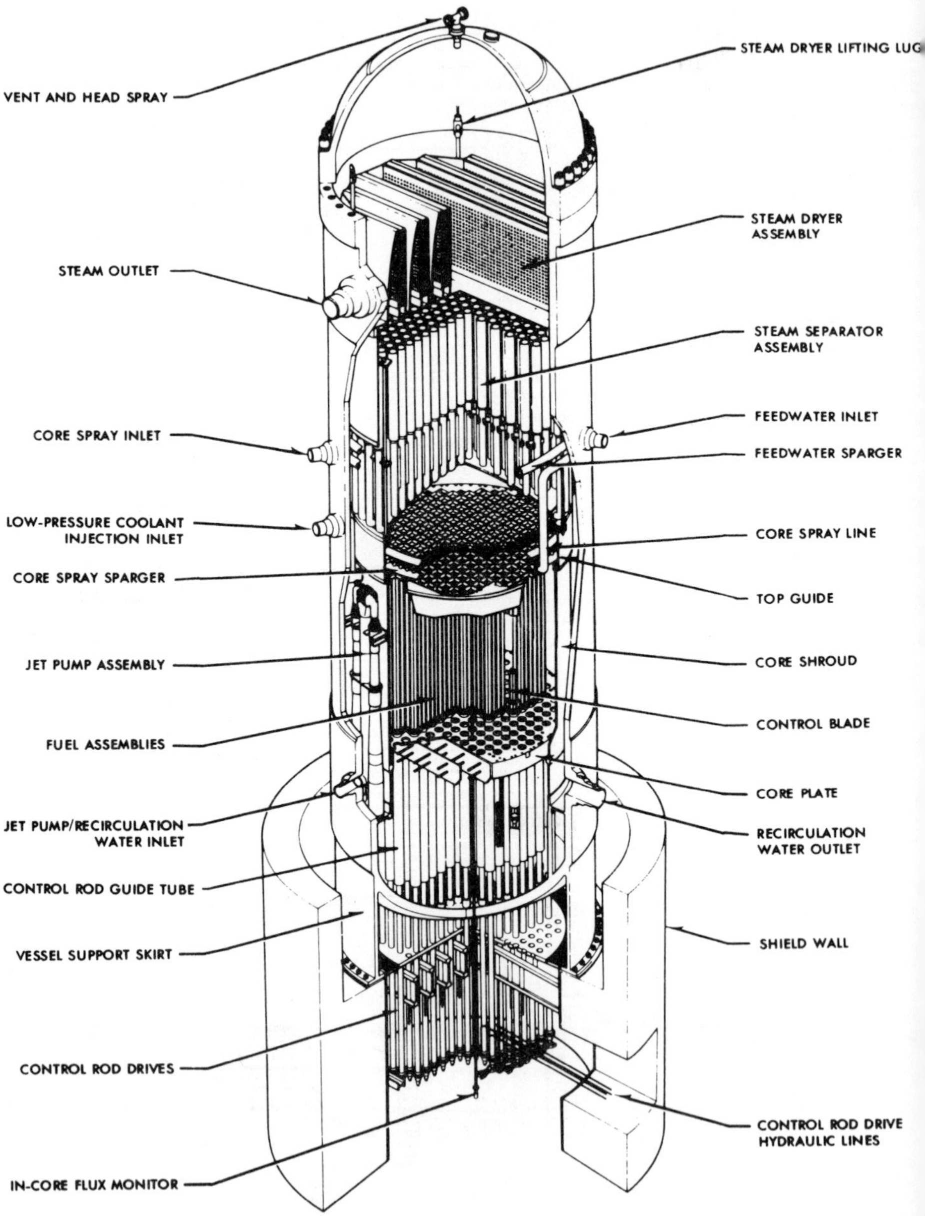

Fig. 1.9. Reactor assembly (BWR/6) [12]

Table 1.6

Design parameters for BWR reactor vessel [1,13]

Base Material	Low Alloy Steel
Cladding Material	Stainless Steel
Design Pressure	1250 psig
Design Temperature	575°F
Operating Pressure	1040 psia
I. D.	238 in.
Wall Thickness, min./max.	5.7 in./6.46 in.
Height	71 ft.
Number Recirculation Loops	2
Number of Jet Pumps	20
Recirculation Line Pipe Sizes, I. D.	24 in.
Recirculation Pump Flow	35,4000 GPM
Number of Steam Separators	263
Steam Separator Pressure Drop	7.3 psi
Flow per Separator	3.99×10^5 lb/hr
Vessel Exit Steam Quality	99.6%
Vessel Weight	975 tons

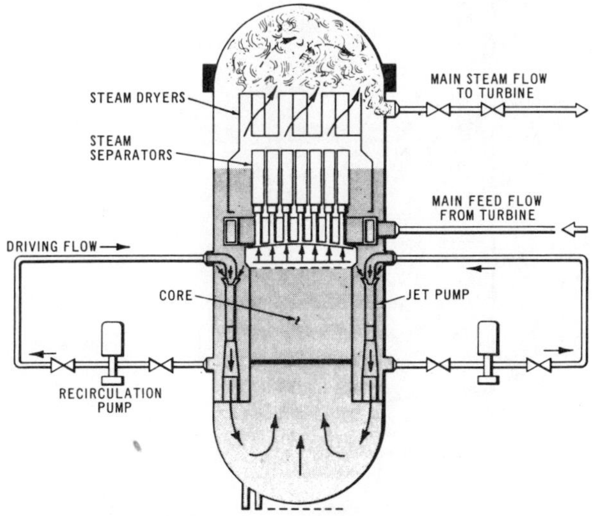

Fig. 1.10. BWR jet pump system. [Courtesy General Electric Co.]

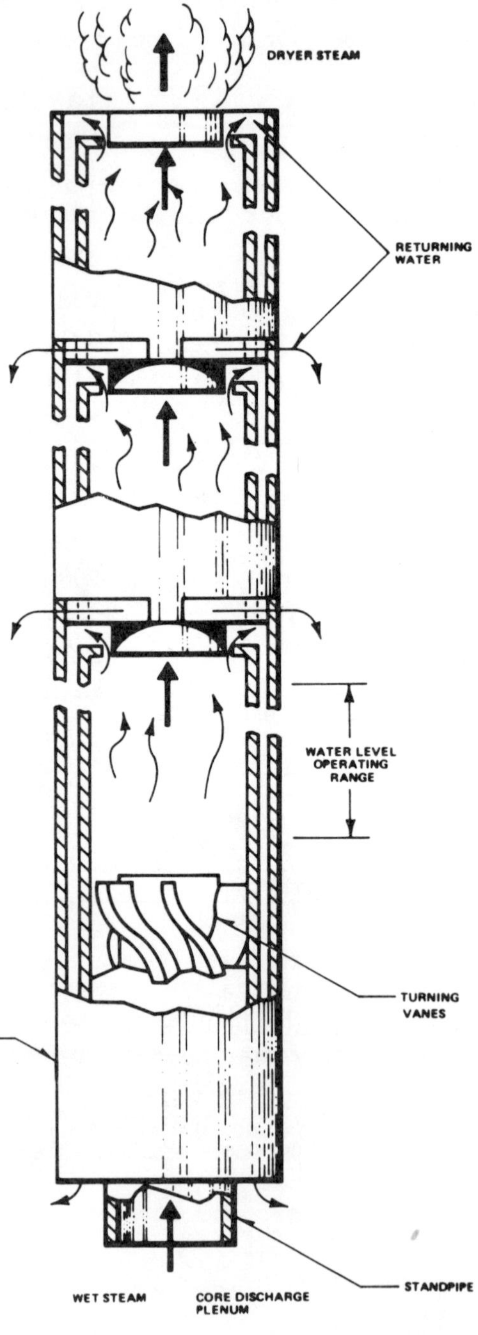

Fig. 1.11. Internal steam separator [12]

The coolant recirculation system consists of two loops external to the reactor vessel. As shown in Figs. 1.9 and 1.10, water is taken from the bottom of the annular downcomer region formed by the core shroud and reactor vessel and enters the suction of the recirculation pumps. Approximately one third of the core flow enters these pumps. The recirculation water pressure is increased by the pumps and is then distributed to a manifold from which connecting pipes are welded to the recirculation water inlet nozzles of the reactor vessel. One inlet nozzle serves two jet pumps.

The jet pumps have no moving parts and provide a continuous internal circulation for a major portion of the core flow. Flow from the recirculation loop enters the jet pump nozzle and is discharged into the throat region. This creates a suction in the throat which draws surrounding water from the downcomer region into the throat where the two flows mix for passage through the jet pump diffuser region. The jet pump discharges into the lower plenum region where flow is then diverted to the fuel assemblies.

Core

A cross section of the reactor core is shown in Fig. 1.12. The irregular shape of the core has a cross sectional area equivalent to that of a circle of 12 ft. diameter. The core consists of 700 or more fuel assemblies with a square cross section 5.52 in. on a side. The core lattice configuration is shown in Fig. 1.13 in which a group of four fuel assemblies are arranged around a cruciform control rod. The core consists of a repetition of this array.

Figure 1.14 illustrates a BWR fuel assembly. Fuel rods are arranged vertically in an 8 x 8 square array. The fuel assembly contains 62 active fuel rods and two hollow "water" rods which are supported by upper and lower tie plates. BWR fuel assemblies differ from those in PWRs because each fuel assembly is enclosed by a Zircaloy-4 channel. The presence of the channel prevents flow between adjacent fuel assemblies. Consequently, the flow to individual fuel assemblies can be orificed so as to maintain an approximately uniform exit quality from all fuel assemblies.

The fuel consists of UO_2 pellets with enrichment varying from 2.2 to 2.7 percent U^{235}. The pellets are 0.416 in. in diameter and 0.41 in. in length. The pellets are contained in a Zircaloy-2 tube with a wall thickness of 0.034 in. The fuel pellets have an average radial clearance from the cladding of 0.0045 inch.

Table 1.7 lists core geometry and thermal-hydraulic characteristics for a BWR/6.

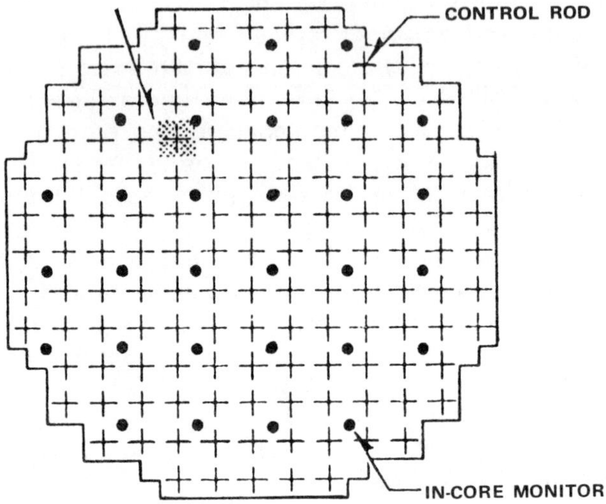

Fig. 1.12. Typical core arrangement (Courtesy General Electric Company)

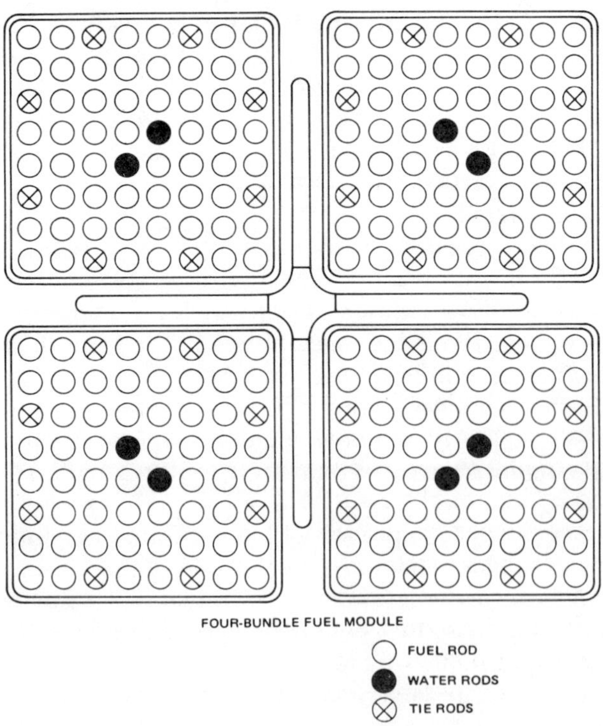

Fig. 1.13. Core lattice (Courtesy General Electric Company)

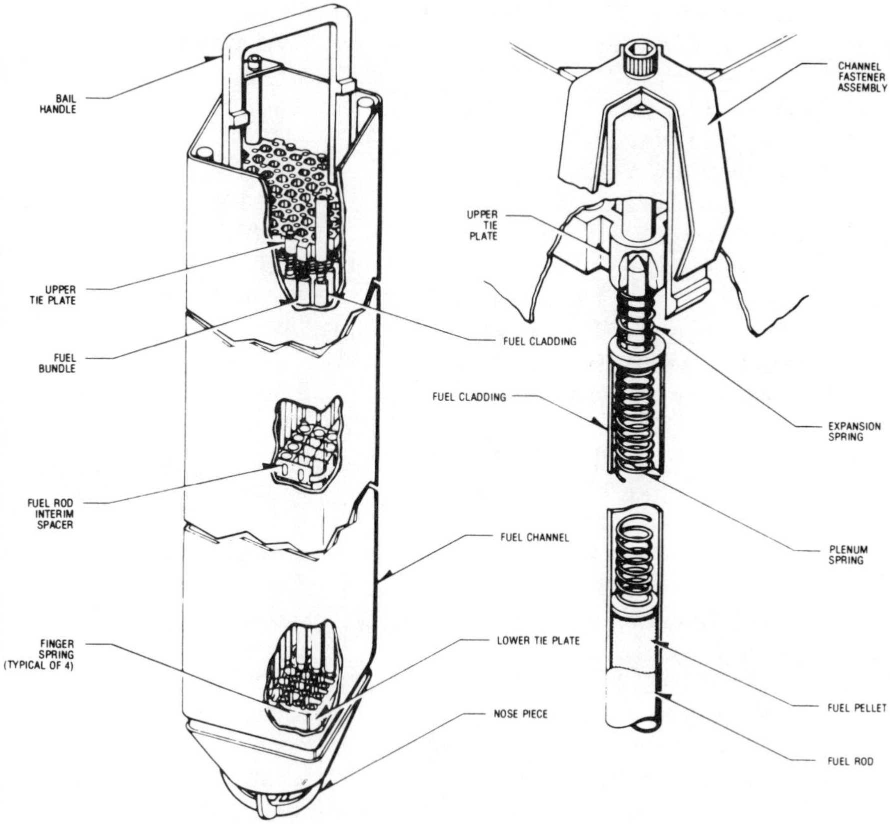

Fig. 1.14. BWR fuel assembly [12]

1.4 High Temperature Gas-Cooled Reactors

An alternative to light water-cooled and moderated reactors is the high temperature gas-cooled reactor (HTGR) which uses graphite as a moderator and helium as the coolant. Figure 1.15 illustrates an HTGR used for electric power generation. The helium coolant enters the reactor at 636°F and exits at 1377°F. Because of these high temperatures it is possible to employ conventional superheat/reheat cycles found in high performance fossil-fueled power plants. The overall plant efficiency for HTGRs is 39 percent which compares quite favorably with the 33 percent efficiency obtained with LWR power plants.

Table 1.7

BWR core geometry and thermal-hydraulic characteristics-BWR/6 [1,5,13]

Thermal Output	3579 MW(t)
Active Height	148 in.
Equivalent Active Diameter	144 in.
Height/Diameter	1.03
Active Core Volume	2260 ft^3
Average Core Power Density	1580 kW/ft^3
Fuel Weight	138,000 kg
Specific Power	25.9 kW/kgU
Burnup	27,500 MWD/MTU
Number of Fuel Assemblies	732
Fuel Element Array	8 x 8
Assembly Dimensions	5.52 in. x 5.52 in.
Assembly Pitch	6.0 in.
Number of Fuel Rods/Assembly	62
Total Number of Fuel Rods	45,384
Fuel Rod O. D.	0.493 in.
Fuel Rod Pitch	0.64 in.
Pitch/Diameter	1.3
Clad Thickness	0.034 in.
Fuel Pellet Diameter	0.416 in.
Pellet-Clad Gap	0.0045 in.
Fuel Enrichment	2.2-2.7%
Pellet Density % Theoretical	~94%
Fission Gas Plenum Length	12 in.
System Pressure	1040 psia
Coolant Flow	1.05 x 10^8 lb/hr
Core Inlet Enthalpy	527.9 Btu/lb
Average Exit Quality	14.6%
Core Average Void Fraction	42.6%
Maximum Exit Void Fraction	76%
Average Inlet Velocity	7.1 ft/sec
Core Pressure Drop	25.9 psi
Average Linear Heat Rate	6.0 kW/ft
Maximum Linear Heat Rate	13.4 kW/ft
Average Heat Flux	159,000 Btu/hr-ft^2
Maximum Heat Flux	354,000 Btu/hr-ft^2
Minimum CHFR	$>$ 1.9
Inlet Temperature (Feedwater)	420°F
Core Inlet Temperature	532°F
Outlet Temperature	547°F
Maximum Fuel Temperature	~3500°F

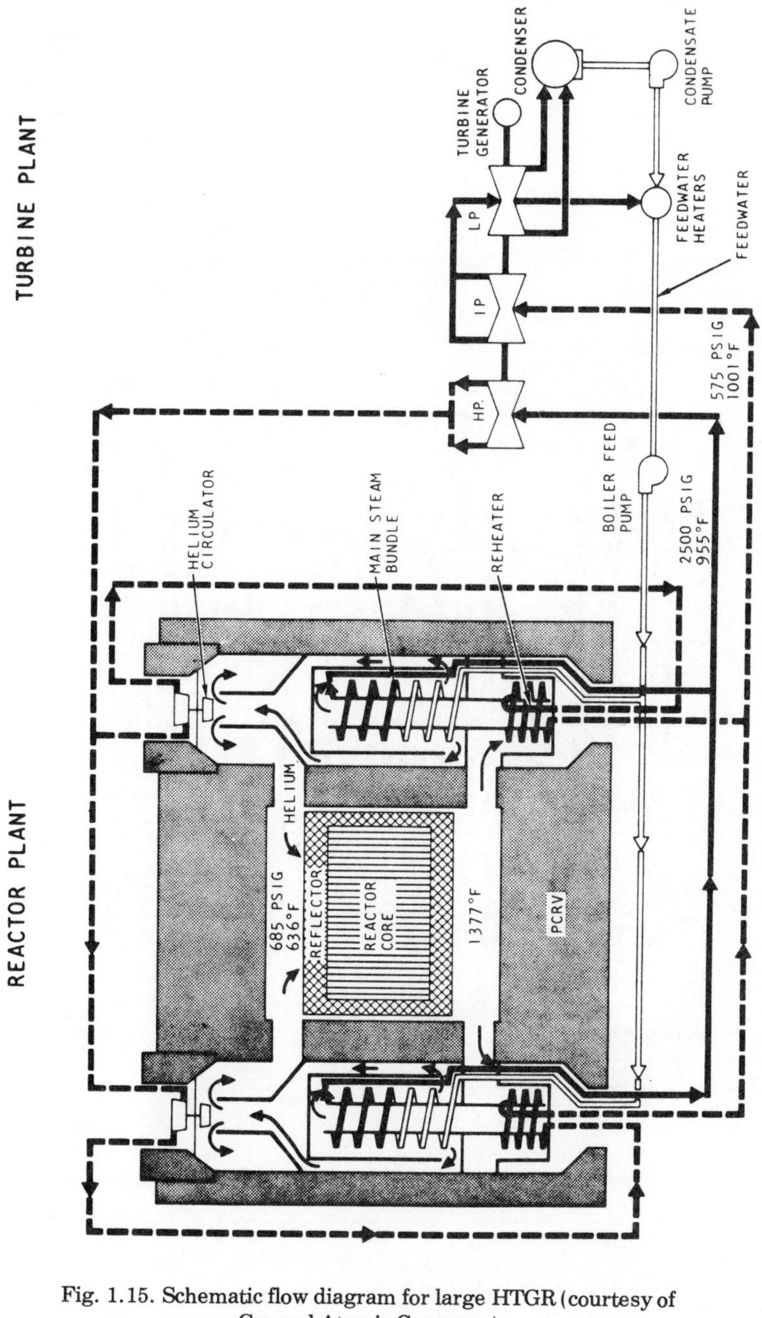

Fig. 1.15. Schematic flow diagram for large HTGR (courtesy of General Atomic Company)

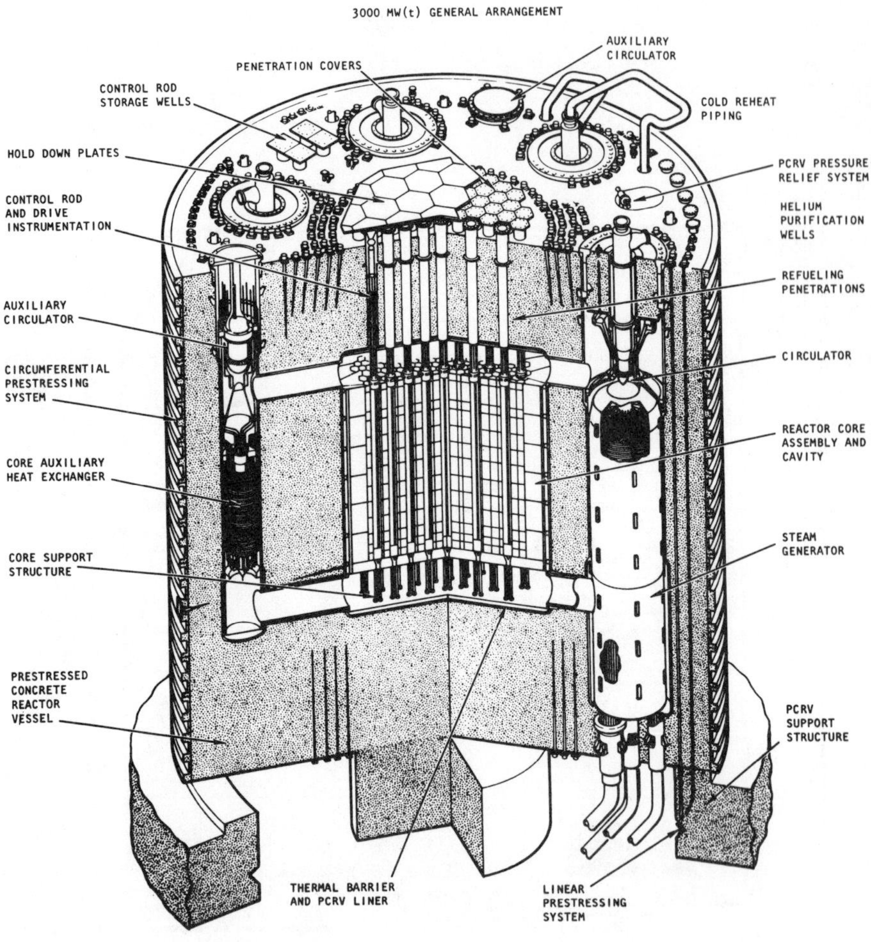

Fig. 1.16. HTGR reactor vessel and internals (Courtesy General Atomic Co.)

The sole supplier of HTGRs in the United States is the General Atomic Company. These reactors have not reached as high a state of development in the United States as LWRs. This may be attributed to these reactors not being suitable for naval applications from which originated the initial impetus for nuclear applications.

In addition to the plant shown in Fig. 1.15, General Atomic has performed design studies on a HTGR employing gas turbines for electric power generation [14]. This plant incorporates a direct cycle because the reactor coolant is expanded through the gas turbine. The turbine-compression units are placed inside the prestressed concrete reactor vessel (PCRV) which enables complete isolation of the primary coolant. An added feature of this type of plant is that dry cooling can be used because of the higher temperatures for heat rejection.

Table 1.8

Design parameters for HTGR [1,5]

Core Thermal Power	3000 MW(t)
Plant Electrical Output	1170 MW
Plant Efficiency	39%
Steam Temperature	950 °F
Steam Pressure	2500 psia
Coolant	Helium
Coolant Pressure	700 psia
Number of Steam Generators	6
Number of Auxiliary Coolers	3
Pressure Vessel Material	reinforced concrete
PCRV Height	91 ft.
PCRV Diameter	100 ft.
PCRV Weight	45,000 tons
Weight of Core	3,000 tons

The reactor vessel for HTGRs is quite different from those employed with LWRs. The reactor vessel is called a prestressed concrete reactor vessel (PCRV). Concrete is continuously poured to form the base of the PCRV and during the pouring reinforcing bars are added to give it additional strength. Four hundred and forty eight steel-lined tubes are placed within the concrete structure to house 160 to 171 quarter inch steel wires. These tendons are stressed to a tension of 1000 tons and then relaxed to 700 tons [15]. This serves to keep the state of the PCRV under compression for which concrete has good properties. Failure testing under air pressure for these vessels causes a small crack to form in the vessel which reseals as the pressure is reduced by venting. Another advantage of this type of pressure vessel is its field fabrication which eliminates transportation of 1000 ton steel pressure vessels.

Helium at 700 psia enters the top of core and flows downward through circular holes in the vertical graphite fuel assemblies. The helium exits from the core and passes through the reheater section of six separate steam generators. The helium then passes through the high pressure steam portion of the steam generators and exits into the suction side of the helium circulators. The six helium circulators are driven by steam turbines which obtain steam from the exhaust from the high pressure turbine. Helium flow is then back to the core inlet. Any one of the six helium loops incorporating a steam generator and helium circulator may be isolated from the core by valves.

In addition to the six primary helium loops, three auxiliary helium loops are available for residual heat removal or in case some emergency situation precludes use of all six primary loops. The auxiliary loops have a water-cooled heat exchanger and helium circulator.

Core

A fuel element for an HTGR is shown in Fig. 1.17. The fuel element is formed from a hexagonal-shaped graphite block 31 in. in height and 14.2 in. across the flats. Holes are bored through the graphite block for 72 coolant channels and inserting 132 graphite fuel rods. Eight fuel elements are stacked on top of each other to form a core stack. Four hundred and ninety three core stacks make up a reactor core which gives a total of 3944 fuel elements in a core. The core is surrounded on all sides by replaceable graphite reflector elements and permanent graphite reflector blocks.

Seventy three core stacks contain control rod fuel elements which have two 4.00 in. diameter and one 3.75 in. diameter holes bored through the graphite block. The 4.00 in. holes accomodate control rod pairs which are motor driven. The 3.75 in. holes are for an emergency shutdown system in which small boron carbide spheres are poured into the reactor. The control rod fuel elements contain 43 coolant channels and 76 fuel rods.

The HTGR fuel, also shown in Fig. 1.17, is in the form of small coated spheres of oxides or carbides of uranium for the fissile material and thorium for the fertile material. The fissile material is 93% enriched U^{235} or recycled U^{233} . The fissile and fertile spheres have different sizes and coatings in order to facilitate their separation during reprocessing. Both fuel kernels are coated with a porous carbon inner layer (buffer zone) to accomodate fuel expansion and provide a region for gaseous fission products. After the buffer zone, the fissile sphere is coated with successive layers of pyrolitic graphite, silicon carbide, and

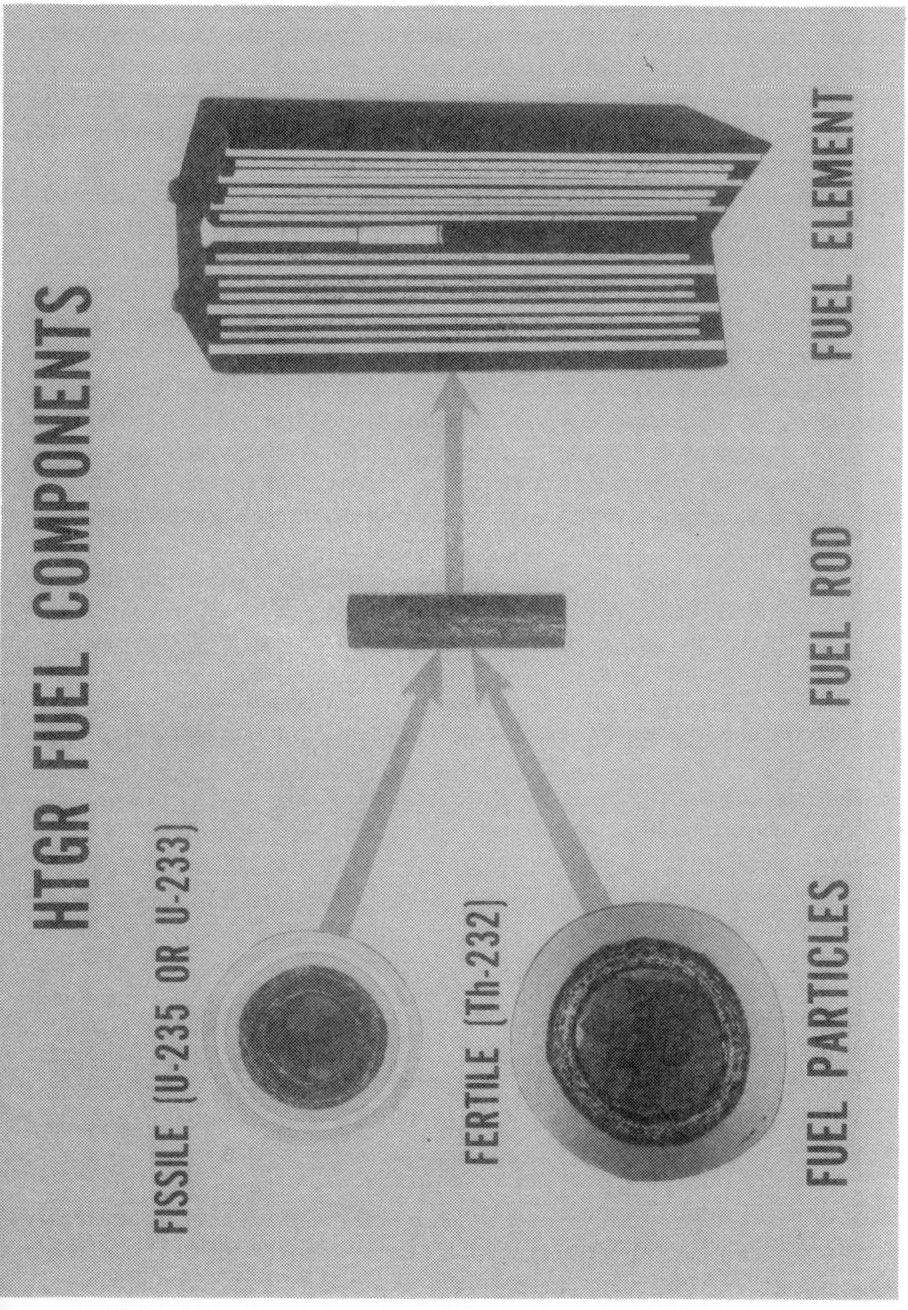

Fig. 1.17. HTGR moderator/fuel assembly (Courtesy of General Atomic Co.)

pyrolitic graphite while the fertile sphere is only covered with a layer of pyrolitic graphite. Because silicon carbide does not burn, this coating aids in separating the two types of particles during reprocessing where the graphite is burned off. Approximate diameters for the particles is 0.3 mm for the fissile sphere and 0.6 mm for the fertile sphere.

The multilayered coatings around the fuel particles serve as small pressure vessels for retention of fission products. In a sense, these coatings are analogous to the cladding surrounding LWR fuels. The fuel particles are formed into rods using a graphite binder. The rods are then inserted into the graphite blocks and the fuel element baked. The fuel rods serve as an additional barrier to fission product loss. Operating experience to high burnups have shown very little fission product loss with these fuel elements.

The HTGR core operates at very high specific powers of the order of 1700 kW/kg of fissile material and high burnups, on an average of 100,000 megawatt-days/metric tonne of U-Th. This is possible because of the large thermal conductivity of graphite, the intimate mixing of fuel and moderator, and the high conversion ratio (approximately 0.7) of the system.

Table 1.9 lists geometry and thermal-hydraulic characteristics for a 3000 MW(t) HTGR core.

Table 1.9

HTGR core geometry and thermal hydraulic characteristics [1,5,6]

Thermal Output	3000 MW(t)
Core Active Height	20.8 ft.
Core Diameter	27.8 ft.
Height/Diameter	0.75
Active Core Volume	12,500 ft^3
Average Core Power Density	240 kW/ft^3
Fuel Type	UC, ThO_2
Fuel Weight (U and Th)	39,000 kg
Uranium Weight	1800 kg
Specific Power	1660 kW/kg U
Average Burnup	98,000 MWD/MTU-Th
Number of Core Stacked Columns	493
Number of Fuel Elements per Column	8
Number of Fuel Elements	3944
Fuel Element Geometry	Hexagonal shape 31 in. high, 14.2 in. across flats

Fuel Rods per Element	132
Fuel Rod Diameter	0.63 in.
Fuel Enrichment	93.5% of U^{235}
System Pressure	700 psia
Coolant Flow	1.1 x 10^7 lb/hr
Coolant Inlet Temperature	636°F
Coolant Outlet Temperature	1377°F
Coolant Channels per Fuel Element	72
Coolant Channel Diameter	0.825 in.
Average Heat Flux	64,800 Btu/hr-ft^2
Maximum Heat Flux	185,000 Btu/hr-ft^2
Maximum Fuel Temperature	2570°F

1.5 Liquid Metal-Cooled Fast Breeder Reactors

In order to assure the availability of a long term supply of fissionable fuels it is essential that breeder reactors be developed in concert with thermal converter reactors already established. Within the United States four types of breeder reactors have been under consideration—**gas-cooled fast breeder reactor** (GCFR) being developed by the General Atomic Company [1,4], **Molten Salt breeder reactor** (MSBR) previously under development by the Oak Ridge National Laboratory [1,3,4], **light-water breeder reactor** (LWBR) under development by DOE's Division of Naval Reactors [1, 4], and the **liquid metal-cooled fast breeder reactor** (LMFBR) under the most active development in the United States and many other nations of the world.

As of this writing the greatest level of breeder reactor activity is directed toward LMFBRs and in particular the Clinch River breeder reactor plant (CRBRP) which is a prototype breeder reactor under development by the Department of Energy, T.V.A., Commonwealth Edison Co., and a consortium of other utilities. The CRBRP employs sodium as a primary and intermediate coolant which has a slight disadvantage because sodium solidifies at 208°F. This problem is easily overcome by wrapping electric heaters around sodium containing components.

Figure 1.18 illustrates a schematic diagram for the CRBRP. Sodium at 995°F exits from the reactor vessel and passes through an intermediate heat exchanger (IHX) where its temperature is reduced to 730°F before the sodium returns to the reactor vessel. In the IHX the primary sodium coolant exchanges heat with an intermediate sodium

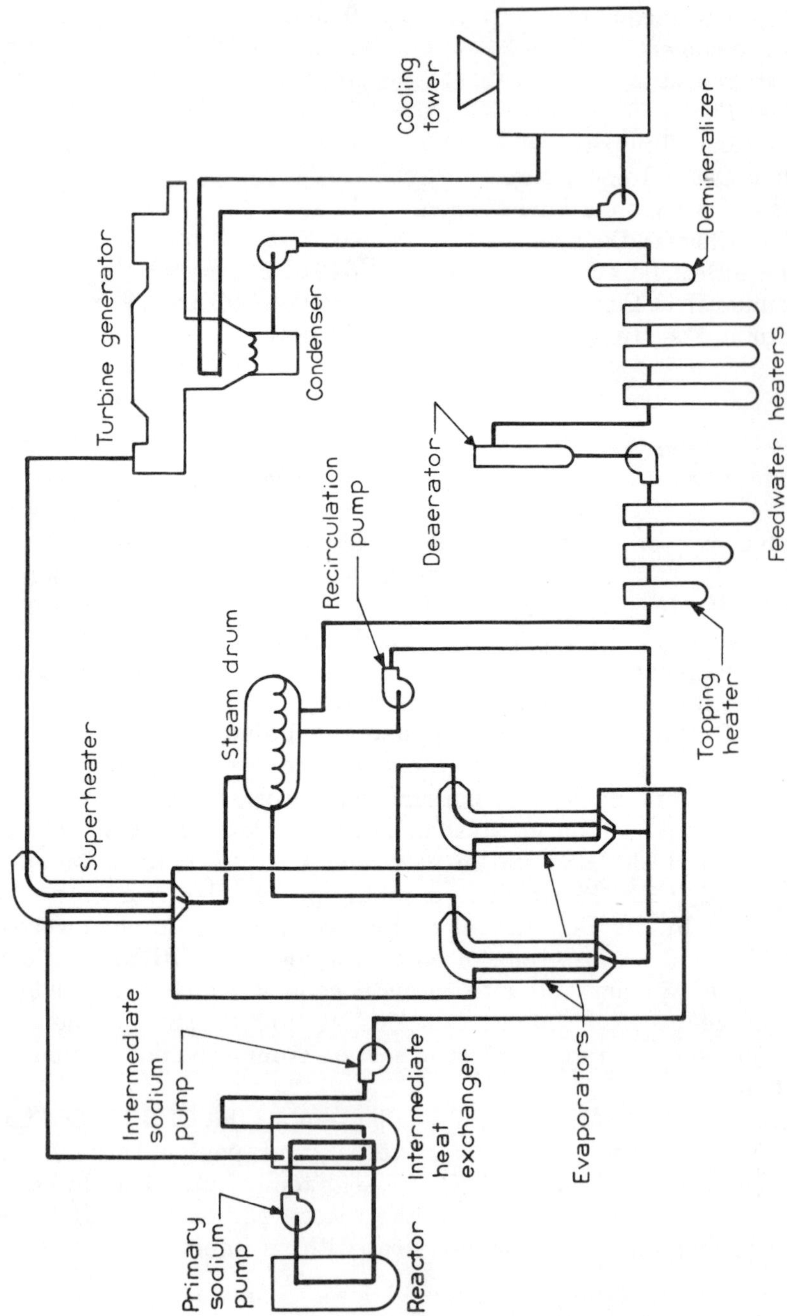

Fig. 1.18. Schematic diagram for Clinch River Breeder Reactor Plant

coolant and raises its temperature from 651°F to 936°F. The intermediate coolant is maintained at a higher pressure than the primary coolant so that any leakage in the IHX is into the primary side. The purpose of the intermediate loop is to prevent a sodium-water reaction with primary coolant in case of a leak in the superheater or evaporators. Three IHXs and steam supply systems are employed on the CRBRP in order to provide redundancy in case one loop is put out of operation.

Sodium from the IHX enters a superheater where its temperature is reduced to 856°F. The flow is then equally split for entry into two evaporators. The sodium exits from the evaporators at 651° F for return to the IHX.

Superheated steam at 905°F and 1550 psia leaves the superheater for passage through high- and low-pressure stages of the turbine. The turbine exhaust is condensed to 101° F and the water pumped by condensate pumps through the demineralizers. Flow continues through low pressure feedwater heaters and then the deaerator. Feedwater pumps draw feedwater from the deaerator and increase the pressure to 2160 psia for passage through two high pressure feedwater heaters and one topping heater. The feedwater enters the steam drum at 468° F.

The thermal output from the reactor is 975 MW and the plant electrical output is 375 MW for an overall thermal efficiency of 38.5 percent.

Reactor Vessel

Figure 1.19 shows the reactor vessel and internals for the CRBRP. The reactor vessel is a vertical, cylindrical shell with a dome-shaped bottom and is capped with a massive closure head attached to an upper flange and support ring. Design parameters for the reactor vessel and closure head are listed in Table 1.10.

The closure head seals the top of the reactor vessel and provides access for control rod assemblies and refueling. The closure head consists of three rotating plugs. The outer, large rotating plug is supported by a riser which is attached to the reactor vessel support ring. Inside and eccentric to the large rotating plug is an intermediate rotating plug. Inside and eccentric to the intermediate rotating plug is the small rotating plug. Rotation of these plugs allows straight-pull access to any fuel assembly or removable component for transfer within the reactor. Plug rotation also allows placing the ex-vessel transfer and storage positions for moving items in and out of the reactor vessel.

Sodium enters three inlet nozzels at the bottom of the reactor vessel and flows upward through 61 inlet modules attached to the core support plate. Flow within each module is distributed to 427 fuel, blanket,

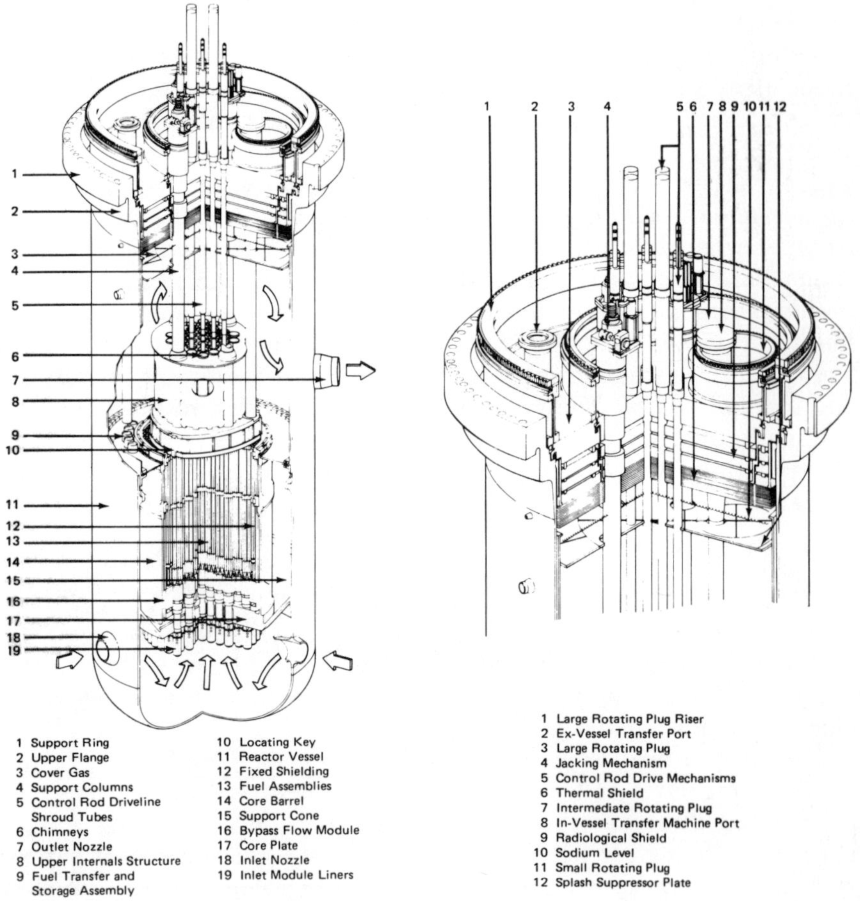

Fig. 1.19. CRBRP reactor and closure head [16]

Table 1.10
Design parameters for CRBRP reactor vessel and closure head [16,17]

Reactor Vessel	
Material	304 Stainless Steel
Height	59 ft.
Inside Diameter	20.25 ft.
Wall Thickness	2 3/8 in.
Weight	505 tons
Sodium Flow Rate	4.15×10^7 lb/hr
Inlet Nozzle (3) Diameter	24 in.
Outlet Nozzle (3) Diameter	36 in.
Inlet to Outlet Pressure Drop	123 psi
Closure Head	
Material	Forged Carbon Steel
Weight	500 tons
Plug Thickness	22 in.
Thermal Shield Thickness	24 in.
Radiological Shield Thickness	29 in.
Diameter of Rotating Plugs	
Large	21.5 ft.
Intermediate	14.67 ft.
Small	5.67 ft.

control rod and radial shield assemblies. The flow continues upward through the various assemblies and exits into the upper internals structure. Chimneys in the upper internals structure guide the sodium into the outlet plenum region where it mixes and exits from three outlet nozzles.

The sodium level is maintained above the sodium splash suppressor plate and a few inches below the closure head. Argon gas is used as a cover gas to provide an inert atmosphere and prevent sodium contact with the head seals.

Core

As of 1979 the exact core configuration for the CRBRP is unclear because of uncertainty in breeder reactor fuel cycles due to the U. S. government's preoccupation with nuclear proliferation. Figure 1.20 shows the reference design core configuration for the CRBRP. The core contains a central region of 108 fuel assemblies with initial cycle enrichment of 18.7 percent and an outer zone of 90 fuel assemblies with an enrichment of 27.1 percent. Within the reactor core are 15 primary control rod assemblies and 4 secondary control rod assemblies of which

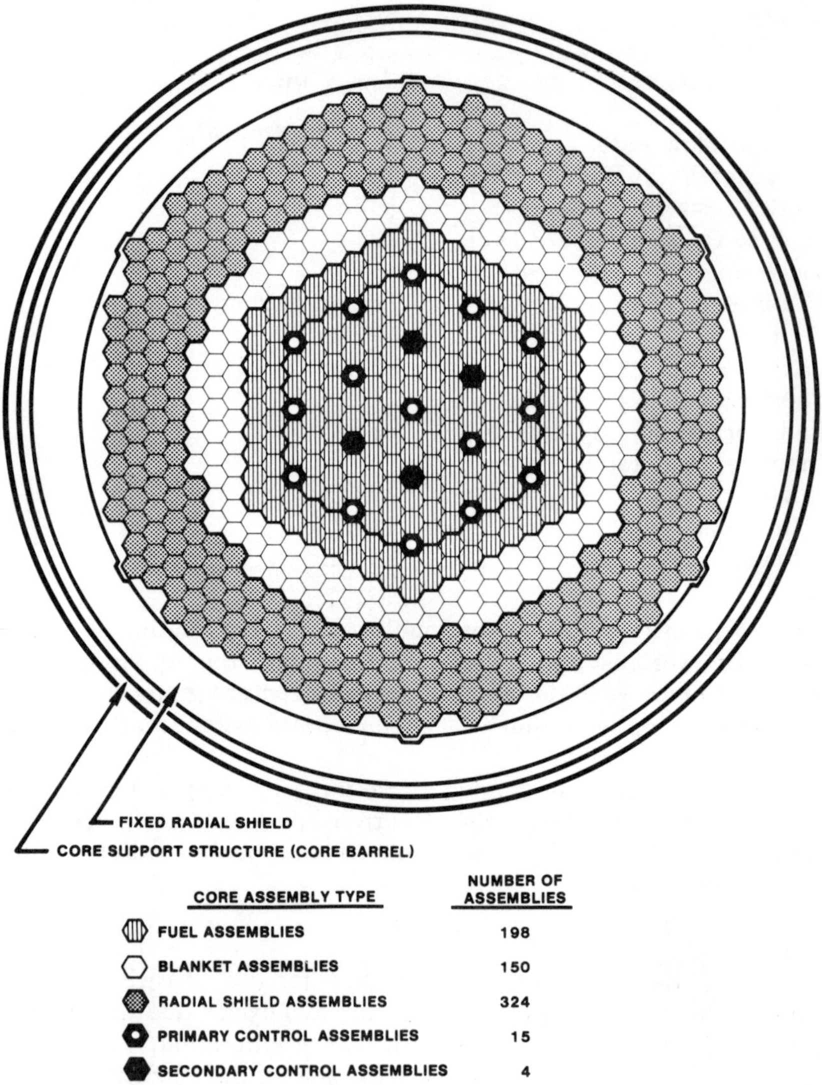

Fig. 1.20. Plan view of CRBRP reactor core [18]

either group can shut down the reactor. Surrounding the core is 150 radial blanket assemblies containing depleted UO_2. Surrounding the blanket is 324 removable radial shield assemblies. Approximately 80 percent of the sodium flow is through the core, 12 percent through the blanket, and 3 percent through the radial shield assemblies.

Figure 1.21 shows a CRBRP fuel assembly. Each fuel assembly is enclosed by a stainless steel hexagonal duct which contains 217 stainless steel, 0.23 in. diameter tubes. Spiral wire spacers are

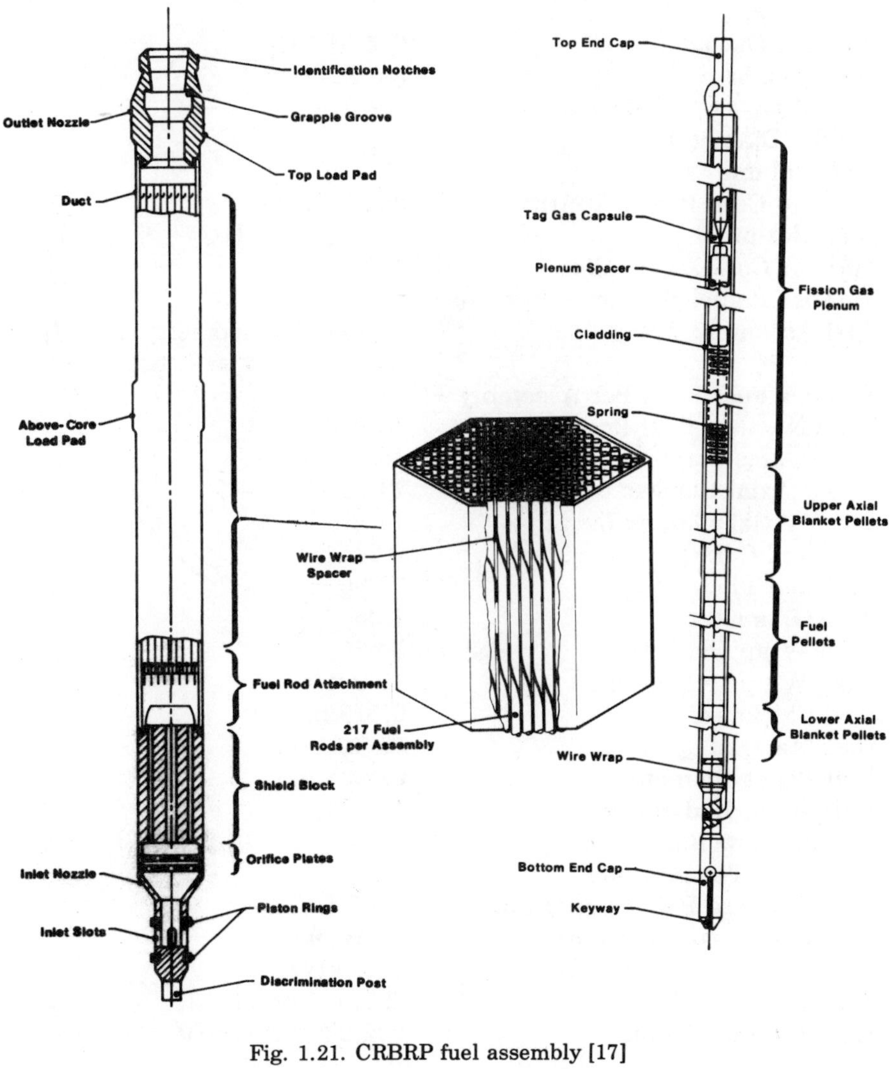

Fig. 1.21. CRBRP fuel assembly [17]

wrapped around the length of each tube to maintain rod spacing. The geometry and thermal-hydraulic characteristics of a fuel assembly are listed in Table 1.11.

Blanket assemblies are the same size as fuel assemblies. However,

Table 1.11

CRBRP core geometry and thermal-hydraulic characteristics [1, 16, 17]

Thermal Output	975 MW(t)
Core Height	3.0 ft.
Core Equivalent Diameter	6.2 ft.
Height/Diameter	0.48
Core Volume	90.6 ft^3
Average Core Power Density	350 kW/liter
Peak Burnup	150,000 MWD/MT
Number Core Assemblies	198
Number Radial Blanket Assemblies	150
Fuel Assembly Array	Hexagonal Geometry with Rods on Triangular Lattice
Number Fuel Rods Per Assembly	217
Total Number Fuel Rods	42,966
Fission Gas Plenum Length	48 in.
Upper Axial Blanket Length	14 in.
Lower Axial Blanket Length	14 in.
Fuel Rod O.D.	0.230 in.
Fuel Rod Pitch	0.290 in.
Pitch/Diameter	1.26
Wire Wrap Diameter	0.056 in.
Wire Wrap Helical Pitch	11.9 in.
Clad Thickness	0.015 in.
Fuel Material	PuO_2–UO_2
Fuel Pellet Diameter	0.1935 in.
Pellet-Clad Radial Gap	0.00325 in.
Sodium Flow Rate	4.15×10^7 lb/hr
Core Inlet Temperature	730°F
Core Average Exit Temperature	995°F
Average Linear Heat Rate	7 kW/ft.
Maximum Linear Heat Rate	14.5 kW/ft
Average Heat Flux	397,000 Btu/hr-ft^2
Maximum Heat Flux	822,000 Btu/hr-ft^2

they contain 61 rods of 0.506 in. diameter which are spaced by spiral wire wraps. The removable radial shield assemblies contain 19 solid stainless steel rods for absorption and reflection of neutrons.

Intermediate Heat Exchanger

Figure 1.22 shows an intermediate heat exchanger for CRBRP. The IHX is a vertical, counterflow, shell-and-tube unit. The lower tube sheet is supported by the straight tube bundle which in turn is supported by the upper tube sheet. Design data for the IHX is given in Table 1.12.

Primary sodium inters the IHX near the middle of the tube bundle where a bypass seal assembly directs flow upward and through a distribution cylinder at the top of the tube bundle. The flow is then directed downward through the shell side to exit from the bottom of the IHX.

Intermediate sodium enters the top of the IHX and is directed through a central downcomer to the bottom of the vessel. Flow is then reversed and directed upward through the tubes to the upper plenum and out the side of the IHX.

Table 1.12

CRBRP intermediate heat exchanger design parameters [16, 17]

Rating	325 MW(t)
Material	304 and 316 Stainless Steel
Number on System	3
Length	52.1 ft.
Shell Diameter	8.8 ft.
Weight (Dry)	230,000 lb.
Weight (Operating)	340,000 lb.
Number of Tubes	2,850
Active Tube Length	26 ft.
Tube O.D.	0.875 in.
Tube Wall Thickness	0.045 in.
Surface Area	11,810 ft^2
Primary Flow	1.38×10^7 lb/hr
Intermediate Flow	1.29×10^7 lb/hr
Primary Inlet Temperature	995°F
Primary Outlet Temperature	730°F
Intermediate Inlet Temperature	651°F
Intermediate Outlet Temperature	936°F

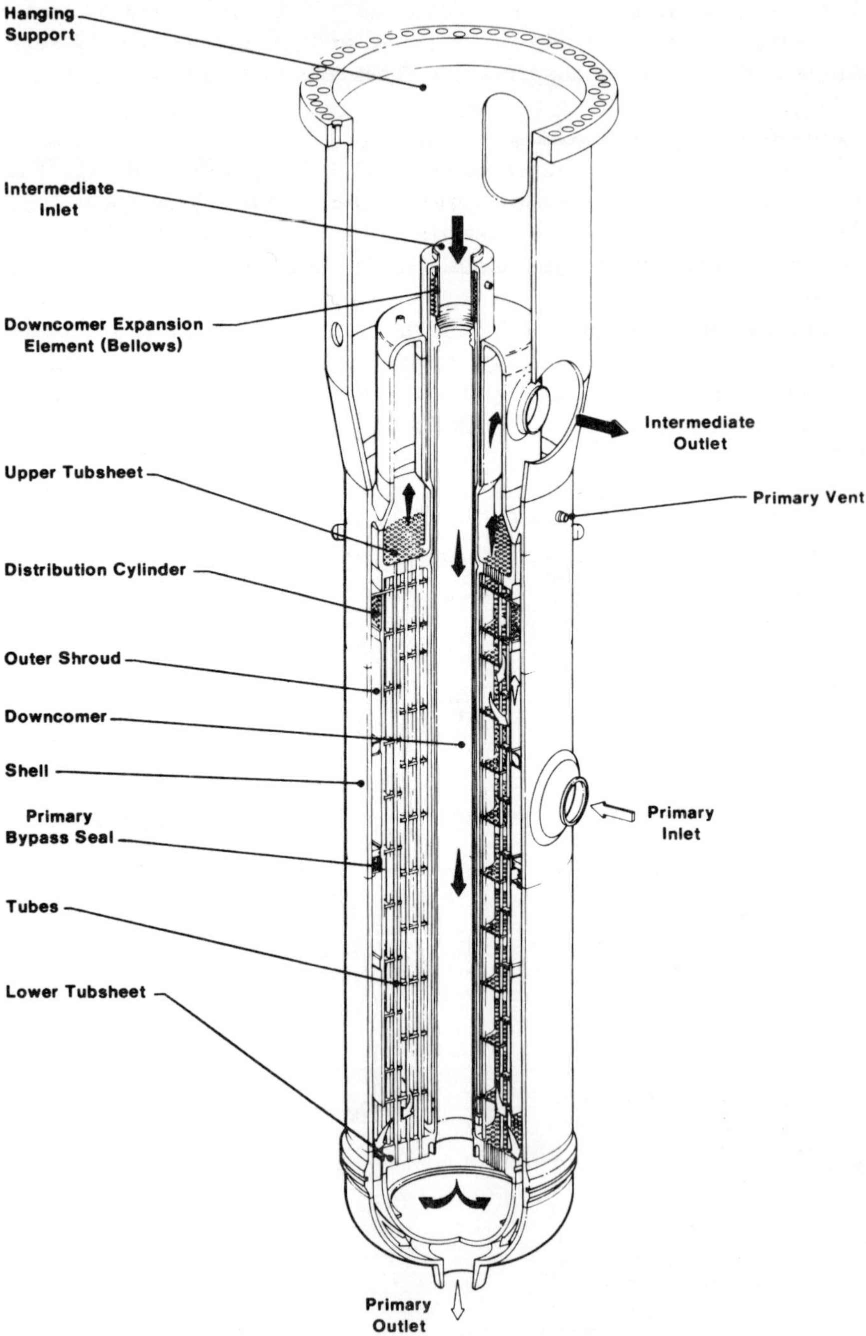

Fig. 1.22. CRBRP intermediate heat exchanger [17]

Steam Generators

The superheaters and evaporators are of identical design which is shown in Fig. 1.23. They are vertical, counterflow, shell-and-tube heat exchangers with fixed tube sheets. The design is called a "hockey stick" because of the 90 degree bend at the top to allow differential expansion between the tube bundle and shell. Design data for the CRBRP steam generators is listed in Table 1.13.

Table 1.13

CRBRP steam generator design data [16, 19]

Weights and Dimensions	
Rating	
Superheater	92 MW(t)
Evaporator	116.5 MW(t)
Material	2 1/4 Cr-1 Mo Steel
Number of Systems	3
Length	65 ft.
Shell Diameter, O. D.	4–4 ½ ft.
Shell Thickness	2 1/4 in.
Number of Tubes	757
Active Tube Length	46 ft.
Tube O.D.	5/8 in.
Tube Wall Thickness	0.109 in.
Dry Weight	106 tons
Functional Performance	
Shell (Sodium) Side	
Sodium Flow Per Loop, Maximum	1.35×10^7 lb/hr
Superheater Inlet Pressure	208 psia
Superheater Sodium Inlet Temperature	936°F
Sodium Friction Pressure Loss, Max.	
Superheater	62 psi
Evaporator	16 psi
Evaporator Inlet Temperature	856°F
Evaporator Outlet Temperature	651°F
Tube (Steam/Water) Side	
Superheater Steam Flow Rate	1.11×10^6 lb/hr
Superheater Outlet Temperature	905°F
Superheater Outlet Pressure	1550 psia
Evaporator Outlet Quality	0.50
Evaporator Steam/Water Flow Rate	1.11×10^6 lb/hr
Recirculating Ratio	2:1
Feedwater Temperature Steam Drum	468°F

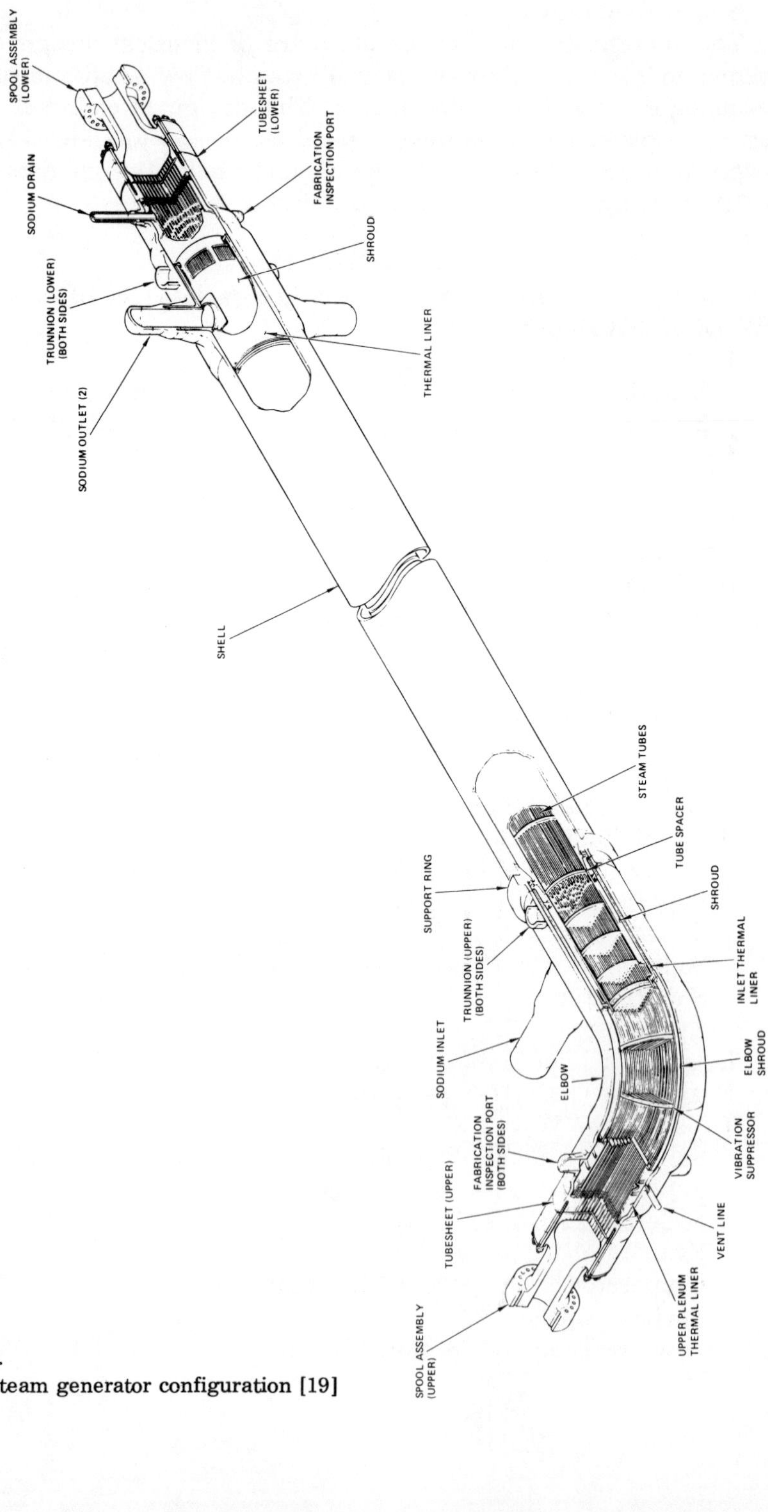

Fig. 1.23.
CRBRP steam generator configuration [19]

Sodium from the IHX enters near the top of the superheater unit, flows down the shell side and exits at the bottom. Two outlets direct sodium to the two evaporators through which the flow is the same as in the superheaters with the exception that there is a single outlet on each unit with discharge piped to a common line for return to the IHX.

Water or steam enters the bottom of the units, flows upward through the tube side and exits at the top.

REFERENCES

1. Nero, A. V., **A Guidebook to Nuclear Reactors**, University of California Press, Berkeley and Los Angeles, California, 1979.
2. Wills, J. G., **Nuclear Power Plant Technology**, Wiley, New York, 1967.
3. Foster, A. R. and R. L Wright, **Basic Nuclear Engineering** (3rd ed), Allyn and Bacon, Boston, 1977.
4. Lamarsh, J. R., **Introduction to Nuclear Engineering**, Addison-Wesley, Reading, Massachusetts, 1975.
5. Graves, H. W., **Nuclear Fuel Management**, Wiley, New York, 1979.
6. Pederson, E. S., **Nuclear Power, Vol. 1 Nuclear Power Plant Design**, Ann Arbor Science, Ann Arbor, Michigan, 1978.
7. **Steam—Its Generation and Use**, 38th ed., Babcock & Wilcox, New York, 1975.
8. Greenwood Units 2 and 3, Preliminary safety analysis report, Detroit Edison Co. (June 4, 1973).
9. The safety of nuclear power reactors (light-water cooled) and related facilities, WASH-1250, USAEC (July 1973).
10. WNP Units 3 and 5, Preliminary safety analysis report, Washington Public Power Supply System (March 5, 1974).
11. Sharon Harris Units 1-4, Preliminary safety analysis report, Carolina Power and Light Co. (September 4, 1971).
12. Lahey, R. T., and F. J. Moody, **The Thermal-Hydraulics of a Boiling Water Nuclear Reactor**, American Nuclear Society, La Grange Park, IL., 1977.
13. Fuller, P. D., et al., General Electric BWR/6 nuclear systems—a performance summary, NEDO-10755, General Electric Company (January 1973).
14. Schoene, T. W., Winkler, E. O., and P. Fortescue, The gas turbine HTGR plant, economical dry cooling or a wet-cooled high efficiency binary configuration, **American Power Conference**, Chicago, Illinois, April 21-23, 1975.
15. Miller, P. G., HTGR—A safer way to nuclear power, **Popular Science** (February 1975).

16. Clinch River breeder reactor plant project, design description, CRBRP-PMC-77-06, Project Management Corporation (1977).
17. Clinch River breeder reactor plant project, 1978 technical progress report, CRBRP-ARD-0230, Westinghouse Electric Corporation (1979).
18. Clinch River breeder reactor plant, technical review Spring 1979, CRBRP-PMC-79-02, Project Management Corporation (April, 1979).
19. Clinch River breeder reactor plant, technical review Fall 1978, CRBRP-PMC-78-04, Project Management Corporation (Oct. 1, 1978).

CHAPTER 2

THERMODYNAMICS OF NUCLEAR POWER PLANTS

2.1 Introduction

In the design of fossil-fueled central station power plants, a large effort is expended to achieve high overall thermal efficiencies (defined as net work output divided by heat input). This is due to fuel costs being of major importance in determining total cost of power. For mobile fossil-fueled power plants used in ships and airplanes, the weight and volume of fuel consumed are large which, consequently, limits carrier range and cargo capacity. Therefore, it is readily evident that high thermal efficiencies are desirable in order to reduce costs and improve range and cargo capacities.

When considering nuclear power plants, overall thermal efficiency is no longer the major parameter. Fuel costs of nuclear plants are normally a smaller fraction of the total power cost when compared to fossil-fueled plants. The weight and volume of fuel are quite small compared to the weight and volume of the power plant. However, these considerations do not mean that high thermal efficiencies are not desirable in nuclear power plants. High thermal efficiencies result in extended core life because fuel element life is limited by energy generated within the element and depletion of fissionable material. In addition, sizes of cores and heat exchangers are dependent upon the amount of heat supplied and, consequently, high thermal efficiencies are important because of a potential reduction in equipment size. Therefore, thermal efficiency is an important parameter in the analysis of nuclear power plants, but only one of a number of important parameters. Usually an extreme effort to produce sizable increases in overall thermal efficiency results in too great an increase in capital costs to justify the effort.

This chapter discusses processes for converting heat to mechanical work that are commonly used in the reactor field. All heat engines of immediate interest use a working fluid which is compressed at low enthalpies, heated, and expanded at high enthalpies. The expansion work is greater than the compression work and external work becomes available. In reactor systems the working fluid flows continuously through the various components. System thermal efficiencies and power requirements are determined by applying steady flow forms of First and Second Laws of Thermodynamics and the Principle of Conservation of Mass.

For purposes of this discussion, English engineering units are employed in describing energy transformations taking place in nuclear systems. These units are:

Mass——pounds mass, lb_m
Force——pounds force, lb_f
Temperature——Fahrenheit or Rankine scales, °F or °R
Energy——British Thermal Unit, Btu or ft–lb_f, 778 ft–lb_f = Btu

In the analysis or reactor systems, one is interested in the conversion of various forms of energy into useful work. The forms of energy of interest are as follows:

Potential Energy

The potential energy of a body is the energy due to its elevation, z, above some datum point. This energy is given by

$$\text{P.E.} = M\frac{g}{g_c} z, \text{ ft–lb}_f \tag{2.1.1}$$

where M is the mass of the body, lb_m, and g is the gravitation acceleration, ft/sec². In the English system of units, the weight of a body at the earth's surface is arbitrarily established as being numerically equal to its mass. As a consequence, at the earth's surface g_c numerically equals g, or $g_c = 32.179\ (lb_m/lb_f)(ft/sec^2)$.

Kinetic Energy

The energy of a body due to its motion (considered with respect to the earth) is given by

$$\text{K.E.} = M\frac{u^2}{2g_c}, \text{ ft–lb}_f \tag{2.1.2}$$

where u is the velocity of the body with respect to the earth, ft/sec.

Internal Energy

A body possesses stored energy due to the internal potential and kinetic energy of its molecules. This energy is designated as

$$\text{Internal Energy} = e, \text{ Btu/lb}_m \tag{2.1.3}$$

Flow Energy

A substance passing across a system boundary possesses an energy quantity entering or leaving the system that is denoted by pv, where p is the pressure of the substance, lb_f/ft^2, and v is the specific volume of the substance, ft^3/lb_m. This energy is due to the substance exerting a force on the system boundary and, therefore, doing work upon the system while entering the system or doing work on the surroundings while leaving the system.

$$\text{Flow Energy} = pv = \frac{p}{\rho}, \frac{\text{ft-lb}_f}{\text{lb}_m} \tag{2.1.4}$$

2.2 Thermodynamics

Steady-Flow Energy Equation

In the thermodynamic analysis of systems, the system is confined by an enclosing boundary so as to produce what is called a control volume. In deriving the Steady-Flow Energy Equation, conservation of energy is applied to the control volume illustrated in Fig. 2.1.

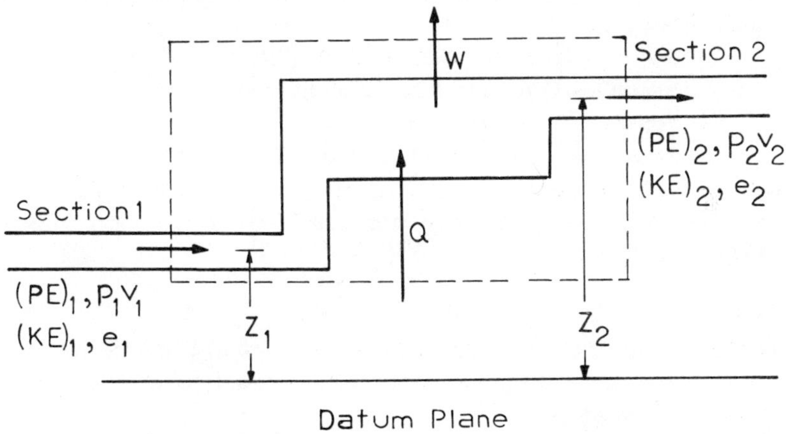

Fig. 2.1. Thermodynamic control volume for steady-flow energy equation

The following assumptions are made in the derivation:

(1) Flow is one-dimensional. This implies the fluid velocity is represented by single vectors, i.e., uniform velocity at entrance and exit.

(2) Equilibrium is achieved; therefore, temperature at any point within the system and work or heat crossing the system boundaries are independent of time.

The assumption of equilibrium implies that the mass rate of flow at (1) equals that of (2). Therefore, the Continuity Equation is

$$\dot{m} = \rho_1 A_1 u_1 = \rho_2 A_2 u_2 = \text{const.} \tag{2.2.1}$$

For the system in Fig. 2.1, under steady-state conditions, conservation of energy requires that

Energy entering = Energy leaving

A more specific form of this general equation is obtained by using the various types of energy previously discussed.

$$\frac{z_1}{J}\frac{g}{g_c} + \frac{u_1^2}{2g_c J} + e_1 + \frac{p_1 v_1}{J} + Q = \frac{z_2}{J}\frac{g}{g_c} +$$

$$\frac{u_2^2}{2g_c J} + e_2 + \frac{p_2 v_2}{J} + W \tag{2.2.2}$$

Q and W are in terms of Btu/lb_m of flowing fluid and J is the Joule conversion factor for conversion of mechanical work to heat, 778 ft-lb_f/Btu.

As shown in Fig. 2.1, heat is added to the system and work extracted from the system. The convention that is used for applying signs to heat and work is as follows:

Positive Q means heat added to system
Negative Q means heat leaving system
Positive W means work leaving system
Negative W means work added to system

Because of their common occurrence in steady-flow problems, the terms e + pv/J are commonly combined into a property known as enthalpy, h defined as

$$h \equiv e + \frac{pv}{J}$$

In addition, potential energy terms are usually negligible in the steady-flow energy equation so that for most practical problems we may write Eq. (2.2.2) as

$$\frac{u_1^2}{2g_c J} + h_1 + Q = \frac{u_2^2}{2g_c J} + h_2 + W \tag{2.2.3}$$

One can define another term, the stagnation enthalpy or total enthalpy

$$h_0 \equiv h + \frac{u^2}{2g_c J}$$

where h is considered the fluid enthalpy as you move with it and h_0 is considered the fluid enthalpy to a stationary observer. Stagnation properties are measured by a stationary probe in a moving fluid.

Consequently, the steady-flow energy equation is rewritten as

$$h_{01} + Q = h_{02} + W \tag{2.2.4}$$

For heating process $W \approx O$ and

$$Q = h_{02} - h_{01} \tag{2.2.5}$$

For turbines, pumps, and compressors, $Q \approx 0$ and

$$W = h_{01} - h_{02} \tag{2.2.6}$$

Equation (2.2.4) is applied to the entire nuclear power plant or any of its components—reactor, heat exchangers, condensers, pumps, or turbines.

Reversibility and Irreversibility

In the thermodynamic analysis of systems we are interested in the various processes of changing energy. It is customary to consider two types of processes—reversible and irreversible processes. In each of these thermodynamic processes there may be a change in the thermodynamic properties of the system and surroundings. A reversible process resulting in different thermodynamic states of a system is a process which could be reversed, returning the system and surroundings back to their respective initial state. As an example, heat might be taken from the surroundings and converted to work by a system. In order for this process to be reversible, the work produced by the system must be able to generate the exact amount of heat taken from the surroundings and return this heat to the surroundings.

An irreversible process is one that is not reversible. In actual practice all processes are irreversible although one can approach reversibility. A process that has heat transferred is irreversible because temperature differences are necessary. Motion of objects usually involves friction with the result that heat transfer takes place. However, the use of the concept of reversibility is not to be ignored; many processes approach reversibility and therefore are treated as such in analysis.

Thermodynamic Properties

So far we have mentioned certain material properties which are used to define thermodynamic states of substances. These properties are pressure, p; temperature, T; density, ρ; internal energy, e; and enthalpy, h. An additional property exists which is called the entropy, s. The entropy is defined for an internally reversible process by

$$ds \equiv \frac{dQ_{Rev}}{T}$$

where s = entropy, Btu/lb$_m$ - ^{0}R
dQ_{Rev} = heat transferred reversibly, Btu/lb$_m$
T = absolute temperature, ^{0}R

In order for dQ to be reversible, T must be the same in both system and surroundings. Integrating gives

$$\Delta s = s_2 - s_1 = \int_1^2 \frac{dQ_{Rev}}{T}$$

For most substances any two independent thermodynamic properties fix all properties of the substance. Consequently, one can make diagrams of thermodynamic states of a substance with two properties as the abscissa and ordinant and a third property as a parameter. For example, the T-s diagram or h-s diagram with pressure as a parameter shown in Fig. 2.2

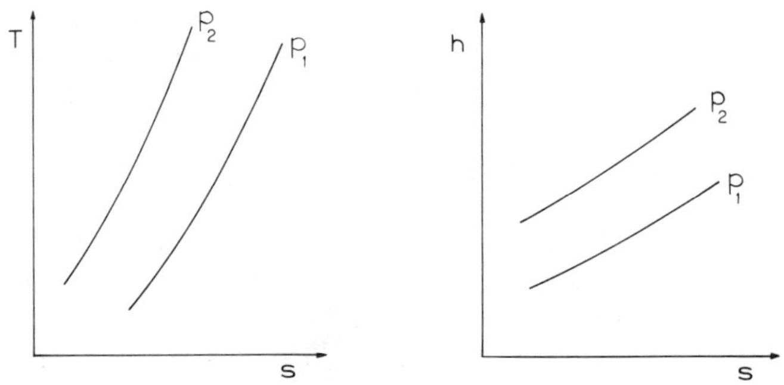

Fig. 2.2 Thermodynamic property diagrams

Diagrams of this nature are most useful in analyzing steam or gas turbine cycles.

Irreversibilities

The steady-flow energy equation deals only with changes in enthalpy and tells nothing about changes in other state variables. Equation 2.6 holds for reversible or irreversible processes, with friction or without. To deal with changes in other variables it is necessary to consider the degree of irreversibility or the "efficiency" of the process. Irreversibility effects are introduced in different ways depending upon the nature of the process under consideration. We will consider several cases.

(1) Pumps and Compressors (Adiabatic Compression)

For pumps handling **incompressible** fluids or compressible fluids with small density changes, the following analysis is performed:

Equation (2.2.2) is differentiated so as to apply to a small elemental flow process

$$dh + \frac{1}{2g_c J} d(u^2) + \frac{g}{Jg_c} dz = dQ - dW \qquad (2.2.7)$$

In addition, for all single component processes

$$T\,ds = dh - \frac{v}{J} dp \qquad (2.2.8)$$

By substituting Eq. (2.2.8) into Eq. (2.2.7) and rearranging we have

$$-dW = (T\,ds - dQ) + \frac{v}{J} dp + \frac{d(u^2)}{2g_c J} + \frac{g}{Jg_c} dz \qquad (2.2.9)$$

The Second Law of Thermodynamics postulates that

$$T\,ds - dQ > 0 \qquad (2.2.10)$$

Therefore, we have

$$-dW = (+) + \frac{1}{J} \left(v\,dp + \frac{d(u^2)}{2g_c} + \frac{g}{g_c} dz\right) \qquad (2.2.11)$$

For an incompressible fluid, Eq. (2.2.11) is integrated to determine the pumping work per lb_m of flowing fluid.

$$-W_p = \int_1^2 (-dW) > \frac{1}{J} \int_1^2 \left(v\,dp + \frac{d(u^2)}{2g_c} + \frac{g}{g_c} dz\right)$$

$$> \frac{1}{J}\left[vp_2 + \frac{u_2^2}{2g_c} + \frac{g}{g_c} z_2\right] - \frac{1}{J}\left[vp_1 + \frac{u_1^2}{2g_c} + \frac{g}{g_c} z_1\right]$$

$$> \frac{1}{J}\left[H_2 - H_1\right] \qquad (2.2.12)$$

where H is defined as the Bernoulli head.

Therefore, the pumping work is greater than the ideal work $(H_2 - H_1)$ and a pump efficiency is defined by

$$\eta_p \equiv \frac{(H_2 - H_1)/J}{W_p} \qquad (2.2.13)$$

η_p usually varies from 0.5 - 0.95 and is a function of many parameters which is discussed in Chapter 4. For large pumps with high flowrates η_p is usually greater than 0.90. Exceptions are electromagnetic pumps in which η_p might be as low as 0.1.

For compressible fluids which undergo large density changes, Eq. (2.2.11) is not easily integrated and another approach is used. Consider the h-s diagram for a flowing compressible fluid shown in Fig. 2.3.

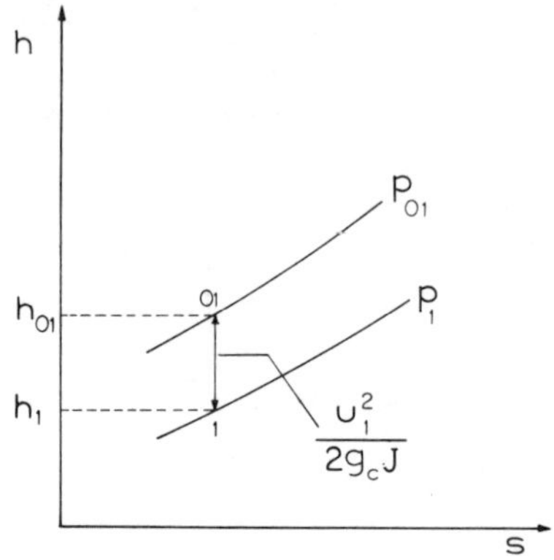

Fig. 2.3 Enthalpy-entropy diagram for flowing fluid

If the state of the moving fluid is known (h_1, p_1), then a point (1) exists on Fig. 2.3 corresponding to that state. Another point (01) a distance $u_1^2/2g_cJ$ vertically from point (1) defines the "stagnation state" of the fluid, which is the state that would be attained if the fluid were decelerated isentropically to rest from (1). We consider only changes in stagnation states in our analysis of cycles; although, in general velocities in systems under consideration are so small that static enthalpy and pressure are essentially equal to stagnation enthalpy and pressure. As an example, let u = 100 ft/sec. Then

$$\frac{u^2}{2g_c J} = \frac{10^4}{64.4 \times 778} = 0.2 \text{ Btu/lb}_m$$

which is usually negligibly small compared to h_1.

When a compressible fluid is compressed from stagnation pressure P_{01} to stagnation pressure P_{02}, the minimum work of compression is that required to accomplish the process isentropically. Let us examine this process on the h-s diagram shown in Fig. 2.4.

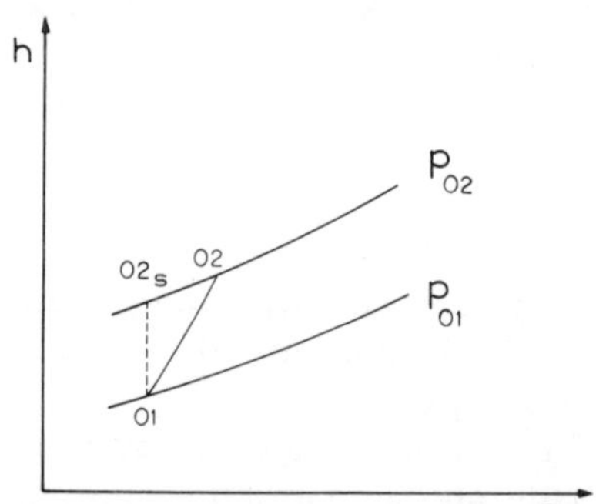

Fig. 2.4 Enthalpy-entropy diagram for compression processes

The isentropic compression to state 02s requires work of compression of $(h_{01} - h_{02s})$. However, in an actual process irreversibilities produce an entropy increase and the final state is 02 which is to the right of 02s on the h-s diagram. The work of compression is now $(h_{01} - h_{02})$ which is greater than the ideal work.

A compressor efficiency is defined by

$$\eta_c \equiv \frac{\text{ideal compressor work}}{\text{actual compressor work}} = \frac{h_{01} - h_{02s}}{h_{01} - h_{02}} \qquad (2.2.14)$$

Compressor efficiencies are usually of the order of 0.8 - 0.9. This figure is a characteristic of the compressor size and flow rate and is supplied by vendors when this number is of use in analysis.

Pumping or compression work is computed by dividing the ideal (isentropic) work by the pump or compressor efficiency. The ideal work is calculated by use of h-s diagrams or thermodynamic property tables for the fluid.

(2) Turbines (Adiabatic Expansion)

Turbines produce the opposite effect of a compressor. The working fluid is expanded from a stagnation pressure p_{01} to a lower stagnation pressure p_{02} to produce work. The h-s diagram shown in Fig. 2.5 is used to indicate the amount of work, W_T, that is produced by this process.

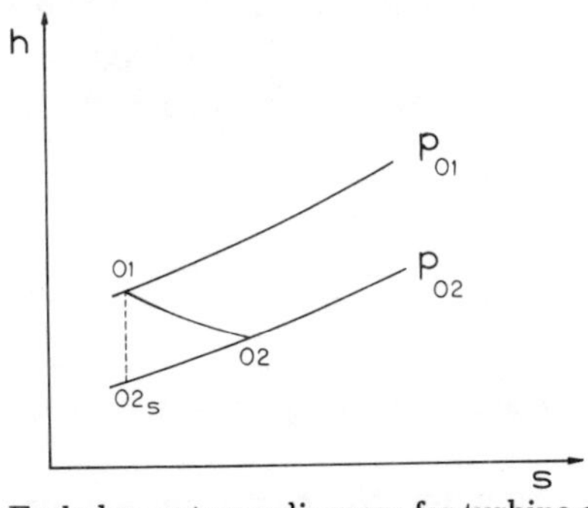

Fig. 2.5. Enthalpy-entropy diagram for turbine expansion

The isentropic expansion between pressures p_{01} and p_{02} produce the maximum amount of shaft work $(h_{01} - h_{02s})$. However, due to irreversibilities the fluid entropy increases and the final state is to the right of 02s. The final fluid enthalpy h_{02} is higher than h_{02s} and, consequently, the output work is reduced to $(h_{01} - h_{02})$. The turbine efficiency is defined by

$$\eta_t \equiv \frac{\text{actual turbine work}}{\text{ideal turbine work}} = \frac{h_{01} - h_{02}}{h_{01} - h_{02s}} \qquad (2.2.15)$$

Turbine efficiencies usually vary from 0.75 to 0.85 for nuclear power plants and 0.8 to 0.9 for fossil-fueled power plants. The reason for the decreased efficiency of turbines used in nuclear power plants is that the moisture content of the steam passing through the turbine is usually higher than in fossil-fueled power plants. Generally, an increase of 1 percent in moisture content results in a decrease of 1 percent in turbine efficiency.

(3) Heat Exchangers

The process of transferring heat results in an irreversibility because temperature differences are involved. The use of primary heat exchangers in power plants reduces plant thermal efficiency because the secondary fluid does not achieve the maximum temperature of the primary coolant. Consequently, less shaft work is possible.

Heat exchanger irreversibilities are defined in terms of the ratio of the amount of heat actually exchanged to that which could have been exchanged under ideal circumstances. A heat exchanger effectiveness is defined in this fashion.

$$\text{H.E. Effectiveness} = \eta_x \equiv \frac{\text{Actual Heat Transfer}}{\text{Ideal Heat Transfer}} \tag{2.2.16}$$

Because of their greater inherent irreversibilities, parallel-flow heat exchangers are rarely used. Counter-flow heat exchangers have more promise for achieving ideal heat transfer conditions. In some gas turbine heat exchangers the mass flow rate of the hot fluid equals the mass flow of the cold fluid and their heat capacities are essentially identical. For the heat exchanger shown in Fig. 2.6,

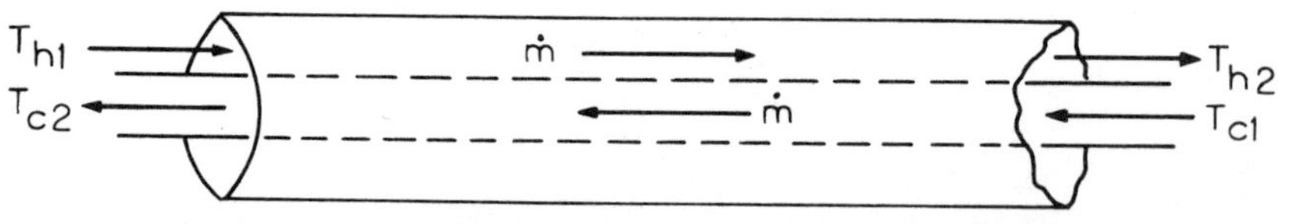

Fig. 2.6. Schematic diagram of a counter-flow heat exchanger

ideal heat transfer occurs when the exit temperature of the cold fluid is equal to T_{h1}. The actual amount of heat being transferred is equal to $\dot{m}c_p(T_{c2} - T_{c1})$ and the ideal heat transfer is $\dot{m}c_p(T_{h1} - T_{c1})$. Therefore, the heat exchanger effectiveness for this example is given by

$$\eta_x = \frac{T_{c2} - T_{c1}}{T_{h1} - T_{c1}} \tag{2.2.17}$$

Heat exchanger effectivenesses vary from 0.5 - 0.9 and are dependent upon the heat transfer area and flow velocities.

Thermodynamic Cycles and Their Efficiencies

A cycle is a series of processes during which a substance starts in a particular state and returns to its initial state. A schematic representation for the power cycle of a reactor system is shown in Fig. 2.7.

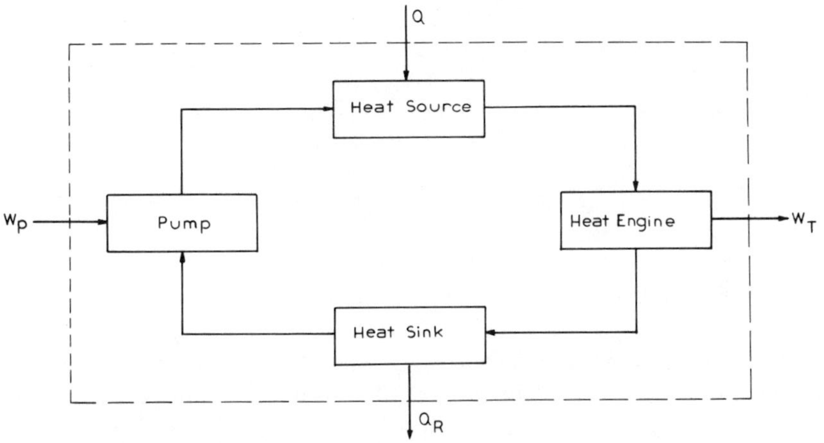

Fig. 2.7. Schematic diagram for a reactor power cycle

The overall thermal efficiency of this cycle is given by

$$\eta = \frac{\text{Net Output Work}}{\text{Input Heat}} = \frac{W_{Net}}{Q} = \frac{W_T - W_p}{Q} \qquad (2.2.18)$$

$$= \frac{Q - Q_R}{Q}$$

For heat engine cycles, the net energy crossing the control volume boundary as heat is equal to the net amount of energy crossing the boundary as work.

For purposes of thermodynamic analysis, the nuclear power plant is broken into two parts—the reactor and its associated coolant loop and the power cycle. In boiling water reactors it is not necessary to separate the reactor cooling system from the power cycle, since reactor heat is supplied directly to the power cycle. In the following sections various types of power cycles are analyzed.

Carnot Cycle

The Carnot cycle is the most efficient heat engine cycle conceivable and is used as the standard of comparison for all heat engines in use today. The Carnot cycle consists of four reversible processes which are illustrated on the temperature-entropy diagram shown in Fig. 2.8.

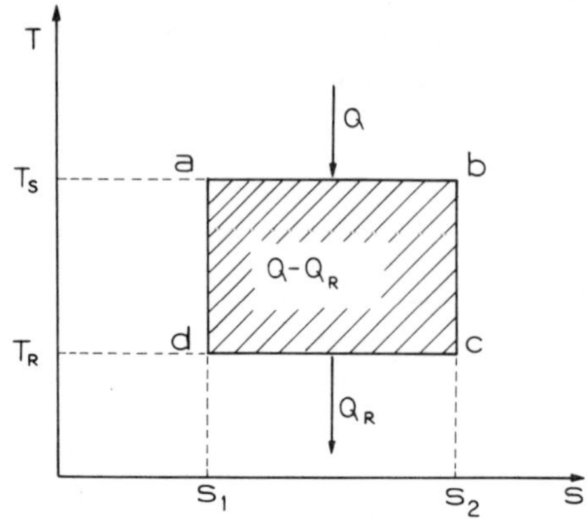

Fig. 2.8. Schematic representation of a Carnot cycle

The four processes consist of (1) isothermal heat addition from a to b, (2) isentropic expansion from b to c, (3) isothermal heat rejection from c to d, and (4) isentropic compression from d to a. The efficiency of this cycle is given by

$$\eta = \frac{Q - Q_R}{Q} = \frac{T_S(s_b - s_a) - T_R(s_c - s_d)}{T_s(s_b - s_a)} = 1 - \frac{T_R}{T_S} \quad (2.2.19)$$

The Carnot cycle efficiency is increased by increasing T_S or decreasing T_R. While no actual power cycle has approached the Carnot cycle efficiency, the basic principle that cycle efficiency is increased by raising the temperature at which energy is received or lowering the temperature at which energy is rejected holds for all heat engine cycles.

2.3 Liquid-Vapor or Rankine Cycle

The liquid-vapor or Rankine cycle is used to produce most of the world's electric power. By condensing vapor after expansion the volume of compressed fluid is made very small. Consequently, the work of compression is almost negligible compared to the turbine output. Therefore, the Rankine cycle engine produces useful work even when component efficiencies are low. This is contrary to gas turbines which need high component efficiencies in order to produce a net work. For Rankine cycles, the conventional working fluid is water, although binary cycles using water and mercury have been built and simple cycles using mercury or potassium are being considered for space applications of nuclear energy.

The simplest liquid-vapor plant (illustrated in Fig. 2.9) consists of a boiler, turbine, condenser, and boiler feed pump. The liquid is vaporized in the boiler at high pressure, expanded in a turbine down to the back pressure, condensed, and pumped up to the boiler pressure again. These processes are conveniently illustrated on p-v, h-s, or T-s diagrams of the working fluid. For convenience, T-s and h-s diagrams are usually used. Figure 2.10 illustrates representative T-s and h-s diagrams for fluids used in liquid-vapor cycles. The solid line illustrates the locus of properties in which the fluid is a saturated liquid or saturated vapor. The dashed lines illustrate the locus of properties in which pressure is constant. The point marked T_c is the location of the critical temperature. At temperatures above the critical temperature, phase changes do not take place in the fluid.

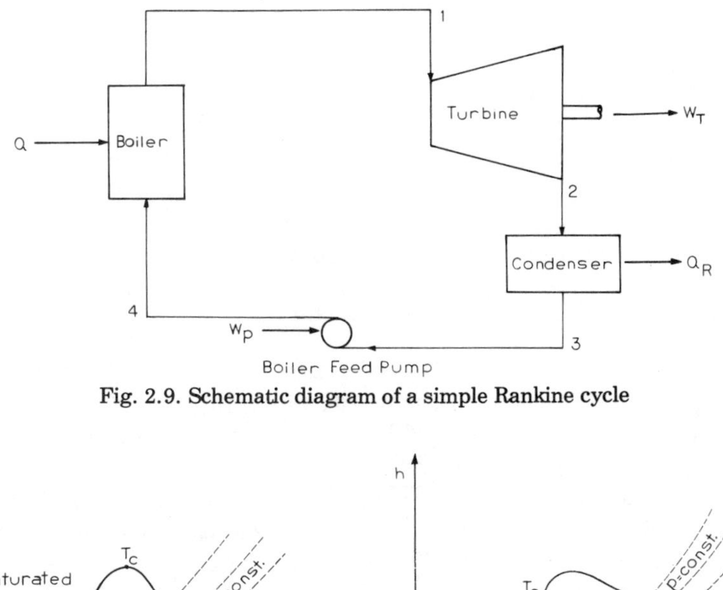

Fig. 2.9. Schematic diagram of a simple Rankine cycle

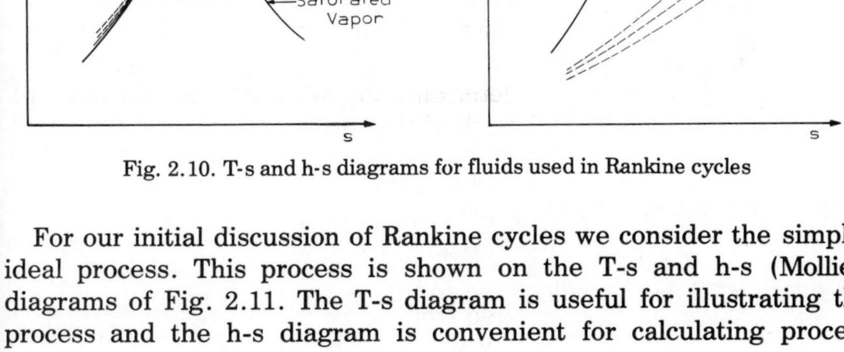

Fig. 2.10. T-s and h-s diagrams for fluids used in Rankine cycles

For our initial discussion of Rankine cycles we consider the simple, ideal process. This process is shown on the T-s and h-s (Mollier) diagrams of Fig. 2.11. The T-s diagram is useful for illustrating the process and the h-s diagram is convenient for calculating process efficiency. In the ideal Rankine cycle, saturated steam (shown at point 1) enters the turbine and expands isentropically to position 2_s. At point 2_s the wet steam enters the condenser where heat is removed until the fluid is condensed to a saturated liquid at point 3. After leaving the condenser the fluid is compressed isentropically from pressure p_2 to the boiler pressure p_1. The high pressure liquid at point 4_s enters the boiler where the fluid is vaporized and emerges as steam at point 1.

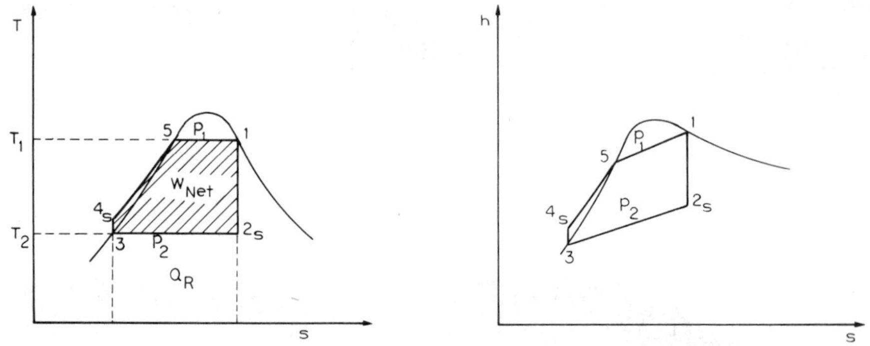

Fig. 2.11. T-s and h-s diagram representations for ideal simple Rankine cycles

Note: We are assuming fluid velocities are zero; i.e., the diagram illustrates stagnation properties.

As seen in Fig. 2.11, in the ideal Rankine cycle heat is rejected at constant temperature T_2. Heat addition is composed of sensible heat between points 4_s and 5 and latent heat at constant temperature T_1 between points 5 and 1. Ordinarily, most of the heat addition is at constant temperature. In the Carnot cycle, all heat addition and rejection is at constant temperature. Consequently, the ideal Rankine cycle closely approaches the Carnot cycle. The shaded area on the temperature-entropy diagram in Fig. 2.11 represents the net work performed by the ideal Rankine cycle. The total area under the cycle curve represents the heat addition and the unshaded area is the heat rejected Q_R.

The h-s diagram is used to determine the net work and efficiency of the Rankine cycle. The shaft work of the turbine is

$$W_T = h_1 - h_{2s} \tag{2.3.1}$$

while the pumping work is

$$W_p = h_{4s} - h_3 \tag{2.3.2}$$

The pump work is also calculated from

$$W_p = \frac{v}{J}(p_1 - p_2) \tag{2.3.3}$$

where v is the average specific volume of the liquid. Note that Eq. (2.3.3) assumes no velocity changes between pump inlet and outlet and a reasonably constant fluid specific volume. For water undergoing pressure changes of 2,000 psi at constant temperature, the change in specific volume is of the order of 10^{-4}cu-ft/lb_m which is less than 1 per cent.

Consequently, the net work per pound of working fluid is

$$W_{Net} = (h_1 - h_{2s}) - (h_{4s} - h_3) = (h_1 - h_{2s}) - \frac{v}{J}(p_1 - p_2) \quad (2.3.4)$$

One should keep in mind that enthalpies and pressures represented in Fig. 2.11 are stagnation enthalpies and pressures. However, for most problems one can use static enthalpies and pressures.

The heat input per pound of fluid is

$$Q = (h_1 - h_{4s}) = (h_1 - h_3) - \frac{v}{J}(p_1 - p_2) \quad (2.3.5)$$

Therefore, the overall thermal efficiency is given by

$$\eta = \frac{W_{Net}}{Q} = \frac{(h_1 - h_{2s}) - (h_{4s} - h_3)}{(h_1 - h_{4s})} = \quad (2.3.6)$$

$$= \frac{(h_1 - h_{2s}) - (v/J)(p_1 - p_2)}{(h_1 - h_3) - (v/J)(p_1 - p_2)}$$

The ideal Rankine cycle is not achieved in practice because of irreversibilities present in the cycle. Some of the irreversibilities are (1) turbine efficiency is less than 100 per cent, (2) condenser subcooling, which increases heat input, (3) pressure drop in the boiler, (4) ducting pressure losses, and (5) pump efficiency is less than 100 per cent. The effect of irreversibilities is shown on the h-s diagram of Fig. 2.12.

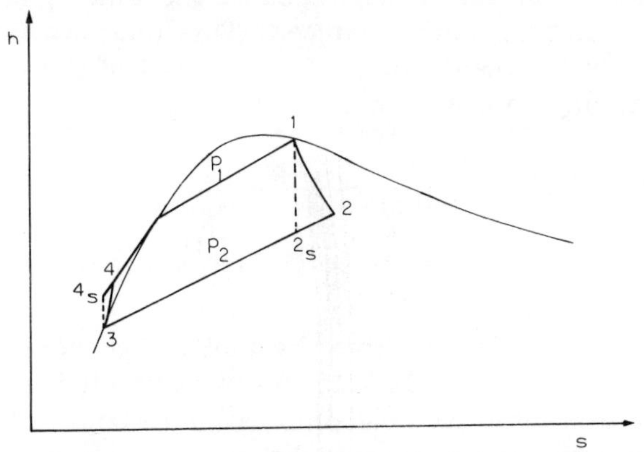

Fig. 2.12. Enthalpy-entropy diagram for an actual simple Rankine cycle

Irreversibilities in pumps and turbines cause an increase in fluid entropy during compression and expansion. The effects of compression work are exaggerated in Fig. 2.12 in order to show detail. In the actual construction of Mollier diagrams for Rankine cycles, the compressor work is barely distinguishable. From the definitions of compressor and turbine efficiencies, we have

$$\eta_p = \frac{h_{4s} - h_3}{h_4 - h_3} \quad , \quad \eta_t = \frac{h_1 - h_2}{h_1 - h_{2s}} \tag{2.3.7}$$

Therefore, the net work per pound of fluid is

$$W_{Net} = \eta_t (h_1 - h_{2s}) - \frac{v\Delta p}{\eta_p J} \tag{2.3.8}$$

and the heat input per pound of fluid is

$$Q = (h_1 - h_4) = (h_1 - h_3) - \frac{v\Delta p}{\eta_p J} \tag{2.3.9}$$

Consequently, the overall thermal efficiency of the practical Rankine cycle is given by

$$\eta = \frac{\eta_t(h_1 - h_{2s}) - (v\Delta p/\eta_p J)}{(h_1 - h_3) - (v\Delta p/\eta_p J)} \tag{2.3.10}$$

Terms which are frequently used as indices of performance for power plants are the **turbine heat rate** and the **net plant heat rate**. The turbine heat rate is the ratio of the heat supplied to the fluid in the boiler to the net work output of the heat engine. Therefore, we may write

$$\text{Turbine Heat Rate} \; \frac{(\text{Btu})}{(\text{kW-hr})} = \frac{Q}{W_{Net}} = \frac{3412\,(h_1 - h_4)}{\eta_t(h_1 - h_{2s}) - (v\Delta p/\eta_p J)} \tag{2.3.11}$$

The net plant heat rate is an index of the total quantity of fuel consumed per unit of work produced. This quantity is of greater significance in power plant evaluation since part of the energy of the consumed fuel may be lost. The net plant heat rate is defined by

$$\text{Net Plant Heat Rate} \; \frac{(\text{Btu})}{(\text{kW-hr})} = \frac{Q_f}{W_{Net}} = \frac{Q_f}{Q} \times \text{Turbine Heat Rate} \tag{2.3.12}$$

where Q_f is energy available from the fuel in Btu per lb_m of flowing fluid. Q_f/Q ranges from 1.10-1.15 for fossil-fueled power plants because 10-15 percent of the heat supplied to the boiler is lost in stack gases and absorption in boiler walls. For nuclear power plants Q_f/Q is approximately 1.0.

Method of Solving Problems

Two simple and convenient methods are available for calculating the work and efficiency of Rankine cycles. Naturally both methods require detailed knowledge of the working fluid thermodynamic properties. For steam cycle computations in the United States, the properties of steam must be taken from the ASME Steam Tables [1].

(1) Graphical Analysis

This method is the type discussed in the last section analyzing the cycle on a Mollier diagram. The work and heat are obtained from enthalpy changes of the working fluid. The Mollier diagram must contain all the thermodynamic states which the working fluid undergoes. This chart is useful for predicting turbine work but its range is frequently not sufficient for analyzing the condensed fluid. Consequently, the use of Steam Tables is usually more suitable and accurate for analyzing power cycles.

(2) Tables of Thermodynamic Properties

The use of thermodynamic property tables for computing cycle efficiencies is best illustrated by working an example problem. In this example we consider producing shaft work by use of heat generated in a boiling water reactor. The approach used in this problem is the same used for the analysis of any Rankine cycle. One must simply know the appropriate thermodynamic properties of the working fluid. Tables of thermodynamic properties of water are given in Appendix B.

Example 2.1: A boiling water reactor produces saturated steam at 1000 psia. The steam passes through a turbine and is exhausted at 1 psia. The steam is condensed to a subcooling of 3°F and then pumped back to the boiler pressure. The turbine efficiency is 0.8 and the pump efficiency is 0.9. Compute the following parameters:

a. Net work per pound of fluid

b. Heat rejected per pound of fluid

c. Reactor heat per pound of fluid

d. Turbine Heat Rate

e. Overall efficiency

Solution:

a. By consulting the ASME Steam Tables it is seen that saturated steam at 1000 psia has the following properties:

$$h_1 = 1192.9 \text{ Btu/lb}_m$$
$$s_1 = 1.3910 \text{ Btu/lb}_m\text{-}^0\text{R}$$

Expanding the steam isentropically in a turbine to 1 psia, the entropy of the exhaust steam is

$$s_{2s} = 1.3910 \text{ Btu/lb}_m\text{-}^0\text{R}$$

At 1 psia pressure the saturated liquid and saturated vapor have the following properties:

Sat. Liquid	Sat. Vapor
$h_f = 69.73\ \mathrm{Btu/lb_m}$	$h_g = 1105.8\ \mathrm{Btu/lb_m}$
$s_f = 0.1326\ \mathrm{Btu/lb_m\text{-}^0R}$	$s_g = 1.9781\ \mathrm{Btu/lb_m\text{-}^0R}$

By comparing the entropy of the expanded fluid with the entropies of saturated liquid and vapor, one sees that the fluid is a mixture of liquid and vapor. It is necessary to determine the fluid vapor-liquid composition in order to calculate its enthalpy. Let y=the amount of liquid (moisture content) in the mixture, then from the known mixture entropy one can compute y from the following relation:

$$s_{2s} = ys_f + (1 - y)s_g \tag{2.3.13}$$

or

$$y = \frac{s_g - s_{2s}}{s_g - s_f}$$

For our example

$$y = \frac{1.9781 - 1.3910}{1.9781 - 0.1326}$$

$$= 0.318$$

The fluid enthalpy from an isentropic expansion is now given by

$$h_{2s} = (0.318)(69.73) + (0.682)(1105.8)$$

$$= 776\ \mathrm{Btu/lb_m}$$

Therefore, the reversible work produced by the turbine is

$$W_{Ts} = h_1 - h_{2s}$$

$$= 1192.9 - 776$$

$$= 417\ \mathrm{Btu/lb_m}$$

However, because of irreversibilities the actual turbine work is given by

$$W_T = \eta_t(h_1 - h_{2s})$$

$$= (0.8)(417)$$

$$= 334\ \mathrm{Btu/lb_m}$$

The work produced is still equal to the change in mixture enthalpy and one can now compute the enthalpy of the fluid exiting from the turbine.

$$\begin{aligned} h_2 &= h_1 - W_T \\ &= 1192.9 - 334 \\ &= 859 \text{ Btu/lb}_m \end{aligned}$$

The isentropic pumping work is given by

$$\begin{aligned} W_{ps} &= \frac{v}{J}(p_1 - p_2) \\ &= \frac{0.0161}{778}(1000 - 1)144 \\ &= 2.96 \text{ Btu/lb}_m \end{aligned}$$

However, the compression is not isentropic and the actual pumping work is given by

$$\begin{aligned} W_p &= \frac{W_{ps}}{\eta_p} \\ &= \frac{2.96}{0.90} = 3.29 \text{ Btu/lb}_m \end{aligned}$$

Therefore, the net work is given by

$$\begin{aligned} W_{Net} &= W_T - W_p \\ &= 334 - 3.29 \\ &= 331 \text{ Btu/lb}_m \end{aligned}$$

b. The heat rejected by the condenser is given by the change in enthalpy between the fluid leaving the turbine and liquid at 1 psia subcooled 3°F.

$$\begin{aligned} Q_R &= h_2 - h_3 \\ &= 859 - 66.73 \\ &= 792 \text{ Btu/lb}_m \end{aligned}$$

c. The heat added by the reactor is given by the sum of the net work and the heat rejected by the condenser.

$$Q = Q_R + W_{Net} \quad \text{or} \quad h_1 - h_4$$
$$= 792 + 331 \quad \text{or} \quad 1192.9 - (66.73 + 3.29)$$
$$= 1123 \text{ Btu/lb}_m \quad \text{or} \quad 1123 \text{ Btu/lb}_m$$

d. The turbine heat rate is given by

$$\text{Turbine Heat Rate} = \frac{Q}{W_{Net}}$$
$$= \frac{(3412)(1123)}{331}$$
$$= 11{,}580 \text{ Btu/(kW-hr)}$$

e. The overall thermal efficiency is given by

$$\eta = \frac{W_{Net}}{Q}$$
$$= \frac{331}{1123}$$
$$= 0.295$$

In this example the steam moisture content is 23.8 per cent at the turbine exhaust. This is unrealistic for practical turbines because a high moisture content decreases turbine efficiency and causes erosion of turbine blades. Moisture contents higher than 10 per cent are undesirable, although with special stellite inserts on rotor blade leading edges these moisture levels can be tolerated. Moisture content in the exhaust fluid is a factor that must be considered in specifying the conditions of a practical Rankine cycle. For the example cited, it would be necessary to have moisture separators between turbine stages in order to reduce the moisture level. The next section discusses various ways to reduce moisture content for a given exhaust pressure and improve upon cycle efficiency.

As the example shows, the pumping work is the order of 1 per cent of the turbine shaft work. Consequently, the overall thermal efficiency of a practical Rankine cycle is approximately equal to the turbine efficiency times the overall efficiency of an ideal Rankine cycle.

Methods of Improving the Rankine Cycle

The expression for overall thermal efficiency can be used as a guide in analyzing factors for improving thermal efficiency.

$$\eta = \frac{W_T - W_p}{Q} \approx \frac{W_T}{Q}$$

The pumping work is quite small for Rankine cycles and, consequently, reducing pumping work usually produced a negligible effect on overall thermal efficiency. In order to improve efficiency one must either increase W_T or decrease Q. We first discuss methods of increasing W_T.

The work produced by the turbine is increased by (1) raising the boiler pressure, (2) lowering the exhaust pressure, (3) using superheat, and (4) using reheat. Factors (1), (3), and (4) effectively increase the temperature of the entering fluid and factor (2) decreases the temperature at which heat is rejected. These trends in inlet and outlet temperature are the same trends used to improve the Carnot cycle efficiency.

The condenser pressure is limited by the temperature of available cooling water, size and cost of the condenser, and size of vacuum pumps required to deaerate the condenser. Consequently, there is a practical lower limit to condenser pressures and existing plants today operate as low as 1″ Hg abs. to 3″ Hg abs. Little room is available for improving cycle efficiency by reducing condenser pressure. Table 2.1 illustrates recommended turbine exhaust pressures as a function of available cooling water temperature. Therefore, improvement in turbine work is essentially achieved by increasing boiler pressure and/or using superheat and reheat. The influence of these factors is best illustrated by temperature-entropy diagrams.

Table 2.1

Recommended turbine exhaust pressures for various cooling water temperatures [2]

Water Temperature	Exhaust Pressure
60 °F	1.0 in Hg. abs.
70 °F	1.5 in Hg. abs.
80 °F	2.0 in Hg. abs.
85 °F	2.5 in Hg. abs.
90 °F	3.0 in Hg. abs.
95 °F	3.5 in Hg. abs.

(1) Raising Boiler Pressure

The effect of increasing boiler pressure on Rankine cycle efficiency is readily shown on the temperature-entropy diagram illustrated in Fig. 2.13. Increasing boiler pressure results in an increase in net work (represented by the shaded area) with a corresponding decrease in heat rejected. However, increasing boiler pressure results in an increase in moisture content of the exhaust steam with its attendant problems in turbine design and efficiency. Consequently, in nuclear power plants there usually exists an optimum boiler pressure which is obtained from

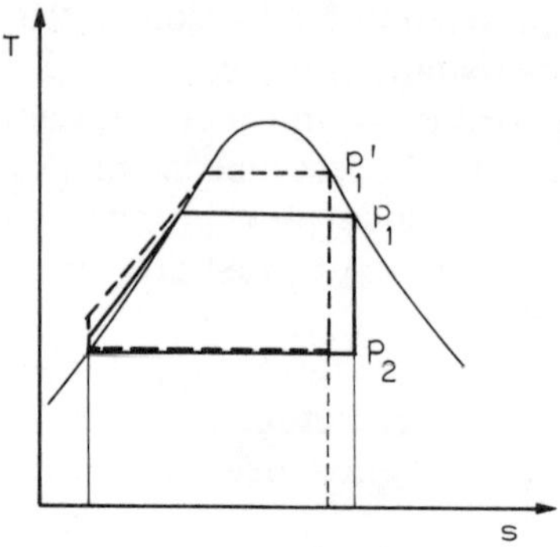

Fig. 2.13. Effects of increasing boiler pressure on the Rankine cycle

a trade-off of increased cycle efficiency for increased power plant reliability and decreased capital costs.

(2) Superheat

Figure 2.14 illustrates the Rankine cycle with superheat. Superheat causes a net increase in temperature at which heat is being received with a resulting improvement in cycle efficiency. Another important factor is that the amount of moisture in the fluid leaving the turbine is reduced which increases turbine efficiency and reduces erosion. However, in order to make use of superheat, one must have a high temperature heat source or reduce boiler pressure.

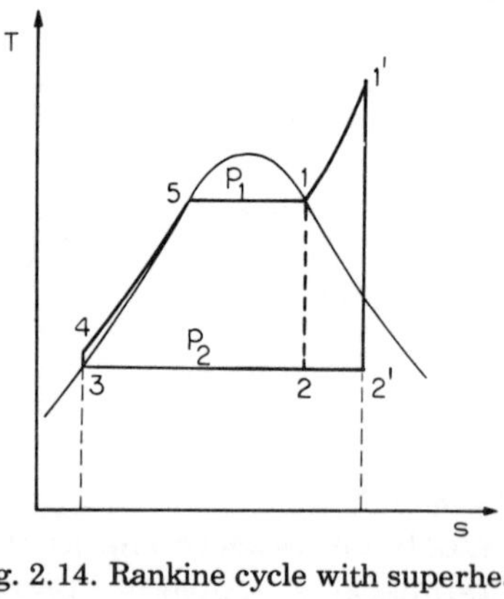

Fig. 2.14. Rankine cycle with superheat

(3) Reheat

The effective temperature of heat addition is increased and the moisture content further reduced by using reheat in the Rankine cycle. A schematic diagram of the power plant and appropriate temperature-entropy diagram is shown in Fig. 2.15. High pressure, superheated steam is expanded in a high-pressure turbine to an intermediate pressure p'_2 and the fluid then returned to a second stage boiler and superheater and reheated to state $1''$. The reheated steam is then expanded in a low-pressure turbine to the final exhaust pressure p''_2. The moisture content of the working fluid is drastically reduced by use of reheat and this approach is used in all fossil-fueled and many nuclear power plants.

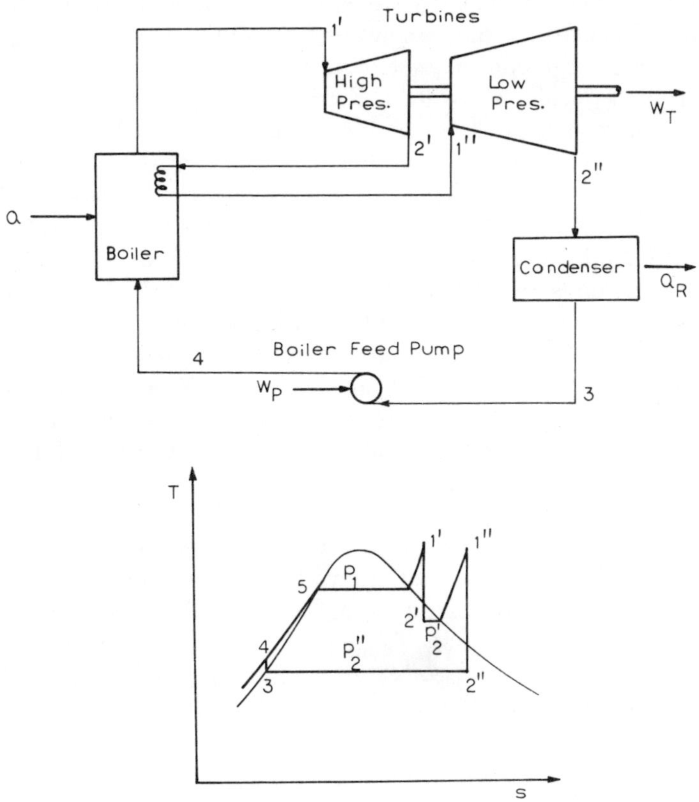

Fig. 2.15. Rankine cycle with reheat

The approach used to compute the work and efficiency of reheat cycles is the same as used in the example problem for the simple Rankine cycle. One calculates the work produced in each turbine separately and the required pumping work. Heat is added to the fluid at two different stages of the cycle and is given by the difference in enthalpy between states $1'$ and 4 and states $1''$ and $2'$.

(4) Regenerative Cycle

The factors mentioned so far for increasing turbine work are employed when practical on nuclear power plants. Another approach for improving cycle efficiency is to use the regenerative cycle which reduces the heat source term, Q, in the expression for overall thermal efficiency.

One of the principal sources of irreversibility in the Rankine cycle is sensible heat addition which occurs over a range of temperatures from the boiler feed pump outlet temperature to the saturation temperature at the boiler pressure. This irreversibility is reduced by regeneration which takes heat from steam expanding through the turbine to provide sensible heat for the compressed liquid.

Ideal regeneration, shown in Fig. 2.16, has compressed liquid between states 4 and 5 receiving heat continuously from vapor expanding between states 1 and 2′ with no temperature difference between the two fluids. This is accomplished, hypothetically, by constructing a turbine with an annular space in its casing and pumping compressed liquid through this spacing in a direction opposite to the steam flow. By constructing a turbine with a thin-walled casing of high thermal conductivity material and having very slow liquid flow, it may be possible for the liquid to be facing vapor at nearly the same temperature and receiving heat from this vapor. Figure 2.17 illustrates a schematic diagram of the ideal regenerative cycle power plant. Under the imposed conditions the heat transfer between the liquid and vapor is approximately reversible.

With ideal regeneration, heat is added to the fluid at constant temperature T_1 and rejected to the surroundings at constant temperature T_2. It is easy to show that the efficiency of the ideal regenerative cycle is the same as a Carnot cycle operating between temperature limits T_1 and T_2. The heat transferred reversibly from the vapor is given by the area under the line 1-2′ in Fig. 2.16. Heat added to the compressed liquid is equal to the area under the line 5-4.

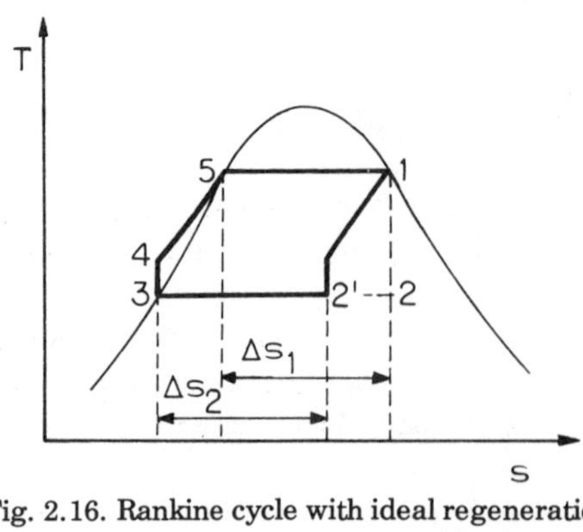

Fig. 2.16. Rankine cycle with ideal regeneration

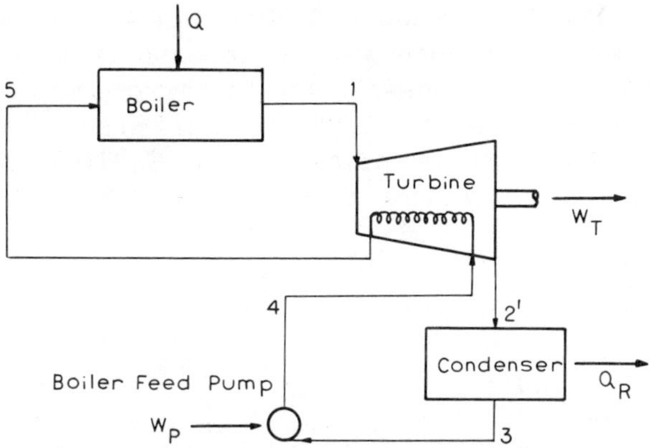

Fig. 2.17. Schematic diagram of a power plant with ideal regeneration

Both of these areas are equal, since it is assumed that no heat is lost to the surroundings, and consequently $\Delta s_1 = \Delta s_2$. Therefore, the overall thermal efficiency of the ideal regenerative Rankine cycle is given by

$$\eta = \frac{\text{heat added} - \text{heat rejected}}{\text{heat added}} = \frac{T_1 \Delta s_1 - T_2 \Delta s_2}{T_1 \Delta s_1} = \frac{T_1 - T_2}{T_1} \tag{2.3.14}$$

which is the same as the Carnot cycle efficiency.

Even if it were possible to build a turbine capable of operating on an ideal regenerative cycle, such a turbine would be very impractical. Ideal regeneration increases vapor moisture content to a prohibitively high level. However, in practice an approximate form of regeneration is used in the manner shown in Fig. 2.18. Steam is withdrawn from the turbine at some point during the expansion and used to heat the condensate. Figure 2.18 shows a single feedwater heater; although, depending upon the economics of the plant, numerous feedwater heaters are used with steam being withdrawn from several points in the turbine.

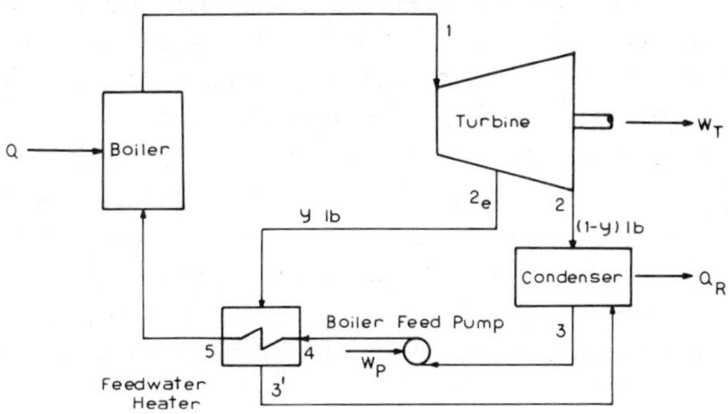

Fig. 2.18. Single heater regenerative cycle

In the analysis of regeneration, one determines the amount of steam that is bled from the turbine to heat the condensate near or to the saturation temperature corresponding to the pressure of the bled steam. A temperature-entropy diagram describing this process for a single-feedwater heater is illustrated in Fig. 2.19. The amount of steam required is found by applying an energy balance to the feedwater

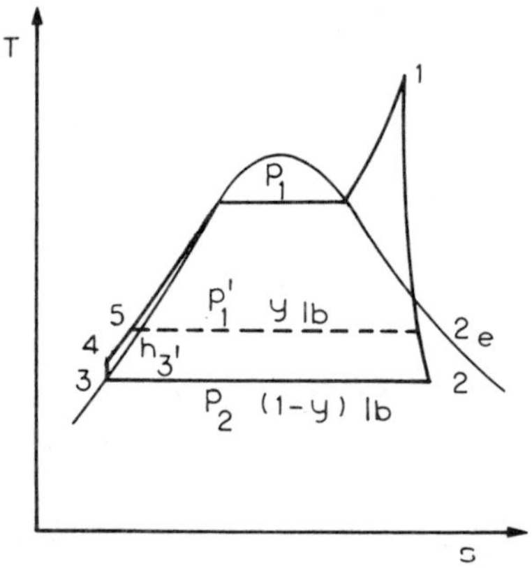

Fig. 2.19. Temperature-entropy diagram for regenerative cycles

heater. Let y equal the amount of bled steam per pound of fluid entering the turbine. Then, because the energy leaving the heater equals the energy entering the heater we have

$$yh_{3'} + h_5 = yh_{2e} + h_4$$

or

$$y = \frac{h_5 - h_4}{h_{2e} - h_{3'}} \tag{2.3.15}$$

Once a value of y is known, it is fairly simple calculating the net work and heat input for the system. One has to remember that the amount of fluid in turbine stages after the point of steam bleeding is reduced by y lb_m per lb_m of fluid entering the turbine.

Steam used for feedwater heating may be withdrawn from the turbine at any point between the turbine inlet and turbine exhaust. Nothing is accomplished by withdrawing steam from the turbine exhaust because, for perfect condensation, the condensate will be at the same temperature as the exhaust steam. At the other extreme, by withdrawing steam from the turbine inlet the heat transferred to the feedwater is at a maximum but no gain in efficiency results because the steam that heats the feedwater serves merely as a heat transfer medium between the heat source and the feedwater. Therefore, for these two extremes the reduction in Q or gain in efficiency is zero, while for some intermediate position the gain is a maximum. The optimum position for steam extraction is found from plots of gain in efficiency versus temperature to which the feedwater is heated. Figure 2.20 illustrates typical plots for various numbers of feedwater heaters.

For a single heater, the maximum reduction in Q occurs when the feedwater is at a temperature about half way between the boiler saturation temperature and condenser outlet temperature. Nuclear power plants employ about five feedwater heaters. Optimization studies using dynamic programming show that for multiple feedwater heaters the maximum efficiency is achieved when the outlet temperature from each feedwater heater is approximately equally spaced between the condenser outlet temperature and the boiler saturation temperature [4].

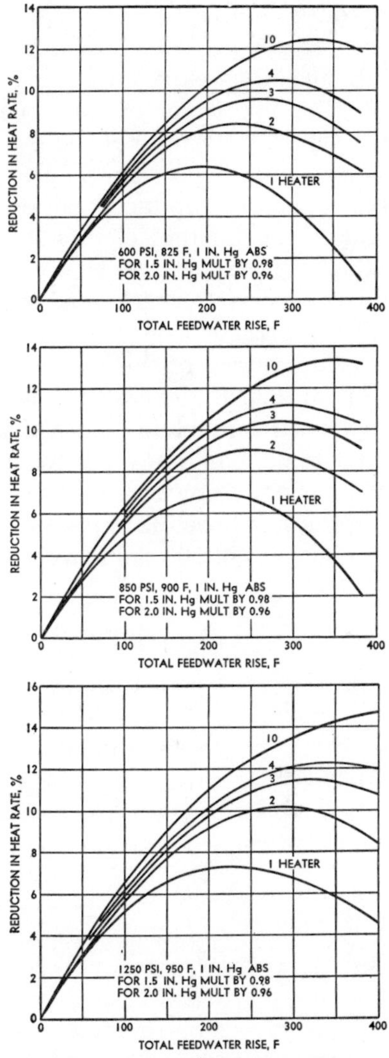

Fig. 2.20. Gain in thermal efficiency of ideal regenerative cycle over Rankine cycle, showing the influence of feed temperature and number of stages of feedwater heating [3].

Heat Addition by Reactor Coolants

The use of superheat and reheat are quite common in fossil-fueled power plants. This is due to the heat source being combustion gases which are at high temperatures. However, with the use of nuclear reactors as heat sources, temperature limitations are sometimes placed upon reactor coolants which makes the use of superheat impractical. Such situations exist for boiling and pressurized water reactors.

Boiling water reactors produce saturated steam and present designs have steam pressures of the order of 1000 psia. Several boiling water reactors employed superheat. The BONUS and Pathfinder boiling water reactors produced superheated steam by having steam pass through special fuel elements on the core periphery. The Elk River boiling water reactor used a coal-fired superheater.

In pressurized water reactors, practical design limitations usually limit the primary coolant pressure to 2300 psia or less. Consequently, the temperature of the primary coolant leaving the reactor is of the order of 620°F. Therefore, it is impractical to produce steam at temperatures much in excess of 580°F which is a limitation. Figure 2.21 illustrates a Rankine cycle for a reactor system in which the primary coolant has a maximum temperature of 620°F. As shown on the diagram, it is possible to produce saturated steam at some temperature in the vicinity of 500°F and a corresponding pressure, p_1. An alternative

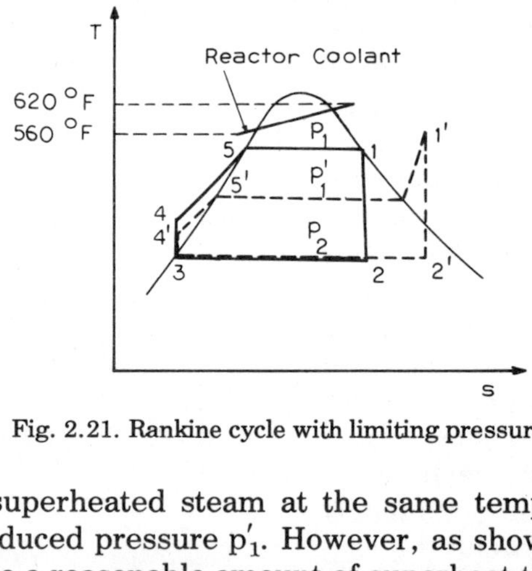

Fig. 2.21. Rankine cycle with limiting pressure

is to produce superheated steam at the same temperature but at a considerably reduced pressure p'_1. However, as shown in Fig. 2.21, in order to produce a reasonable amount of superheat the temperature at which the fluid receives most of its heat, the temperature at which vaporization occurs, is considerably reduced with a net result that the use of superheat reduces the Rankine cycle efficiency. Consequently, superheat is usually not used with pressurized water reactors. An exception is systems manufactured by the Babcock & Wilcox Company using once-through steam generators which produce steam with about 50°F superheat.

In the Rankine cycle shown in Fig. 2.21, excessive moisture is a problem and this is relieved in several ways. Reheat is used to shift the cycle to the right and reduce moisture content or moisture separators are employed. The latter approach is used in practically every water-cooled nuclear power plant. Moisture separation is accomplished internally by having grooves on the turbines blades which by centrifugal action cause the moisture to flow to the turbine casing from which it can be drained. Moisture is removed externally by passing the full steam flow through an external moisture separator. Fig. 2.22 illustrates a PWR steam cycle employing reheat by use of high pressure and bleed steam, moisture separation, and six stages of feedwater heaters using bleed steam and moisture [5].

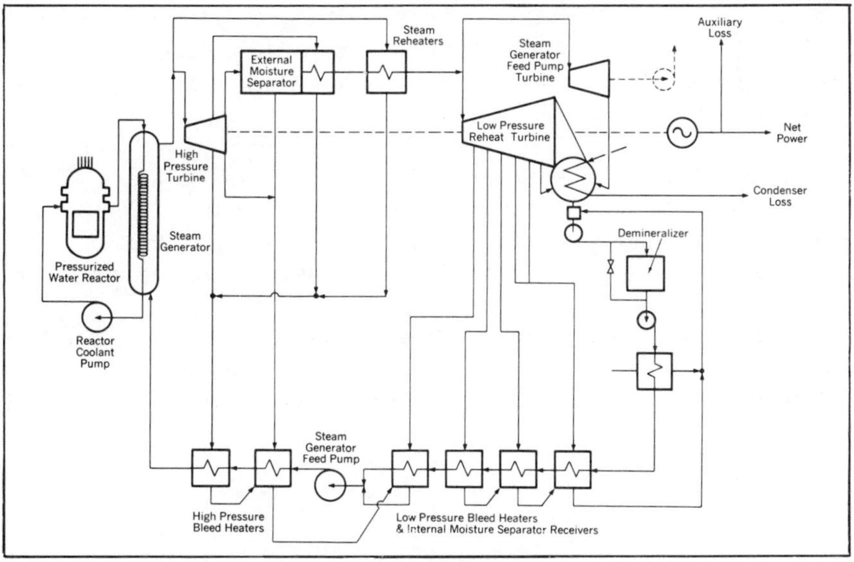

Fig. 2.22. Power cycle diagram, nuclear fuel—reheat by bleed and high pressure steam, moisture separation, and 6-stage regenerative feed heating—900 psia, 566/503F steam [5].

One can employ superheat with PWR systems by using fossil-fueled superheaters. This method was used with the Indian Point I Reactor. However, fossil-fueled superheaters are more expensive when incorporated with nuclear reactors than when used with conventional boilers. Consequently, recent nuclear power plants have not used fossil-fueled superheaters.

In boiling water reactors, heat is added directly to the working fluid and either saturated or superheated steam passes to the turbine. The steam pressure can be varied up to the practical design limitations of pressure vessels. However, in other types of reactor systems which use a primary coolant and intermediate heat exchangers, the steam pressure and temperature is now controlled by the temperature of the coolant entering and leaving the reactor. The working fluid temperature can not exceed the primary coolant temperature at any point in the heat exchanger and is usually kept 30-100°F less in PWRs and 300°F less in HTGRs in order to enable reasonable sized heat exchangers. Because of temperature limitations placed upon many coolants, the use of primary coolants frequently places great restrictions upon power plant thermodynamic cycles.

If the primary coolant (i.e. water, organic liquids, gases, or liquid metals) does not undergo phase changes in the primary loop, then the processes taking place in the primary heat exchanger can be represented by temperature-% heat exchanged diagrams similar to Fig. 2.23. A point of minimum temperature difference between the primary coolant and working fluid is called a **pinch point**. Two pinch points may occur in a heat exchanger. The first point is located at the position where vaporization commences and, consequently, limits the steam pressure. The second pinch point occurs at the end of a superheater and limits maximum steam temperature. Temperature differences at pinch points are design criteria and are usually kept at least 20-30°F in order to limit heat exchanger sizes to reasonable values.

A superheater is frequently worthwhile only if a substantial degree of superheat is obtained. In pressurized water reactors the coolant temperature rise is small and, therefore, the available temperature

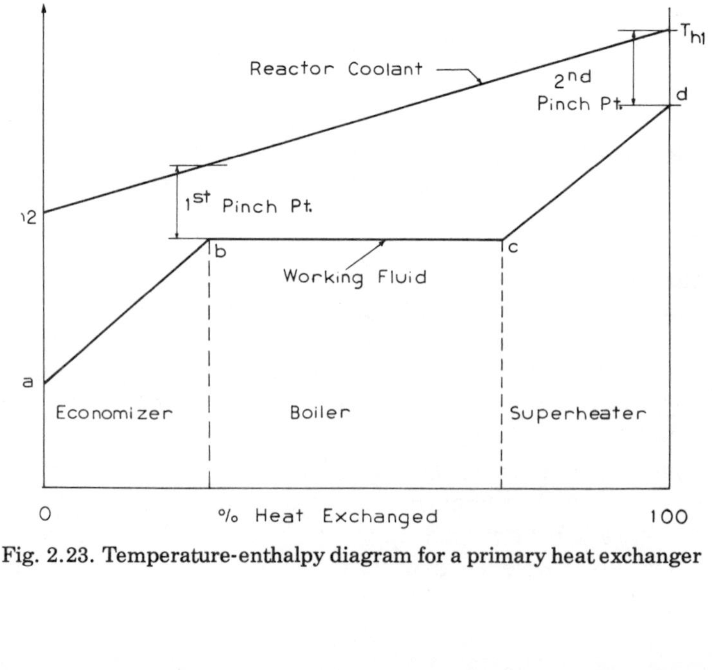

Fig. 2.23. Temperature-enthalpy diagram for a primary heat exchanger

difference varies little along the heat exchanger and superheat can accomplish little. However, in gas-cooled and liquid-metal-cooled reactors, the primary coolant outlet temperature is high and changes in coolant temperature substantial and, consequently, large temperature differences are available for superheating.

Reactor coolant temperatures are usually specified from heat transfer, corrosion, and nuclear considerations in the core. Thermodynamically it is desirable to have the highest permissible outlet temperature for reactor coolants. Pinch point temperature differences are specified from economic considerations of the heat exchanger. Therefore, with reactor coolant temperatures and pinch point temperature differences fixed, an analysis of the heat exchanger determines working fluid temperatures and pressure. Pinch point positions are determined by trial-and-error. Using the nomenclature illustrated in Fig. 2.23, the procedures used in the calculations are as follows:

(1) T_{h1}, T_{h2}, and pinch point temperature differences are specified.

(2) Assume the temperature for the first pinch point (T_b) which gives the per cent heat at the pinch point, steam pressure, and h_b.

(3) Calculate the sensible heat added to the working fluid or $h_b - h_a$.

(4) Calculate the total heat added to the working fluid $h_d - h_a$ from

$$h_d - h_a = \frac{h_b - h_a}{\%\ \text{heat at pitch point}} \tag{2.3.16}$$

This calculation gives h_d.

(5) If it is desired to produce only saturated steam, h_d should be the enthalpy of saturated steam corresponding to the pressure determined in part 2. If superheat is desired, then the temperature of the exiting working fluid, T_d, is calculated from steam tables.

(6) The assumed pinch point temperature is varied until h_d and/or T_d come into the desired range. With superheat T_d is determined by specifying a second pinch point.

Binary Vapor Cycles

The conventional working fluid in vapor-cycle plants is water because of its well-understood technology. Because the critical temperature of water is 706°F, from the standpoint of efficiency it is not an ideal fluid for use in vapor-cycle engines. The average temperature at which heat is added is usually considerably less than 700°F and, consequently, this poses a restriction on cycle thermal efficiency. Because of extreme pressures involved in steam cycles, a maximum superheat temperature of 1000°F is imposed upon plants because of the "metallurgical limit" of materials. With vaporization of water occurring at temperatures of the order of 600°F, the average temperature of heat addition is of the order of 650-700°F or less which is well under the so-called "metallurgical limit" of the containing materials. The maximum thermal efficiency of the highest performance steam plants today is about 42 per cent. Efficiencies for nuclear power plants range from 0.33 for water-cooled reactors to about 0.40 for HTGRs and LMFBRs.

A means of increasing average temperature of heat addition in vapor-cycle systems and, therefore, increasing thermal efficiencies is to employ the binary vapor cycle. In a binary cycle, heat from the heat source is used to vaporize a fluid with a high critical temperature and a reasonably high boiling point. This vapor passes through a turbine to a condenser where its heat of condensation is used to vaporize water. The steam in turn passes through a second turbine to another condenser where its heat of condensation is rejected to the surroundings. Water and mercury were employed in binary vapor-cycle systems with overall efficiencies of the order of 46-48 per cent.

Figure 2.24 illustrates a schematic diagram of a water-mercury binary cycle and the appropriate temperature-entropy diagram which is

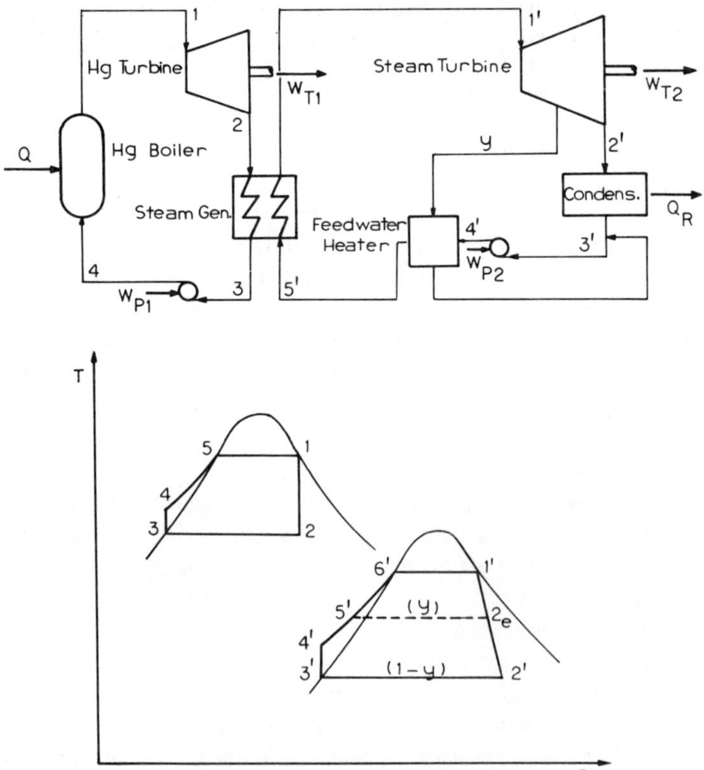

Fig. 2.24. Binary mercury-water cycle

used in its analysis. The analysis of this system is no different from the analysis used in the preceding sections on the Rankine cycle. One requirement is that flow of the high temperature fluid must be sufficient to enable complete vaporization of the low temperature fluid. For the plant described, it is convenient to perform the analysis on the basis of one pound of water. Therefore, from an energy balance on the steam generator, the flow of mercury per pound of water is given by

$$\dot{m}_{Hg} = \frac{h_{1'} - h_{5'}}{h_2 - h_3}, \frac{lb_m\,(Hg)}{lb_m\,(H_2O)} \tag{2.3.17}$$

The net work performed by this plant on the basis of one pound of water is

$$W_{Net} = W_{T2} - W_{p2} + \dot{m}_{Hg}(W_{T1} - W_{p1}) \tag{2.3.18}$$

and the overall efficiency is

$$\eta = \frac{W_{Net}}{Q} = \frac{W_{T2} - W_{p2} + \dot{m}_{Hg}(W_{T1} - W_{p1})}{Q} \tag{2.3.19}$$

A Mollier diagram for mercury is given in Appendix B.

The use of mercury as a second working fluid has been abandoned in stationary power plants which may be due to its extreme toxicity and difficulty in handling. However, organic materials such as diphenyl and diphenyl oxide (Dowtherm) were considered for nuclear applications because these materials were used as reactor coolants.

2.4 Gas Turbine or Brayton Cycles

In gas turbines the working fluid is a vapor throughout the entire cycle and, therefore, the work of compression is quite significant. In order for gas turbines to be practical as energy converters, high component efficiencies are essential. Such efficiencies were not attained in compressors until the 1930s. Today we see turbines completely superceding reciprocating engines for high-speed aircraft.

In order to achieve reasonable thermal efficiencies, gas turbines require high turbine inlet temperatures and, consequently, the use of gas turbines with nuclear reactors has awaited the development of high temperature gas-cooled reactors. At present, gas turbines have not been used for stationary power plants. However, General Atomic Company has performed many design studies using gas turbines with HTGRs [6]. One advantage of this cycle is that dry cooling can be used for heat rejection which eliminated locating a power plant near an abundant supply of water.

A considerable freedom of choice exists in the selection of working fluids for gas turbines. However, material, nuclear, and thermal design considerations have limited the desirable gases to hydrogen, helium, carbon dioxide, nitrogen, and air. Helium and carbon dioxide are the most desirable fluids because of their inertness, low absorption cross sections, and reasonable heat transfer characteristics. The departure of CO_2 from perfect gas laws results in less work of compression, and therefore, slightly better cycle performance than helium.

Four possible systems can be considered when using gas turbines with nuclear reactors—(1) direct open-cycle, (2) indirect open-cycle, (3) direct closed-cycle, and (4) indirect closed-cycle. Figure 2.25 shows schematic diagrams of these systems.

Each of the systems in Fig. 2.25 has its own particular set of advantages and disadvantages. The direct open-cycle is simple and compact; however, it rejects radioactive air to the atmosphere. The indirect open-cycle eliminates the radioactivity problem, but air is not the best working fluid for turbines. The direct closed-cycle enables use of any gas for a working fluid, however, there is risk of contamination of this gas by neutron activation and fission product release. The indirect closed-cycle is the safest of these systems, but also the most complex. The use of intermediate heat exchangers in indirect cycles also results

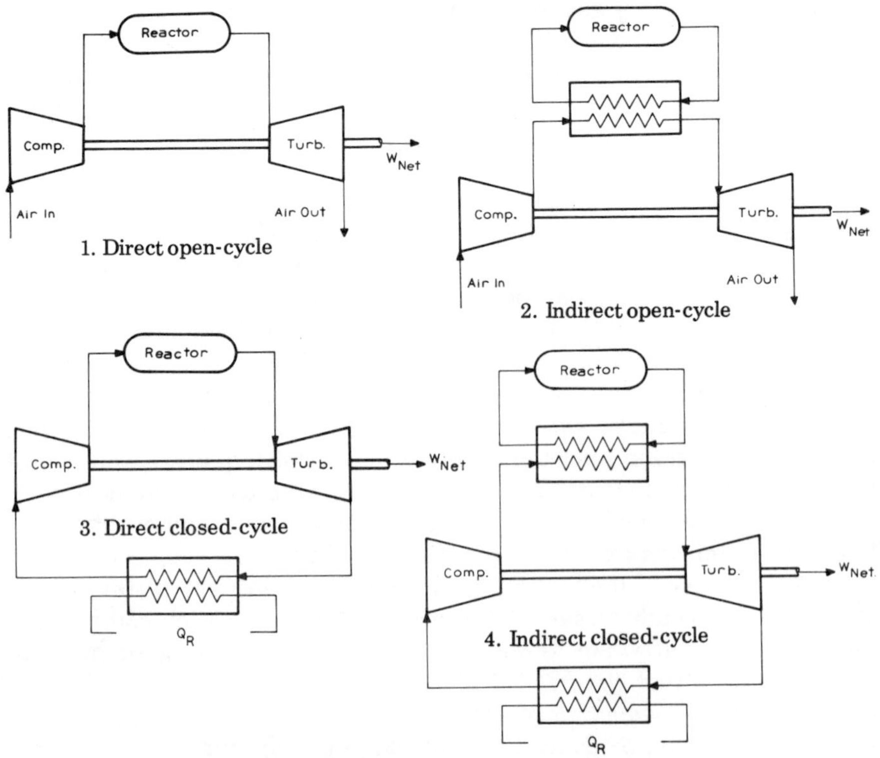

Fig. 2.25. Nuclear powered gas turbine systems

in lower efficiencies because of reduced working fluid temperatures.

It is difficult to justify the use of gas turbines over steam turbines on the basis of efficiency alone. Steam turbines usually have higher efficiencies even when gas turbine working fluids are above 1200°F. However, gas turbines have an advantage in that they enable more compact systems because the lowest working fluid pressure in closed-cycle gas turbines is much higher than in steam plants, i.e., gas pressures are of the order of 100 psia as opposed to steam back pressures of the order 1 psia. In addition, erosion of gas turbine blades is an insignificant problem compared to possible erosion in steam turbines.

The Ideal Brayton Cycle

The simple ideal Brayton cycle is illustrated on p-v, T-s, or h-s diagrams and for purposes of calculations, T-s or h-s diagrams are the most convenient. Figure 2.26 illustrates a simple Brayton cycle plant and its corresponding h-s diagram. For ideal gases, h-s diagrams are equivalent to T-s diagrams because enthalpy is proportional to temperature.

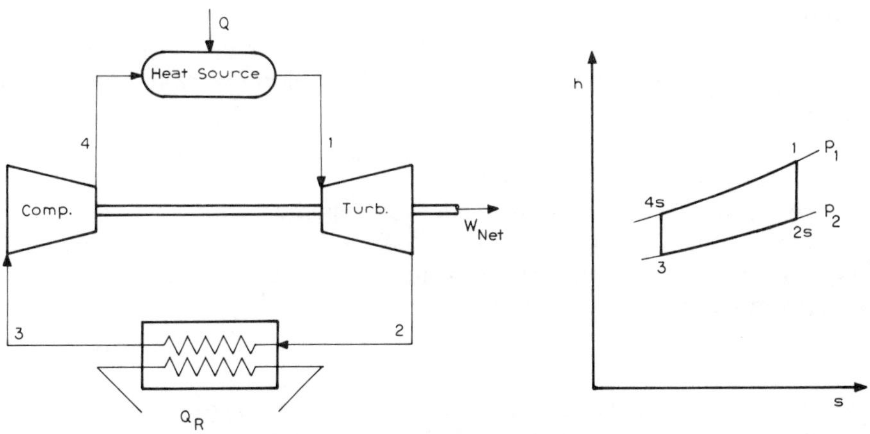

Fig. 2.26. Ideal simple Brayton cycle

In the ideal Brayton cycle, a high pressure, high temperature gas at state 1 undergoes an adiabatic and isentropic expansion through a turbine to a lower pressure and temperature state 2s. Heat is removed from the fluid at constant pressure to state 3, the fluid is compressed adiabatically and isentropically to state 4s , and heated at constant pressure to state 1 and the process repeated.

Assuming negligible potential and kinetic energy changes in the fluid; the turbine work is given by the fluid enthalpy change.

$$W_{Ts} = h_1 - h_{2s} \tag{2.4.1}$$

The compressor work is given by

$$W_{cs} = h_{4s} - h_3 \tag{2.4.2}$$

and the heat added to the fluid is given by

$$Q = h_1 - h_{4s} \tag{2.4.3}$$

Therefore, the overall thermal efficiency of the ideal Brayton cycle is

$$\eta = \frac{W_{Net}}{Q} = \frac{(h_1 - h_{2s}) - (h_{4s} - h_3)}{h_1 - h_{4s}} \tag{2.4.4}$$

Gas Tables published by Keenan and Kaye [7] list thermodynamic properties for many common gases. By specifying necessary conditions at states 1, 2, 3, and 4—two independent thermodynamic properties at each state—one readily computes overall thermal efficiencies.

However, in order to use Gas Tables, the pressures p_1 and p_2 must be specified as well as temperatures at states 3 and 1. The temperature and pressure at state 1 is usually specified by materials and design limitations. However, the pressure p_2 is determined by further analysis.

By assuming a perfect gas, enthalpy differences are written in terms of temperature differences. Therefore, the turbine work is rewritten in the form

$$W_{Ts} = c_p(T_1 - T_{2s}) = c_p T_1 \left(1 - \frac{T_{2s}}{T_1}\right) \tag{2.4.5}$$

where c_p is the gas specific heat at constant pressure (assumed constant) and T is the absolute temperature. It is assumed that expansions through turbines are adiabatic and isentropic. For such expansions a perfect gas relationship is

$$pv^\gamma = \text{const} \tag{2.4.6}$$

where γ is the ratio of the specific heat at constant pressure to that at constant volume. This relationship is transformed by use of the perfect gas law to give

$$\frac{T_{2s}}{T_1} = \left(\frac{p_2}{p_1}\right)^{(\gamma-1)/\gamma} \tag{2.4.7}$$

Therefore, the turbine work is rearranged to give

$$W_{Ts} = c_p T_1 \left[1 - \frac{1}{(p_1/p_2)^{(\gamma-1)/\gamma}}\right] \tag{2.4.8}$$

Similarly the compressor work is

$$W_{cs} = c_p(T_{4s} - T_3) = c_p T_{4s}\left(1 - \frac{T_3}{T_{4s}}\right)$$

$$= c_p T_{4s}\left[1 - \frac{1}{(p_1/p_2)^{(\gamma-1)/\gamma}}\right] \tag{2.4.9}$$

The pressure ratios across the turbine and compressor are assumed equal. In actual cycles the turbine pressure ratio is less than that across the compressor because of pressure losses in the heating and cooling processes. The compressor pressure ratio is defined by

$$r_p \equiv \frac{p_1}{p_2} \tag{2.4.10}$$

Consequently, the net work is given by

$$W_{Net} = W_{Ts} - W_{cs} = c_p(T_1 - T_{4s})\left[1 - \frac{1}{(r_p)^{(\gamma-1)/\gamma}}\right] \tag{2.4.11}$$

The term $c_p(T_1 - T_{4s})$ is the heat added to the cycle. Therefore, the thermal efficiency is given by

$$\eta = 1 - \frac{1}{(r_p)^{(\gamma-1)/\gamma}} \tag{2.4.12}$$

It should be noted that Eq. (2.4.12) does not contain temperatures.

As previously stated, materials limitations usually specify an upper limit to the maximum fluid temperature T_1. Consequently, it is necessary to consider what is the best pressure ratio for optimum power plant performance. For maximum thermal efficiency, it is desirable to expand the gas down to the lowest permissible temperature T_3 and also compress the gas to the maximum possible temperature T_1. The pressure ratio is given by

$$r_p = \left(\frac{T_1}{T_3}\right)^{(\gamma-1)/\gamma} \tag{2.4.13}$$

This cycle yields a Carnot cycle efficiency

$$\eta = 1 - \frac{T_3}{T_1} \tag{2.4.14}$$

and is illustrated in Fig. 2.27.

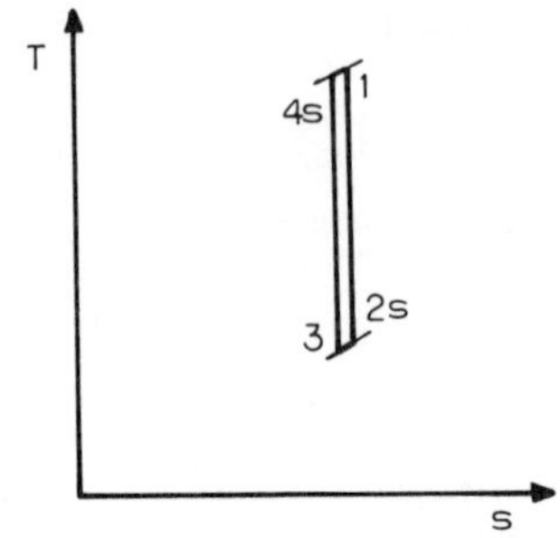

Fig. 2.27. Maximum efficiency Brayton cycle

As shown in Fig. 2.27, for maximum thermal efficiency the net work per pound of fluid is zero and, consequently, an infinite flow is required for a finite work.

The equation for net work is rearranged slightly to put it in terms of maximum and minimum fluid temperatures—T_1 and T_3.

$$W_{Net} = c_p\left[T_1 - T_3\,(r_p)^{(\gamma-1)/\gamma}\right]\left[1 - \frac{1}{(r_p)^{(\gamma-1)/\gamma}}\right] \tag{2.4.15}$$

$$= c_p T_3\left\{\left[1 - (r_p)^{(\gamma-1)/\gamma}\right] + \frac{T_1}{T_3}\left[1 - \frac{1}{(r_p)^{(\gamma-1)/\gamma}}\right]\right\}$$

The thermal efficiency and specific work, $W_{Net}/(c_p T_3)$, are plotted in Fig. 2.28 as a function of $(r_p)^{(\gamma-1)/\gamma}$ and a given value of T_1/T_3. For given values of T_1/T_3, such curves are universal and apply to all perfect gases.

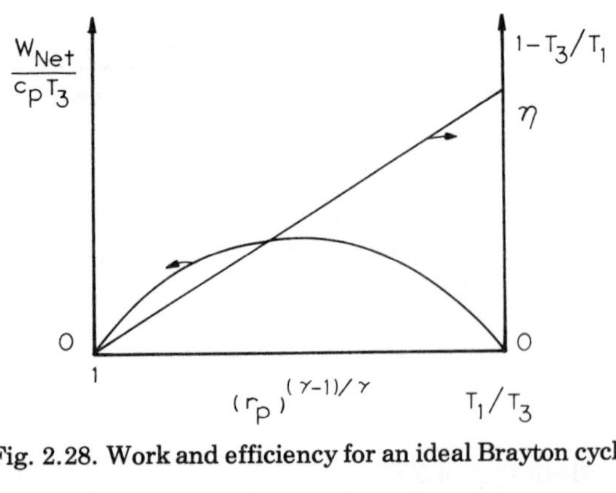

Fig. 2.28. Work and efficiency for an ideal Brayton cycle

As shown in Fig. 2.28, there exists an optimum pressure ratio for a maximum value of W_{Net}. This ratio is obtained by differentiating the expression for net work with respect to r_p and equating the derivative to zero. The result is

$$(r_p)_{opt} = \left(\frac{T_1}{T_3}\right)^{\gamma/2(\gamma-1)} \tag{2.4.16}$$

and therefore

$$\sqrt{\frac{T_1}{T_3}} = (r_p)_{opt}^{(\gamma-1)/\gamma} = \frac{T_{4s}}{T_3} \tag{2.4.17}$$

Consequently, for maximum work, the overall efficiency is given by

$$\eta = 1 - \frac{1}{(r_p)_{opt}^{(\gamma-1)/\gamma}} = 1 - \sqrt{\frac{T_3}{T_1}} \tag{2.4.18}$$

In addition, the fluid temperature after compression is given by

$$T_{4s} = \sqrt{T_1 T_3} \tag{2.4.19}$$

The properties of some common gases are listed in Table 2.2

Table 2.2

Physical properties of gases (68 °F) [7]

	c_p Btu/lb_m -- °F	γ c_p/c_v
H_2	3.42	1.405
He	1.25	1.659
CO_2	0.202	1.290
Air	0.240	1.40
N_2	0.248	1.40

The quantity $\gamma/2(\gamma - 1)$ decreases as γ increases. Therefore, for fixed maximum and minimum cycle temperatures, the optimum pressure ratio for monatomic gases (He) is in general lower than for polyatomic gases (CO_2 , air, etc.). Therefore, for closed-cycle turbines, higher turbine exhaust pressures are permissible and plant sizes are reduced.

The Actual Brayton Cycle

Brayton cycles are more susceptible to irreversibility effects than Rankine cycles. Pressure losses in ducts have a substantial influence on overall efficiency because compressor pumping work is large. Turbines and compressors have efficiencies less than one which also contribute to a reduction in overall efficiency.

Figure 2.29 illustrates an h-s diagram for an actual simple Brayton cycle. The dashed lines illustrate an ideal simple Brayton cycle operating with the same turbine pressures as the actual Brayton cycle. As seen, for the actual Brayton cycle, the pressure ratio in the compressor is higher than the pressure ratio in the turbine. The differences in pressure ratios are exaggerated in this diagram.

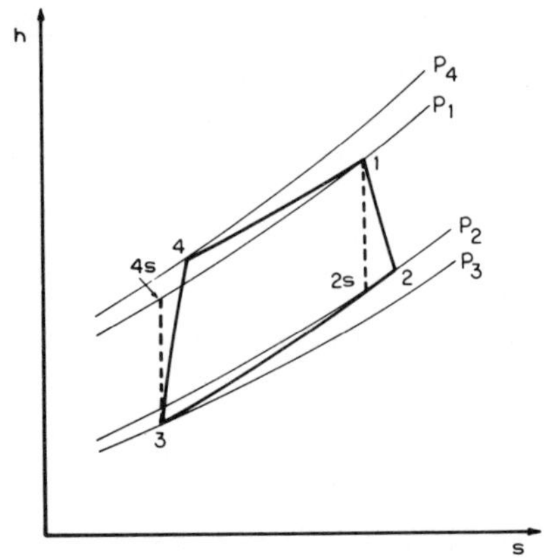

Fig. 2.29. Actual simple Brayton cycle

The same concept of compressor and turbine efficiencies is applied to Brayton cycles as is applied to Rankine cycles. Neglecting fluid velocity and height changes the compressor efficiency for adiabatic compression is given by

$$\eta_c = \frac{\text{Ideal Work}}{\text{Actual Work}} = \frac{h_{4s} - h_3}{h_4 - h_3} \tag{2.4.20}$$

For ideal gases, we have

$$\eta_c = \frac{T_{4s} - T_3}{T_4 - T_3} \tag{2.4.21}$$

Similarly, for an adiabatic expansion the turbine efficiency is given by (Neglecting velocity and height changes)

$$\eta_t = \frac{\text{Actual Work}}{\text{Ideal Work}} = \frac{h_1 - h_2}{h_1 - h_{2s}} \tag{2.4.22}$$

and for ideal gases

$$\eta_t = \frac{T_1 - T_2}{T_1 - T_{2s}} \tag{2.4.23}$$

Consequently, on the basis of the defined efficiency, the compressor work for an ideal gas with adiabatic compression is given by

$$W_c = c_p (T_4 - T_3) = \frac{c_p}{\eta_c} (T_{4s} - T_3) = \frac{c_p T_3}{\eta_c} \left(\frac{T_{4s}}{T_3} - 1 \right) \quad (2.4.24)$$

Since the ideal compression is adiabatic and isentropic, the temperature ratio is given in terms of pressures at states 3 and 4 and Eq. (2.2.24) is written

$$W_c = \frac{c_p T_3}{\eta_c} \left[\left(\frac{p_4}{p_3} \right)^{(\gamma-1)/\gamma} - 1 \right] = \frac{c_p T_3}{\eta_c} \left[(r_p)^{(\gamma-1)/\gamma} - 1 \right] \quad (2.4.25)$$

where r_p is the compressor pressure ratio.

Similarly, the turbine work is given by

$$W_T = c_p(T_1 - T_2) = \eta_t c_p(T_1 - T_{2s}) = \eta_t c_p T_1 \left(1 - \frac{T_{2s}}{T_1} \right) \quad (2.4.26)$$

Because the ideal turbine expansion is adiabatic and isentropic, the temperature ratio of the working fluid is written in terms of the turbine pressure ratio.

$$W_T = \eta_t c_p T_1 \left[1 - \left(\frac{p_2}{p_1} \right)^{(\gamma-1)/\gamma} \right] \quad (2.4.27)$$

The turbine pressure ratio is rewritten as

$$\left(\frac{p_2}{p_1} \right)^{(\gamma-1)/\gamma} = \left(\frac{p_3}{p_4} \right)^{(\gamma-1)/\gamma} \times \left(\frac{p_4}{p_1} \times \frac{p_2}{p_3} \right)^{(\gamma-1)/\gamma}$$

$$= \frac{1}{(r_p)^{(\gamma-1)/\gamma}} \times \beta \quad (2.4.28)$$

where β is a pressure loss ratio defined by

$$\beta \equiv \left(\frac{p_4}{p_1} \cdot \frac{p_2}{p_3} \right)^{(\gamma-1)/\gamma} \quad (2.4.29)$$

The turbine work is now given by

$$W_T = \eta_t c_p T_1 \left[1 - \frac{\beta}{(r_p)^{(\gamma-1)/\gamma}} \right] \quad (2.4.30)$$

The term β is evaluated from pressure losses in the ducts between the compressor and turbine. Let Δp_1 be the pressure loss between the turbine inlet and compressor outlet and Δp_2 be the loss between the turbine exhaust and compressor inlet, then β is given by

$$\beta = \left[\left(1 + \frac{\Delta p_1}{p_1} \right) \left(1 + \frac{\Delta p_2}{p_3} \right) \right]^{(\gamma-1)/\gamma} \quad (2.4.31)$$

For gas turbine systems, the pressures at the turbine inlet and exhaust may be quite large and pressure losses small compared to these pressures. Consequently, since the exponent on the pressure terms is considerably less than one, β will be close to one. Increasing ducting pressure losses increase β and therefore reduce the work output of the turbine. As γ increases, smaller percentage pressure losses are required to produce the same effect on cycle performance. However, as γ increases, the optimum compressor ratio for the Brayton cycle decreases with a result of higher pressures throughout the system. This means higher gas densities and lower velocities with a result of smaller component pressure losses. Consequently, β remains approximately constant for a given application with most gases.

The net work is given by

$$W_{Net} = c_p T_3 \left\{ \eta_t \frac{T_1}{T_3} \left[1 - \frac{\beta}{(r_p)^{(\gamma-1)/\gamma}} \right] - \frac{1}{\eta_c} \left[(r_p)^{(\gamma-1)/\gamma} - 1 \right] \right\} \quad (2.4.32)$$

and the heat input is

$$Q = c_p(T_1 - T_4) = c_p\left(T_1 - T_3 - \frac{W_c}{c_p}\right)$$

$$= c_p T_3 \left\{ \left(\frac{T_1}{T_3} - 1 \right) - \frac{1}{\eta_c} \left[(r_p)^{(\gamma-1)/\gamma} - 1 \right] \right\} \quad (2.4.33)$$

Therefore, the overall thermal efficiency is

$$\eta = \frac{W_{Net}}{Q} = \frac{\eta_t (T_1/T_3) \left[1 - \frac{\beta}{(r_p)^{(\gamma-1)/\gamma}} \right] - (1/\eta_c) \left[(r_p)^{(\gamma-1)/\gamma} - 1 \right]}{(T_1/T_3) - 1 - (1/\eta_c) \left[(r_p)^{(\gamma-1)/\gamma} - 1 \right]} \quad (2.4.34)$$

The efficiency of ideal Brayton cycles is independent of temperature and increases with increasing pressure ratio. However, as indicated by Eq. (2.4.34), the efficiency of actual cycles is strongly dependent upon gas temperatures. In addition, for a given set of temperatures and component efficiencies, the overall thermal efficiency is zero for small compressor pressure ratios, increases to a maximum with increasing compressor pressure ratios, and then decreases to zero as the compressor pressure ratio is further increased. A similar behavior exists for the net work produced by the turbine. Consequently, optimum compressor pressure ratios exist for maximum overall thermal efficiencies and maximum net work per pound of flowing gas. Unfortunately, the optimum pressure ratios are not the same for these two parameters and a compromise is necessary in design.

For given values of η_c, η_t, β, and T_1/T_3, universal plots are made of overall thermal efficiency and $W_{Net}/(c_p T_3)$ as a function of $(r_p)^{(\gamma-1)/\gamma}$. Figure 2.30 illustrates a plot for $\eta_c = \eta_t = 0.85$, $\beta = 1.05$, and $T_1/T_3 = 3$ (i.e., $T_1 = 1340°F$ and $T_3 = 140°F$). As seen in Fig. 2.30, the performance of a simple Brayton cycle is quite poor. A simple Rankine cycle operating between the same temperature limits would have an overall thermal efficiency far greater than 0.30.

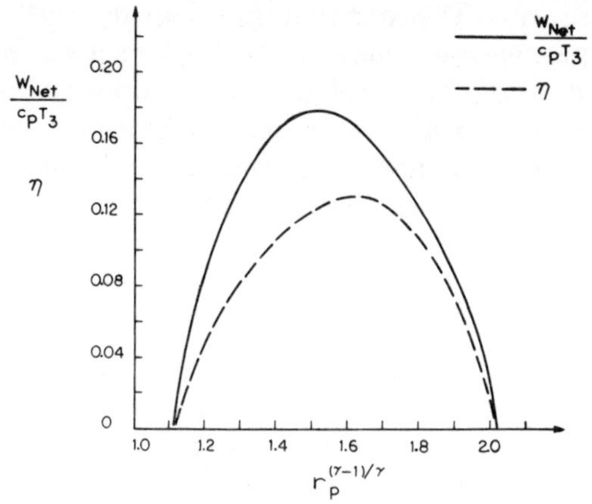

Fig. 2.30. Net work and efficiency of an actual Brayton cycle

Methods for Improving Brayton Cycle Efficiency

Brayton cycle efficiencies are improved by techniques similar to those used to improve Rankine cycle efficiencies. Reheat is used to increase cycle work output, regenerative heat exchange between the turbine exhaust gas and compressor outlet reduces heat input, and intercooling between compressor stages reduces compressor work. Figure 2.31 illustrates a schematic diagram of a gas turbine power plant

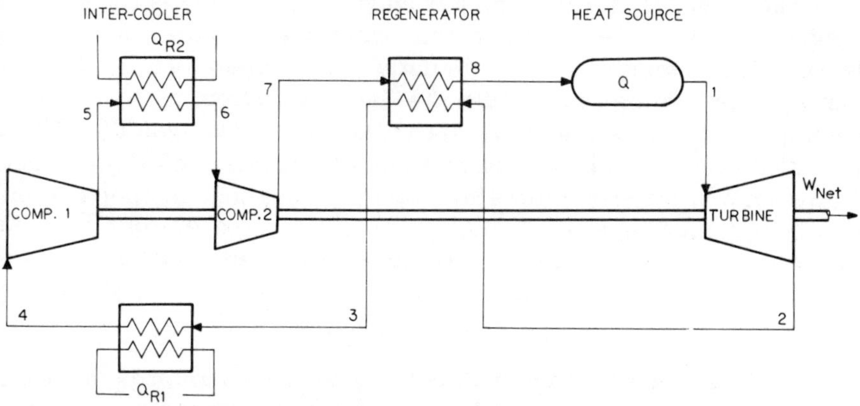

Fig. 2.31. Brayton cycle with regeneration and intercooling

using a regenerator and two-stage compression with intercooling to reduce compressor work.

The analysis of Fig. 2.31 is most easily understood by examining the temperature-entropy diagram shown in Fig. 2.32. The hot gas at pressure p_1 and temperature T_1 expands through a turbine to pressure p_2 and temperature T_2 . The exhaust gas then passes through a regenerator where it preheats the compressed gas before it enters the reactor or heat source. The exhaust gas leaving the regenerator at temperature T_3 passes into the cooling heat exchanger for rejecting heat to the surroundings. The cold gas at temperature T_4 now enters a two-stage compressor with intercooling between stages and is compressed to pressure p_7 and temperature T_7. From the compressor, the gas passes through the regenerator and heat source to complete the cycle.

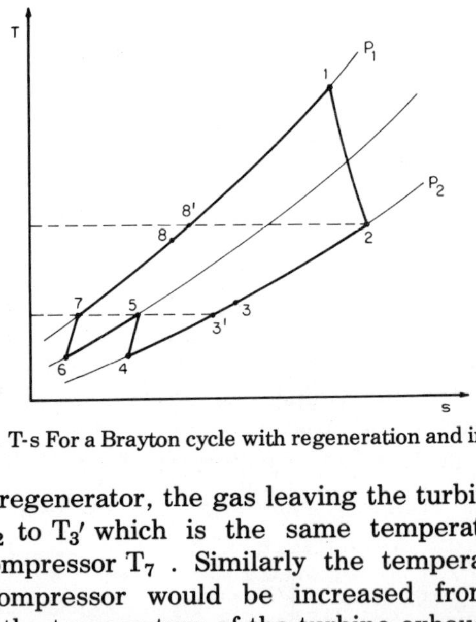

Fig. 2.32. T-s For a Brayton cycle with regeneration and intercooling

For an ideal regenerator, the gas leaving the turbine is cooled from temperature T_2 to T_3' which is the same temperature as the gas leaving the compressor T_7 . Similarly the temperature of the gas leaving the compressor would be increased from T_7 to T_8' which corresponds to the temperature of the turbine exhaust gas T_2 . These conditions correspond to a regenerator effectiveness of 100 percent. However, actual regenerator effectiveness is less that 100 percent and the compressed gas is heated to a lower temperature T_8 . As previously defined, the heat exchanger effectiveness is given by

$$\eta_x = \frac{\text{Actual Heat Transfer}}{\text{Ideal Heat Transfer}}$$

Since the mass flow through both sides of the regenerator is the same and the heat capacities of both gases are approximately equal, the regenerator effectiveness is given by

$$\eta_x = \frac{T_8 - T_7}{T_2 - T_7} \tag{2.4.35}$$

The method of analysis for improved Brayton cycles is the same as used for simple Brayton cycles. Work and heat requirements for each component are computed and appropriate summations formed to calculate Q and W_{Net} . Optimum compressor pressure ratios in terms of minimum cost of power or maximum specific work or efficiency are then determined.

If the gas leaving the first stage of compression is cooled back to T_4 before entering the second stage, then the compressor work for each stage is given by

$$W_{c1} = \frac{c_p T_4}{\eta_c}\left[(p_5/p_4)^{(\gamma-1)/\gamma} - 1\right]$$

$$W_{c2} = \frac{c_p T_4}{\eta_c}\left[(p_7/p_6)^{(\gamma-1)/\gamma} - 1\right] \tag{2.4.36}$$

Therefore, the total compressor work is given by

$$\begin{aligned} W_c &= W_{c1} + W_{c2} \\ &= \frac{c_p T_4}{\eta_c}\left[(p_7/p_6)^{(\gamma-1)/\gamma} + (p_5/p_4)^{(\gamma-1)/\gamma} - 2\right] \end{aligned} \tag{2.4.37}$$

For compression from pressure p_4 to p_7 in two stages, the minimum compressor work occurs when $p_5/p_4 = p_7/p_6$. Therefore, the minimum compressor work is given by

$$W_c = \frac{2c_p T_4}{\eta_c}\left[(r_p)^{(\gamma-1)/\gamma} - 1\right] \tag{2.4.38}$$

where r_p equals p_5/p_4 the compressor pressure ratio for each stage. For "n" compressor stages the minimum compressor work occurs when the pressure ratio across each stage is the same. Consequently, Eq. (2.4.38) is suitable for calculating the total compressor work by substituting "n" for the 2 in the equation.

The turbine work is given by

$$W_T = \eta_t c_p T_1 \left[1 - \left(\frac{p_2}{p_1}\right)^{(\gamma-1)/\gamma}\right] \tag{2.4.39}$$

The turbine pressure ratio is expressed in terms of system pressure ratios as

$$\begin{aligned} \left(\frac{p_2}{p_1}\right)^{(\gamma-1)/\gamma} &= \left(\frac{p_2}{p_3}\cdot\frac{p_3}{p_4}\cdot\frac{p_4}{p_5}\cdot\frac{p_5}{p_6}\cdot\frac{p_6}{p_7}\cdot\frac{p_7}{p_8}\cdot\frac{p_8}{p_1}\right)^{(\gamma-1)/\gamma} \\ &= \beta\left(\frac{1}{r_p}\right)^{2(\gamma-1)/\gamma} \end{aligned} \tag{2.4.40}$$

where β is the overall pressure loss ratio and $r_p = p_5/p_4 = p_7/p_6$

$$\beta = \left(\frac{p_2}{p_3}\cdot\frac{p_3}{p_4}\cdot\frac{p_5}{p_6}\cdot\frac{p_7}{p_8}\cdot\frac{p_8}{p_1}\right)^{(\gamma-1)/\gamma} \tag{2.4.41}$$

Consequently, Eq. (2.4.39) is written as

$$W_T = \eta_t c_p T_1 \left[1 - \frac{\beta}{(r_p)^{2(\gamma-1)/\gamma}}\right] \tag{2.4.42}$$

The net work, obtained from subtracting Eq. (2.4.38) from Eq. (2.4.42)

$$W_{net} = c_p T_4 \left\{\eta_t \frac{T_1}{T_4}\left[1 - \frac{\beta}{(r_p)^{2(\gamma-1)/\gamma}}\right] - \frac{2}{\eta_c}\left[(r_p)^{(\gamma-1)/\gamma} - 1\right]\right\} \tag{2.4.43}$$

Heat added by the reactor or heat source is given by

$$Q = c_p(T_1 - T_8) = c_p T_4 \left(\frac{T_1}{T_4} - \frac{T_8}{T_4}\right) \tag{2.4.44}$$

The temperature of the compressed gas leaving the regenerator, T_8 , is found from further analysis. By using the known value of the regenerator effectiveness, η_x , we have

$$T_8 = T_7 + \eta_x (T_2 - T_7) \tag{2.4.45}$$

The temperature of the gas exiting from the turbine and compressor, T_2 and T_7 , is expressed as

$$T_2 = T_1 - \frac{W_T}{c_p}; \quad T_7 = T_6 + \frac{W_{c2}}{c_p} \tag{2.4.46}$$

Since T_6 equals T_4 and W_{c2} equals $W_c/2$, Eq. (2.4.45) is written

$$T_8 = T_4 + \frac{W_c}{2c_p} + \eta_x\left[T_1 - \frac{W_T}{c_p} - \left(T_4 + \frac{W_c}{2c_p}\right)\right]$$

or

$$\frac{T_8}{T_4} = 1 + \frac{W_c}{2c_p T_4} + \eta_x\left[\frac{T_1}{T_4} - \frac{W_T}{c_p T_4} - \left(1 + \frac{W_c}{2c_p T_4}\right)\right] \tag{2.4.47}$$

Therefore, substituting Eq. (2.4.47) into Eq. (2.4.44) gives for Q

$$Q = c_p T_4 \left[(1 - \eta_x)\left(\frac{T_1}{T_4} - 1 - \frac{W_c}{2c_p T_4}\right) + \eta_x \frac{W_T}{c_p T_4}\right] \tag{2.4.48}$$

The overall thermal efficiency is obtained by dividing Eq. (2.4.43) by Eq. (2.4.48). Both equations are in terms of component efficiencies, η_c and η_t ; heat exchanger effectiveness, η_x ; temperature ratio, T_1/T_4 ;

pressure loss ratio, β ; and pressure ratio, r_p . Therefore, for a given set of values of η_c , η_t , η_x , β , and T_1/T_4 , universal plots of η and W_{Net}/c_pT_4 as a function of $(r_p)^{(\gamma-1)/\gamma}$ are obtained. Figure 2.33 illustrates the quantities for the same values of β, η_c, η_t, and T_1/T_4 used in Fig 2.30 and η_x equal to 0.90. Comparing Fig. 2.33 with Fig. 2.30 shows that the overall thermal efficiency is almost doubled by using regeneration and compressor intercooling. The efficiency now approaches efficiencies of steam plants. Consequently, proposed gas cycle plants for stationary or maritime use will most likely incorporate a regenerator and multi-stage compression with intercooling.

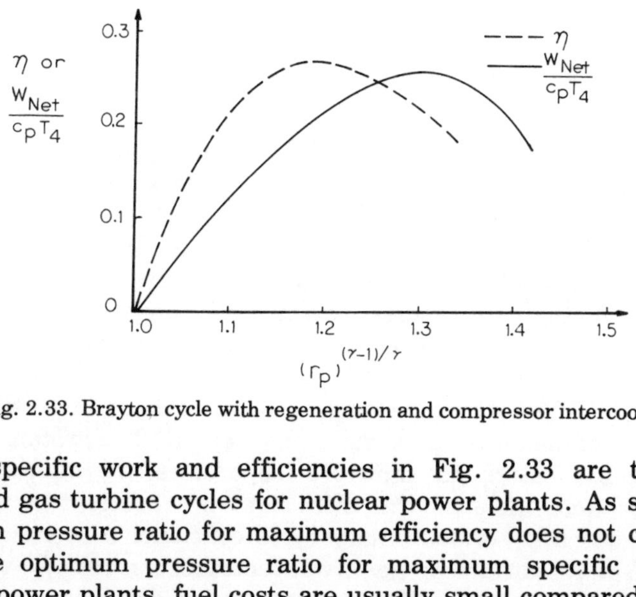

Fig. 2.33. Brayton cycle with regeneration and compressor intercooling

The specific work and efficiencies in Fig. 2.33 are typical for proposed gas turbine cycles for nuclear power plants. As shown, the optimum pressure ratio for maximum efficiency does not correspond with the optimum pressure ratio for maximum specific work. For nuclear power plants, fuel costs are usually small compared to capital costs and, consequently, the effort in design is toward reducing capital costs. Therefore, from cost considerations, the optimum pressure ratio is probably close to the pressure ratio for maximum specific work.

Power costs for nuclear power plants are divided into three categories—capital, fuel, and operation. For nuclear power plants using gas turbines, operating costs should be approximately constant and, thus, independent of compressor pressure ratio. A plot of the various power costs as a function of compressor pressure ratio enables finding the optimum compressor pressure ratio for minimum power cost. Figure 2.34 illustrates a typical plot of cost vs. compressor pressure ratio.

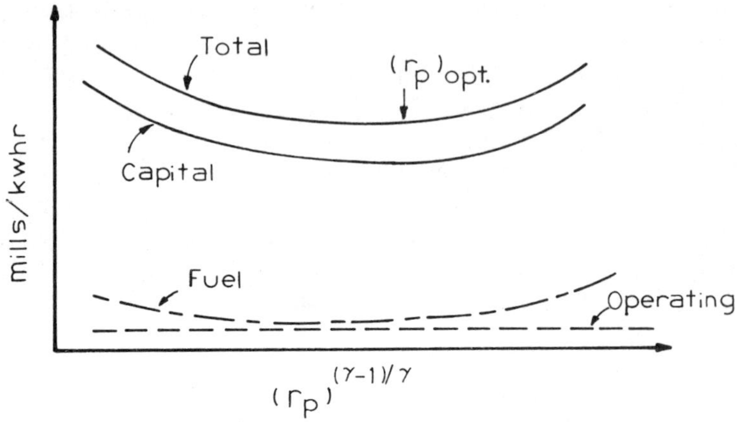

Fig. 2.34. Cost vs. pressure ratio for gas turbine power plants

2.5 Binary Cycles

A high efficiency energy conversion scheme suitable for use with HTGRs is a binary cycle. A binary cycle employs a gas turbine cycle which rejects its waste heat to a Rankine cycle for further energy conversion. Several studies were performed on binary cycles using high temperature gas-cooled reactors as the heat source [8, 9, 10]. These studies showed high efficiencies are achieved; however, optimum plant configurations and operating conditions for maximizing overall efficiencies were not found.

Figure 2.35 illustrates a generalized binary cycle which is analyzed for purposes of maximizing overall efficiency [4]. The system employs M feedwater heaters, L moisture separators, N intercoolers, a regenerator, and includes the effect of superheating. A model is developed which describes binary cycle efficiencies in terms of the following set of variables: gas cycle compressor ratio, number of intercoolers, regenerator effectiveness, boiler exit pressure, superheated steam exit temperature, number of feedwater heaters, and feedwater heater inlet saturation temperatures. The binary cycle efficiency is then optimized with respect to these variables in the presence of engineering design constraints such as minimum temperature differences in heat exchangers and component pressure losses.

As part of the optimization procedure, the combined system is partitioned into two components, a top cycle (Brayton cycle) and bottom cycle (Rankine cycle) by fixing a set of coupling variables. The two uncoupled cycles are then optimized separately to give a maximum binary cycle efficiency for the given coupling variables. The procedure is repeated for different values of coupling variables. The global optimum binary cycle efficiency is chosen from the best value through the entire search. Dynamic programming is used to optimize the Rankine cycle and an enumerative search technique is used for the Brayton cycle.

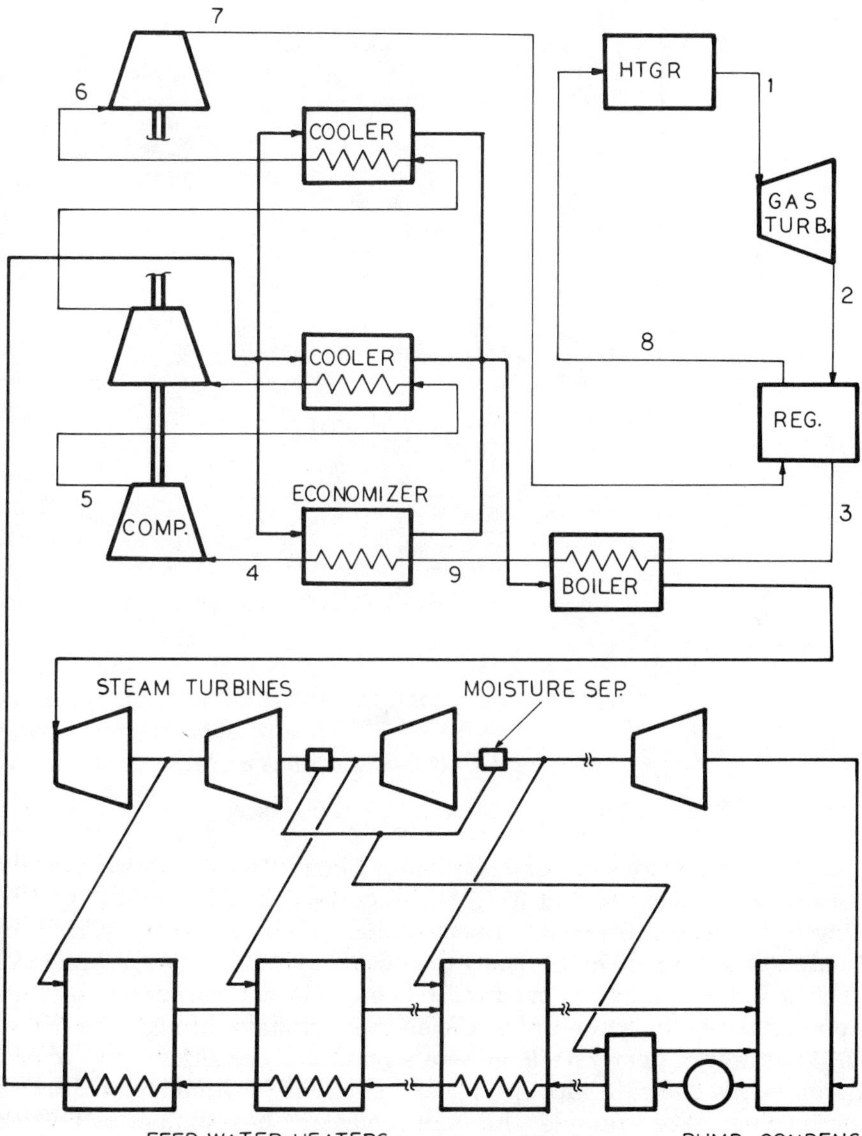

Fig. 2.35. Binary cycle plant schematic diagram

Figure 2.36 illustrates optimum binary cycle efficiencies at several boiler saturation temperatures as a function of reactor exit temperature. Binary cycle efficiencies increase almost linearly with reactor exit temperature and increase with increasing boiler saturation temperature.

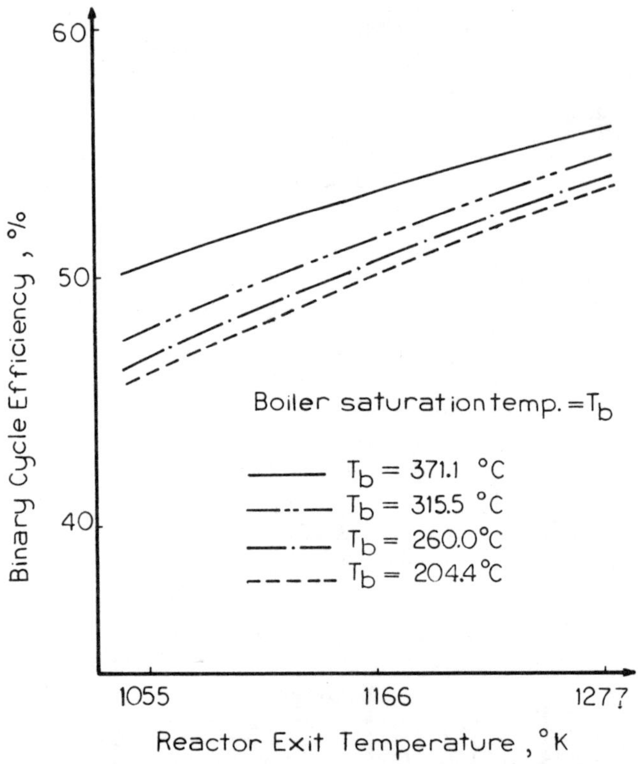

Fig. 2.36. Binary cycle efficiency vs. reactor exit temperature

For the conditions studied, optimum binary cycles require a boiler saturation temperature of 371.1°C, superheat of 56°C or more, and single stage compression. Even though nine or more stages of feedwater heaters is numerically the optimum number, employment of such a large number of feedwater heaters is not necessary because approximately the same cycle efficiency is obtained through the use of few feedwater heaters. In general, optimum conditions for binary cycles do not coincide with optimized conditions for Rankine or Brayton cycles alone. For example, the final feedwater heater inlet saturation temperature of an optimal binary cycle is much lower than the corresponding temperature of an optimum Rankine cycle and the optimal gas compression ratio for the Brayton cycle portion of an optimum binary cycle is much lower than the optimum compression ratio of a Brayton cycle. A final conclusion is that binary cycles optimized to maximize overall plant efficiency tend to have simple plant configurations.

REFERENCES

1. Meyer, C.A., et al., **Thermodynamic and Transport Properties of Steam**, A.S.M.E., New York, 1967.
2. Keenan, J. H., **Thermodynamics**, Wiley, New York (1941).
3. **Steam—Its Generation and Use**, Babcock & Wilcox Co., New York, 1975.
4. Kim, R. L., **Optimization of Brayton-Rankine Binary Cycles**, Ph.D. Thesis, Georgia Institute of Technology, January 1978.
5. **Steam—Its Generation and Use**, Babcock & Wilcox Co., New York, 1963.
6. Koutz, S. L., et al., HTGR gas turbine power plant preliminary design, **Trans. ASME International Gas Turbine Conference**, Zurich, Switzerland (March 31-April 4, 1974).
7. Keenan, J. H. and Kaye, J., **Gas Tables**, John Wiley & Sons, Inc., New York, 1945.
8. Kilparti, S. R. and Nagib, M. M., Combined helium and steam cycle for nuclear power plants, **Mechanical Engineering** 93, 14-18 (August 1971).
9. McCracken, K. W., Rust, J. H., and Miller, R., Improved binary cycles for HTGRs, **Trans. Amer. Nucl. Soc.** 15, 371-372 (June 1972).
10. Schoene, T. W., Winkler, E. O., and Fortescue, P., The gas turbine HTGR plant, economical dry cooling or wet-cooled high efficiency binary configuration, **American Power Conference**, Chicago, Illinois (April 21-23, 1975).

PROBLEMS

2.1 Steam at 1600 psia and 1000 °F is expanded in a turbine to a pressure of 1 psia and a moisture content of 10 percent. Water leaves the condenser at 90° F. If the boiler feed pump has a compression efficiency of 85 percent, what is (1) turbine efficiency, (2) ideal Rankine cycle efficiency, and (3) actual cycle efficiency?

2.2 A steam turbine for a BWR has saturated steam at 1000 psia enter the turbine. The steam is expanded down to 100 psia in a high pressure turbine where steam is bled for use in a feedwater heater. The remaining steam is expanded to 1 psia in a low pressure turbine. The feedwater is heated to 320°F and the condensed steam is cooled to 10°F above the temperature of the entering feedwater. The high pressure turbine efficiency is 0.85 and the low pressure turbine efficiency is 0.8. Find the overall plant efficiency and the heat rate.

2.3 A power reactor concept considered by several countries is the heavy water moderated, organic cooled reactor. This type reactor uses an organic coolant (Santowax OM) which is thermally separated from the D_2O moderator by an insulated tube. The organic coolant permits higher coolant temperatures and, thus, better steam conditions with much lower coolant pressure than possible if heavy water were used as a coolant.

You are asked to investigate the steam cycle of this system. Five percent of the reactor power is deposited in the D_2O moderator and 95% of the power in the organic coolant. The D_2O is used to preheat the high pressure water from the boiler feed pump. The organic coolant inlet temperature to the reactor core is 670° F and the core outlet temperature is 720° F. The specific heat of Santowax OM is 0.58 Btu/lb_m-°F. The turbine and pump efficiency is 85% and condenser subcooling 10°F. Pump work on organic and D_2O loops is 1% the turbine output.

A. For a boiler pinch point temperature differential 50°F, a pinch point difference in the superheater of 20° F, and condenser pressure of 1 psia, what is the steam flow rate in lb_m steam/lb_m organic?

B. Calculate the power cycle thermal efficiency?

C. What is the efficiency if steam at 100 psia is bled from the turbine for a single stage feedwater heater? The feedwater is heated to within 10° F of the bled steam and the condensed bled steam is cooled to within 10° F of the feedwater inlet temperature.

2.4 Saturated steam at 1000 psia enters a high pressure turbine of 85% efficiency and expanded down to 100 psia. The steam then enters a moisture separator where all entrained moisture is removed. Saturated steam at 100 psia enters a low pressure turbine of 85% efficiency and is exhausted to a condenser at 1 psia. Saturated liquid from the moisture separator is sent to a feedwater heater where it is cooled to 110° F before it is sent to the condenser. The pump efficiency is 85%. Calculate the overall plant efficiency.

2.5 In order to relieve moisture problems in turbines, nuclear power plants use reheaters employing primary steam. Primary steam is condensed and subcooled in the reheater and sent to act as a feedwater heater. The plant schematic is given below. Saturated steam at 900 psia leaves the steam generator and part passes through a high pressure turbine of 0.83 efficiency. The expansion is down to 100 psia and the wet mixture is sent to the reheater where the mixture is superheated to 500° F. This steam passes through a low pressure turbine of 0.9 efficiency where it exhausts at 1 psia.

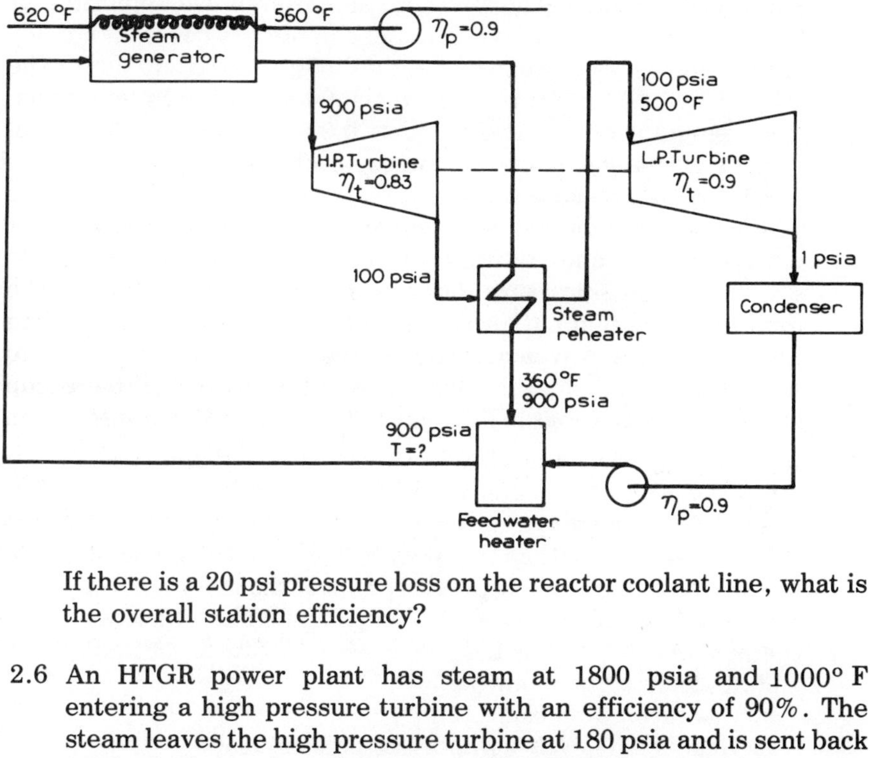

If there is a 20 psi pressure loss on the reactor coolant line, what is the overall station efficiency?

2.6 An HTGR power plant has steam at 1800 psia and 1000° F entering a high pressure turbine with an efficiency of 90%. The steam leaves the high pressure turbine at 180 psia and is sent back to the boiler to be reheated to 1000°F. This steam leaves the boiler at 160 psia. The full flow steam is used to drive the helium circulators with a turbine of 100% efficiency. The steam then passes to a low pressure turbine of 85% efficiency where it is expanded to an exhaust pressure of 1 psia. The helium circulator has an efficiency of 90% and the helium average density is 0.04 lb/ft^3 during the compression of 20 psi. Helium has a heat capacity of 1.25 Btu/lb-° F and a temperature rise of 600° F in the reactor.

a. Calculate the plant overall efficiency.

b. In 50 words or less, explain how you change working this

problem if the efficiency of the turbine driving the helium circulator is 90%.

2.7 The Marine Gas Cooled Reactor proposed by General Atomic used a closed-cycle gas turbine to produce power. Helium is heated in a nuclear reactor to 1300°F and then fed to a gas turbine. The helium leaving the turbine passed in series through a regenerator, a cooler, a low pressure compressor, an intercooler, a high pressure compressor, and back through the regenerator before entering the reactor. The following data is assumed applicable for this system:

Efficiencies: Compressors $\eta_c = 0.85$

Turbine $\eta_t = 0.90$

Regenerator effectiveness $\eta_x = 0.90$

Heat exchanger pressure losses (each side) = 2 percent of entering pressure

Reactor pressure loss = 5 percent of entering pressure

Inlet suction temperature each compressor = 100°F

Assume equal pressure ratios in the two compressors.

Find the maximum thermal efficiency for this cycle and the compression ratio which produces the maximum efficiency. At what temperature does the helium enter the reactor? If the reactor produces 100 thermal MW, what is the helium flow rate?

2.8 Helium at 900 psia and 1500°F leaves an HTGR and enters a gas turbine where the helium is expanded down to 1000°F. The helium rejects heat in a boiler where it produces saturated steam at 1000 psia. The helium is cooled to 300°F in the boiler and then enters a compressor. The pressure loss of the helium in the boiler and HTGR is 10 psi in each component. The efficiency of all components is 0.9. For helium c_p = 1.25 Btu/lb–°F and γ = 1.66. Calculate the following:

a. Gas turbine pressure ratio
b. Gas turbine cycle compressor pressure ratio
c. Net work from steam cycle in Btu/lb_m
d. Plant overall efficiency.

CHAPTER 3

FLUID FLOW CONCEPTS

3.1 Introduction

Understanding the analytical methods used in thermal-hydraulic analyses of nuclear reactor systems requires comprehension of many basic concepts of fluid mechanics. This task in itself is formidable and this chapter only serves to provide a rapid review of a few areas of importance. Readers desiring a more detailed coverage should consult some of the references listed at the end of this chapter.

3.2 Physical Properties

In order to perform a fluid flow analysis of reactor systems, it is essential to have accurate values of certain fluid physical properties. These properties are usually viscosity, density, surface tension, and vapor pressure. Reference [1] contains a summary of properties of common reactor coolants which are divided into four classes: (1) aqueous and organic liquids, (2) gases, (3) liquid metals, and (4) fused inorganic liquids. Reference [2] gives a detailed treatment of methods for evaluating some of these properties. Treatment in depth is beyond the scope of this book and interested persons are urged to seek high-level books on fluid mechanics or physical chemistry. Properties of interest for several reactor coolants are tabulated in Appendix C.

Viscosity

Viscosity is a measure of the resistance of a fluid to deformations produced by flows near boundaries. Near boundaries, the velocity of a fluid varies with distance from the boundary in a manner similar to that shown in Fig. 3.1. For the case of streamline (laminar) flow, over any plane 1-1 which is parallel to the direction of flow, a shear stress exists due to the differences in velocity at successive points normal to the direction of flow. The value of the shear stress is given by

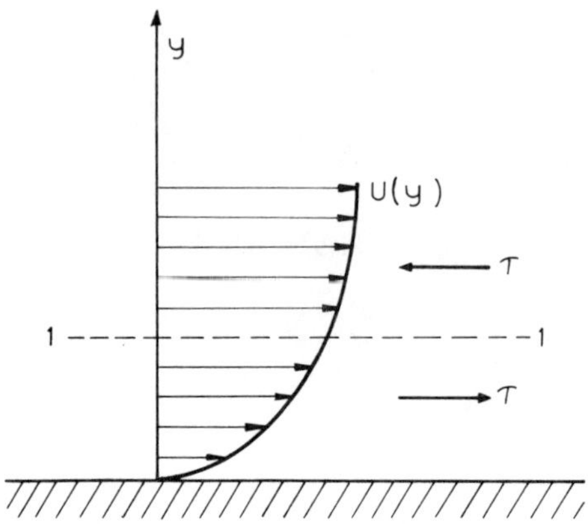

Fig. 3.1. Shear stress in a fluid motion

$$\tau_{yx} = -\frac{\mu}{g_c}\frac{du}{dy} \tag{3.2.1}$$

The proportionality factor μ is the viscosity and Eq. (3.2.1) is used to define viscosity. Any fluid in which μ does not depend upon τ or du/dy is called a "Newtonian fluid." The majority of fluids such as water and gases are Newtonian fluids. For most fluids, viscosity is a function of temperature, the nature of the fluid, and is independent of pressure, except at extremely high pressures. The viscosity of liquids decreases with increasing temperature; the viscosity of gases increases roughly with the square root of its absolute temperature.

The system of units for viscosity which is most convenient for engineering calculations is

$$\mu = \left[\frac{lb_m}{ft\text{-}sec}\right]$$

The normal units given in viscosity tables are the poise or centipoise (1/100 poise). The conversion from metric units to English engineering units is

$$\frac{lb_m}{ft\text{-}sec} = 6.72 \times 10^{-4} \times \text{centipoise}$$

Another type of viscosity frequently used is the kinematic viscosity which is defined by

$$\nu \equiv \frac{\mu}{\rho}$$

Density

Densities of most materials are tabulated in the literature as a function of temperature and pressure. For many engineering uses, one may assume liquids incompressible and gases obeying the ideal gas law

$$\rho = \frac{p}{RT} \tag{3.2.2}$$

where R is the gas law constant and T is the absolute temperature. A better representation for the gas law is

$$\rho = \frac{p}{ZRT} \tag{3.2.3}$$

where Z is the compressibility factor which is given as a function of reduced pressure and reduced temperature in most physical chemistry books.

The standard English engineering units for density are

$$\rho = \left[\frac{lb_m}{ft^3}\right]$$

Vapor Pressure and Surface Tension

Vapor pressure is important for predicting whether boiling takes place in a liquid. Boiling may be undesirable in many nuclear reactor applications.

Surface tension is of use in the analysis of vapor-liquid mixtures, i.e., boiling. Surface tension is a measure of the force acting on a vapor-liquid interface and causes a liquid to assume the smallest surface area for a given volume. The standard English engineering units for surface tension are

$$\sigma = \left[\frac{lb_f}{ft}\right]$$

3.3 Review of Fluid Mechanics

Two concepts of flow theory are essential for understanding fluid mechanics and, as will be shown later, for understanding convection heat transfer. These fundamental concepts are (1) concept of the boundary layer and (2) concept of turbulence.

For flow along a solid surface, the fluid velocity is zero at the interface. Consequently, the velocity distribution near a boundary is similar to that shown in Fig. 3.2.

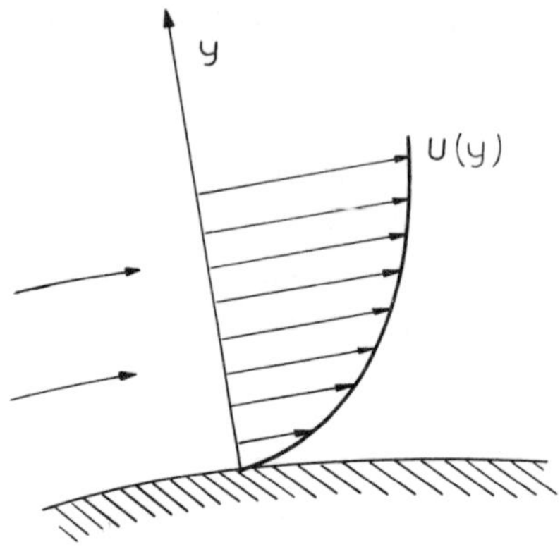

Fig. 3.2. Velocity profile in a fluid near a boundary

In this section, we consider flow in two important cases: (1) flow along a flat plate and (2) flow through a channel.

Flow Along a Smooth Flat Plate [External Boundary Layer]

Let us examine the velocity profile along the submerged smooth flat plate illustrated in Fig. 3.3. The fluid upstream from the plate has a uniform velocity U_∞ parallel to the plate surface. In the vicinity of the plate leading edge, a laminar and external boundary layer develops along the plate with a velocity distribution similar to that shown in Fig. 3.3. The laminar boundary layer thickness at any point on the plate is given by δ_L which is calculated by [3].

$$\frac{\delta_L}{x} = \frac{5.0}{(U_\infty x\rho/\mu)^{0.5}} = \frac{5.0}{(N_{Re,x})^{0.5}} \tag{3.3.1}$$

where $N_{Re,x}$ is the Reynolds number based upon distance from the plate leading edge. Flow outside the boundary layer is maintained at the free stream velocity U_∞.

Numerical solutions exist for velocity profiles in laminar boundary layers [3]. However, a more important result is the shear stress on the plate surface which enables calculating forces exerted on the plate by the fluid motion. The wall shear stress is evaluated from solutions for the velocity profile by

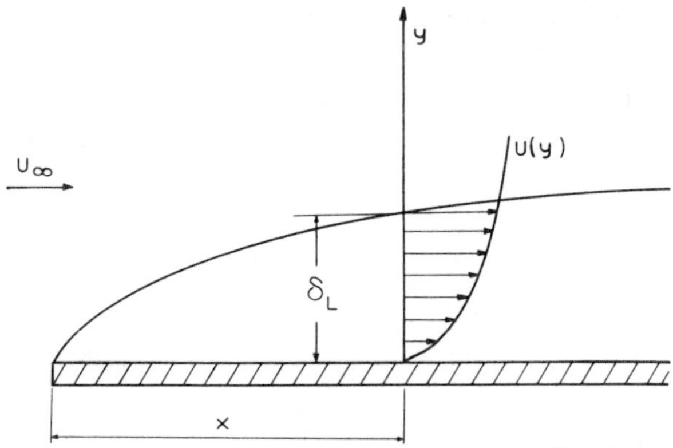

Fig. 3.3. Laminar boundary layer along a flat plate

$$\tau_w(x) = \frac{\mu}{g_c}\left[\frac{\partial u(x)}{\partial y}\right]_{y=0} \tag{3.3.2}$$

For flows without a pressure gradient in the free stream, the wall shear stress is given by [3]

$$\tau_w(x) = 0.332\ U_\infty \mu (U_\infty/\nu x)^{0.5}/g_c \tag{3.3.3}$$

Experiments appear to indicate that Eq. (3.3.3) is valid if the free stream velocity is turbulent [4]. However, free stream pressure gradients alter velocity profiles [4] and Eq. (3.3.3) should not be used with these types of flows.

In the discussion of boundary layer flows it is customary to define a **local drag coefficient** or skin-friction coefficient by

$$C_f(x) = \frac{\tau_w(x)}{\rho U_\infty^2/2g_c} \tag{3.3.4}$$

Consequently, for laminar boundary layers without a free stream pressure gradient the drag coefficient is given by

$$C_f(x) = \frac{0.664}{(U_\infty x/\nu)^{0.5}} \tag{3.3.5}$$

The results for flow past a flat plate can be used for flow past curved surfaces providing the radius of curvature of the surface is very large compared to the boundary layer thickness.

At some distance along the plate, the flow becomes unstable and the laminar boundary layer becomes turbulent. Figure 3.4 illustrates regions of various types of boundary layer flows. Immediately downstream from the laminar boundary layer is a boundary layer region called the transition zone in which the flow is intermittently laminar or turbulent. Downstream from the transition zone is an external boundary layer in which the flow is completely turbulent. However, adjacent to the surface in the turbulent boundary layer is a thin sublayer in which the flow is considered laminar.

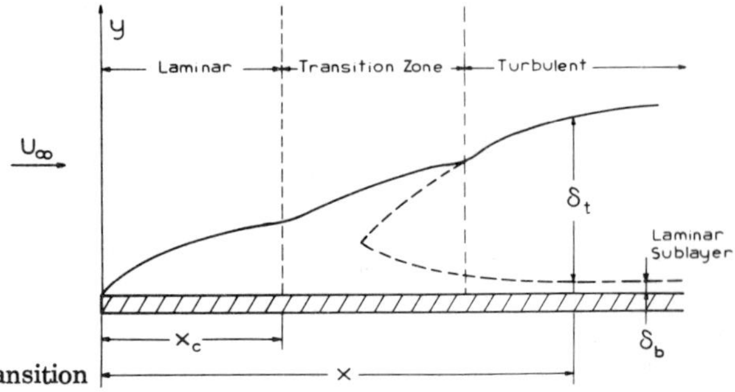

Fig. 3.4. Boundary-layer transition

Experience has shown that there is a critical Reynolds number which characterizes the change from laminar to turbulent boundary layers. The range of critical Reynolds numbers is quite large and depends upon the plate surface conditions, shape of the leading edge, vibration of the plate, and the nature of the upstream flow. The critical Reynolds number is based upon the distance downstream from the leading edge and falls in the range

$$80{,}000 \leqslant (U_\infty x_c \rho)/\mu = N_{Re,c} \leqslant 5{,}000{,}000$$

For engineering design estimates without prior knowledge of flow conditions, the transition from a laminar to a turbulent boundary layer may be assumed when $N_{Re,x} = 3 \times 10^5$.

Due to the random nature of turbulence, the turbulent boundary layer thickness at any given location varies randomly with time. An approximate average thickness of the turbulent external boundary layer without a free stream pressure gradient based upon distance from the plate leading edge is given by [3]

$$\frac{\delta_t}{x} = \frac{0.374}{(U_\infty x\rho/\mu)^{0.2}} \qquad (3.3.6)$$

Next to the plate surface in a turbulent boundary layer is a laminar sublayer whose thickness is approximated by

$$\frac{\delta_b}{x} = \frac{29}{(U_\infty x\rho/\mu)^{0.9}} \qquad (3.3.7)$$

Note that δ_b increases slowly with one-tenth root of distance from the leading edge.

The local wall shear stress or drag coefficient for turbulent boundary layers is determined by experiment. Probably the best expression for the drag coefficient is proposed by Schultz-Grunow [5] which is reported to be applicable up to length Reynolds numbers of 10^9.

$$C_f(x) = 0.370 \ (\log N_{Re,x})^{-2.58} \tag{3.3.8}$$

The local drag coefficient is unaffected by roughened surfaces whose height of asperities is less than the thickness of the laminar sublayer.

Velocity profiles in turbulent external boundary layers are determined experimentally since solutions to the fluid flow equations have not been found. On the basis of data obtained by Schultz-Grunow, a universal velocity profile is proposed in which the boundary layer is divided into four distinct regions—laminar sublayer, buffer zone, turbulent core, and turbulent wake [6]. The parameters used in the universal velocity profile are a dimensionless velocity and a dimensionless distance measured in a normal direction from the plate surface. The dimensionless velocity is given by

$$u^+ = u/\sqrt{\tau_w g_c/\rho} \tag{3.3.9}$$

and the dimensionless distance is given by

$$y^+ = \frac{y\sqrt{\tau_w g_c/\rho}}{\nu} \tag{3.3.10}$$

The local wall shear stress used in these dimensionless parameters is evaluated from Eq. (3.3.8).

The formulation for the turbulent boundary layer velocity profile without a pressure gradient is as follows [6]:

(1) Laminar Sublayer. Because of a lack of experimental data in this region, the velocity profile in the laminar sublayer is assumed the same as for channel flows. Therefore, for $0 \leqslant y^+ \leqslant 5$

$$u^+ = y^+ \tag{3.3.11}$$

(2) Buffer Zone. The same form of velocity profile used in channel flows is also assumed for the buffer zone. The extent of the buffer zone is based upon extrapolation of data from the exterior turbulent core. Consequently, for $5 \leqslant y^+ \leqslant 18.2$

$$u^+ = -3.05 + 5.0 \ln y^+ \tag{3.3.12}$$

(3) Turbulent Core. The turbulent core adjacent to the buffer zone

extends over approximately one-third the turbulent boundary layer. The formulation for the velocity profile is based entirely upon data taken by Schultz-Grunow. For $18.2 \leqslant y^+ \leqslant 360$

$$u^+ = 4.4 + 2.43 \ln y^+ \tag{3.3.13}$$

(4) Turbulent Wake. The turbulent wake extends from the turbulent core to the free stream which is approximately two-thirds the thickness of the turbulent boundary. For $y^+ > 360$, the velocity data are approximately correlated by

$$u^+ = -4.4 + 3.96 \ln y^+ \tag{3.3.14}$$

Figure 3.5 illustrates a comparison of the universal flat-plate velocity profile with experimental data.

The results presented so far are only applicable for external boundary layer flows without a free stream pressure gradient. For boundary layer flows with a pressure gradient, such as found in channel flows, more sophisticated approaches are required to determine velocity profiles and local drag coefficients. Some of the procedures for attacking these types of problems are given in Refs. [3], [7], and [8].

The concept of a boundary layer thickness is somewhat arbitrary since in theory the fluid velocity is influenced by the plate for an infinite distance in a direction perpendicular to the plate. The boundary layer thickness given by the preceding equations represent distances at

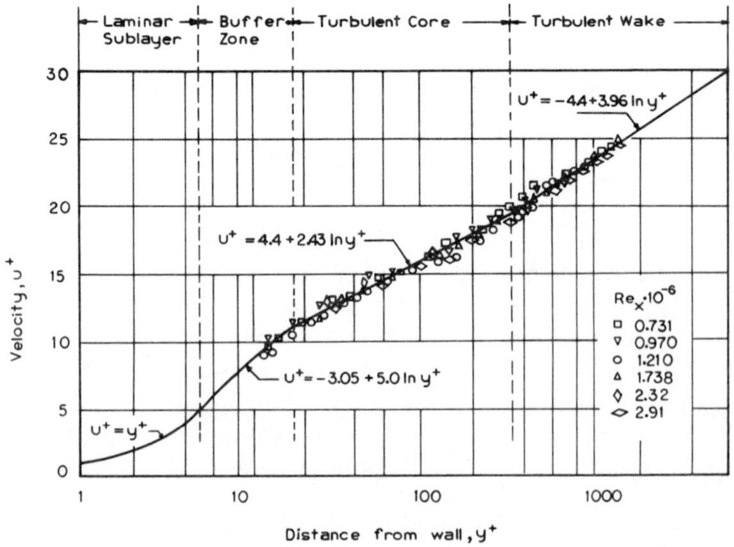

Fig. 3.5. Comparison of universal flat-plate velocity profile with experimental data [6]

which the fluid velocity is approximately 99 percent of the free stream velocity U_∞. Another type of boundary layer thickness is the displacement thickness. The displacement thickness is found by equating the area under the velocity profile to an equal area rectangle of length given by the free stream velocity. The height of the rectangle is the displacement thickness δ^*. A physical interpretation of the displacement thickness is that it is the distance the solid boundary would have to be displaced in an outward direction in order to keep the flow field unchanged if the fluid was frictionless and, consequently,

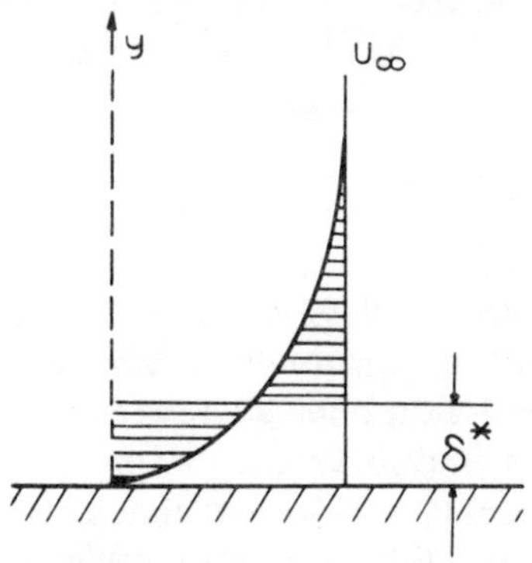

Fig. 3.6. Displacement thickness of the boundary layer

there was no boundary layer. Approximate values for the displacement thickness are given for the laminar boundary layer by[9]

$$\delta_L^* \approx \frac{\delta_L}{3} \tag{3.3.15}$$

and for the turbulent boundary layer by [9].

$$\delta_t^* \approx \frac{\delta_t}{8} \tag{3.3.16}$$

Flow in a Channel

The flow conditions at a channel inlet are similar to parallel flow past a plate. Figure 3.7 illustrates development of velocity profiles at a channel entrance. Boundary layers form immediately on the channel wall and increase in size until they meet at a downstream distance L_e. At L_e the velocity profile is a parabola or arched curve depending upon whether the flow in the channel is laminar or turbulent. Further downstream the velocity profile does not change shape and the flow is in a region called **fully-developed flow**. For fully-developed flow the velocity distribution is independent of the coordinate in the direction of flow.

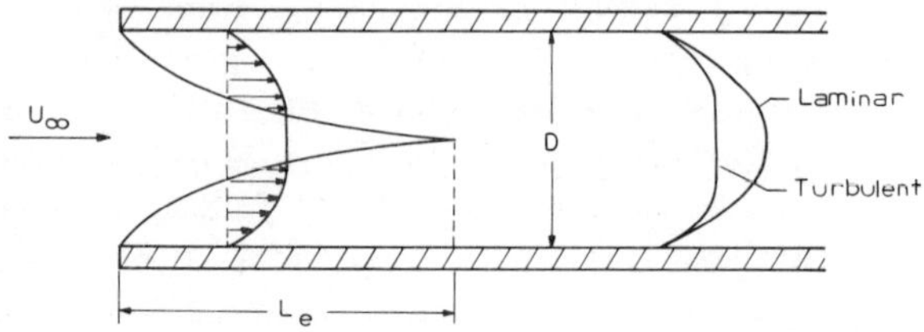

Fig. 3.7. Development of the velocity profile at a channel inlet

For laminar flow, the entrance length for merging of boundary layers is calculated from an analysis of the differential equations governing the flow. The entrance length is given by [9]

$$\frac{L_e}{D} = 0.0288 \frac{u_m D\rho}{\mu} = 0.0288\, N_{Re} \qquad (3.3.17)$$

where u_m is the average fluid velocity in the channel and N_{Re} is a Reynolds number based upon channel diameter.

For turbulent flow, no simple expression exists for determining L_e. However, the entrance length decreases because of the more rapid buildup of boundary layers. The entrance length decreases with increasing Reynolds number to a minimum L_e/D of 40 and then increases with increasing Reynolds number. Experimental values of L_e equal to L_e/D ratios as high as 120 are observed.

Experience has shown that for isothermal fully-developed flow, the transition from laminar to turbulent flow occurs at a critical Reynolds number of

$$N_{Re,c} = \frac{D u_m \rho}{\mu} \approx 2300 \qquad (3.3.18)$$

Carefully conducted experiments with an object of increasing critical Reynolds number have succeeded in producing critical Reynolds numbers as high as 5 x 10^4. For ordinary design purposes, a critical Reynolds number of 2300 is satisfactory for channel flows.

For heated flows of liquids, the onset of turbulence takes place at a lower critical Reynolds number which is probably due to natural convection being superimposed on the forced flow. For horizontal tubes, the critical Reynolds number is given by [10]

$$600 \leqslant N_{Re,c} \leqslant 800 \qquad (3.3.19)$$

and for vertical tubes [10]

$$400 \leqslant N_{Re,c} \leqslant 600 \qquad (3.3.20)$$

Mixed flow (combined forced and natural convection) is not well understood and the exact establishment of a critical Reynolds number probably requires experimental measurements.

For fully-developed, isothermal, laminar flow in a circular pipe, the velocity profile is dependent only upon radial distance and is symmetric with respect to the pipe axis. The velocity profile is given by the well-known parabolic form

$$u = \frac{g_c (p_1 - p_2)}{4\mu L} (r_w^2 - r^2) \qquad (3.3.21)$$

where $(p_1 - p_2)/L$ is the pressure loss due to friction over a length L of pipe and r_w is the pipe radius. Under the same conditions the equation for the laminar velocity distribution in a wide, flat channel is given by

$$u = \frac{g_c (p_1 - p_2)}{2\mu L} (yb - y^2) \qquad (3.3.22)$$

where b is the channel spacing.

For fully-developed, isothermal, turbulent flow in channels, no simple relation exists for determining velocity distributions. A theoretical treatment is hampered because little is understood about the nature of turbulence. Consequently, it is necessary to rely upon experiments to provide information about velocity distributions.

Many empirical turbulent flow velocity distributions are proposed for channel flows. The best-known correlation is that of von Karman [11] who proposed that velocity distributions in turbulent flow consist of three regions—a laminar sublayer, a buffer zone, and a turbulent core. The von Karman velocity profile is written in terms of the non-dimensional velocity parameter, u^+, and non-dimensional distance paramenter, y^+. The wall shear stress for channel flows is given by

$$\tau_W = \frac{f_W}{4} \frac{\rho u_m^2}{2g_c} \qquad (3.3.23)$$

where f_W is the Darcy-Weisbach friction factor which is discussed in the next chapter and u_m is the average channel velocity.

Adjacent to the wall the flow is thought to be laminar and the velocity distribution in terms of non-dimensional parameters is given by

$$u^+ = y^+ \qquad (3.3.24)$$

for $0 \leqslant y^+ \leqslant 5$. The buffer zone next to the laminar sublayer has a velocity distribution given by

$$u^+ = -3.05 + 5.0 \ln y^+ \qquad (3.3.25)$$

for a range $5 \leqslant y^+ \leqslant 30$. Outside the buffer zone, the flow is fully turbulent and the velocity distribution is given by

$$u^+ = 5.5 + 2.5 \ln y^+ \qquad (3.3.26)$$

for $30 < y^+$.

The main difference in turbulent velocity profiles between channel flows and external boundary layers occurs away from the wall. This is to be expected, since away from the wall the free stream conditions are different for these two examples. Near the wall, where wall effects predominate, the two types of flow should be similar.

Figure 3.8 illustrates the channel flow velocity distribution in terms of non-dimensional parameters. There is good experimental agreement with the von Karman velocity profile for flows at Reynolds numbers above 30,000. This profile is not satisfactory for predicting velocities at the axes of pipes or channels because the velocity gradient is not zero (a necessary condition in actual symmetric flows).

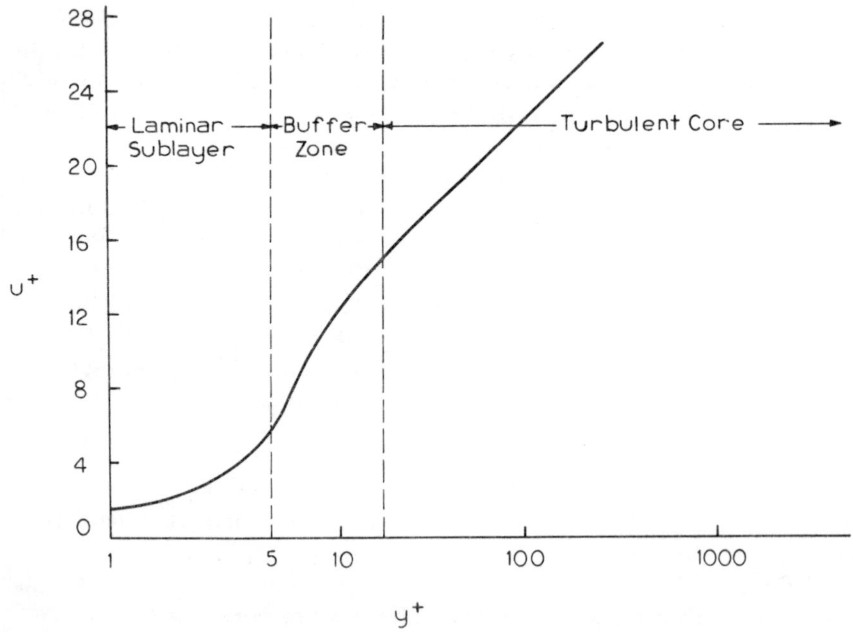

Fig. 3.8. Universal velocity profile for channel flows [9]

Velocity distributions for pipe flows are shown in Fig. 3.9 for laminar and turbulent flow. As the Reynolds number increases the velocity distribution becomes more uniform and, consequently, the ratio of maximum velocity to average velocity u_m approaches unity.

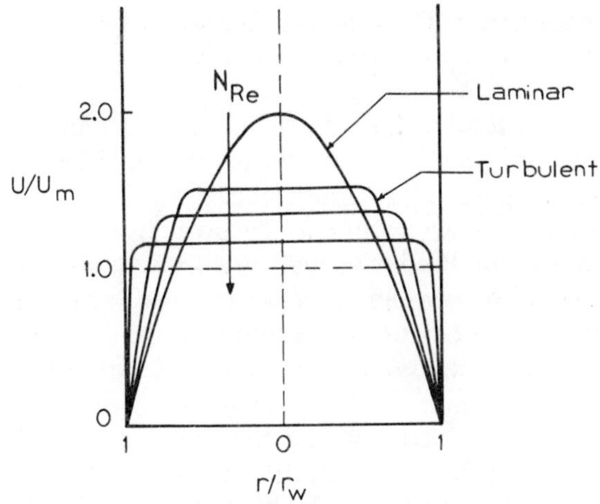

Fig. 3.9. Velocity distributions in a smooth pipe

Streamline Flow

A streamline is a line in a flow path in which fluid particles do not cross. A flow field in which streamlines exist is called streamline flow. Naturally, fluid velocity vectors are tangent to streamlines.

The fundamental equations of fluid flow are the continuity equation, momentum equations, and energy equation. A simple form of these equations is derived for streamline flow by examining flow in a "stream tube." A stream tube is generated by enclosing a portion of the flow volume with a surface made of streamlines as shown in Fig. 3.10.

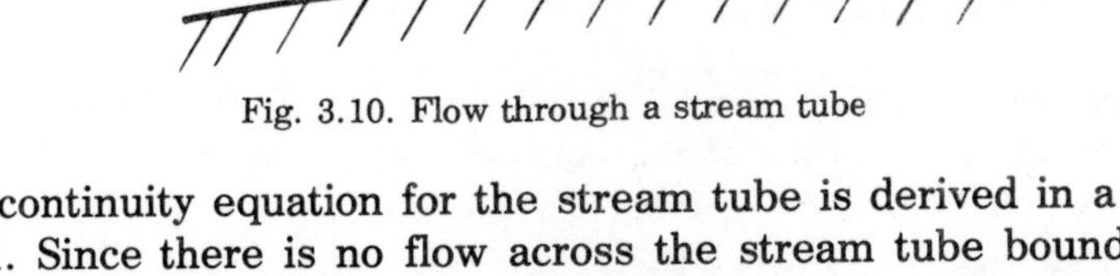

Fig. 3.10. Flow through a stream tube

The continuity equation for the stream tube is derived in a simple fashion. Since there is no flow across the stream tube boundary, in steady flow the mass flow rate entering the stream tube equals the mass flow rate leaving the stream tube. Therefore, if u_1 and u_2 are the mean fluid velocities entering and leaving the tube at positions 1 and 2, the continuity equation is written

$$\dot{m} = \rho_1 A_1 u_1 = \rho_2 A_2 u_2 \qquad (3.3.27$$

For an imcompressible fluid, Eq. (3.3.27) simplifies to

$$A_1 u_1 = A_2 u_2 \tag{3.3.28}$$

The momentum equation for steady flow in a stream tube is derived by applying Newton's Second Law to the stream tube, or

$$\text{Rate of change of momentum} = g_c \vec{F} \tag{3.3.29}$$

where $\vec{F}$ is the resultant of all external forces acting on the stream tube. Applying Eq. (3.3.29) over an interval of time Δt yields the equality between impulse and momentum change, $\vec{F} g_c \Delta t = m \Delta \vec{u}$, where $\Delta \vec{u}$ is the vectorial change in velocity. For a steady flow process

$$\vec{F} g_c = \Delta \vec{u} \frac{m}{\Delta t} = \dot{m} \Delta \vec{u} \tag{3.3.30}$$

where $\dot{m}$ is the mass flow rate into the stream tube, lb_m /sec. However, the mass flow rate is given by Eq. (3.3.27) and we may therefore write for the resultant force acting between positions 1 and 2

$$\vec{F} g_c = \rho_1 A_1 u_1 (\vec{u}_2 - \vec{u}_1) \tag{3.3.31}$$

The simplest form of the energy equation is the Bernoulli equation which is applied to the steady flow of a frictionless, incompressible, isothermal fluid. For a stream tube the Bernoulli equation is given by

$$\frac{g}{g_c} z_1 + v p_1 + \frac{u_1^2}{2g_c} = \frac{g}{g_c} z_2 + v p_2 + \frac{u_2^2}{2g_c} \tag{3.3.32}$$

It should be noted that Eq. (3.3.32) implies the fluid velocity is uniform at the inlet and exit of the stream tube. This is due to the average kinetic energy of a fluid not being given by the square of the mean velocity for non-uniform velocity distributions.

For a steady, frictionless, compressible flow at variable temperature, the energy equation is given by

$$\frac{g}{g_c} z_1 + v_1 p_1 + e_1 + \frac{u_1^2}{2g_c} = \frac{g}{g_c} z_2 + v_2 p_2 + e_2 + \frac{u_2^2}{2g_c} \tag{3.3.33}$$

where e_1 and e_2 are the internal energies of the fluid at positions 1 and 2 in the stream tube. By making use of the definition of enthalpy, $h = e + pv$, Eq. (3.3.33) is rewritten

$$\frac{g}{g_c} z_1 + h_1 + \frac{u_1^2}{2g_c} = \frac{g}{g_c} z_2 + h_2 + \frac{u_2^2}{2g_c} \tag{3.3.34}$$

These examples illustrate the fundamental fluid flow equations in extremely simple forms. In more general terms, the fluid flow equations are a set of partial differential equations which are difficult to solve without some simplification. The next section briefly treats these general equations.

Differential Equations of Fluid Flow

The differential equations which govern the motion of and energy transfer within a flowing fluid are derived by examining a stationary elemental control volume in the flow field. A derivation of these equations is found in many books on heat transfer or fluid mechanics [3,7,12].

(1) Continuity or Conservation of Mass Equation

Using vector notation, the general equation for the conservation of mass is given by

$$\nabla \cdot \rho u + \frac{\partial \rho}{\partial t} = 0 \qquad (3.3.35)$$

For an incompressible fluid, Eq. (3.3.35) simplifies to

$$\nabla \cdot \vec{u} = 0 \qquad (3.3.36)$$

which is written in rectangular coordinates as

$$\frac{\partial u_x}{\partial x} + \frac{\partial u_y}{\partial y} + \frac{\partial u_z}{\partial z} = 0 \qquad (3.3.37)$$

where u_x, u_y, and u_z are velocity components in the x–, y–, and z– directions, respectively. In tensor notation Eq. (3.3.37) is written

$$\frac{\partial u_i}{\partial x_i} = 0 \qquad (3.3.38)$$

(2) Momentum or Navier-Stokes Equations

The momentum equations are simply a statement of Newton's Second Law or

$$\Sigma \vec{F} = m\vec{a} \qquad (3.3.39)$$

where $\vec{F}$ represents external forces acting upon a fluid particle and $\vec{a}$ is the acceleration of the fluid particle. An equation is written for each of the coordinate directions.

In the language of fluid mechanics, the term mass x acceleration is called the inertial force. Consequently, the inertial force is equal to the sum of external forces acting on a fluid particle. The external forces are field forces (gravity, magnetic fields, etc.), normal forces (pressure), and shear or tangential forces (forces due to resistance of the fluid to deformation).

For rectangular coordinates, the momentum equation for the x-direction is given by

$$\rho \frac{Du_x}{Dt} = \rho B_x - \frac{\partial p}{\partial x} + \frac{\partial}{\partial x}\left\{\mu\left(2\frac{\partial u_x}{\partial x} - \frac{2}{3}\nabla\cdot\vec{u}\right)\right\}$$

$$+ \frac{\partial}{\partial y}\left[\mu\left(\frac{\partial u_y}{\partial x} + \frac{\partial u_x}{\partial y}\right)\right] + \frac{\partial}{\partial z}\left[\mu\left(\frac{\partial u_z}{\partial x} + \frac{\partial u_x}{\partial z}\right)\right] \quad (3.3.40)$$

where B_x is a general formulation for field forces acting in the x-direction and D/Dt is called the substantial derivative which is defined by

$$\frac{D}{Dt} \equiv u_x \frac{\partial}{\partial x} + u_y \frac{\partial}{\partial y} + u_z \frac{\partial}{\partial z} + \frac{\partial}{\partial t} \quad (3.3.41)$$

Equation (3.3.40) assumes the fluid is Newtonian and, consequently, is not applicable for other classes of fluids.

Similar equations exist for the y- and z- directions. One may also use different coordinates such as cylindrical or spherical coordinates to describe the momentum equations.

The momentum equations are difficult to solve in their rigorous form and it is necessary to make simplifications in order to reduce their complexity. Assuming the fluid is incompressible and the viscosity constant, the x-direction momentum equation is now written

$$\rho \frac{Du_x}{Dt} = \rho B_x - \frac{\partial p}{\partial x} + \mu\left(\frac{\partial^2 u_x}{\partial x^2} + \frac{\partial^2 u_x}{\partial y^2} + \frac{\partial^2 u_x}{\partial z^2}\right) \quad (3.3.42)$$

Further simplifications are made to the Navier-Stokes equations by analyzing the order of magnitude or relative importance of each term in the equations. This type of simplification is used by Prandtl in his boundary layer theory [7] which leads to a system of equations solvable for laminar flow.

With the use of tensor notation, the three momentum equations are represented by a single equation. The Navier-Stokes equations are given by

$$\rho \frac{Du_i}{Dt} = \rho B_i - \frac{\partial p}{\partial x_i} + \mu \frac{\partial^2 u_i}{\partial x_j \partial x_j} \quad (3.3.43)$$

(3) Energy Equation

The energy equation is a formulation of the First Law of Thermodynamics which states that for a fluid element the rate of change of internal energy equals the rate of internal heat generation within the element plus the heat transfer rate to the element minus the rate of work done by the element. For rectangular coordinates the energy equation is given by

$$\rho \frac{De}{Dt} = q''' + \frac{\partial}{\partial x}\left(k \frac{\partial T}{\partial x}\right) + \frac{\partial}{\partial y}\left(k \frac{\partial T}{\partial y}\right) + \frac{\partial}{\partial z}\left(k \frac{\partial T}{\partial z}\right) + \frac{p}{\rho}\frac{D\rho}{Dt} + \Phi \quad (3.3.44)$$

where q''' is the volumetric heat generation rate in the fluid, Btu/hr–ft^3, and Φ is a complicated term [12] due to viscous dissipation.

The internal heat generation term in Eq. (3.3.44) is due to chemical reactions, nuclear reactions, or passing an electric current through a conducting fluid. For the majority of fluid flow problems, internal heat generation is nonexistent. However, for flow problems involving the circulation of liquid nuclear reactor fuels (such as in the Molten Salt Reactor) this term can be quite significant.

For an incompressible fluid with constant thermal conductivity and no internal heat generation, the energy equation is written

$$\frac{\partial T}{\partial t} + u_x \frac{\partial T}{\partial x} + u_y \frac{\partial T}{\partial y} + u_z \frac{\partial T}{\partial z} = \frac{k}{\rho c_p} \nabla^2 T + \frac{\Phi}{\rho c_p} \quad (3.3.45)$$

Equation (3.3.45) is the form of the energy equation frequently used in theoretical analysis. Further simplifications are obtained by neglecting terms which are small in comparison to the rest of the terms in the equation. For most flows the viscous dissipation term Φ is neglected. However, Φ is not negligible in the treatment of high speed flows which occur in some aerodynamic problems.

(4) Fundamental Equations for Turbulent Flow

A considerable difference exists between laminar and turbulent flows. Turbulent flow is characterized by a random unsteadiness to the flow. A fast response velocity sensor at a point shows the following time dependence for the velocity component in the direction of mean flow.

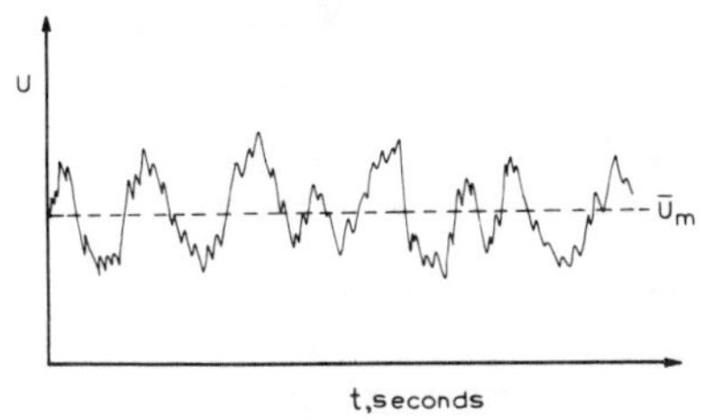

Fig. 3.11. Time dependence of turbulent flow

$\overline{u}_m$ represents the time-averaged velocity or mean velocity. Turbulent velocity components exist in all coordinant directions. However, their mean values are zero in directions transverse to a mean flow.

The differential equations of fluid mechanics given by Eqs. (3.3.35, 3.3.40, and 3.3.45) are applicable for both laminar or turbulent flows. However, for turbulent flows these equations are quite awkward because, in addition to their spatial dependence, velocity, pressure, and temperature vary randomly with time. For compressible fluids, density also varies randomly with time.

For the treatment of steady flows, a technique for simplifying the fluid flow equations is to represent the random variables as the sum of a time-independent mean and a random quantity. Consequently, for steady, incompressible turbulent flows, at any point in space the velocity, pressure, and temperature are given by

$$\begin{aligned} u_x &= \overline{u}_x + u_x' \\ u_y &= \overline{u}_y + u_y' \\ u_z &= \overline{u}_z + u_z' \\ p &= \overline{p} + p' \\ T &= \overline{T} + t' \end{aligned} \qquad (3.3.46)$$

where the mean quantities are given by

$$\overline{u}(\vec{r}) = \lim_{\mathcal{T}\to\infty} \frac{1}{\mathcal{T}} \int_0^{\mathcal{T}} u(\vec{r},t)\, dt \qquad (3.3.47)$$

Naturally, from Eq. (3.3.47) the time-average of the random quantities denoted by a prime is zero.

The expressions represented by Eq. (3.3.46) are substituted into the fluid flow equations and a time-average taken of each term in the equations. For the continuity equation this yields

$$\lim_{\mathcal{T}\to\infty} \frac{1}{\mathcal{T}} \int_0^{\mathcal{T}} \left[\frac{\partial(\overline{u}_x + u_x')}{\partial x} + \frac{\partial(\overline{u}_y + u_y')}{\partial y} + \frac{\partial(\overline{u}_z + u_z')}{\partial z} \right] dt = 0 \qquad (3.3.48)$$

or

$$\frac{\partial \overline{u}_x}{\partial x} + \frac{\partial \overline{u}_y}{\partial y} + \frac{\partial \overline{u}_z}{\partial z} = 0 \qquad (3.3.49)$$

which is the same result previously obtained, but now in terms of local mean velocities.

The same operation indicated by Eq. (3.3.47) is performed on the x-direction momentum equation and, by making use of Eq. (3.3.49) and performing some arrangement, we have

$$\rho\left[\bar{u}_x\frac{\partial \bar{u}_x}{\partial x} + \bar{u}_y\frac{\partial \bar{u}_x}{\partial y} + \bar{u}_z\frac{\partial \bar{u}_x}{\partial z}\right] = \rho B_x - \frac{\partial \bar{p}}{\partial x}$$

$$+ \frac{\partial}{\partial x}\left[\mu\frac{\partial \bar{u}_x}{\partial x} - \rho\overline{u_x'^2}\right] + \frac{\partial}{\partial y}\left[\mu\frac{\partial \bar{u}_x}{\partial y} - \rho\overline{u_x'u_y'}\right]$$

$$+ \frac{\partial}{\partial z}\left[\mu\frac{\partial \bar{u}_x}{\partial z} - \rho\overline{u_x'u_z'}\right] \qquad (3.3.50)$$

Equations similar to Eq. (3.3.50) are derived for the y- and z-directions. Using tensor notation the three momentum equations are given by

$$\rho\bar{u}_j\frac{\partial \bar{u}_i}{\partial x_j} = \rho B_i - \frac{\partial \bar{p}}{\partial x_i} + \frac{\partial}{\partial x_j}\left[\mu\frac{\partial \bar{u}_i}{\partial x_j} - \rho\overline{u_i'u_j'}\right] \qquad (3.3.51)$$

The three equations represented by Eq. (3.3.51) are the Reynolds equations for mean velocities in turbulent flow. The momentum equations are only altered by the appearance of terms containing the time-average of products of fluctuating velocity components. Equation (3.3.51) is written in a form in which the additional terms appear as stresses due to turbulent motion. The turbulent stress components are called Reynolds stresses. There are nine Reynolds stresses; however, only six of the components are independent.

Normal Stresses: $-\rho\overline{u_x'^2}$, $-\rho\overline{u_y'^2}$, $-\rho\overline{u_z'^2}$

Shear Stresses: $-\rho\overline{u_x'u_y'}$, $-\rho\overline{u_x'u_z'}$, $-\rho\overline{u_y'u_z'}$

Substituting the relations given by Eq. (3.3.46) into the energy equation for turbulent, incompressible flow and performing a time-average yields

$$\rho c_p\left[\bar{u}_x\frac{\partial \bar{T}}{\partial x} + \bar{u}_y\frac{\partial \bar{T}}{\partial y} + \bar{u}_z\frac{\partial \bar{T}}{\partial z}\right] = q''' + \frac{\partial}{\partial x}\left[k\frac{\partial \bar{T}}{\partial x} - \rho c_p\overline{u_x't'}\right]$$

$$+ \frac{\partial}{\partial y}\left[k\frac{\partial \bar{T}}{\partial y} - \rho c_p\overline{u_y't'}\right] + \frac{\partial}{\partial z}\left[k\frac{\partial \bar{T}}{\partial y} - \rho c_p\overline{u_z't'}\right] \qquad (3.3.52)$$

Viscous dissipation is considered negligible in formulating Eq. (3.3.52).

Equation (3.3.52) is similar to the original energy equation with the exception of the appearance of terms containing time-averages of products of fluctuating velocities and temperature. In the arrangement of Eq. (3.3.52), the new terms appear as heat fluxes which are due to the convective action of the turbulent motion.

The fundamental fluid flow equations given by Eqs. (3.3.49, 3.3.51, and 3.3.52) constitute a set of five equations containing fourteen unknowns. Naturally this system of equations is unsolvable and it is necessary to have available expressions which describe the spatial dependence of the time-averaged turbulent components.

Phenomenological Approach to Turbulence

The present theoretical approaches used to describe the fluctuating velocities and temperatures in turbulent flows are lumped into two general classes: (1) **Statistical** approach and (2) **Phenomenological** approach. Since the statistical approach at this time has not yielded results suitable for engineering design applications, this approach is not treated in this book. However, it is thought that if complete understanding of turbulence should ever be forthcoming, it will require use of statistical approaches.

The phenomenological approach develops models describing turbulent motion and then use experimental data to supply constants for the model equations so as to force agreement between experiment and the semi-empirical theory. Most phenomenological approaches make use of considerations of fluid motion in a manner similar to that proposed by Prandtl in his **mixing-length theory** [7].

The present thinking about turbulent flow is that fluid motion is in the form of lumps or eddies which are continuously being formed and destroyed. These eddies travel randomly in all directions, but may be carried along by a mean-motion superimposed on the flow. Prandtl came to these conclusions because he thought the increased friction pressure loss in turbulent flow is due to the exchange of momentum between fluid particles.

Let us examine the turbulent flow field shown in Fig. 3.12 in which the mean motion is in the x-direction and only a function of y.

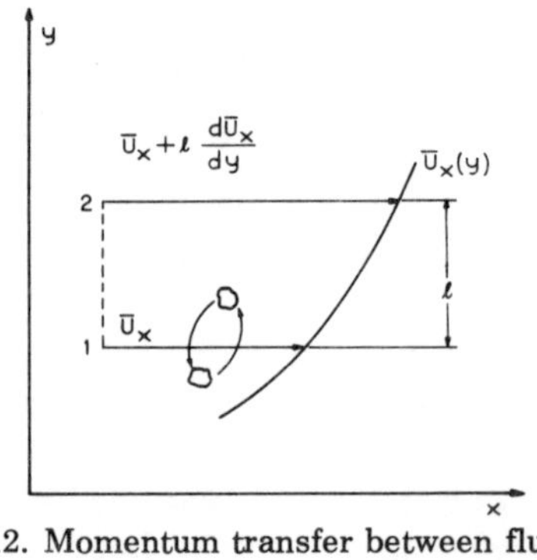

Fig. 3.12. Momentum transfer between fluid particles

The difference between mean velocities in two nearby fluid layers 1 and 2 is $\ell d\bar{u}_x/dy$. At layer 2, the fluid moves with a faster velocity than at layer 1. Consequently, a fluid particle or eddy moving from layer 2 to layer 1 is slowed down while at the same time this particle momentarily tends to speed up the flow at layer 1. Conversely, a fluid particle moving from layer 1 to layer 2 acts as a drag in layer 2 and also has its speed increased. The distance ℓ between the fluid layers 1 and 2 is the **mean mixing length** which Prandtl defined as the distance a fluid particle moves in a direction transverse to the mean flow before its identity is lost.

Since the transverse motion of fluid particles produces instantaneous changes in fluid velocity, the fluctuations in velocity may be described in a manner similar to that used for the change in mean velocity. Consequently, the instantaneous velocity fluctuation in the x-direction is given by

$$u_x' = \ell_x' \frac{d\bar{u}_x}{dy} \tag{3.3.53}$$

where ℓ_x' is the instantaneous mixing length for the x-direction velocity component. With this approach the x-direction velocity fluctuation is equal to the difference in mean velocities of two layers separated by a distance ℓ_x'.

The instantaneous velocity fluctuation in the y-direction is written as

$$u_y' = \ell_y' \frac{d\bar{u}_x}{dy} \tag{3.3.54}$$

where ℓ_y' is the instantaneous mixing length for the y-direction velocity component.

The shear stress acting on a fluid surface parallel to the mean flow $\bar{u}_x$ is broken into two components for turbulent flow.

$$\begin{aligned} \tau &= \tau_{\text{molecular}} + \tau_{\text{turbulent}} \\ &= \tau_L + \tau_t \end{aligned} \tag{3.3.55}$$

where the molecular (laminar flow) shear stress is given by

$$\tau_L = \mu \frac{d\bar{u}_x}{dy} \tag{3.3.56}$$

and the turbulent shear stress is given by

$$\tau_t = \rho \overline{u_x' u_y'} \tag{3.3.57}$$

By substituting instantaneous velocity components given by Eqs.

(3.3.53) and (3.3.54) into Eq. (3.3.57) and taking a time-average, we have

$$\tau_t = -\rho\overline{\ell_x'\ell_y'}\left(\frac{d\bar{u}_x}{dy}\right)\left(\frac{d\bar{u}_x}{dy}\right) \tag{3.3.58}$$

The term $u_x' u_y'$ has on the average a sign that is the negative of the sign on $d\bar{u}_x/dy$. For positive $d\bar{u}_x/dy$, when u_y' is positive u_x' is negative and vice versa. The same type of argument also holds for negative $d\bar{u}_x/dy$. Consequently, by combining mixing length we have

$$\tau_t = \rho\ell^2\left|\frac{d\bar{u}_x}{dy}\right|\frac{d\bar{u}_x}{dy} \tag{3.3.59}$$

Experiments offer some support for the mixing length approach. Velocity fluctuations increase in regions with larger velocity gradients and $u_x' u_y'$ is negative in regions of positive velocity gradients. One problem that occurs in channel flows is that at a point of symmetry, $d\bar{u}_x/dy = 0$; however, measurements show that $u_x' u_y'$ is not zero which is what intuitively one would expect.

By combining $\ell^2|d\bar{u}_x/dy|$ into one term, an eddy viscosity or eddy diffusivity for momentum transfer is defined

$$\epsilon_m \equiv \ell^2\left|\frac{d\bar{u}_x}{dy}\right| \tag{3.3.60}$$

Therefore, the total shear stress in a fluid with a one-directional turbulent flow is given by

$$\tau = \rho(\nu + \epsilon_m)\frac{d\bar{u}_x}{dy} \tag{3.3.61}$$

By making assumptions about the shear stress spatial dependence in the flow field and using the universal velocity profile equations, it is possible to develop analytical expressions for ϵ_m.

The approach presented is quite simplified and is only applicable for a plane, parallel motion. Hinze has treated the use of eddy viscosities in a more general manner and has shown that the eddy viscosity is a fourth-order tensor, $\epsilon_{M,ijkl}$ [13]. However, it is customary in the treatment of flows with mean velocity components in two or three directions to assume the eddy viscosity a scalar.

An approach analogous to that just illustrated is applied to heat transfer in turbulent flows. Once again we assume a flow field in which the mean flow is only in the x-direction. For a given x-location, the temperature distribution is assumed described in the manner shown by Fig. 3.13.

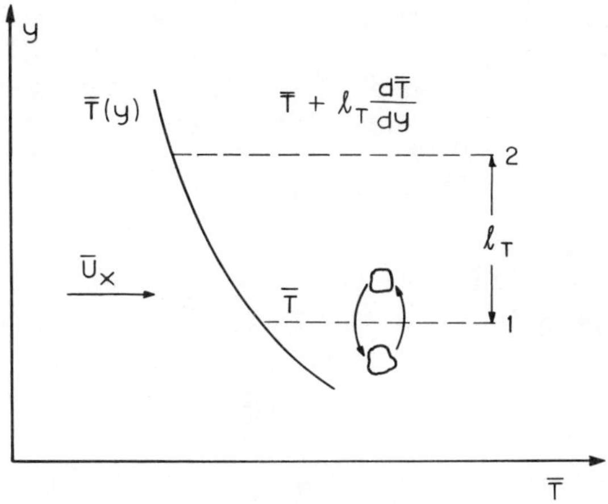

Fig. 3.13. Heat transfer in turbulent flow

The difference in mean temperatures between two fluid layers 1 and 2 is $\ell_T \partial\overline{T}/\partial y$, where ℓ_T is the mean thermal mixing length.

The heat transfer rate per unit area in the y-direction is due to conduction and turbulent convection, or

$$\begin{aligned} q''_y &= q''_y(\text{conduction}) + q''_y(\text{turbulent}) \\ &= q''_{yc} + q''_{yt} \end{aligned} \tag{3.3.62}$$

The heat transfer rates due to conduction and turbulent convection are obtained from Eq. (3.3.52) as

$$q''_{yc} = -k\frac{\partial\overline{T}}{\partial y} \tag{3.3.63}$$

and

$$q''_{yt} = \rho c_p \overline{u'_y t'} \tag{3.3.64}$$

By assuming instantaneous temperature fluctuations are described in a similar manner as velocity fluctuations, we have

$$t' = \ell'_T \frac{\partial\overline{T}}{\partial y} \tag{3.3.65}$$

where ℓ'_T is the instantaneous thermal mixing length. By substituting Eqs. (3.3.54) and (3.3.65) into Eq. (3.3.64) and taking a time-average, we have

$$q''_{yt} = \rho c_p \overline{\ell'_y \ell'_T}\left(\frac{\partial\overline{T}}{\partial y}\right)\left(\frac{\partial\overline{u}_x}{\partial y}\right) \tag{3.3.66}$$

The term $u'_y t'$ will on the average be the negative of $\partial \bar{T}/\partial y$ because a positive u'_y causes a positive t' for a region in which $\partial \bar{T}/\partial y$ is negative. The time-average of the two mixing lengths in Eq. (3.3.66) is combined into a single term to give

$$q''_{yt} = -\rho c_p \ell_T^2 \left| \frac{d\bar{u}_x}{dy} \right| \frac{\partial \bar{T}}{\partial y} \tag{3.3.67}$$

An eddy diffusivity for heat transfer is defined in the same manner as the eddy diffusivity for momentum transfer.

$$\epsilon_H \equiv \ell_T^2 \left| \frac{d\bar{u}_x}{dy} \right| \tag{3.3.68}$$

Therefore, the heat transfer rate in the transverse direction for a one-directional flow is given by

$$q''_y = -\rho c_p (a_t + \epsilon_H) \frac{\partial \bar{T}}{\partial y} \tag{3.3.69}$$

where a_t is the thermal diffusivity given by $k/\rho c_p$.

The use of a mixing length model to describe temperature fluctuations has some experimental support, since the magnitude of temperature fluctuations increases in regions of increasing temperature gradients [14]. For the treatment presented, if ℓ_t were equal to ℓ_x, then $\epsilon_H = \epsilon_M$. This assumption is made in a considerable amount of analytical work involving the energy equation. However, making such an assumption is incorrect in theory since in the most general sense the eddy diffusivity for heat transfer is a second-order tensor, $\epsilon_{M,ij}$. In addition, experiments with liquid metals indicate that when employing the scalar representations for the eddy diffusivities, $\epsilon_M >> \epsilon_H$.

Solutions of the Fluid Flow Equations

In problems of determining velocity distributions and pressure losses in flow systems, a rigorous approach requires simultaneous solution of the continuity and momentum equations. If the flow is non-isothermal and fluid physical properties dependent upon temperature, then the energy equation must also be included in the set of equations. The fundamental equations presented in the previous section are insolvable in their general form because of extreme non-linearity and presence of more unknowns than equations for turbulent flow. Consequently, the approach used to find solutions is to assume most of the terms in the equations are negligible and attempt to find solutions to a vastly simplified set of equations. For turbulent flow, the simplified fundamental equations are still insolvable because of the inability to find mathematical expressions which describe random fluctuations of velocity and temperature with time. Consequently, in order to find solutions to turbulent flow equations, further assumptions are made about the nature of turbulence and empirical expressions formulated which describe these assumptions.

3.4 Dimensional Analysis

As mentioned in the preceding section, due to the complexity of the fluid flow equations, exact solutions are found for only a few special cases. Consequently, in the analysis of fluid flow and convection heat transfer it is necessary to rely heavily upon experiment. The necessity of using experimental data also creates special problems. One must know what variables to study in the experiment and how to successfully correlate experimental data. An additional problem is the scale-up of data from small-size experiments to full-size systems. Dimensional analysis provides a powerful technique for handling problems of this nature. However, caution should be used in applying dimensional analysis, since it is necessary that all dimensional variables are known.

In the study of fluid mechanics, the dimensions of most variables are given in terms of three fundamental quantities—Mass (M), Length (L), and Time (t). One can also consider Force (F) as a fundamental dimension. However, force is expressed in terms of the first three quantities as ML/t^2. With energy transfer, one can consider two more fundamental quantities—Temperature (T) and Energy (E). Energy (E) is also equivalent to work (FL) and, consequently, it is not necessary to use energy as a fundamental dimension. However, in applying dimensional analysis to engineering problems it is convenient to use energy as a fundamental dimension. Therefore, most variables used in engineering analysis are expressed in terms of five fundamental dimensions—M, L, t, T, and E.

Similitude

In the use of dimensional analysis, it is necessary to consider similarity between different systems. For the scale-up of experimental data from pilot plants, geometric similarity must exist between systems. An additional requirement for scale-up of systems with flowing fluids is dynamic similarity in the flow fields. The conditions required for geometric and dynamic similarity between systems are as follows [3]:

(1) Geometric Similarity

Let the solid boundaries of a system by completely described by a number of length segments L_1, L_2, L_3,...L_n. If these segments are divided by L_1, the system is then described by L_1, r_2, r_3,...r_n, where

$$r_2 = L_2/L_1$$
$$r_3 = L_3/L_1$$
$$\vdots$$
$$r_n = L_n/L_1 \tag{3.4.1}$$

Geometric similarity exists between two bodies if the ratios r_2, r_3,...r_n are the same for each system.

(2) Dynamic Similarity

Dynamic similarity considers the relationship between various forces acting on systems. These forces are inertial, normal, shear, and field forces.

In geometrically similar systems, dynamic similarity exists at the same relative point in each system if the following conditions are satisfied:

$$\frac{\text{Inertial Force}_{\text{I}}}{\text{Viscous Force}_{\text{I}}} = \frac{\text{Inertial Force}_{\text{II}}}{\text{Viscous Force}_{\text{II}}}$$
$$\frac{\text{Inertial Force}_{\text{I}}}{\text{Gravity Force}_{\text{I}}} = \frac{\text{Inertial Force}_{\text{II}}}{\text{Gravity Force}_{\text{II}}} \tag{3.4.2}$$

The ratio of inertial force to viscous force is the Reynolds number. This is shown by examining dimensions on each term. The inertial force is given by mass x acceleration or $V\rho(du/dt)$ where V is the volume under consideration. In terms of dimensions, the inertial force is given by

$$L^3\rho u/t = L^2\rho u^2 \tag{3.4.3}$$

The viscous force is given by

$$A\tau = A\mu\,(du/dy) \tag{3.4.4}$$

where A is an area and τ a shear stress. In terms of dimensions, the viscous force is represented by

$$L^2\mu u/L = L\mu u \tag{3.4.5}$$

Therefore, the ratio of these forces is

$$\frac{L^2\rho u^2}{L\mu u} = \frac{\rho u L}{\mu} = N_{Re} \tag{3.4.6}$$

The ratio of inertial force to gravity force is obtained in the same manner. The inertial force was shown to be represented by $L^2\rho u^2$. The gravity force is given by mass x acceleration due to gravity or $V\rho g$. In terms of dimensions, the gravity force is represented by $L^3\rho g$. Therefore, the ratio of inertial force to gravity force is given by

$$\frac{L^2\rho u^2}{L^3\rho g} = \frac{u^2}{Lg} \tag{3.4.7}$$

Equation (3.4.7) is a dimensionless quantity called the Froude number, or

$$N_{Fr} = \frac{u^2}{Lg} \tag{3.4.8}$$

For flow along solid boundaries, gravity forces are usually negligible and equal Reynolds numbers are the criteria for dynamic similarity of geometrically similar systems. For flow along free surfaces, for example the top of the ocean, viscous forces are negligible and equal Froude numbers are the important criteria for dynamic similarity of geometrically similar systems.

Methods of Dimensional Analysis

The three methods of applying dimensional analysis examined in this section are (1) analysis of differential equations, (2) Rayleigh method, and (3) Buckingham method. The foundation for all of these methods is Buckingham's theorem which states "If an equation is dimensionally homogeneous, it can be reduced to a relationship among a complete set of dimensionless products" [15].

(1) Analysis of the Differential Equations

The simplified equations for steady, laminar flow in a boundary layer adjacent to a flat plate are given by [7]

(continuity) $$\frac{\partial u_x}{\partial x} + \frac{\partial u_y}{\partial y} = 0 \tag{3.4.9}$$

(momentum) $$u_x\frac{\partial u_x}{\partial x} + u_y\frac{\partial u_x}{\partial y} = \nu\frac{\partial^2 u_x}{\partial y^2} \tag{3.4.10}$$

(energy) $$u_x\frac{\partial T}{\partial x} + u_y\frac{\partial T}{\partial y} = a\frac{\partial^2 T}{\partial y^2} \tag{3.4.11}$$

By making the momentum equation dimensionless, fluid physical properties appear in the equation in the form of the Reynolds number. The boundary layer equations are made dimensionless by dividing all velocities by the free stream velocity, U_∞, and all length terms by x, the distance from the plate leading edge. The new variables are defined by

$$U_x \equiv u_x/U_\infty \qquad U_y \equiv u_y/U_\infty$$
$$X \equiv x/x \qquad Y \equiv y/x \tag{3.4.12}$$

By using the variables defined in Eq. (3.4.12), the continuity and momentum equations are written as

$$\frac{\partial U_x}{\partial X} + \frac{\partial U_y}{\partial Y} = 0 \tag{3.4.13}$$

and

$$U_x \frac{\partial U_x}{\partial X} + U_y \frac{\partial U_x}{\partial Y} = \frac{\nu}{U_\infty x} \frac{\partial^2 U_x}{\partial Y^2}$$

or

$$U_x \frac{\partial U_x}{\partial X} + U_y \frac{\partial U_x}{\partial Y} = \frac{1}{N_{Re}} \frac{\partial^2 U_x}{\partial Y^2} \tag{3.4.14}$$

By inspection of Eq. (3.4.14) it is apparent that for any position along the plate the velocity U_x is given by

$$U_x = f(Y, N_{Re}) \tag{3.4.15}$$

The energy equation is written in dimensionless form by using the dimensionless variables defined in Eq. (3.4.12) and a dimensionless temperature defined by

$$\theta = \frac{T_w - T}{T_w - T_\infty} \tag{3.4.16}$$

where T_∞ is the free stream temperature and T_w the wall temperature. Both T_w and T_∞ are assumed constant. Substituting the new variables gives

$$U_x \frac{\partial \theta}{\partial X} + U_y \frac{\partial \theta}{\partial Y} = \frac{a}{U_\infty x} \frac{\partial^2 \theta}{\partial Y^2} \tag{3.4.17}$$

The term $U_\infty x/a$ is a dimensionless number called the Peclet number, N_{Pe}. The Peclet number is also the product of the Reynolds and Prandtl numbers. Therefore, Eq. (3.4.17) is written as

$$U_x \frac{\partial \theta}{\partial X} + U_y \frac{\partial \theta}{\partial Y} = \frac{1}{N_{Pe}} \frac{\partial^2 \theta}{\partial Y^2} \tag{3.4.18}$$

An examination of Eq. (3.4.18) shows that θ is a function of Y, U_x, U_y, and N_{Pe}. However, the velocity terms were previously shown to be a function of Y and N_{Re}. Therefore, the functional dependence of θ is given by

$$\theta = f_\theta (Y, N_{Re}, N_{Pe}) \tag{3.4.19}$$

and since $N_{Pe} = N_{Pe} . N_{Pr}$, Eq. (3.4.19) is also written

$$\theta = f_\theta (Y, N_{Re}, N_{Pr}) \tag{3.4.20}$$

For fluids with $N_{Pr} = 1$, the dimensionless momentum and energy equations are of identical form. Consequently, the solution of one equation is used as the solution for the other providing T_w and T_∞ are constant.

The heat-transfer coefficient for a flat plate is given by

$$h \equiv \frac{q_w''}{T_w - T_\infty} = \frac{-k\,(\partial T/\partial y)_w}{T_w - T_\infty} \tag{3.4.21}$$

where q_w'' is the wall heat flux, Btu/hr-ft². By rearrangement in terms of dimensionless quantities, Eq. (3.4.21) is written

$$\frac{hx}{k} = \left(\frac{\partial \theta}{\partial Y}\right)_w \tag{3.4.22}$$

The dimensionless temperature is a function of Y, N_{Re}, and N_{Pr} and, consequently, its derivative must also be a function of these quantities. Since we are evaluating the derivative of θ at the wall or Y = 0, the derivative is a function of N_{Re} and N_{Pr}. Therefore, the heat-transfer coefficient is given by

$$\frac{hx}{k} = N_{Nu} = f'_\theta (N_{Re}, N_{Pr}) \tag{3.4.23}$$

where N_{Nu} is a dimensionless number called the Nusselt number.

(2) Rayleigh Method [15]

The Rayleigh method consists of establishing a relationship between some quantity and its known dependent variables through dimensional homogeneity.

Let Q_1, Q_2, Q_3,...Q_n be "n" dimensional variables upon which a given physical process is dependent. Let r_2, r_3,...r_m be the dimensionless length ratios required to geometrically describe the solid boundaries of the system. The functional relationship between all these variables is

$$F_1(Q_1, Q_2, \ldots Q_n \,;\, r_2, r_3, \ldots r_m) = 0 \tag{3.4.24}$$

For geometrically similar systems, the relationship becomes

$$F_2(Q_1, Q_2, \ldots Q_n) = 0 \tag{3.4.25}$$

The Rayleigh method removes one variable from Eq. (3.4.25) and states

$$Q_1 = F_3(Q_2, Q_3, \ldots Q_n) \qquad (3.4.26)$$

Dimensional homogeneity is achieved by having the dimensions of the variables Q_2, $Q_3, \ldots Q_n$ taken to such powers that the dimensions of their product are the same as Q_1. Otherwise, in terms of dimensions of each variable

$$[Q_1] = [Q_2]^a [Q_3]^b [Q_4]^c, \text{ etc.} \qquad (3.4.27)$$

The unknown exponents in Eq. (3.4.27) are evaluated by dimensional analysis.

The application of the Rayleigh method is easily understood by working a simple problem. Let us find a correlation for predicting the pressure loss in a smooth tube. It is apparent that the pressure loss Δp is dependent upon fluid mean velocity u_m, fluid density ρ, fluid viscosity μ, pipe length L, and pipe diameter D. The units of these terms must satisfy the relationship

$$[\Delta p] = [\rho]^a [u_m]^b [L]^c [\mu]^d [D]^e \qquad (3.4.28)$$

By using mass, length, and time system of units, Eq. (3.4.28) is written

$$\frac{ML}{L^2 t^2} = \frac{M}{Lt^2} = \left(\frac{M}{L^3}\right)^a \left(\frac{L}{t}\right)^b (L)^c \left(\frac{M}{Lt}\right)^d (L)^e \qquad (3.4.29)$$

In order to satisfy dimensional homogeneity, the exponents on each fundamental dimension must satisfy the following relationships:

For M, $\qquad 1 = a + d \qquad (3.4.30)$

For L, $\qquad -1 = -3a + b + c - d + e \qquad (3.4.31)$

For t, $\qquad -2 = -b - d \qquad (3.4.32)$

From these equations three of the unknown exponents are eliminated, or

$$a = 1 - d \; ; \quad b = 2 - d \; ; \quad c = -d - e \qquad (3.4.33)$$

These relationships are substituted into Eq. (3.4.28) to give

$$\begin{aligned} [\Delta p] &= [\rho]^{1-d} [u_m]^{2-d} [L]^{-d-e} [\mu]^d [D]^e \\ &= [\rho u_m^2] [\mu/\rho u_m L]^d [D/L]^e \\ &= [\rho u_m^2] [N_{Re,L}]^{-d} [D/L]^e \end{aligned} \qquad (3.4.34)$$

It should be noted that the Reynolds number in Eq. (3.4.35) is based upon pipe length (L) instead of pipe diameter (D). Experience has shown that pressure loss in a pipe is a function of a Reynolds number based upon pipe diameter. Equation (3.4.35) is multiplied by $(D/D)^d$ and one obtains

$$[\Delta p] = [\rho u_m^2]\,[\rho u_m D/\mu]^{-d}\,[D/L]^{e+d} \qquad (3.4.35)$$

$$= [\rho u_m^2]\,[N_{Re}]^{-d}\,[D/L]^{e+d}$$

In terms of dimensionless quantities, Eq. (3.4.36) should be interpreted as

$$\frac{\Delta p}{\rho u_m^2} = F_3\,(D/L, N_{Re}) \qquad (3.4.36)$$

If the product form is accepted as the functional relationship for Q_1 and its independent variables, experimental data can be used to determine the unknown exponents and a proportionality constant. However, this conclusion can be quite erroneous since the Rayleigh method only implies a relationship exists among dimensionless quantities and does not indicate anything about the form of this functional dependence.

(3) Buckingham Method [16]

The Buckingham method also makes use of the relationship between variables in a physical process given by Eq. (3.4.25), or

$$F_2\,(Q_1, Q_2, Q_3, \ldots Q_n) = 0 \qquad (3.4.37)$$

Buckingham's π theorem states that if k is the maximum number of dimensional quantities which cannot be combined to form a dimensionless term, then the equation $F_2 = 0$ can be rearranged into a new equation containing n-k dimensionless terms, or

$$F_2\,(\pi_1, \pi_2, \ldots\ \pi_{n-k}) = 0 \qquad (3.4.38)$$

The dimensionless groups π are given by the following equations:

$$\pi_1 = Q_1^{a1}\,Q_2^{b1} \ldots Q_k^{k1}\,Q_{k+1}$$

$$\pi_2 = Q_1^{a2}\,Q_2^{b2} \ldots Q_k^{k2}\,Q_{k+2}$$

$$\vdots$$

$$\pi_{n-k} = Q_1^{an-k}\,Q_2^{bn-k} \ldots Q_k^{kn-k}\,Q_n \qquad (3.4.39)$$

The dimensional quantities $Q_1, Q_2, \ldots Q_k$ must between them contain all of the fundamental dimensions; i.e., M, L, T, t, E, etc.

Usually k is equal to the number of fundamental dimensions. It can never exceed this value, although it may be less. The dimensional quantities that make up $Q_1, Q_2, \ldots Q_k$ are selected arbitrarily. Generally, they are independent variables in an experiment such as density, diameter, etc. For convenience, it is desirable to have quantities which are varied in an experiment, like flow rate or pressure loss, to appear in the dimensionless groups without being taken to some power. In addition, it is convenient to select $Q_1, Q_2, \ldots Q_k$ as terms with few dimensions.

As an example, let us apply the Buckingham method to our previous problem for the pressure loss in a smooth pipe. The equation for the functional dependence among variables is written as

$$F_2(\rho, \mu, L, D, u_m, \Delta p) = 0 \tag{3.4.40}$$

In Eq. (3.4.40), n = 6 and the number of fundamental dimensions is M, L, and t or 3. By inspection of the variables it is easy to verify that k = 3. Therefore, the number of dimensionless terms required in the transformed equation is 3, or

$$F'_2(\pi_1, \pi_2, \pi_3) = 0 \tag{3.4.41}$$

Let us take u_m, ρ, and D as the variables which may appear with exponents. Therefore,

$$Q_1 = \rho \qquad Q_2 = u_m \qquad Q_3 = D \tag{3.4.42}$$

The fundamental dimensions are now expressed in terms of these variables as follows:

$$\rho = M/L^3 \quad \text{or} \quad M = L^3\rho = D^3\rho$$

$$u_m = L/t \quad \text{or} \quad t = L/u_m = D/u_m$$

$$D = L \quad \text{or} \quad L = D \tag{3.4.43}$$

π_1, π_2, π_3, contain μ, L, and Δp, respectively. Therefore, by using the expressions for the fundamental dimensions given by Eq. (3.4.43), the π's are written by inspection.

$$\mu = \frac{M}{Lt} = \frac{D^3\rho}{DD/u_m} = \rho u_m D \quad \text{or} \quad \pi_1 = \frac{\mu}{\rho u_m D} = \frac{1}{N_{Re}} \tag{3.4.44}$$

$$L = L = D \qquad \text{or} \qquad \pi_2 = L/D \tag{3.4.45}$$

$$\Delta p = \frac{M}{Lt^2} = \frac{D^3 \rho}{D(D/u_m)^2} = u_m^2 \rho \quad \text{or} \quad \pi_3 = \frac{\Delta p}{u_m^2 \rho} \qquad (3.4.46)$$

The functional relationship is

$$F_2' \left(N_{Re}, L/D, \Delta p/u_m^2 \rho\right) \qquad (3.4.47)$$

which is the same result previously obtained by application of the Rayleigh method. The Buckingham method has an advantage in that it is a simpler and faster method of dimensional analysis.

Caution should be used in applying the Rayleigh and Buckingham methods of dimensional analysis. It is absolutely essential to consider all dimensional variables applicable to a particular problem or an erroneous result occurs. The problem of carelessly omitting pertinent variables does not occur when applying dimensional analysis to the fundamental differential equations which govern a system.

Occasionally additional information is obtained from dimensional analysis by allowing for differences in the fundamental length dimension for the x-, y-, and z-directions. This approach is known as vector or directional dimensional analysis [17].

REFERENCES

1. Waggener, J. P., Friction factors for pressure-drop calculations, **Nucleonics** 19, No. 11, 145-147 (November 1961).
2. Bonilla, C. F., ed., **Nuclear Engineering**, McGraw-Hill, New York, 1957.
3. Knudsen, J. G. and Katz, D. L., **Fluid Dynamics and Heat Transfer**, McGraw-Hill, New York, 1958.
4. Junkhan, G. H. and Serovy, G. K., Effects of free-stream turbulence and pressure gradient on flat-plate boundary-layer velocity profiles and heat transfer, **Trans. of the ASME** 89, Series C, 169-176 (May 1967).
5. Schultz-Grunow, F., NACA TMI 86, Washington, D. C. (1941).
6. Reynolds, W. C., Kays, W. M., and Kline, S. J., Heat transfer in the turbulent incompressible boundary layer. I-constant wall temperature, NASA MEMO 12-1-58W (1958).
7. Schlichting, H., **Boundary Layer Theory**, 6th ed, McGraw-Hill, New York, 1968.
8. Rotta, J. C., Recent developments in calculation methods for turbulent boundary layers with pressure gradients and heat transfer, **ASME Journal of Applied Mechanics** 33, 429-437 (June 1966).

9. Eckert, E. R. G., **Introduction to the Transfer of Heat and Mass**, McGraw-Hill, New York, 1950.
10. Metais, B., Heat transfer laboratory TR No. 51, University of Minnesota (1963).
11. von Karman, T., The analogy between fluid friction and heat transfer, **Transactions of the ASME** 61, 705 (November 1939).
12. Kays, W. M., **Convective Heat and Mass Transfer**, McGraw-Hill, New York, 1966.
13. Hinze, J. O., **Turbulence**, McGraw-Hill, New York, 1959.
14. Rust, J. H. and Sesonske, A., Turbulent temperature fluctuations in mercury and ethylene glycol in pipe flow, **Int. Journal of Heat Mass Transfer** 9, 215-227 (March 1966).
15. Bridgman, P. W., **Dimensional Analysis**, Yale University Press, New Haven, CT, 1931.
16. Buckingham, E., **Transactions of the ASME** 35, 263 (1915).
17. Huntley, **Dimensional Analysis**, MacDonald, London, 1952.

PROBLEMS

3.1 Water at 60^0 F flows past a smooth flat plate of dimensions 6 ft. width by 4 ft. length. If the free stream velocity is uniform at 2 ft/sec, what is the drag on the plate? If the plate surface is uniformly roughened with particles of 0.001 in. height, how is the drag affected? Hint: Solution to this problem is made easier if the following expression is used for the turbulent flow drag coefficient:

$$C_f(x) = \frac{0.0584}{\left(\frac{\rho U_\infty x}{\mu}\right)^{0.2}}$$

3.2 A. Write out in fully expanded form the three momentum equations which are given in tensor form by Eq. (3.4.43).

B. Consider the steady, laminar flow of an incompressible fluid between two infinite parallel plates. Starting with the equations from part A, derive the momentum equations necessary for finding the velocity distribution between the plates. Carefully orient your coordinate system and discuss all assumptions made.

C. Solve the equations from Part B for the velocity distribution between the parallel plates.

3.3 By using the expressions for the von Karman universal velocity profile, derive an analytical expression for the eddy diffusivity for momentum transfer for fully-developed pipe flows.

Hint: In order to do this problem it is necessary to make an assumption about the distribution of the shear stress in the fluid. It is up to the student to come up with a logical assumption based upon physical reasoning. Let the shear stress at the wall be given τ_w. Put expression in terms of ϵ_m/ν. Sketch distribution of ϵ_m/ν as a function of distance from the wall, use y^+ as the distance parameters.

3.4 Show that the eddy diffusivity for heat transfer must be a second-order tensor and give a physical interpretation for this result.

3.5 The variables involved in determining the characteristics of a centrifugal pump are as follows:

$$F_2(Q, \rho, \omega, \mu, d, H) = 0$$

where

Q = volumetric flow rate, ft^3/sec
ρ = density, lb_m/ft^3
ω = impeller speed, rad/sec
μ = viscosity, $lb_m/ft\text{-}sec$
d = impeller diameter, ft
H = total head, $ft\text{-}lb_f/lb_m$

Find the relationship of dimensionless variable by use of the Rayleigh and the Buckingham methods.

Chapter 4

FLUID SYSTEM ANALYSIS

4.1 Introduction

As shown in preceding chapters, different fluids are employed as reactor coolants and working fluids for energy conversion systems. As a consequence, a large variety of parameters must be considered when selecting fluids for use in reactor systems. For fluids which pass through reactor cores, one must consider its nuclear, physical, and economic characteristics before a selection is made. For out-of-core fluids, the physical and economic characteristics are of immediate interest and one may also have to consider nuclear characteristics when safety analyses are performed on the system.

In selecting fluids for reactor systems, the following characteristics are usually considered desirable [1]:

(i) Nuclear

a. Low neutron absorption cross section
b. Moderating power suitable for the reactor type
c. Low induced radioactivity (as an alternative, short half-lived activity for those present)
d. Good radiation stability

(ii) Physical

a. Low melting point so that the fluid will not solidify during shutdown
b. Vapor pressure suitable for the application of the fluid—low vapor pressures when the fluid is used as a coolant so as to avoid high pressurization
c. Compatibility with containment materials—fuel cladding, ducts, heat exchangers, pumps, valves, etc. It is desirable to avoid exotic containment materials such as niobium, etc., which are expensive and difficult to fabricate
d. Good thermal stability—decomposition of coolants at high temperatures is undesirable and a serious problem with organic fluids

(iii) Economic

a. Low initial cost
b. Reasonable availability—availability and low cost are coupled, since rare materials are usually expensive
c. Low pumping power requirements—desirable materials have low viscosities, reasonable densities, and large heat capacities
d. Large heat-transfer coefficients—desirable materials have large thermal conductivities and small viscosities.

Usually, for a given application, the choice of coolants is quickly reduced because of these factors. For fast reactors, coolants containing hydrogen are unsuitable because of these materials' moderating power. In natural uranium reactors, coolants other than gases or heavy water are eliminated because of excessive neutron absorption. For high temperature use in radiation fields, the present organic coolants are undesirable because of thermal and radiation instability.

No single fluid possesses all desirable characteristics. In selecting fluids for reactor systems, compromises must be made between conflicting requirements, i.e., increasing fuel enrichment so that water can be used as a coolant instead of heavy water.

The thermal-hydraulic analysis of nuclear power plants requires consideration of both steady-state and transient operation. These studies include component pressure losses, flow distribution through the reactor core, and pumping power requirements. In addition, for two-phase flows one must consider flow stability and critical (maximum) flow. In this chapter we restrict ourselves to steady-state conditions; Chapter 9 will cover several types of transients.

4.2 Single Phase Flow Pressure Loss

The ability to predict pressure losses in reactor system components is of great importance because of the necessity for determining pumping power requirements, pump sizes and capacities, component diameters and lengths from economic considerations, system stability to flow rate perturbations, and optimum flow rates from economic and heat transfer criteria. Pressure losses are calculated by the mechanical-energy balance form of the steady-flow energy equation

$$-\int_{p_1,\,v_1}^{p_2,v_2} v\,dp = (z_2 - z_1)\frac{g}{g_c} + F + \frac{a_2 u_2^2 - a_1 u_1^2}{2g_c} \qquad (4.2.1)$$

where F is the pressure loss due to friction.

Note: In the practical application of Eq. (4.2.1), it is customary to evaluate terms in the following manner: z is evaluated at the center of gravity of the cross section of the stream considered and u is the average velocity of the fluid flowing through the cross-sectional area at the point in question. Usually values found for these terms are substantially correct in theory; however, the kinetic energy term is usually theoretically incorrect if average velocities are used. The term a is a factor that corrects the average kinetic energy flow through a cross section when it is written in terms of average velocity. Various values of a are given in Table 4.1. For turbulent flow, the parameter f_W is the Darcy-Weisbach friction factor, which is discussed later in this section.

TABLE 4.1
Kinetic energy correction factor, a [2]

		Pipe	Annulus, radii $=R_1, R_2$ R_2/R_1			Wide Flat Channel	Square Channel
			3	2	1.2		
Laminar Flow		2	1.568	1.552	1.543	1.543	2.2
Slug Flow		1.0	1.0	1.0	1.0	1.0	1.0
Turbulent Flow	f_W						
	0.04	1.135				1.229	
	0.02	1.073				1.055	
	0.01	1.040				1.022	

Generally, as p decreases, v remains constant or increases along a curve of moderate curvature. Therefore, to a good approximation

$$-\int_{p_1,v_1}^{p_2,v_2} v\,dp \approx v_{avg}\,(p_1 - p_2) \tag{4.2.2}$$

where v_{avg} is the arithmetic mean of the terminal specific volumes. For isothermal flow of most liquids, Eq. (4.2.2) is precise. For isothermal flow of gases, the error is very small if p_1/p_2 is less than two [3]. Therefore, for the practical solution of most problems concerning liquids and many problems concerning gases, the mechanical-energy balance equation is written

$$v_{avg}\,(p_1 - p_2) = (z_2 - z_1)\,\frac{g}{g_c} + F + \frac{a_2 u_2^2 - a_1 u_1^2}{2g_c} \tag{4.2.3}$$

In applying the mechanical-energy balance equation to flow systems, pressure losses are computed step-by-step, from component to component, and summed up over the several components in order to find the pressure loss across a portion of the system. In addition, in the analysis of closed loops the sum of the pressure losses in all components due to variations in heights is zero. Therefore, for many cases of closed-loop analysis, the mechanical-energy balance equation is written

$$v_{avg}(p_1 - p_2) = \frac{a_2 u_2^2 - a_1 u_1^2}{2g_c} + F \tag{4.2.4}$$

In the analysis of many reactor components, the contribution to Eq. (4.2.4) from variations in fluid velocity is zero or small compared to the friction loss. Consequently, a major problem in analyzing system pressure losses is determining the friction loss F.

It should be emphasized that Eq. (4.2.1) and all approximation equations that follow are energy equations. Consequently, pressure losses due to fluid momentum changes due to changes in flow direction, such as would occur in flow through an elbow, are incorporated in the friction loss term F.

In order to apply Eq. (4.2.1), etc., a flow channel with well-defined boundaries must be specified. The selection of surfaces which make up the enclosing sides of a channel is arbitrary. Such surfaces may be a fluid-solid interface or an imaginary surface passing through the fluid. We will define a **closed channel** as a flow channel whose side boundaries consist entirely of solid surfaces. An **open channel** is defined as a flow channel whose side boundaries are partially or entirely a fluid surface. Naturally, boundary selection is made to facilitate analysis; for flow through pipes it is obvious that one selects the pipe walls for the flow channel boundary.

For flow through closed channels, the friction loss F is due to shear stress exerted by the fluid on the solid boundary. In open channels the friction loss F is due to two factors: (1) shear stresses at fluid-solid interfaces and (2) shear stresses on imaginary fluid boundaries. The shear stress on a fluid surface is due to viscous effects and momentum exchange due to turbulence. In some types of analysis, equations similar to Eq. (3.3.61) are used to calculate shear stresses on fluid surfaces.

Pressure Loss With Flow Through Closed Channels

For steady laminar or turbulent flow in smooth or rough closed channels, the gradient of the pressure loss due to friction is frequently expressed as

$$\frac{dp_F}{dx} = \rho \frac{dF}{dx} = f_W \frac{\rho u_m^2}{2g_c} \frac{1}{D_e} \tag{4.2.5}$$

where x represents the axial location in the channel, u_m is the average fluid velocity in the channel, f_W is the local Darcy-Weisbach friction factor, and D_e is the channel equivalent diameter. The friction factor is dependent upon many parameters and, consequently, a universal expression does not exist for determining its value. Some of the parameters which friction factors depend are fluid density, viscosity, average channel velocity, channel cross section geometry, surface roughness, and whether flow is developing or fully-developed and isothermal or non-isothermal.

(1) Fully-Developed Isothermal Flow Through Circular Channels

For fully-developed isothermal laminar flow through circular channels, the friction factor is given by [4]

$$f_W = 64/N_{Re} \tag{4.2.6}$$

where the Reynolds number is based upon the pipe diameter. Ordinarily, laminar flow is not encountered in nuclear reactor systems because of the large velocities employed. The critical Reynolds number for transition of isothermal flow from laminar to turbulent is of the order of 2300.

Flow in a Reynolds number range from 2000 to 4000 is not well understood because flow is intermittently laminar or turbulent. Since for the same fluid velocity, friction factors for turbulent flow are larger than for laminar flow, flow in the transition range is subject to oscillations because the wall shear stress intermittently increases and decreases as the flow changes from laminar to turbulent and back. Consequently, design of system components which operate in the transition Reynolds number range is not advisable because flow oscilations may produce vibrations or undesirable heat transfer effects.

A vast quantity of friction pressure loss data exists for turbulent flow. For fully-developed, incompressible, isothermal flow in smooth pipes, von Karman proposes the following empirical equation [5]:

$$(f_W)^{-1/2} = 2 \log_{10} [0.398\, N_{Re}(f_W)^{1/2}] \tag{4.2.7}$$

Because of its transcendental nature, Eq. (4.2.7) is solved by trial-and-error which makes it cumbersome to use in engineering analysis. For given Reynolds number ranges, several approximations are suggested for Eq. (4.2.7) which lead to greater mathematical simplicity.

For a range of Reynolds numbers from 4×10^3 to 3×10^6 , the following equation fits the von Karman relationship to $\pm$ 5% [3]:

$$f_W = 0.0056 + 0.50(N_{Re})^{-0.32} \tag{4.2.8}$$

For the entire range of Reynolds numbers, the von Karman relationship is generally in the lower half of these tolerance limits.

McAdams [6] proposed the relation

$$f_W = 0.184(N_{Re})^{-0.20} \tag{4.2.9}$$

which is in good agreement with the von Karman relationship for Reynolds numbers from 3×10^4 to 2×10^6

The friction factor equations discussed so far are applicable only for flow through clean, smooth pipes. Colebrook [7] proposed an empirical equation which is applicable for both smooth and rough pipes. His equation is given by

$$\frac{1}{f_W^{1/2}} = -2\log_{10}\left[\frac{\epsilon/D}{3.70} + \frac{2.51}{N_{Re}f_W^{1/2}}\right] \tag{4.2.10}$$

where ϵ/D is the relative roughness of the pipe inner surface. Values of ϵ/D are tabulated in many handbooks [3,8] for various types of pipes. Typical values of ϵ are as follows: drawn tubing, 0.000,005 ft; commercial steel pipe, 0.00015 ft; galvanized iron pipe, 0.0005 ft; cast iron, 0.00085 ft; and riveted pipe, 0.003 to 0.03 ft [2]. Pipe roughness is measured with special instruements that are capable of measuring microinch depressions in materials; ϵ is the average value of the measured depressions.

A graphical representation of the Colebrook equation is the well-known Moody diagram [9] shown in Fig. 4.1.

(2) Fully-Developed Isothermal Flow Through Non-Circular Channels

Most friction factor data are from experiments in circular tubes. In calculating pressure losses for non-circular geometries, certain modifications are made to the friction factor and pressure loss equations. For ducts that have sufficiently smooth walls and no "stagnant" zones, the pressure loss in **turbulent** flow is usually calculated with equations for circular tubes, providing the equivalent diameter is taken as the characteristic cross section dimension of the duct. The equivalent diameter is defined by

$$D_e \equiv \frac{4 \text{ x Flow Area}}{\text{Wetted Perimeter}} \tag{4.2.11}$$

For ducts of complicated cross section, the use of circular tube equations may lead to erroneous results.

In the analysis of flow in nuclear reactor systems, non-circular geometry in important components is the rule rather than the exception. Fuel elements in reactor cores usually consist of plates which have flow areas of narrow rectangular channels or rods which have flow areas of square or triangular subchannels with rods located at the vertices. Flow in reactor cores is further complicated by use of wire or grid spacers to maintain required separation between plates or rods. Figure 4.2 illustrates typical fuel element geometries encountered in reactor cores. For rod bundles, the distance between rod centers is called the pitch (P).

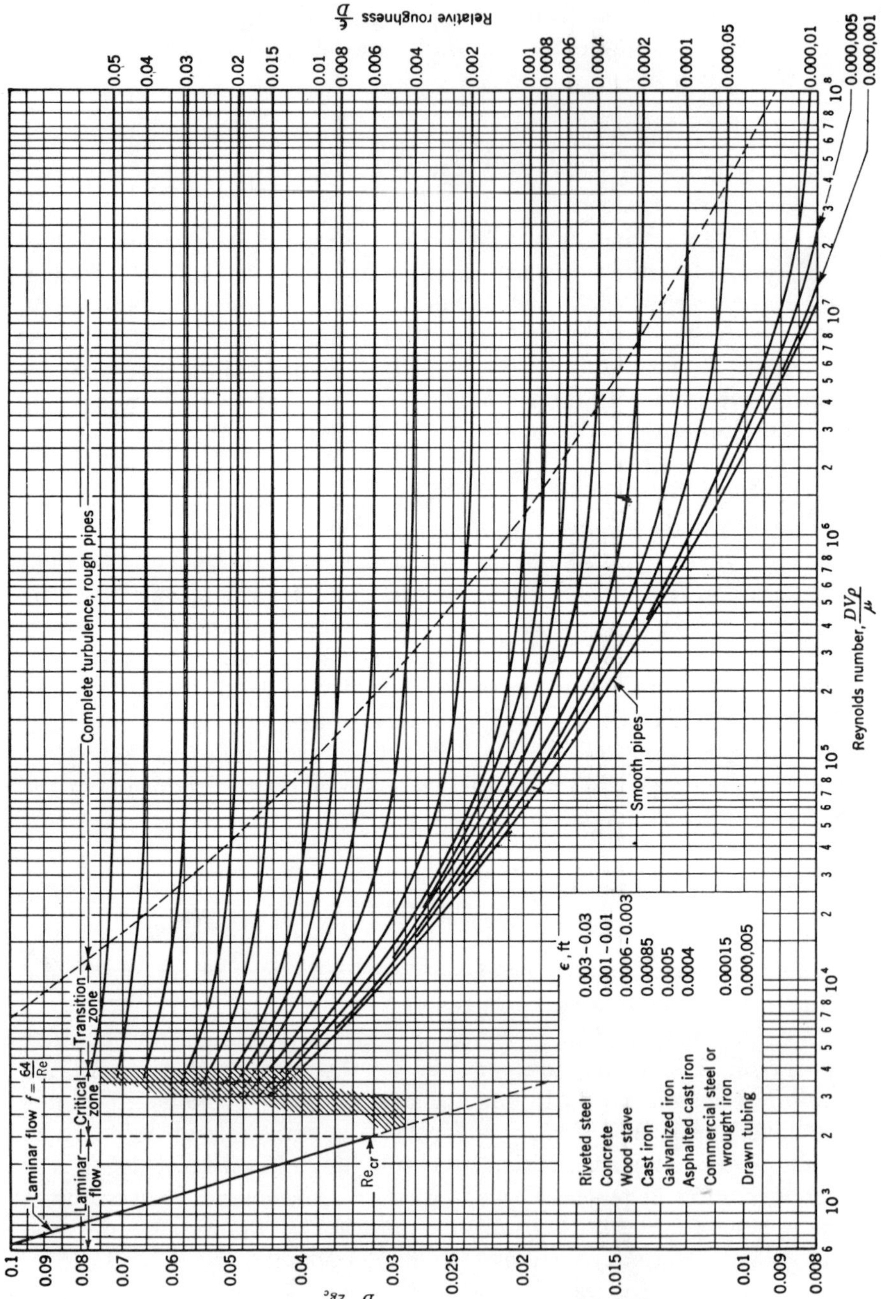

Fig. 4.1. Moody diagram for friction factors [1].

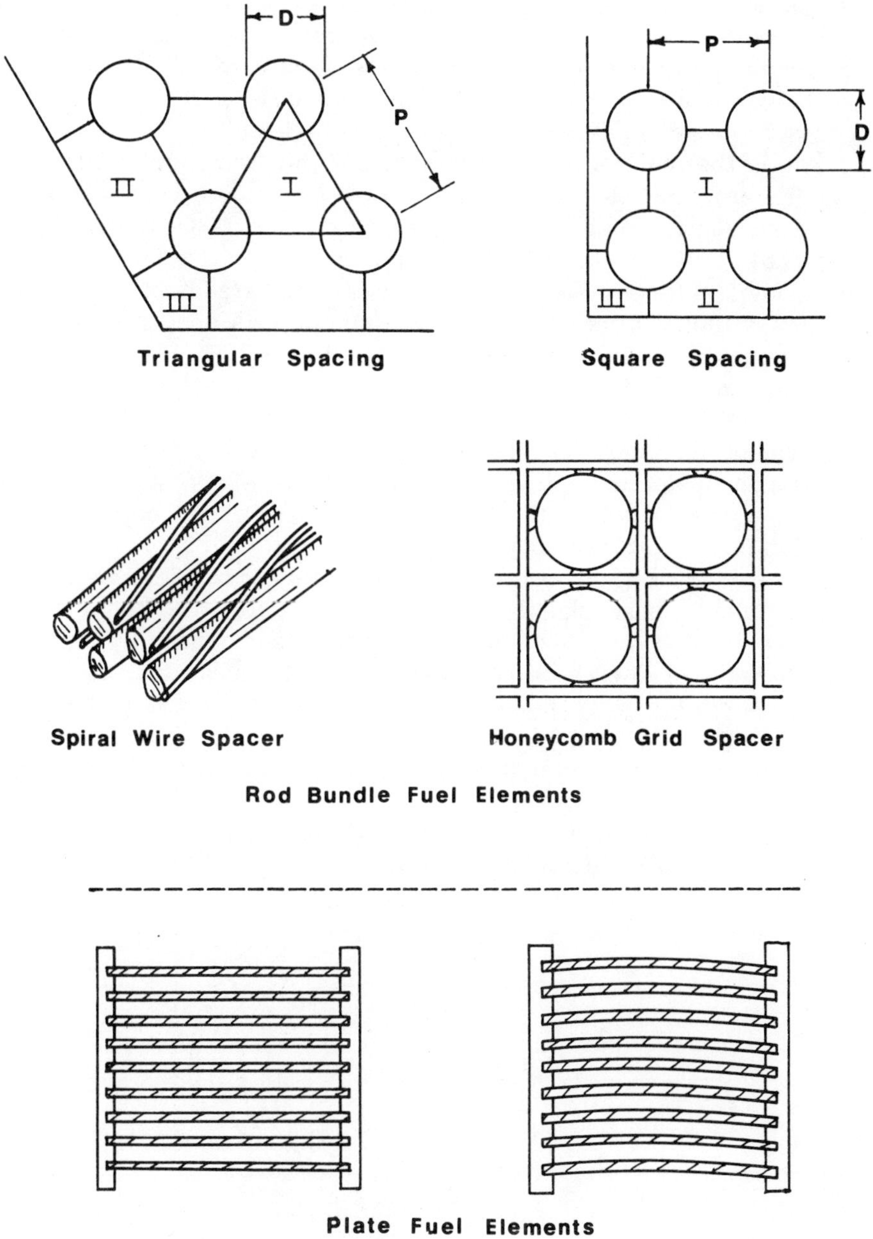

Fig. 4.2. Fuel element configurations.

Heat exchangers, which usually consist of tube bundles enclosed by a shell, are another important reactor component with complicated flow patterns. Depending upon baffle arrangements, the flow in the shell side of heat exchangers may be in-line or cross flow to the tube bundle, or a combination of both. Pressure losses through the shell side of heat exchangers are usually intractable to analysis and, consequently, pressure loss characteristics are supplied by vendors.

For isothermal turbulent flow through thin rectangular channels, experiments indicate that friction factors for smooth channels are in good agreement with smooth pipe friction factors from the Moody diagram [10]. However, for roughened rectangular channels, experimental friction factors are not in as good agreement with friction factors obtained from the Colebrook equation or Moody diagram with the same relative roughness [10,11]. Depending upon Reynolds number and relative roughness, rectangular channel friction factors may be greater or less than friction factors for pipes with the same equivalent diameter and relative roughness. For a relative roughness less than 0.002 and Reynolds numbers in excess of 5×10^5, rectangular channel friction factors are smaller than corresponding friction factors for circular tubes [10].

In the analysis of rod bundles, it is necessary to consider a variety of subchannels whose shapes are dependent upon whether a rod is adjacent to a flat- or curved-surface shroud or surrounded by rods. The subchannels of interest in square- and triangular-lattice rod bundles are shown in Fig. 4.2. The equivalent diameter for each subchannel is calculated by Eq. (4.2.10). Because of the complexity in analysis of open-channel flows, rod bundles are treated in greater detail in a later section.

Other flow geometries used or considered for nuclear reactors are concentric annuli and packed beds of coated fuel spheres. Figure 4.3

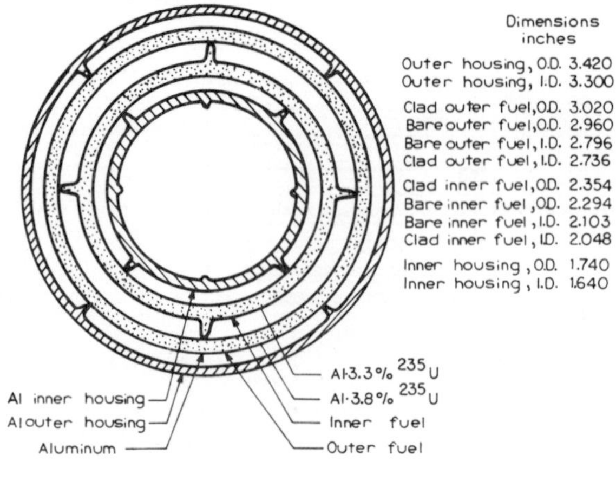

Fig. 4.3. Cross section of high flux fuel assembly [13].

illustrates the cross section of a fuel assembly for the Savannah River Laboratory High Flux Reactor. For turbulent flow the friction factor is calculated using standard equations or Moody diagram, provided the Reynolds number is based upon the annulus equivalent diameter [12].

Although turbulent flow friction factors for circular tubes is used to calculate pressure losses for other geometries, this simplification does not exist for laminar flow. One must consult the literature for expressions giving laminar flow friction factors.

(3) Flow Not Fully-Developed

At the entrance and for some distance downstream in a channel, the flow is not fully-developed and, consequently, fully-developed flow friction factors are not applicable for calculating pressure losses. Pressure losses in entrance regions are larger than in fully-developed regions because of two factors: (i) at channel entrances, velocity profiles are more nearly uniform and resulting steeper velocity gradients near boundaries produce larger wall shear stresses; and (ii) the increase in momentum flux as the velocity profile adjusts from nearly uniform to a parabolic or arched curve distribution $(a > 1.0)$.

Very little work has been done on flow development in channel entrances for situations in which fully-developed flow is turbulent. This problem is complicated because boundary layer development at channel entrance is initially laminar. Some of the techniques discussed in Ref. [14] can be used for calculating entrance pressure losses, but not without great difficulty. For channels in which L/D_e is greater than 100, one can probably assume fully-developed flow over the entire channel length for estimating pressure losses.

(4) Non-Isothermal Flow

The preceding friction factor equations are based upon analytical and experimental data taken from unheated channels. This insures a reasonable constancy in fluid physical properties. However, in the analysis of nuclear reactor systems, heat transfer occurs in components of major importance, i.e., core and heat exchangers. With heat transfer, appreciable temperature gradients may exist in the fluid with a result that fluid physical properties are subject to large variations. The density of liquids does not change substantially with temperature; however, viscosity is subject to substantial variations with small temperature changes. In the flow of gases, both viscosity and density are temperature dependent. Therefore, the isothermal friction factor equations are not applicable for calculating pressure losses in reactor cores or heat exchangers.

The approaches used for calculating pressure losses in fluids with heating or cooling are not as well developed as for isothermal flows. Present approaches employ empirical friction factor equations in which fluid physical properties are evaluated at bulk fluid temperatures, wall temperatures, or defined film temperatures which are combinations of bulk and wall temperatures, i.e., $T_f = (T_w + T_b)/2$. Because of the considerable difference in temperature dependence of fluid properties for gases and liquids it is probably advisable to use different friction factor equations for these classes of fluids.

For the flow of heated oils, McAdams [6] recommends multiplying isothermal friction factor by the Seider-Tate correlation used to correct heat-transfer coefficients with variable viscosity flows. This equation is given by

$$f_W = f_{W,iso}\left(\frac{\mu_w}{\mu_b}\right)^{0.14} \tag{4.2.12}$$

where μ_b is the bulk viscosity, μ_w is the viscosity at the tube wall, and $f_{W,iso}$ is the isothermal friction factor evaluated at the bulk fluid temperature. Equation (4.2.12) is used to calculate local or channel average f_W. Local f_W are evaluated at local T_w and T_b and channel average f_W are evaluated at average T_b and T_w.

Rohsenow and Clark [15] examined Eq. (4.2.12) for water flowing at pressures from 1500 to 2000 psia. They concluded that the exponent in Eq. (4.2.12) should be 0.60 for the friction pressure loss and 0.14 for both friction and momentum pressure losses.

For water heated in rectangular channels and rod bundles at 2000 psia, measurements have shown that the friction factor is correlated by [10,16]

$$f_W = (1 - 0.0019\Delta T_f)f_{W,iso} \tag{4.2.13}$$

where ΔT_f is a film temperature drop in $^\circ$F evaluated from

$$\Delta T_f = q''_w/h \tag{4.2.14}$$

where q''_w is the wall heat flux, Btu/hr–ft^2, and h is the heat-transfer coefficient given by

$$h = 0.023\frac{k}{D_e}(N_{Re})_b^{0.8}(N_{Pr})_b^{0.4} \tag{4.2.15}$$

All fluid physical properties in Eqs. (4.2.13) and (4.2.15) are evaluated at the bulk temperature. In contradiction to Eq. (4.2.13), Gambill [11] reported that applying a temperature correction to Moody diagram friction factors for rectangular channels increased the discrepancy between experimental and calculated friction factors. His data indicated that rectangular channel friction factors are independent of heating.

Although pressure loss for flows of non-isothermal gases has been studied extensively, no single expression for determining friction factors is generally accepted. For turbulent flow in pipes, Taylor [17] proposes that local or average friction factors be calculated from

$$f_W = \left[0.0028 + \frac{0.25}{(N_{Re})_s^{0.32}}\right]\left(\frac{T_b}{T_w}\right)^{0.5} \tag{4.2.16}$$

where $(N_{Re})_s$ is a modified surface Reynolds number in which the fluid density is evaluated at the bulk temperature and viscosity evaluated at the channel wall temperature. Temperatures are given in absolute units. Equation (4.2.16) correlated data examined to within ± 10 percent for modified surface Reynolds numbers greater than 3000, $0.35 < T_w/T_b < 7.35$, and length-to-diameter ratios from 21 to 200. An examination of pressure loss data for flow between heated parallel plates indicates that Eq. (4.2.16) is suitable for use with this geometry.

For treating non-isothermal turbulent flow of gases, Humble, Loudermilk, and Desmon recommend a modified form of the von Karman equation for calculating friction factors [18]

$$\frac{1}{\left(f_W \dfrac{T_f}{T_b}\right)^{1/2}} = 2.0 \log_{10}\left[(N_{Re})_b \frac{\nu_b}{\nu_f}\left(f_W \frac{T_f}{T_b}\right)^{1/2}\right] - 0.8 \tag{4.2.17}$$

where the film temperature, T_f, is defined by $T_f = (T_w + T_b)/2$; ν_f and ν_b are kinematic viscosities evaluated at film and bulk temperatures, respectively; and the Reynolds number is evaluated with bulk fluid properties. Equation (4.2.17) appears to be satisfactory for conditions in which Reynolds numbers are greater than 20,000, surface-to-bulk fluid temperature ratios T_w/T_b are less than 2.5, and channel length-to-diameter ratios are greater than 60. On the basis of analytical and experimental studies of turbulent flow of air through tubes and along rods, Deissler proposes [19]

$$f_W = 0.1159(N_{Re})_b^{-0.2776} \tag{4.2.18}$$

In Eq. (4.2.18) all fluid properties are evaluated at the bulk temperature.

Investigations by Magee, et al., indicate that friction factors for non-isothermal turbulent flow of gases are dependent upon length of the thermal entrance region [20]. The data examined are correlated by

$$\frac{f_W}{f_{KN}} = \left(\frac{T_W}{T_b}\right)^n \tag{4.2.19}$$

where

$$n = 1/2 - 1/6 \ln(x/D)$$

$$\text{for } 5 < x/D < 55$$

and

$$n = 1/6 \quad \text{for} \quad x/D > 55$$

The distance x is measured from the start of the heated region and f_{KN} is the von Karman friction factor (Eq. 4.2.7) evaluated at the local bulk Reynolds number.

Because of uncertainty in determining friction factors for non-isothermal flows, it is recommended for analyses that requires precise calculation of pressure losses that the literature be reviewed to obtain data which are suitable for the fluid, pressure range, temperature range, and geometry under consideration.

(5) Pressure Loss for Abrupt Changes in Cross Section

When flow channel cross sections change abruptly, as at the inlet grid plates of reactor cores, there is a friction pressure loss added to the potential and kinetic energy terms of the steady-flow energy equation. Figure 4.4 illustrates the entrance pressure drop and exit pressure rise which occur for sudden contractions and expansions at inlets and exits of heat exchangers or reactor cores.

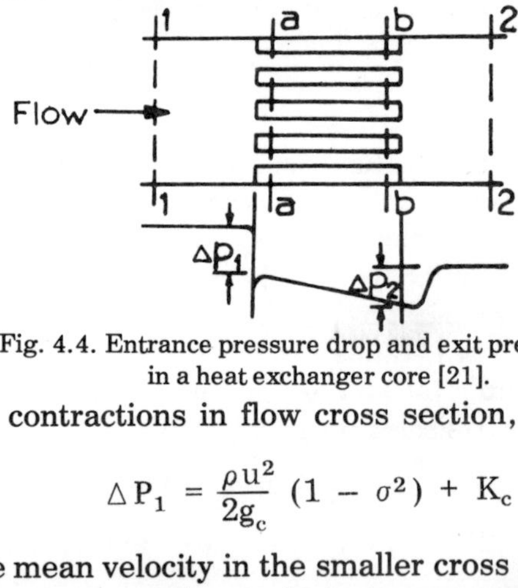

Fig. 4.4. Entrance pressure drop and exit pressure rise in a heat exchanger core [21].

For sudden contractions in flow cross section, the pressure loss is given by [21]

$$\Delta P_1 = \frac{\rho u^2}{2g_c}(1 - \sigma^2) + K_c \frac{\rho u^2}{2g_c} \tag{4.2.20}$$

where u is the mean velocity in the smaller cross section duct, σ is the ratio of smaller cross section to larger cross section, and K_c is the contraction pressure loss coefficient which is dependent upon Reynolds number and shape of the contracting surface. Figure 4.5 illustrates K_c for a sharp-edged entrance to a multiple-circular-tube heat exchanger.

For rounded or conical-shaped entrances, K_c is reduced by a considerable amount and may be as small as 0.05 for turbulent flow in the smaller duct. The literature should be consulted for the treatment of gradual expansions and contractions or flow cross sectional area inlet and exit geometries different from that shown in Fig. 4.5 [3, 21].

The exit pressure rise is given by [21]

$$\Delta P_2 = \frac{\rho u^2}{2g_c}(1 - \sigma^2) - K_e \frac{\rho u^2}{2g_c} \tag{4.2.21}$$

where all terms are the same as in Eq. (4.2.20) with the exception of K_e which is the exit pressure-loss coefficient also shown in Fig. 4.5.

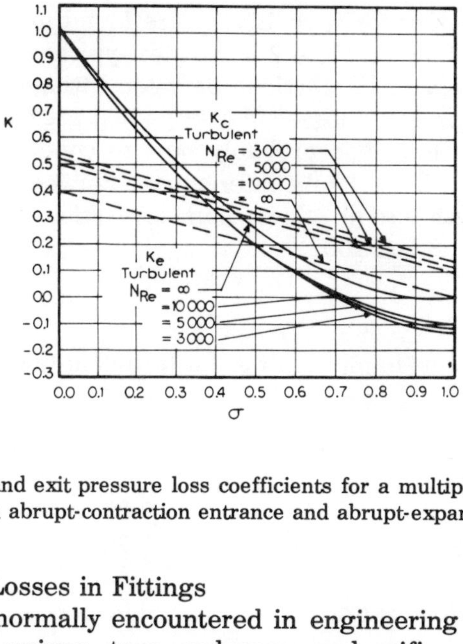

Fig. 4.5. Entrance and exit pressure loss coefficients for a multiple-circular-tube heat exchanger core with abrupt-contraction entrance and abrupt-expansion exit [21].

(6) Pressure Losses in Fittings

Pipe fittings normally encountered in engineering applications are valves, elbows, unions, tees, reducers, and orifice plates. Pressure losses in fittings cannot be reliably calculated because various fittings are not highly reproducible, they vary from manufacturer to manufacturer, and they affect each other when not widely separated by lengths of straight pipe.

Two equivalent methods are used for predicting fitting pressure losses: (1) express the loss in terms of the number (K) of pipe average-velocity heads, or (2) assign a fictitious or equivalent length, L_e, to the fitting and calculate the pressure loss using the straight-pipe friction loss equation, Eq. (4.2.5).

In turbulent flow, irregularities in most fittings are analogous to large values of surface roughness and, consequently, pressure losses are generally proportional to the square of the mean velocity, u_m^2. Therefore, values of K or L_e are generally independent of fluid velocity or Reynolds number. Consequently, friction pressure losses due to fittings are expresses as

$$(p_1 - p_2)_F = K \frac{\rho u_m^2}{2g_c} \tag{4.2.22}$$

or

$$(p_1 - p_2)_F = f_w \frac{\rho u_m^2}{2g_c} \frac{L_e}{D_e} \tag{4.2.23}$$

where u_m is the average fluid velocity entering the fitting.

Approximate values for K or L_e/D_e are found in a variety of books which treat fluid flow. Table 4.2 illustrates typical values for common fittings [3]. Because of interactions between fittings, Table 4.2 is not used to calculate pressure losses in fittings that are closely spaced. For this type of condition, it is probably best to conduct experiments to determine losses. Most likely interactions between fittings will not occur when length-to-equivalent diameter ratios of pipes or ducts between fittings exceed 100.

TABLE 4.2
Friction in pipe fittings for turbulent flow [3]

Fitting	K	L_e/D_e
Elbows		
45° Standard	0.35	15
45°Long Radius	0.20	10
90° Standard	0.82	35
90° Long Radius	0.53	20
Bends		
180° Close	2.0	90
180° Medium	1.2	55
Valves		
Open 0° Gate	0.16	7
Open 0° Globe	6	270
Open 0° Swing Check	2	90
Open 90° Angle	3	135
Couplings, Union, 0°	0.04	2
Tees		
0° Through	0.4	20
90° Side Inlet	1.5	70
90° Side Outlet	1.3	60

(7) Pressure Losses in Curved Channels

Bends are employed in the primary piping of nuclear power plants because pipe diameters are too large for conventional fittings. Flow through curved channels presents greater problems than flow through straight channels because of fluid momentum changes due to changes in flow direction. Secondary flows, illustrated in Fig. 4.6, are generated

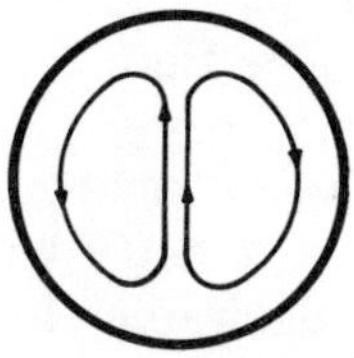

Fig. 4.6. Secondary flow in curved channels.

in directions transverse to the mean flow. These effects cause changes in wall shear stresses which result in larger friction factors for curved channels of the same dimensions.

There has not been a great deal of work done on flow through curved channels; some experimental friction factor data is correlated in the following form [22]: For fully-developed laminar flow

$$\frac{f}{f_{\text{straight tube}}} = \frac{1}{1 - \left[1 - \left(\dfrac{11.6}{N_{Re}\sqrt{r_0/R}}\right)^{0.45}\right]^{2.22}} \tag{4.2.24}$$

where r_0 is the tube radius and R is the radius of curvature of the tube centerline. Equation (4.2.24) is valid for

$$11.6 < N_{Re}\sqrt{r_0/R} < 2000.$$

For fully-developed turbulent flow

$$\frac{f}{f_{\text{straight tube}}} = \left[N_{Re}(r_0/R)^2\right]^{0.05} \tag{4.2.25}$$

for

$$N_{Re}(r_0/R)^2 > 6$$

Pressure Loss with Incompressible Flow Through Open Channels

The analytical approaches presented for flow through closed channels are rather simplified and careful calculations are expected to yield fairly accurate results. Unfortunately, due to the difficulty of the problems involved, the procedures for the analysis of open-channel flows are not at this time in as great a state of development.

Open channels constitute a common flow geometry in nuclear reactor cores. Because of the large surface areas per unit volume and ease of fuel element fabrication, rod bundles are used almost exclusively in power reactor cores. The basic configurations are the square and triangular lattices shown in Fig. 4.2. As seen, various subchannels are defined which depend upon the shroud arrangement.

For the majority of nuclear reactors, rod dimensions vary from 0.2 - 0.5 inch in diameter and lengths from 1 to 16 feet with spacings from 0.02 to 0.2 inch.

Subchannel configurations are usually specified in order to have some degree of geometrical symmetry with regard to the entire flow geometry. It is probably desirable for fluid-fluid interfaces on subchannel boundaries to be in regions where velocity gradients normal to the interface are zero or small.

Because of the variety of subchannels, flow in rod bundles is not uniform. For rod bundles enclosed in a shroud, there is usually a reduced flow in outer channels because of larger ratio of wetted perimeter to flow cross sectional area. This creates problems in design of power reactors. The coolant enthalpy rise in reduced flow subchannels is larger than if a uniform enthalpy rise occurred in the rod bundle. However, interchannel mixing due to turbulence and rod supports helps reduce the importance of this effect.

The maintenance of proper spacing between rods in a reactor rod bundle is usually achieved by use of fins, wires, or grids. Figures 4.2 and 4.7 illustrate some spacer arrangements employed in power reactors. The use of wires or fins produces a cross flow between subchannels which greatly enhances mixing between subchannels. The engineering analysis of rod bundles with spacers is very complex.

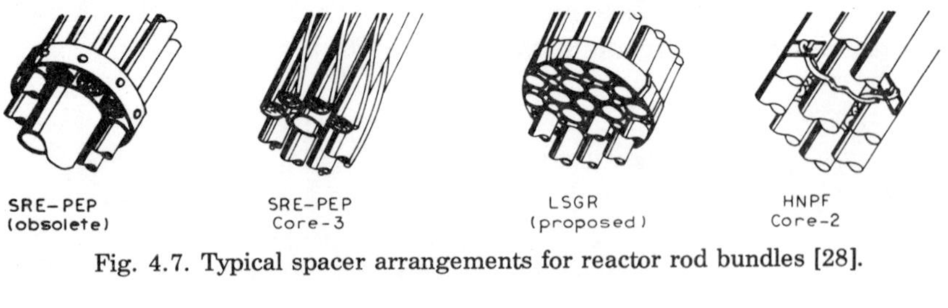

Fig. 4.7. Typical spacer arrangements for reactor rod bundles [28].

(1) Experimental-Pressure Loss in Rod Bundles

Theoretical analysis and experiments show that for close-packed rod bundles (P/D = 1) there are stagnant zones adjacent to the line of contact between rods [23]. Consequently, wall shear stress varies with angular position on the rods. Experimental data published by Subbotin and co-workers [23] indicate that friction factors for close-packed rod bundles with either triangular or square lattices are 63 percent of circular tube friction factors.

There is considerable discrepancy in friction factor data for rod bundles with P/D greater than one. In general, there is a tendency toward increasing friction factors with increasing P/D. This is probably due to cross flow between channels in the rod bundle and turbulent shear stress at fluid-fluid interfaces. Deissler [24] published data for flow past rod bundles arranged in square and triangular lattices without spacers. His data indicate that, for triangular lattices (P/D = 1.12), friction factors are about 3 percent below the Moody line for smooth tubes. Data for rod bundles arranged in square lattices (P/D = 1.12 and 1.2) fell as much as 10 percent below the Moody line for smooth pipes at Reynolds numbers of the order of 10^4. However, for Reynolds numbers in excess of 7×10^4, friction factors are higher than the smooth pipe Moody curve.

Tests on a 37-rod bundle of 3 foot heated length with 0.625 inch diameter rods arranged on a triangular lattice with P/D = 1.46 correlated the friction factor as [25]

$$f_W = \frac{0.296}{(N_{Re})^{0.2}} \tag{4.2.26}$$

These results are about 60 percent higher than the Moody curve for smooth pipes. Later experiments on a 6 x 6 rod bundle of 11.5 foot length with 0.59 inch diameter rods with P/D = 1.3 yield [26]

$$f_W = \frac{0.337}{(N_{Re})^{0.24}} \tag{4.2.27}$$

Equation (4.2.27) correlated data with a maximum deviation of $\pm 10\%$ and agrees with the Moody curve for a relative roughness $\epsilon/D = 0.0002$ within $\pm 8\%$ over a Reynolds number range of 10^5 to 10^6.

(2) Pressure Loss in Rod Bundles with Spacers

In the design of rod bundle fuel elements, spacers are usually placed between adjacent rods to prevent vibrations and bowing. These spacers are divided into two broad categories: (i) spiral-spacer wires wrapped around individual rods and (ii) transverse spacer grids of various shapes.

In calculating pressure losses across rod bundle cores of complicated geometry, the core designer considers the following components:

1. Pressure loss at core entrance and exit, Δp_E.
2. Pressure loss across the rod bundle without spacers, Δp_R.
3. Pressure loss across supporting devices, i.e., spacers, Δp_s.

Entrance and exit pressure losses are calculated using Kay s [21] entrance and exit pressure-loss coefficients, K_e and K_c, previously discussed.

In the most general treatment, pressure loss due to spacers is calculated by considering successive contractions and expansions of the coolant as it passes through various spacers. One method of calculating spacer pressure losses is using the Kays and London [21] abrupt section-change coefficients and Moody's friction factor in the spacing device. This approach gives satisfactory results for predicting spacer pressure losses for some specific types of transverse grids and widely discrepant results for other types of grids [27]. However, because of the constant flow cross section, this method is not suitable for analyzing spiral-wire spacer systems since the pressure loss varies with wire pitch.

A. N. deStordeur [27] developed a method for calculating spacer pressure losses using experimental drag coefficient correlations. In order to correlate experimental pressure loss data, the total pressure loss Δp_T is broken into three components: Δp_E at the rod bundle entrance and exit, Δp_R calculated for the rod bundle without spacers, and Δp_s produced by the spacers. Therefore, the spacer pressure loss is given by

$$\Delta p_s = \Delta p_T - (\Delta p_E + \Delta p_R) \tag{4.2.28}$$

In order to evaluate Δp_s from experimental total pressure losses, the entrance and exit losses are determined experimentally or calculated using abrupt section-change coefficients from Kays and London and Δp_R calculated using Moody's friction factor for round pipes of the same equivalent diameter and surface roughness. From the experimental Δp_s, a drag coefficient C_s is calculated using the following definition:

$$C_s = \frac{2g_c A}{u_s^2 S} \Delta p_s \tag{4.2.29}$$

where

C_s = experimental drag coefficient, dimensionless
u_s = average fluid velocity in the spacer region, ft/sec
S = spacer projected frontal area, ft^2
g_c = constant, $32.2(lb_m/lb_f)(ft/sec^2)$
A = unrestricted flow area of the bundle away from spacer, ft^2
Δp_s = experimental spacer pressure loss, $ft\text{-}lb_f/lb_m$

Drag coefficient correlated by de Stordeur for several types of spacer geometries are illustrated in Fig. 4.8 as a function of spacer Reynolds numbers based upon spacer frontal dimensions. For spiral-wire spacers this dimension is the wire diameter and for grid spacers this dimension is the grid equivalent thickness. The grid equivalent thickness is found by dividing the spacer projected frontal area(s) by the total projected spacer length in the subchannel under consideration. The velocity is the fluid velocity in the spacer region. The type of spacer arrangements used in developing the correlations are shown in Fig. 4.7. Later measurements by Brindley [28] of spacer drag-coefficients for transverse grid and spiral-wire spacers show good agreement with de Stordeur's correlations.

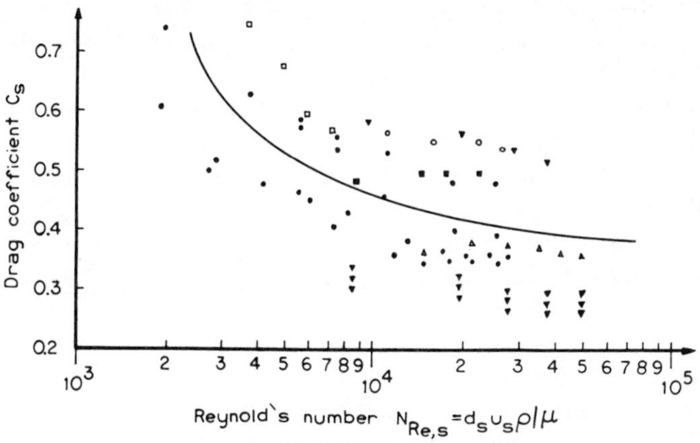

Spiral-wire Spacers

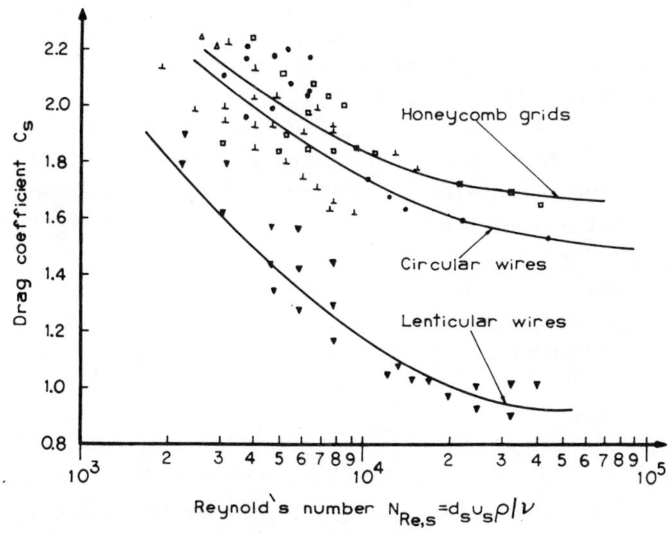

Transverse Grid Spacers

Fig. 4.8. Drag coefficients for rod bundle spacers [27].

Drag coefficients for spiral wires are considerably smaller than for transverse grids because they offer a smaller immediate frontal area resistance to flow. Of the three types of transverse grids studied, the lenticular grid has smaller drag coefficients due to streamlining of the trailing edge. The honeycomb grid has slightly larger drag coefficients than circular wire grids because it has the least amount of streamlining. The proposed drag coefficients for transverse grid spacers correlated experimental data to within $\pm 15\%$, while the spiral-wire spacer drag coefficients correlated data to within $\pm 40\%$ [27].

The use of spacer drag coefficients yields reasonable estimates of total pressure loss in rod bundles with spacers. On the basis of calculations, the spacer pressure loss in rod bundles is usually 30–70% of the total pressure loss. Therefore, due to the states uncertainty in the proposed drag coefficients, the uncertainty in pressure loss calculations for rod bundles is of the order of 10–30% for spiral-wire spacers and 5–10% for transverse grid spacers. This accuracy is adequate for preliminary design calculations.

Example 4-1

The use of de Stordeur's method of calculating rod bundle pressure losses is best illustrated by working an example problem. Consider a rod bundle with spiral-wire spacers having the following characteristics [27]:

1. 121 rod bundle on a square pitch, each rod having a wire wrap;
2. Rod diameter, D = 0.193 in. = 0.0161 ft;
3. Spiral-wire diameter, d_s = 0.025 in. = 0.00208 ft;
4. Pin pitch, P = 0.2205 in. = 0.01838 ft
   [Note: $P \neq D + d_s$ because of the oversized containment can. The clearance between the bundle and can is assumed equally distributed between rods because of the usually slightly wavey shape of the rods.];
5. Rod length, L = 2 ft;
6. Helical spacer wire pitch, h = 0.5 ft;
7. Fluid is water at 190°F ($\nu = 3.6 \times 10^{-6}$ ft²/sec).

Flow Distribution

The first step in the calculations is determining the flow distribution within the rod bundle on the basis of its subchannel geometries. Three types of coolant subchannels, which are illustrated in Fig.4.2 for the square-lattice rod bundle, are considered in this problem. The number of nominal or central coolant subchannels, n_I, is given by

$$n_I = 100 \text{ (81 interior rods + 36 half-peripheral rods + 4 quarter-corner rods)} \quad (4.2.30)$$

The number of peripheral or wall coolant subchannels associated with peripheral rods is given by n_{II} and n_{III}, where

$$n_{II} = 40 \text{ (40 half-peripheral rods)} \tag{4.2.31}$$

and

$$n_{III} = 4 \text{(4 quarter-corner rods)} \tag{4.2.32}$$

The area of a central flow channel without a spacer wire, A_I, is given by

$$A_I = P^2 - \frac{\pi}{4} D^2 = 1.344 \times 10^{-4} \text{ ft}^2 \tag{4.2.33}$$

Therefore, the equivalent diameter of a central flow channel without a spacer wire is

$$D_{e,I} = \frac{4A_I}{\pi D} = 1.062 \times 10^{-2} \text{ ft} \tag{4.2.34}$$

The area of a central flow channel with a spacer wire, A'_I, is given by

$$A'_I = A_I - \frac{\pi}{4} d_s^2 = 1.3104 \times 10^{-4} \text{ ft}^2 \tag{4.2.35}$$

and the equivalent diameter of this channel is

$$D'_{e,I} = \frac{4A'_I}{\pi D + \pi d_s} = 9.183 \times 10^{-3} \text{ ft} \tag{4.2.36}$$

For the peripheral channel II, the flow area with a spacer wire is given by

$$\begin{aligned} A'_{II} &= P(P - D/2) - \frac{\pi}{8} (D^2 + d_s^2)^* \\ &= 0.859 \times 10^{-4} \text{ ft}^2 \end{aligned} \tag{4.2.37}$$

The equivalent diameter for the peripheral channel II with spacer wire, $D'_{e,II}$, is given by

$$D'_{e,II} = \frac{4A'_{II}}{P + \frac{\pi}{2}(D + d_s)} = 7.32 \times 10^{-3} \text{ ft} \tag{4.2.38}$$

In the corner channel III, the flow area with a spacer wire is given by

$$\begin{aligned} A'_{III} &= (P - D/2)^2 - \frac{\pi}{16} (D^2 + d_s^2) \\ &= 0.547 \times 10^{-4} \text{ ft}^2 \end{aligned} \tag{4.2.39}$$

---

*Remarks: Each rod has a wire-wrap spacer. Consequently, every channel which has a fraction of a rod associated with its geometry is assumed to have the same fraction of spacer wire in the channel.

and the equivalent diameter of a corner channel is

$$D'_{e,III} = \frac{4A'_{III}}{2(P - D/2) + \frac{\pi}{4}(D + d_s)} = 6.28 \times 10^{-3} \text{ ft} \quad (4.2.40)$$

The total pressure loss across all coolant channels is the same and, therefore, one can write

$$\Delta p_I = \Delta p_{II} = \Delta p_{III} \quad (4.2.41)$$

where Δp_I , Δp_{II} , and Δp_{III} are the total pressure losses across channels I, II, and III, respectively. A pressure imbalance between adjacent channels at a given axial location produces a cross flow between channels which relieves the pressure difference. Therefore, it is assumed the friction pressure loss across a rod bundle channel of length L is the same in all rod bundle channels. In addition, it is assumed the friction pressure loss is represented by the Darcy-Weisbach equation and, therefore, we can write

$$\Delta p_I = f_{W,I} \frac{L}{D'_{e,I}} \frac{\rho u_I^2}{2g_c}$$

$$\Delta p_{II} = f_{W,II} \frac{L}{D'_{e,II}} \frac{\rho u_{II}^2}{2g_c}$$

$$\Delta p_{III} = f_{W,III} \frac{L}{D'_{e,III}} \frac{\rho u_{III}^2}{2g_c} \quad (4.2.42)$$

Since the channel volumetric flow rate is given by the product of the average fluid velocity and channel cross sectional area, we can write

$$f_{W,I}\left(\frac{L}{D'_{e,I}}\right)\left(\frac{Q_I}{A'_I}\right)^2 = f_{W,II}\left(\frac{L}{D'_{e,II}}\right)\left(\frac{Q_{II}}{A'_{II}}\right)^2 = f_{W,III}\left(\frac{L}{D'_{e,III}}\right)\left(\frac{Q_{III}}{A'_{III}}\right)^2 \quad (4.2.43)$$

where Q_I , Q_{II} , and Q_{III} are the respective volumetric flow rates in the central and peripheral channels. By assuming the friction factors are equal we may write

$$\frac{Q_{II}}{Q_I} = \frac{A'_{II}}{A'_I}\left[\frac{D'_{e,II}}{D'_{e,I}}\right]^{1/2} = K_1 = 0.542 \quad (4.2.44)$$

and

$$\frac{Q_{III}}{Q_I} = \frac{A'_{III}}{A'_I}\left[\frac{D'_{e,III}}{D'_{e,I}}\right]^{1/2} = K_2 = 0.32 \quad (4.2.45)$$

The total flow through the entire rod bundle, Q_T, is given by

$$\begin{aligned} Q_T &= n_I Q_I + n_{II} Q_{II} + n_{III} Q_{III} \\ &= (n_I + n_{II} K_1 + n_{III} K_2) Q_I \\ &= 123 Q_I \end{aligned} \tag{4.2.46}$$

Finally, the average velocity in the central flow channels is represented by

$$u_I = \frac{Q_I}{A_I'} = \frac{Q_T}{123 A_I'} \tag{4.2.47}$$

Pressure Loss Without Spacers

The pressure loss due to rods is found using the standard friction loss equation

$$\Delta p_R = f_{W,I} \left(\frac{L}{D_e} \right) \frac{u_I^2}{2g_c} \tag{4.2.48}$$

The equivalent diameter for the rods is given by $4A_I/\pi D$, which is rearranged to give $D_{e,I} A_I'/A_I$. Therefore, Eq. (4.2.48) is written as

$$\Delta p_R = f_{W,I} \left(\frac{L}{D_{e,I}} \right) \frac{u_I^2}{2g_c} \frac{A_I}{A_I'} \tag{4.2.49}$$

and by use of Eq. (4.2.47), we have

$$\Delta p_R = f_{W,I} \left(\frac{L}{D_{e,I}} \right) \frac{Q_T^2}{2g_c} \frac{A_I}{(123)^2 A_I'^3} \tag{4.2.50}$$

The friction factor, $f_{W,I}$, is taken from Moody's correlation with a Reynolds number*

$$N_{Re,R} = \frac{D_{e,I} Q_I / A_I'}{\nu} = \frac{D_{e,I} Q_T}{123 A_I' \nu} \tag{4.2.51}$$

The values for ν and rod bundle dimensions are inserted into Eqs. (4.2.50) and (4.2.51) to give

---

*Note: The author disagrees with this definition of Reynolds number because it is not based upon the same equivalent diameter used in Eq. (4.2.48) for the rod pressure loss. However, since experimental drag coefficients are calculated in this manner, it is necessary to follow de Stordeur's method if his drag coefficient correlations are used. If the example problem used a grid spacer instead of wire wraps, then equivalent diameters and flow areas are calculated in positions of unrestricted flow.

$$\Delta p_R = f_{W,I} Q_T^2 \times 1.16 \times 10^4, \text{ ft-lb}_f/\text{lb}_m$$
$$N_{Re,R} = Q_T \times 1.83 \times 10^5 \tag{4.2.52}$$

Spacer Pressure Loss

The pressure loss due to spacers is given by

$$\Delta p_s = C_s \frac{u_I^2}{2g_c} \frac{S}{A_I'} = C_s \frac{[Q_T/(123A_I')]^2}{2g_c} \frac{S}{A_I'} \tag{4.2.53}$$

where S is the spacer projected frontal area. For this example, with each turn of the spiral wire, S is an annulus of thickness equal to the spacer wire diameter. Therefore, for the entire rod length, S is given by

$$S = \pi(D + d_s)d_s \frac{L}{h} \tag{4.2.54}$$
$$= 4.48 \times 10^{-4} \text{ ft}^2$$

Upon substituting the appropriate dimensions into Eq. (4.2.53), the spacer loss equation is simplified to give

$$\Delta p_s = 205 C_s Q_T^2, \text{ ft-lb}_f/\text{lb}_m \tag{4.2.55}$$

The drag coefficient, C_s , is found from the appropriate curve in Fig. 4.8 for a Reynolds number of

$$N_{Re,s} = \frac{d_s u_I}{\nu} = \frac{d_s Q_T/(123A_I')}{\nu} \tag{4.2.56}$$
$$= 3.59 \times 10^4 Q_T$$

Finally, the total pressure loss across the rod bundle is determined by summing individual losses. By neglecting entrance and exit contraction and expansion losses,

$$\Delta p_T = \Delta p_R + \Delta p_s$$

Table 4.3 lists values of total pressure loss as a function of total volumetric flow rate.

As a comparison between experimental and calculated pressure losses shows, calculated pressure losses using drag coefficients agreed within ±12% . This agreement is quite adequate for design purposes. Most rod bundle reactor cores use transverse spacer grids. The approach used for calculating pressure losses for grid spacers is similar to that illustrated for spiral-wire spacers—one determines flow distributions throughout the rod bundle and then calculates pressure loss contributions due to grid spacers and rod bundle without spacers.

Predicting pressure losses in reactor cores of complicated geometry is not considered a precise calculation. At present, the most reliable

TABLE 4.3
Spiral-wire spacer pressure loss calculations

Q_T (ft^3/sec)	0.223	0.3345	0.5575	0.669
$N_{Re,R}$	40,800	61,200	102,000	122,500
$f_{W,I}$	0.21	0.0197	0.0177	0.0171
Δp_R (ft–lb_f/lb_m)	12.1	25.6	64.0	89.0
$N_{Re,s}$	8,000	12,000	20,000	24,000
C_s (Fig. 4.8)	0.49	0.44	0.405	0.395
Δp_s (ft–lb_f/lb_m)	5.0	10.1	25.9	36.3
Δp_T (ft–lb_f/lb_m)				
Calculated	17.1	35.6	89.9	125.3
Experimental	16.8	36.4	94.2	134.8

method of determining pressure losses in reactor channels is a direct experimental measurement of pressure losses in the channels or their models.

A later publication by Rehme [29] gives a more general, easier to apply, and possibly more accurate approach for calculating spacer pressure losses. The grid spacer pressure loss is given by

$$\Delta p_s = C_v \epsilon^2 \frac{\rho u_B^2}{2g_c} \tag{4.2.57}$$

where

C_v = modified drag coefficient

ϵ = ratio of projected grid cross section to undisturbed flow cross section

u_B = average bundle fluid velocity

Figure 4.9 illustrates modified drag coefficients for grid spacers as a function of rod bundle Reynolds number defined as

$$N_{Re,B} = \frac{\rho u_B D_e}{\mu} \tag{4.2.58}$$

where $D_e = 4\,A/S$ with A being the total unrestricted flow cross section and S the wetted perimeter including channel wall. Eight different types of grid spacers are tested in developing Fig. 4.9 and this approach appears to have produced a great deal of generality among the various spacers.

For wire-wrapped rod bundles, the pressure loss depends upon P/D, helical spacer wire pitch, H, and number of rods. Rehme [29] suggests that rod bundle pressure loss be calculated by a modified form of the friction factor approach instead of assuming the loss is due to the sum of losses from the bare rod and spacer wire. The pressure loss is given by [29]

$$\Delta p = f' \frac{S_b}{S_t} \frac{\rho u_{eff}^2}{2g_c} \frac{L}{D_e} \tag{4.2.59}$$

where

f' = modified friction factor

S_b/S_t = ratio of wetted perimeter of rods and wires to total wetted perimeter

u_{eff} = effective velocity accounting for swirl flow due to wire wraps

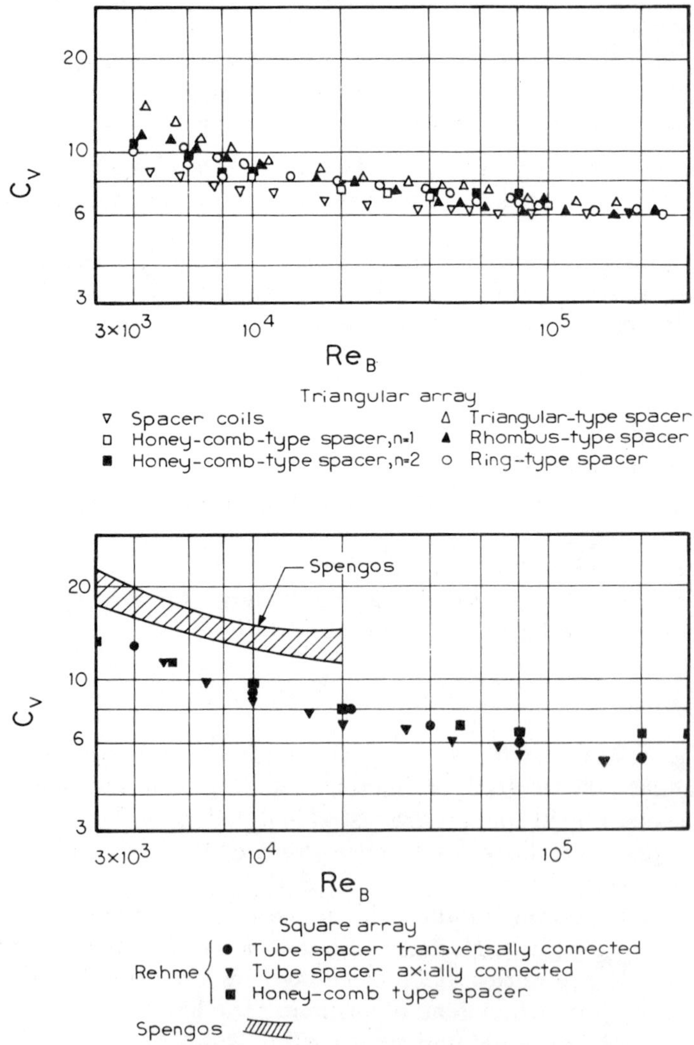

Fig. 4.9. Modified drag coefficients [29].

The effective velocity depends upon rod bundle geometry and is given by

$$\left(\frac{u_{eff}}{u_B}\right)^2 = \left(\frac{P}{D}\right)^{0.5} + \left[7.6\,\frac{D}{H}\left(\frac{P}{D}\right)^2\right]^{2.16} \qquad (4.2.60)$$

where H is the helical pitch of the wire wraps. The modified friction factor depends upon a modified Reynolds number

$$N'_{Re} = \frac{\rho u_{eff} D_e}{\mu} \qquad (4.2.61)$$

and is correlated as

$$f' = \frac{64}{N'_{Re}} + \frac{0.0816}{(N'_{Re})^{0.133}} \qquad (4.2.62)$$

For $N'_{Re} > 2 \times 10^4$, 92% of ~ 1400 data points fell within ± 5% of Eq. (4.2.62). The experiments covered a wide variety of geometrical parameters—P/D = 1.125 to 1.417, H/D = 6 to 45, and number of rods from 7 to 61.

Example 4.2

The use of Rehme's method of calculating rod bundle pressure losses is illustrated by repeating example problem 4.1. The coolant volumetric flow rate is assumed 0.5575 ft³/sec

The first step in the calculations is to compute the rod bundle equivalent diameter which is given by $4A_t/S_t$ where A_t is total flow area and S_t is total wetted perimeter. Rehme recommends taking into account that the wire cross section perpendicular to the rods is an ellipse; however, this additional complexity is ignored.

$$A_t = (12P - D)^2 - 121\pi(D^2 + d_m^2)/4 = 0.0168 \text{ ft}^2$$

$$S_t = 4(12P - D) + 121\pi(D + d_m) = 7.72 \text{ ft.}$$

$$D_e = 4A_t/S_t = 0.0087 \text{ ft}$$

$$u_B = Q/A_t = 33.2 \text{ ft/sec}$$

and from Eq. (4.2.60) the square of the effective velocity is

$$\left(\frac{u_{eff}}{u_B}\right)^2 = \left(\frac{P}{D}\right)^{0.5} + \left[7.6\,\frac{D}{H}\left(\frac{P}{D}\right)^2\right]^{2.16} = 1.154$$

or

$$u_{eff} = 35.7 \text{ ft/sec}$$

The modified Reynolds number is then

$$N'_{Re} = \frac{\rho u_{eff} D_e}{\mu} = 85{,}900$$

The modified friction factor is given by Eq. (4.2.62)

$$f' = \frac{64}{N'_{Re}} + \frac{0.0816}{(N'_{Re})^{0.133}} = 0.0188$$

The wetted perimeter of the rods and wires is

$$S_b = 121\pi(D + d_m) = 6.91 \text{ ft}$$

and $S_b/S_t = 0.894$. From Eq. (4.2.59) the rod bundle pressure loss is given by

$$\Delta p = f'\frac{S_b}{S_t}\frac{u_{eff}^2}{2g_c}\frac{L}{D_e} = 76.6 \text{ ft–lb}_f/\text{lb}_m$$

For the geometry and flow conditions given, the measured pressure loss is (94.2 ft–lb_f/lb_m [27], which is considerably higher than calculated. For this example, the technique cited in Ref. [27] gives a better agreement between calculated and measured pressure losses.

4.3 Pressure Loss With Compressible Fluids

For flow of gases through constant cross-sectional area channels, fluid velocity changes occur because of fluid density changes. Consequently, the form of the steady-flow energy equation used for incompressible fluids is not suitable because of continually changing fluid velocity or kinetic energy along the channel. For isothermal flow of gases, the incompressible fluid form of the steady-flow energy equation gives good approximate results providing pressure losses are small or p_2/p_1 is not substantially less than unity [3].

For isothermal flow of gases, experiments show that standard friction factor equations or the Moody diagram can be used for predicting friction pressure losses at any point in a channel. Consequently, a differential form of the Darcy-Weisbach friction loss equation can be substituted into the differential form of the steady-flow energy equation.

$$\frac{g}{g_c}dz + v\,dp + \frac{\alpha u\,du}{g_c} = \delta W - f_W\frac{u^2}{2g_c}\frac{dx}{D_e} \tag{4.3.1}$$

where x is channel length and z is elevation. Equation (4.3.1) is integrated over a given channel length in order to find the pressure loss. However, the integration is not performed in the simple manner used for incompressible fluids because density and velocity are a function of channel position.

In the treatment of compressible flows, it is convenient to eliminate velocity terms from the steady-flow energy equation by substituting terms using mass flow rate and density. From conservation of mass, the mass flow rate through a channel is constant. Therefore, at any channel cross section, the average fluid velocity is given by

$$u = \frac{\dot{m}}{A} v = Gv \tag{4.3.2}$$

where $\dot{m}$ is the mass flow rate in lb_m/sec, A is the channel cross sectional area in ft^2, and G is the mass velocity in $lb_m/ft^2\text{-}sec$. Since G is constant for channels of constant cross sectional area

$$du = G\,dv \tag{4.3.3}$$

By assuming the work term is zero and using Eqs. (4.3.2) and (4.3.3), Eq. (4.3.1) is rearranged to give

$$-dp = \frac{1}{v}\frac{g}{g_c}\,dz + \frac{aG^2\,dv}{g_c} + f_W\frac{G^2 v}{2g_c}\frac{dx}{D_e} \tag{4.3.4}$$

Equation (4.3.4) is a difference equation well suited for numerical computations. Since ρu is constant and the fluid viscosity is essentially independent of pressure, for isothermal flows the Reynolds number is constant and f_W will not vary along channel length.

The potential energy term is expressed in terms of distance along the channel by using

$$dz = K\,dx \tag{4.3.5}$$

where K is the sine of the channel angle of inclination at position dx. When G and K are constant, the steady-flow energy equation is readily integrated because the variables are separated. Integration gives

$$-\int_{p_1}^{p_2} dp = \int_0^L \frac{g}{g_c}\frac{K}{v}\,dx + \int_{v_1}^{v_2} \frac{aG^2}{g_c}\,dv + \int_0^L \frac{f_W G^2 v}{2g_c D_e}\,dx \tag{4.3.6}$$

or

$$p_1 - p_2 = K\frac{g}{g_c}\int_0^L \rho\,dx + \frac{aG^2}{g_c}(v_2 - v_1) + \frac{f_W G^2}{2g_c D_e}\int_0^L v\,dx$$

If pressure losses are not too large, for isothermal flow the fluid density or specific volume changes with small curvature along the channel length and, consequently, the integrals in Eq. (4.3.6) are replaced by average values for fluid density and specific volume. Therefore

$$p_1 - p_2 = \rho_{avg}\frac{g}{g_c}KL + \frac{G^2}{g_c}\left[a(v_2 - v_1) + \frac{f_W v_{avg} L}{2D_e}\right] \tag{4.3.7}$$

where $\rho_{avg} = (\rho_1 + \rho_2)/2$ and $v_{avg} = (v_1 + v_2)/2$. Note that $\rho_{avg} \neq 1/v_{avg}$. In addition, the sign on K depends upon whether the flow is up or down. For upflow, the angles of inclination vary from 0 to 90° while for downflow, the angles vary from $180°$ to $270°$.

Solutions to Eq. (4.3.7) require trial-and-error procedures because p_2 is needed to evaluate ρ_{avg} and v_{avg}. A few trials are sufficient for predicting $(p_1 - p_2)$ accurately enough for design purposes. It should be kept in mind that solutions for pressure losses that use Eqs. (4.3.6) or (4.3.7) alone require the flow be isothermal. This is necessary because pressure alone can not be used for determining densities or specific volumes. Consequently, Eq. (4.3.7) is used for isothermal flow situations which most likely would exist for adiabatic flows with small pressure losses.

Unless the system pressure is high and, consequently, the gas density of significance, the potential energy term in Eqs. (4.3.6) and (4.3.7) is usually negligible compared to other terms.

For flows in which pressure losses are large, which would occur in high speed flows, the fluid temperature changes as it expands down the channel. Equation (4.3.6) is still valid for analyzing pressure losses; however, it along with the equation of state (i.e., gas law) consist of a set of two equations containing three unknowns—p, v, T. In order to treat this problem, it is necessary to bring in further relations from fundamental thermodynamics.

Equations (4.3.6) and (4.3.7) are momentum equations obtained by applying Newton's Laws of Motion to a constant cross section channel. Applying the First Law of Thermodynamics to a volume element in the channel yields

$$J\,dh + d\left(\frac{\alpha u^2}{2g_c}\right) + \frac{g}{g_c}\,dz - J\,dQ = 0 \qquad (4.3.8)$$

For adiabatic flow $dQ = 0$ and for simplicity we will assume $dz = 0$; therefore, Eq. (4.3.8) becomes

$$J\,dh + d\left(\frac{\alpha u^2}{2g_c}\right) = 0 \qquad (4.3.9)$$

Substituting Eq. (4.3.2) into Eq. (4.3.9), assuming $\alpha = 1.0$, and integrating between axial locations (1) and (2) yields

$$Jh_2 + G^2\frac{v_2^2}{2g_c} = Jh_1 + G^2\frac{v_1^2}{2g_c} \qquad (4.3.10)$$

Equation (4.3.10) represents an infinite number of relations, each corresponding to different values of G, passing through the state corresponding to h_1 and v_1. Figure 4.10 illustrates several lines representing Eq. (4.3.10) on an h-s diagram. These lines are called

Fanno lines. Accordingly, adiabatic flow through constant cross section channels is called **Fanno flow.**

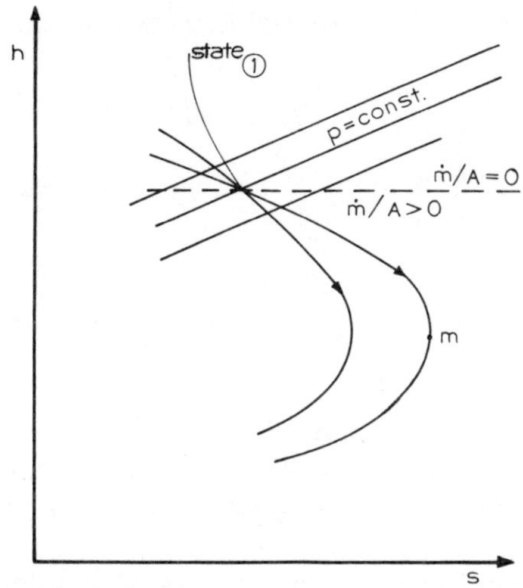

Fig. 4.10. Fanno lines for a compressible fluid expanding through a constant area channel

The isenthalpic Fanno line through state 1 corresponds to zero (negligible) flow. For a given mass flow rate, positions on the Fanno line move toward higher entropies as the enthalpy and pressure decreases. At some pressure, the entropy reaches a maximum (point m in Fig. 4.10) and then decreases with any further change in pressure. However, an expansion at constant area cannot proceed beyond point m since a change to a lower entropy for an adiabatic process violates the Second Law of Thermodynamics. If the chamber pressure into which the pipe exhausts is lower than the pressure at state m, then the pressure drop from state m to the exhaust pressure occurs outside the pipe where the stream is not restricted to constant area.

The fluid velocity at state m is the velocity of sound corresponding to conditions at m. For a perfect gas, the velocity of sound at state m is given by

$$u_s = \sqrt{\gamma g_c p_m v_m} \tag{4.3.11}$$

$$u_s = \sqrt{\gamma g_c R T_m}$$

where $\gamma = c_p/c_v$ and the subscript m implies conditions at state m.

For constant cross section channels, if the fluid velocity initially is subsonic, then nowhere in the channel can velocity exceed the speed of sound corresponding to local channel conditions.

In summary, pressure losses with compressible flow without restrictions requires simultaneous solution of four equations. These equations are (1) energy equation, (2) momentum equation, (3) equation of state, and (4) a relation between enthalpy and temperature and pressure. For steady, **adiabatic** flow of a perfect gas in a constant cross section channel, in differential form these equations are given by

$$J\,dh + G^2 d\left(\frac{\alpha v^2}{2g_c}\right) + \frac{g}{g_c} K\,dx = 0 \tag{4.3.12}$$

$$-dp = \frac{1}{v}\frac{g}{g_c} K\,dx + \frac{\alpha G^2}{2g_c} dv + f_W \frac{G^2 v}{2g_c} \cdot \frac{dx}{D_e} \tag{4.3.13}$$

$$pv = RT \tag{4.3.14}$$

$$dh = c_p\,dT \tag{4.3.15}$$

For a given mass flow rate and initial conditions, Eqs. (4.3.12-4.3.15) contain four unknowns—p, v, T, and h. In differential form these equations are well suited for numerical step-wise computation down a channel.

Doing hand calculations with Eqs. (4.3.12-4.3.15) requires more ingenuity than a simple successive elimination of unknowns. One approach is to substitute Eq. (4.3.15) into Eq. (4.3.12) and integrate between states 1 and 2. For convenience, we assume the potential energy term is negligible and this gives

$$Jc_p(T_2 - T_1) + G^2 \frac{\alpha v_2^2}{2g_c} = G^2 \frac{\alpha v_1^2}{2g_c} \tag{4.3.16}$$

By assuming a value for T_2 , the specific volume at state 2 is evaluated from Eq. (4.3.16). The pressure at state 2 is then evaluated from the gas law, Eq. (4.3.14). Equation (4.3.13) is integrated to determine the channel length corresponding to state 2. In order to integrate Eq. (4.3.13), it is convenient to divide by v, therefore

$$-\int_1^2 \frac{dp}{v} = \frac{\alpha G^2}{g_c} \ln\frac{v_2}{v_1} + f_W \frac{G^2}{2g_c}\frac{L}{D_e} \tag{4.3.17}$$

The integral on the left is evaluated graphically or integrated directly by solving for v in terms of p using Eqs. (4.3.16) and (4.3.14).

Employing this approach yields values of channel length L corresponding to assumed values of p_2 . A plot can then be made of channel pressure loss ($p_1 - p_2$) as a function of channel length L. If initial channel velocities are large, a check should be made to insure that sonic velocities do not appear in any downstream portion of the channel. If such a situation is predicted, for a given applied pressure difference the mass flow rate would decrease to a point where all channel velocities are subsonic.

In gas-cooled nuclear power plants, non-adiabatic flows occur in reactor cores and heat exchangers. Consequently, the heat addition term is retained in the energy equation. This term is given by

$$\dot{m}\, dQ = q''_w P_h\, dx \tag{4.3.18}$$

where q''_w is the local wall heat flux and P_h is the channel heated perimeter. Since q''_w is highly space dependent in reactor cores, it is best to use numerical techniques for simultaneous solution of the four basic equations.

Generally in nuclear reactors the differential heat addition term is quite large compared to the kinetic energy and potential energy terms in Eq. (4.3.8). Consequently, as an approximation

$$dh \approx dQ = \frac{q''_w P_h\, dx}{\dot{m}} \tag{4.3.19}$$

This assumption is not valid for the analysis of nuclear rocket engines. Therefore, assuming a perfect gas and integrating, the temperature difference between two points in a channel is determined.

$$\begin{aligned} c_p(T_2 - T_1) &= \frac{1}{\dot{m}} \int_1^2 q''_w P_h\, dx \\ &= \frac{q''_{avg} P_h L}{\dot{m}} \end{aligned} \tag{4.3.20}$$

The momentum equation, Eq. (4.3.6), and equation of state, Eq. (4.3.14), are employed along with Eq. (4.3.20) to solve for pressure loss along a channel length.

For a uniform heat flux, an approximate method is developed for calculating $p_1 - p_2$. With this restraint, the fluid mean temperature at any point in the channel is given by

$$\begin{aligned} T &= T_1 + \frac{q''_w P_h x}{\dot{m} c_p} \\ &= T_1 \left[1 + \left(\frac{T_2 - T_1}{T_1}\right) \frac{x}{L}\right] \\ &= T_1 \left[1 + \frac{\Delta T}{T_1} \frac{x}{L}\right] \end{aligned} \tag{4.3.21}$$

In addition, it is assumed that pressure change in the channel also varies linearly and, consequently

$$\begin{aligned} p &= p_1 + \frac{p_2 - p_1}{L} x \\ &= p_1 \left[1 + \frac{\Delta p}{p_1} \frac{x}{L}\right] \end{aligned} \tag{4.3.22}$$

Substituting Eqs. (4.3.14), (4.3.21), and (4.3.22) into Eq. (4.3.6) gives

$$p_1 - p_2 = \frac{Kgp_1}{g_c RT_1} \int_o^L \frac{\left[1 + \frac{\Delta p}{p_1}\frac{x}{L}\right] dx}{1 + \frac{\Delta T}{T_1}\frac{x}{L}} +$$

$$\frac{G^2}{g_c}\left[\frac{RT_2}{p_2} - \frac{RT_1}{p_1} + \frac{RT_1}{2D_e p_1}\int_o^L f_w \frac{\left[1 + \frac{\Delta T}{T_1}\frac{x}{L}\right] dx}{1 + \frac{\Delta p}{p_1}\frac{x}{L}}\right] \tag{4.3.23}$$

Equation (4.3.23) is readily integrated. The friction factor is assumed constant and is evaluated from Eqs. (4.2.16) or (4.2.17) using average wall and fluid bulk temperatures. Integration gives

$$p_1 - p_2 = \frac{KgLT_1}{g_c v_1 \Delta T}\left[\frac{\Delta p}{p_1} + \left(1 - \frac{\Delta p}{p_1}\frac{T_1}{\Delta T}\right) \ln\left(1 + \frac{\Delta T}{T_1}\right)\right]$$

$$+ \frac{G^2 v_1}{g_c}\left[\frac{T_2/T_1}{1 + \Delta p/p_1} - 1\right.$$

$$\left. + \frac{f_W p_1 L}{2D_e \Delta p}\left\{\frac{\Delta T}{T_1} + \left(1 - \frac{\Delta T p_1}{T_1 \Delta p}\right) \ln\left(1 + \frac{\Delta p}{p_1}\right)\right\}\right]$$

(4.3.24)

At first glance (Eq. 4.3.24) appears to be rather formidable for calculating pressure losses. However, for gas-cooled reactors, design pressures are high and, consequently, $|\Delta p/p_1| << 1.0$. Therefore, in solving for $p_1 - p_2$, one can assume a reasonable value for $\Delta p/p_1$ like -0.1, and solve for $p_1 - p_2$. A check is then made on the assumed value for $\Delta p/p_1$. If the agreement is not satisfactory, the calculation is repeated using the new calculated value for $\Delta p/p_1$.

In order to check the validity of Eq. (4.3.24), a comparison is made of a solution to a sample problem using this equation and a fourth-order Runge-Kutta numerical solution to the basic differential equations. The problem involves helium flowing through a 5/8 inch diameter tube with a heated length of 15.5 feet. The helium inlet temperature is 760°F and the inlet pressure is 600 psia. These conditions are representative of conditions in a HTGR. Both solution methods yield identical results which are shown in Fig. 4.11.

In gas-cooled reactors, pressure losses can be quite high with a result that pumping power requirements are large. As seen in Eq. (4.3.24), pressure loss is roughly proportional to G^2 and v; and, as discussed later, the pumping power is roughly proportional to G^3. One method for reducing core pressure losses is to split the core into two regions and introduce the coolant at the center of the flow channels. This reduces G by a factor of two in each split section. Therefore, the pumping power requirements for the core are reduced by almost a factor of eight. This method for reducing pressure loss is quite suitable for large gas-cooled reactors and was used in the Brookhaven natural uranium, graphite-moderated reactor.

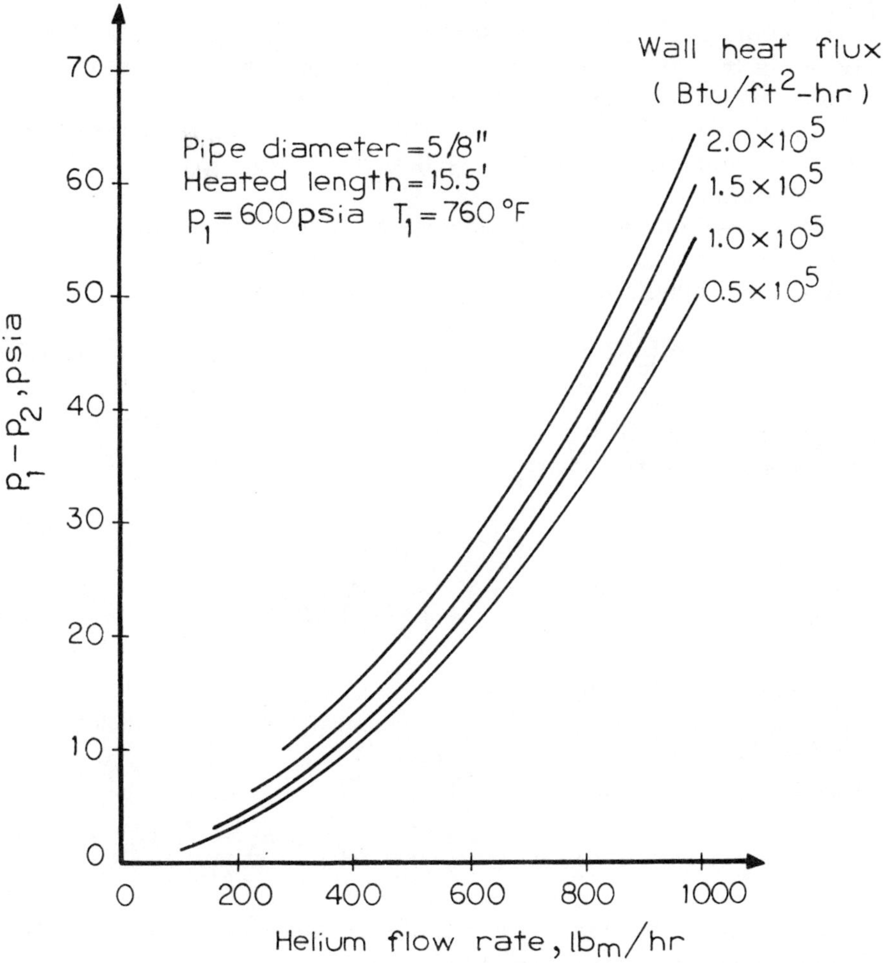

Fig. 4.11. Pressure loss vs. flow rate for helium in a heated circular channel

4.4 Flow With Liquid Thermal Buoyancy

In the flow of liquids through heated or cooled channels, changes in fluid density due to temperature changes produce buoyancy forces which may aid or oppose flow. For flow in horizontal channels, buoyancy forces are perpendicular to the direction of flow and produce a warping of fluid velocity profile which only affects wall shear stress. However, for flow in inclined or vertical channels, buoyancy forces are either in the same or opposite direction of flow. With heat addition to the fluid, buoyancy forces are upward and aid upflow or oppose downflow. Consequently, a flow instability due to buoyancy effects is possible for downflow in channels with heat addition.

Downflow is not used in either water-cooled or liquid-metal-cooled power reactors. However, it is used in swimming pool research reactors because of upward flow of irradiated water would cause a radiation hazard due to induced N^{16} activity. Therefore, it may be necessary to examine stability of the flow at low flow rates in the safety analysis of pool reactors.

The steady-flow energy equation is applicable for analyzing pressure loss with heat addition affecting liquid density. Equation (4.3.4) is written for channels with constant cross section area

$$-dp = \frac{g}{g_c} K\rho \, dx + \frac{G^2}{g_c}\left[a \, dv + \frac{f_W v}{2D_e} \, dx\right] \tag{4.4.1}$$

where for upflow K = 1.0 and downflow K = -1.0. The variable densities and specific volumes are represented to first order by

$$\rho = \rho_1 \left[1 + \left(\frac{1}{\rho}\frac{d\rho}{dx}\right)_1 dx\right] \tag{4.4.2}$$

$$v = v_1 \left[1 + \left(\frac{1}{v}\frac{dv}{dx}\right)_1 dx\right]$$

where the subscript 1 means evaluation at the channel inlet. Making use of the definition of volumetric coefficient of thermal expansion β, $(d\rho/\rho)_1$ is given by $-\beta \, dT$ and $(dv/v)_1$ is given by $\beta \, dT$. Substituting these relations and Eq. (4.4.2) into Eq. (4.4.1) gives

$$-dp = \rho_1 \frac{g}{g_c} K(1 - \beta \, dT) \, dx + \frac{v_1 G^2}{g_c}\left[a\beta \, dT + \frac{f_W(1 + \beta \, dT)}{2D_e} dx\right] \tag{4.4.3}$$

In finite difference form, Eq. (4.4.3) is well-suited for step-wise numerical calculations up a channel.

Equation (4.4.3) is integrated over the channel length $x = 0$ to $x = L$ and $T = T_1$ to $T = T_2$ which gives

$$(p_1 - p_2) - \rho_1 \frac{g}{g_c} KL = -\rho_1 \frac{g}{g_c} K\beta \int_{x=0}^{x=L} \int_{T=T_1}^{T} dT\, dx$$

$$+ \frac{v_1 G^2}{g_c} \left[a\beta(T_2 - T_1) + \frac{f_W L}{2D_e} + \frac{f_W \beta}{2D_e} \int_{x=0}^{x=L} \int_{T=T_1}^{T} dT\, dx \right] \qquad (4.4.4)$$

If the temperature rise in the fluid is linear, approximately true for reactor cores unless the neutron flux distribution is considerably warped in the axial direction, the integrals of the temperature distribution are represented by $L(T_2 - T_1)/2$. Therefore, Eq. (4.4.4) is written

$$(p_1 - p_2) - \rho_1 \frac{g}{g_c} KL = -\rho_1 \frac{g}{g_c} K\beta \frac{(T_2 - T_1)}{2} L$$

$$+ \frac{v_1 G^2}{g_c} \left[a\beta(T_2 - T_1) + \frac{f_W L}{4D_e} (2 - \beta(T_2 - T_1)) \right] \qquad (4.4.5)$$

The first term on the right of Eq. (4.4.5) is the pressure change due to change in fluid density. With downflow ($K = -1.0$) and heat addition, this results in a pressure loss. The second term contains pressure losses due to friction and increasing fluid kinetic energy with heat addition. For channels without substantial temperature changes, $\beta(T_2 - T_1)$ is small compared to 2. In addition, for long channels with small equivalent diameters, the friction loss term is large compared to $a\beta(T_2 - T_1)$. Consequently, in many situations the pressure loss is due to contributions from the change in fluid density and friction. The ratio of the buoyancy term and friction loss for vertical channels ($K = \pm 1$) is given by

$$\frac{\rho_1 \dfrac{g}{2g_c} \beta(T_2 - T_1)L}{\dfrac{v_1 G^2 f_W L}{2g_c D_e}} = \frac{\dfrac{\rho_1^2 \beta g(T_2 - T_1)D_e^3}{2\mu^2}}{\dfrac{f_w G^2 D_e^2}{2\mu^2}}$$

$$= \frac{N_{Gr}}{\dfrac{f_W}{2} N_{Re}^2} = \frac{N_{GR}}{N_{Ka}} \qquad (4.4.6)$$

where N_{Gr} is the Grashof number employing $(T_2 - T_1)/2$ as the usual temperature difference and N_{Ka} is the Karman number.

Reactor cores usually contain many parallel coolant channels with common inlet and outlet plenum regions. Consequently, the left-hand side of Eq. (4.4.5) is the same for each coolant channel as well as the right-hand side. Two limiting cases involving Eq. (4.4.6) are considered. If $N_{Gr}/N_{Ka} << 1.0$, say of the order of 0.05, then buoyancy effects are minor and pressure loss is treated as if the fluid is incompressible. If $N_{Gr}/N_{Ka} >> 1.0$, of the order of 20 or more, then friction is negligible and the flow distributes itself in each channel so that $(T_2 - T_1)$ is the same for all channels. This is a useful technique for maintaining uniform temperatures and low thermal stresses in heated systems by employing natural convection cooling. Naturally, since flow is small, heat rates must be small.

Temperatures are eliminated from Eq. (4.4.5) by using a heat balance on the flow channel $Q = \dot{m}c_p(T_2 - T_1)$, where Q is the rate of heat added to the fluid in the channel. Equation (4.4.5) is written

$$(p_1 - p_2) - \rho_1 \frac{g}{g_c} KL = \frac{v_1 G^2 f_W L}{2g_c D_e} - \frac{\rho_1 g\beta KLQ}{2g_c \dot{m} c_p} + \frac{v_1 G^2 \beta Q}{g_c \dot{m} c_p}\left[a + \frac{f_W L}{4D_e}\right] \tag{4.4.7}$$

For a given reactor core geometry, pressure loss $(p_1 - p_2)$, and power level, Eq. (4.4.7) is used to predict channel mass flow rate $\dot{m}$. If the total mass flow rate, $\dot{m}_T$, is given through a reactor core—this is a common specification from thermodynamic analysis—Eq. (4.4.7) is used to compute flow through individual channels. A convenient method of solving problems involving a specified total mass flow rate is to assume reasonable values for $(p_1 - p_2)$ and calculate flow through individual channels. A plot of $(p_1 - p_2)$ as a function of the sum of the flow through all channels, $\Sigma\dot{m}$, yields the required pressure loss when $\Sigma\dot{m} = \dot{m}_T$. Of course this example is somewhat simplified since entrance, exit, and spacer pressure losses must also be accounted for if applicable to a problem.

For downflow in vertical channels with heat addition, a flow instability is possible because of buoyancy effects. In Eq. (4.4.7), $G = \dot{m}/A$ where A is the channel cross section area. Inspection of Eq. (4.4.7) shows that, for constant heat addition and decreasing mass flow rate, the friction pressure loss decreases with the square of the mass flow rate and the buoyancy pressure loss increases with the reciprocal of the decreasing mass flow rate. Consequently, with decreasing flow rates a situation develops in which a further decrease in flow rate produces an increase in pressure loss. With downflow, this condition defines a flow instability since either flow stoppage or flow reversal occurs.

Mass flow rates at the onset of flow instabilities are obtained by setting the derivative of Eq. (4.4.7) with respect to $\dot{m}$ equal to zero. By assuming the far right term of Eq. (4.4.7) is negligible, the critical mass velocity is

$$G_{crit} = \left[\frac{-K\rho_1^2 g\beta(T_2 - T_1)D_e}{2f_W}\right]^{1/2} \tag{4.4.8}$$

Substituting Eq. (4.4.8) into Eq. (4.4.6) yields a criteria for predicting flow instabilities, if

$$\frac{N_{Gr}}{N_{Ka}} \geqslant -\frac{2}{K} = 2 \tag{4.4.9}$$

then downflow is unstable. Consequently, downflow in reactor systems should be viewed with caution unless $N_{Gr}/N_{Ka} << 2$.

The problem of flow instability is best illustrated by a diagram of total pressure loss as a function of flow rate. Figure 4.12 illustrates a typical diagram of $(p_1 - p_2)$ at low flow rates for 10 percent and 100 percent reactor power. At 100 percent power, a reduction in flow rate reduces the pressure loss; however, as flow rate decreases, the fluid exiting from the channel reaches its saturation temperature and bulk boiling commences. The rapid change in fluid density with bulk boiling invalidates the use of Eq. (4.4.7) in the analysis. For the mass flow rates considered at 100 percent full power, the flow is stable since an increase in $(p_1 - p_2)$ accelerates the fluid to the higher flow rate which corresponds to the increased pressure loss.

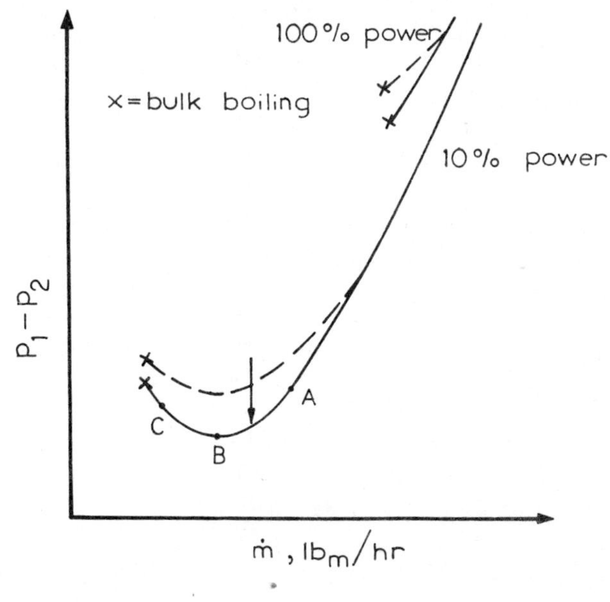

Fig. 4.12. Pressure loss with downflow in a reactor coolant channel at varying percentages of rated power (arrow corresponds to maintaining constant $T_2 - T_1$ for rated power and flow)

If full power flow rate is maintained at lower power levels, then $(T_2 - T_1)$ decreases, and consequently, N_{Gr}/N_{Ka} decreases and the stability of the flow increases. However, to decrease pumping power and thermal stresses, it may be desirable to maintain $(T_2 - T_1)$ constant during low power operation by decreasing the flow rate proportionally with the decreasing power level. As a consequence, N_{Ka} decreases and N_{Gr}/N_{Ka} increases and an instability in downflow is more likely. Figure 4.12 illustrates the pressure loss versus flow rate curve for a reactor operating at 10 percent full power. The arrow on the curve marks the flow rate that has the same $(T_2 - T_1)$ that exists at full power with rated flow. Operation at flow rates lower than point B on the 10 percent power curve is unstable. For example, at point C an increase in $(p_1 - p_2)$ accelerates the flow; however, increased flow requires a smaller pressure loss and, consequently, the flow rate increases past point B to some stabilizing point near A. As a converse, a slight decrease in $(p_1 - p_2)$ at C causes a decrease in flow which requires a larger value of $(p_1 - p_2)$ to maintain the flow and, consequently, the flow rate decreases further until a fuel element failure or flow reversal occurs.

Consequently, in the operation of power reactors with downflow, it is undesirable to reduce flow rates and power levels to a point that is near the point of instability, B, on a pressure loss versus flow rate curve. For power reactors, the flow resistance and heat generation vary from channel to channel and, therefore, average values do not accurately represent stability of all channels. If it is desirable to keep the header temperatures, T_1 and T_2, constant during all levels of power operation, the stability of the system is increased by adding flow resistances such as nozzles in series with the channels. This technique is discussed in more detail in the section on two-phase flow, since flow stability is a more severe problem with two-phase flow with heating.

The method of calculating pressure losses with thermal buoyancy, Eq. (4.4.7), is not precise because of difficulty in calculating friction pressure losses in non-isothermal flows. The superimposed buoyancy effects change velocity distributions in the fluid and, consequently, alter wall shear stresses. With heating in most liquids, the fluid viscosity decreases and the friction pressure loss or wall shear stress is reduced. Another important factor neglected in deriving Eq. (4.4.7) is that, with decreasing mass flow rate with heating, the fluid approaches its saturation temperature and subcooled nucleate boiling takes place. Friction pressure losses with subcooled nucleate boiling are larger than without nucleate boiling. This effect tends to add to the instability of the system since, with constant heat input, decreasing mass flow rates increase the amount of subcooled nucleate boiling. An estimate of the trend of this neglected factor is shown on Fig. 4.12 by dashed lines. As discussed in the section on two-phase flow, estimating pressure losses with subcooled nucleate boiling is not as precise as with single-phase flow.

4.5 Two-Phase Flow

With the advent of boiling water and high power density pressurized water reactors, the analysis of fluids consisting of liquid-vapor mixtures has become extremely important. In order to analyze reactor systems employing liquid-vapor mixtures, it is necessary to predict liquid-vapor density, pressure drop across a given channel length, flow stability, maximum flow rates, and heat transfer rates. All of these topics are discussed in this section with the exception of heat transfer which is treated in Chapter 6.

A sound theoretical treatment of the flow of two-phase mixtures is not possible because of the complexity of the physical processes. Consequently, in order to reduce two-phase flow problems to tractable proportions, it is necessary to combine some forms of analysis with experimental results to obtain correlations of desired quantities as functions of known parameters.

Various flow patterns exist in two-phase flow which are primarily dependent upon volume flow rates, direction of flow, and channel orientation. Channel dimensions, for channels large enough that capillary forces are not significant, and physical properties of either phase do not appear to have a significant influence in determining the flow pattern in two-phase flow. Figure 4.13 illustrates various flow regimes experimentally shown to exist in adiabatic two-phase flows. These configurations are observed by passing liquid-gas mixtures through transparent tubes. The order of the regimes is in terms of increasing gas flow for a fixed liquid flow. Since definitions of various flow regimes are arbitrary and the transition between them is uneven, it is not possible to distinguish between regimes with much precision. Data do exist for estimating the type of flow regimes for two-phase mixtures in terms of liquid and vapor velocities or vapor volume fraction and average mixture velocity.

With heat addition, as in nuclear reactors, two-phase flow is complicated even further since the added heat vaporizes more fluid and, consequently, the vapor volume fraction continually increases along the channel. Figure 4.14 illustrates various flow regimes which are thought to exist in a heated vertical channel. For reactors these regimes are called:

Bubbly Flow - small bubbles exist in a continuous liquid phase; regime occurs only with small void fractions.

Slug Flow - vapor bubbles agglomerate and the flow is characterized by slugs of liquid and vapor following each other; this appears to be the transition between bubbly and annular flow and occurs at moderate

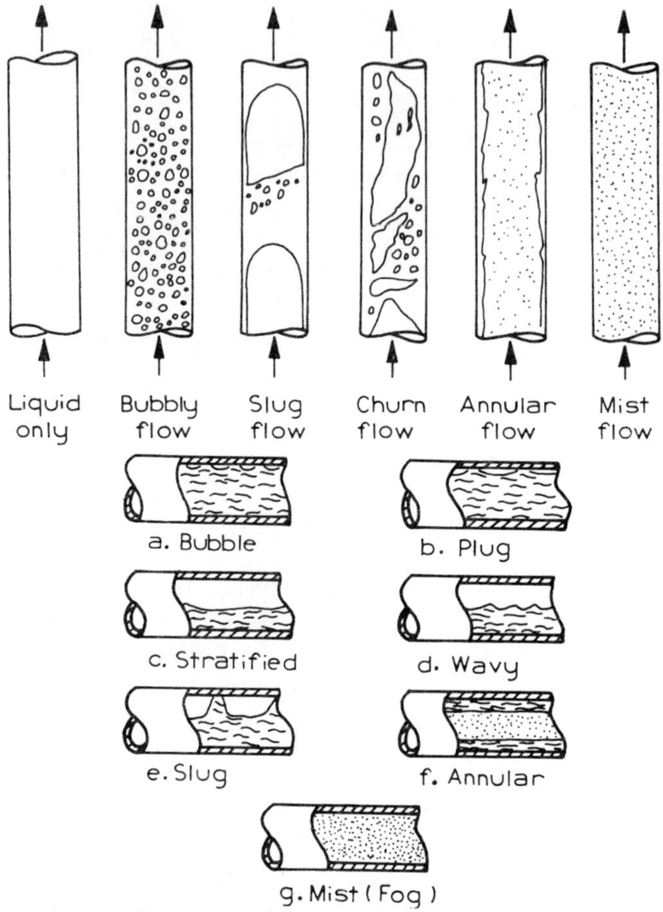

Fig. 4.13. Flow patterns in two-phase flow.

void fractions and low velocities.

Annular Flow - phases separate into a vapor core and an annular liquid phase which flows along the channel walls; regime occurs at high velocities and void fractions.

Pressure drop in two-phase flow is affected by flow patterns as given by the distribution of phases. However, correlations for each flow pattern that are consistently reliable have not, in general, been developed. Martinelli, et al. [30] reported that, for steady state flow and no radial pressure gradients, the pressure loss may be independent of flow pattern. Most investigators disagree with this prediction.

Classifications based on visual observations are made by numerous investigators [31, 32, 33, 34, 35] for horizontal and vertical upflow in rectangular channels and tubes. The patterns for vertical down flow have received minimal attention, undoubtedly due to the limited

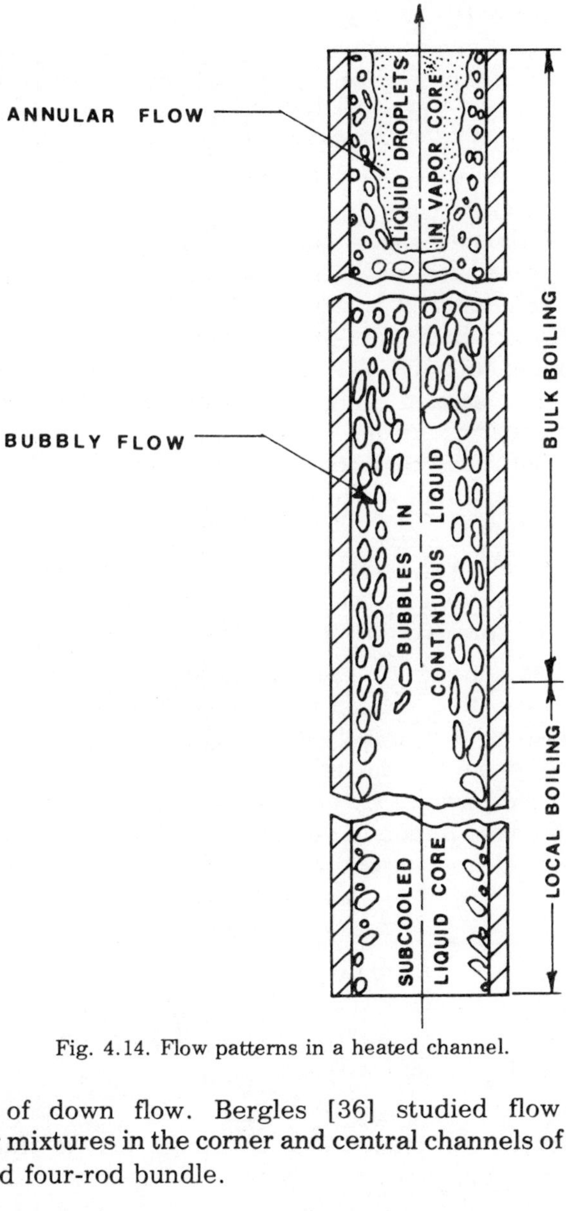

Fig. 4.14. Flow patterns in a heated channel.

application of down flow. Bergles [36] studied flow regimes of steam-water mixtures in the corner and central channels of an unheated and a heated four-rod bundle.

Fundamental Concepts

In the analysis of boiling systems it is necessary to predict subchannel void fraction and mixture density in order to perform nuclear calculations, determine pressure loss and flow stability, and evaluate core heat transfer.

Contemplation of the boiling channel shown in Fig. 4.14 leads one to conclude that at a given channel location fluid conditions continually change with time. At one instant in time the fluid may be vapor while at another time the fluid may be liquid, consequently, describing fluid conditions at a point in a channel requires taking time-averages in a manner similar to that used to describe single-phase turbulence discussed in Sec. 3.3

The use of time-averaged point-wise properties to describe flow conditions in two-phase flows is too cumbersome for most methods of analysis. Consequently, it is customary to describe flow conditions in terms of cross-sectional averages taken across the channel flow cross section. Keeping in mind that the flow properites being developed involve flow cross section averages of time-averaged local properties, we now proceed to define a few basic concepts and develop relationships between these concepts. A more detailed description of this process in contained in Ref. [37].

(1) Void Fraction, a

The void fraction in a vapor-liquid mixture is defined by

$$a \equiv \frac{\text{volume of vapor in mixture}}{\text{total volume of liquid–vapor mixture}} \tag{4.5.1}$$

The term void is somewhat misleading, since there is no void in boiling systems; however, the vapor in a liquid-vapor mixture essentially acts as void because of its relatively low density.

The void fraction can be considered over a channel volume or reduced to a "local" value by taking the average over a channel cross section. For the void region in the differential length element shown in Fig. 4.15 the void fraction is given by

$$a = \frac{\Delta Z \iint_{A_g} dA}{\Delta Z \iint_{A} dA} = \frac{A_g}{A_\ell + A_g} \tag{4.5.2}$$

One should keep in mind that the void regions shown in Fig. 4.15 vary randomly with position and time. Consequently, A_g in Eq. (4.5.2) represent a time-averaged void flow area.

(2) Quality, x_s , x , x_e

Three qualities are used in the treatment of two-phase mixtures which can have widely varied meaning. It is essential that one be very clear as to the type of quality under consideration when using quality in two-phase flow analysis.

The first, quality is the **static quality** defined by

$$x_s \equiv \frac{\text{mass of vapor in mixture}}{\text{total mass of mixture}} \tag{4.5.3}$$

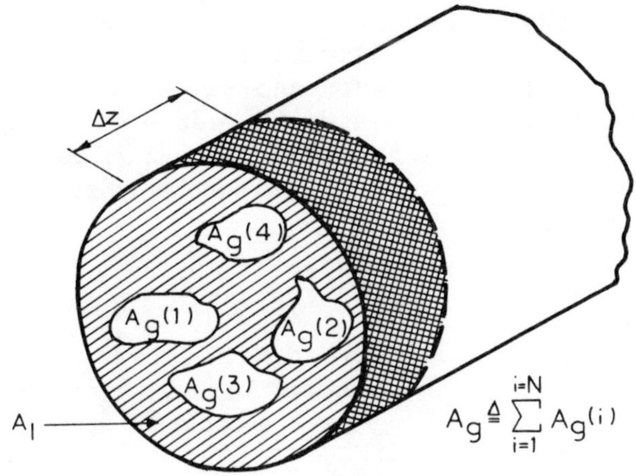

Fig. 4.15. Void distributions [37]

For the void region shown in Fig. 4.15, the static quality is given by

$$x_s = \frac{\rho_g A_g}{\rho_\ell A_\ell + \rho_g A_g} \tag{4.5.4}$$

The static quality is applicable to nonflow systems or systems in which the average velocity of both phases is the same. The later situation describes a type of flow called **homogeneous flow**.

The second quality is the **flow quality** which is defined by

$$x \equiv \frac{\text{mass flow rate of vapor}}{\text{mass flow rate of mixture}} \tag{4.5.5}$$

If u_g and u_ℓ are cross-sectional average velocities of the vapor and liquid phases, then from Fig. 4.15 the flow quality is given by

$$x = \frac{\rho_g u_g A_g}{\rho_\ell u_\ell A_\ell + \rho_g u_g A_g} \tag{4.5.6}$$

Flow qualities range from 0 to 1 and are of great interest during subcooled boiling and post-dryout boiling conditions where thermodynamic equilibrium between phases does not exist [37]. For nonflow conditions or homogeneous flow x and x_s are the same. The flow quality is usually the quality used in two-phase flow and heat transfer correlations discussed in this section and Chapter 6.

The final quality is the thermodynamic **equilibrium quality** defined by

$$x_e = \frac{h - h_f}{h_{fg}} \tag{4.5.7}$$

where h is the mixture enthalpy and h_f and h_{fg} are saturated liquid enthalpy and latent heat of vaporization, respectively. Equilibrium qualities can be negative or positive and exceed unity. This quality is the easiest to evaluate because it is determined by a simple energy balance on a heated channel. For closed channels (or open channels in which energy exchange between adjacent subchannels is neglected) the fluid enthalpy is given by

$$\dot{m}(h - h_{in}) = \int_0^z q''_w P_h \, dz \tag{4.5.8}$$

where q''_w is the wall heat flux, Btu/hr-ft^2 P_h is the channel heated perimeter, ft, and z is channel position. As shown later in this section, for bulk boiling conditions x_e and x become the same.

(3) Slip Ratio, S

As discussed in the treatment of flow quality, we can describe average velocities for the liquid and vapor phases. If the total mass flow rate for the mixture is given by $\dot{m}(lb_m/hr)$, then the vapor mass flow rate is $x\dot{m}$ and the liquid mass flow rate is $(1 - x)\dot{m}$. Applying the continuity equation to each phase at a given channel cross section gives for the liquid phase velocity

$$u_\ell = \frac{(1 - x)\dot{m}}{\rho_\ell A_\ell} \tag{4.5.9}$$

and the vapor phase velocity

$$u_g = \frac{x\dot{m}}{\rho_g A_g} \tag{4.5.10}$$

Under most circumstances the phase velocities are not the same with u_g being the larger. The difference between the vapor phase and liquid phase velocities is called the **slip velocity**. The **slip ratio**, S, is defined as the ratio of vapor phase to liquid phase velocities. Taking the ratio of Eq. (4.5.9) to Eq. (4.5.10) gives for S

$$S = \frac{u_g}{u_\ell} = \frac{x}{1 - x} \frac{A_\ell \rho_\ell}{A_g \rho_g} \tag{4.5.11}$$

(4) Basic Relations

Rearranging Eqs. (4.5.2) and (4.5.4) gives a simple relationship between void fraction and static quality

$$\alpha = \frac{1}{1 + [(1 - x_s)/x_s]\, v_\ell / v_g} \tag{4.5.12}$$

where v_ℓ and v_g are respective specific volumes of the liquid and vapor phases. For low static qualities void fractions are relatively large.

Consider a steam water mixture at 1000 psia with $x_s = 0.02$. The specific volumes are $v_\ell = 0.02159\ \mathrm{ft^3/lb_m}$ and $v_g = 0.4496\ \mathrm{ft^3/lb_m}$. Therefore

$$a = \frac{1}{1 + [(1 - 0.02)/0.02]\ 0.02159/0.44596} \tag{4.5.13}$$

$$= 0.297 \text{ or } 29.7\% \text{ void}$$

Substituting Eqs. (4.5.9) and (4.5.10) into Eq. (4.5.4) for x_s gives

$$x_s = \frac{x\dot{m}/u_g}{x\dot{m}/u_g + (1 - x)\dot{m}/u_\ell} \tag{4.5.14}$$

By rearrangement we have

$$\frac{1 - x_s}{x_s} = \frac{u_g}{u_\ell}\,\frac{1 - x}{x} \tag{4.5.15}$$

At low pressures the vapor phase normally moves faster than the liquid phase and, consequently, the static quality is smaller than the flowing quality. Experiments have shown that S decreases with increasing system pressure and volumetric flow rate and increases with increasing power density. S is also found to increase with increasing quality at high pressures but decrease with increasing quality at low pressures [1).

Substituting Eq. (4.5.15) into Eq. (4.5.12) gives the relationship between a and x including the effect of slip ratio

$$a = \frac{1}{1 + [(1 - x)/x]\ (v_\ell/v_g)S} \tag{4.5.16}$$

For a given quality, increasing slip ratios cause the void fraction to decrease. Figure 4.16 illustrates calculated void fractions as a function of flow quality and slip ratio for steam-water mixtures at 1000 psia.

Experimental slip ratios are found from void fraction measurements in two-phase flows. There is a considerable amount of data for S and a as a function of various operating parameters for flowing steam-water mixtures. Figures 4.17 and 4.18 illustrate some data for unheated channels. u_i in these illustrations is the average mixture velocity of the channel inlet and is calculated using the continuity equation at the inlet

$$u_i = \dot{m}/\rho_i A \tag{4.5.17}$$

where ρ_i is inlet mixture density and A is the channel flow cross section area.

The mixture density is found by considering Fig. 4.15 which yields

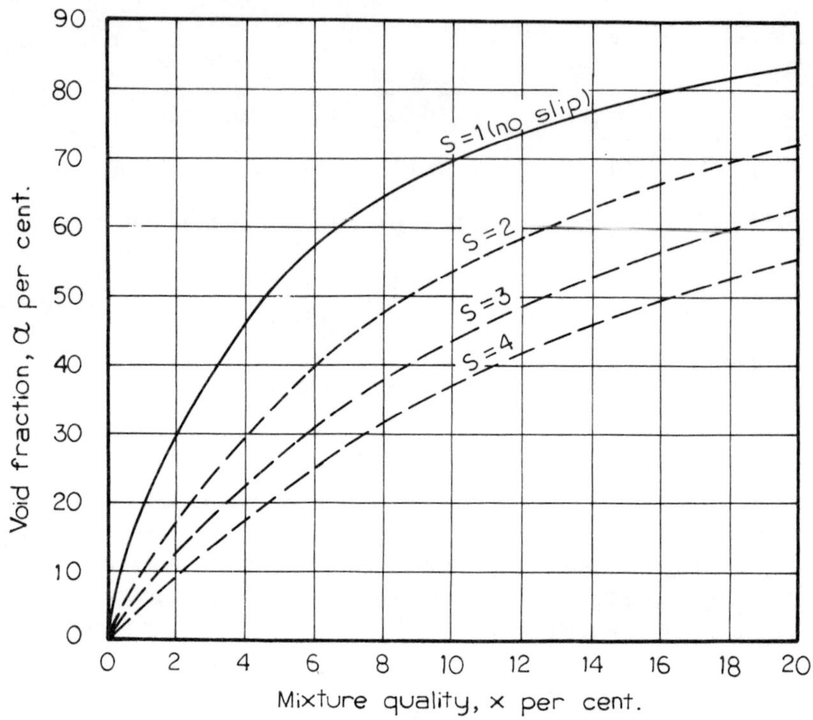

Fig. 4.16. α vs. x for water at 1,000 psia and various slip ratios

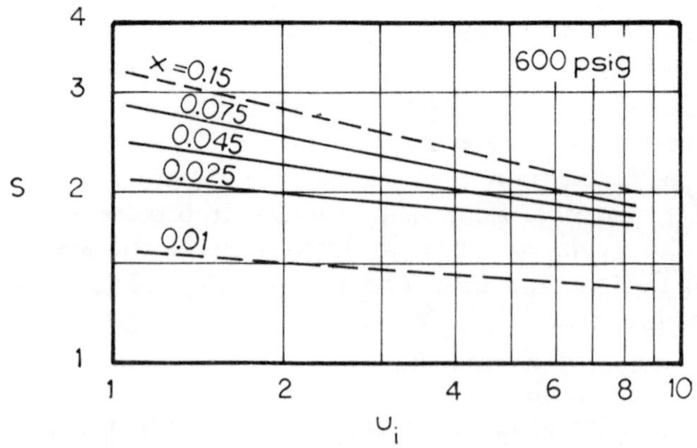

Fig. 4.17. Slip ratios as a function of u_i and x at 600 psig [1]

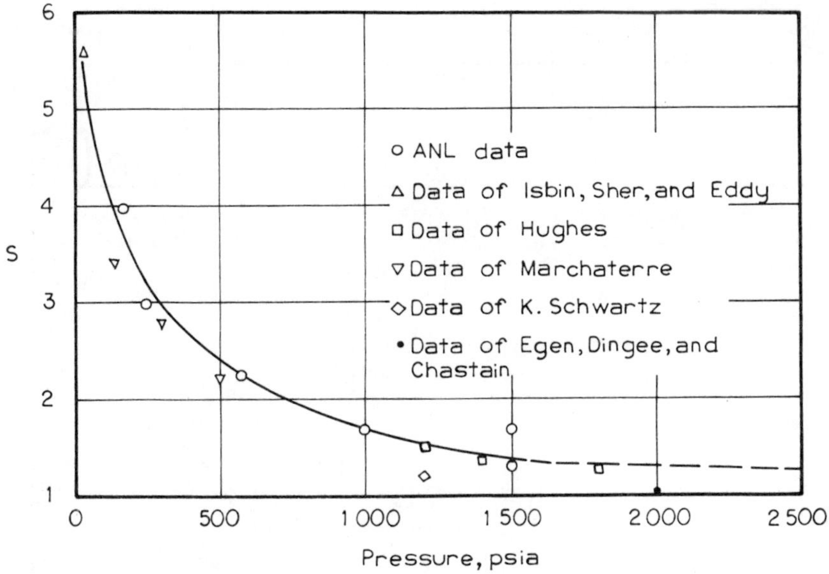

Fig. 4.18. Slip ratio as a function of pressure for $u_i \approx 2$ ft/sec and $x \approx 0.05$ [1]

$$\bar{\rho} = \frac{\Delta z \iint_{A_g} \rho_g \, dA + \Delta z \iint_{A_\ell} \rho_\ell \, dA}{\Delta z \iint_A dA} \tag{4.5.18}$$

$$= \frac{\rho_g A_g + \rho_\ell A_\ell}{A_g + A_\ell}$$

From Eq. (4.5.2), $\bar{\rho}$ becomes

$$\bar{\rho} = a\rho_g + (1 - a)\rho_\ell \tag{4.5.19}$$

Void Fraction

The distribution of void in a heated channel is shown in Fig. 4.19. For subcooled boiling the void is divided into region I in which the voidage is thin and attached to the wall and region II of lower subcooling in which bubbles detach themselves from the wall and are slowly condensed further downstream. The void fraction in region I is quite small and is usually considered negligible in analysis.

Several techniques are available for predicting the boundaries for regions I and II. The location of the beginning of the subcooled boiling region (z_n) is called the onset of nucleate boiling. This position is

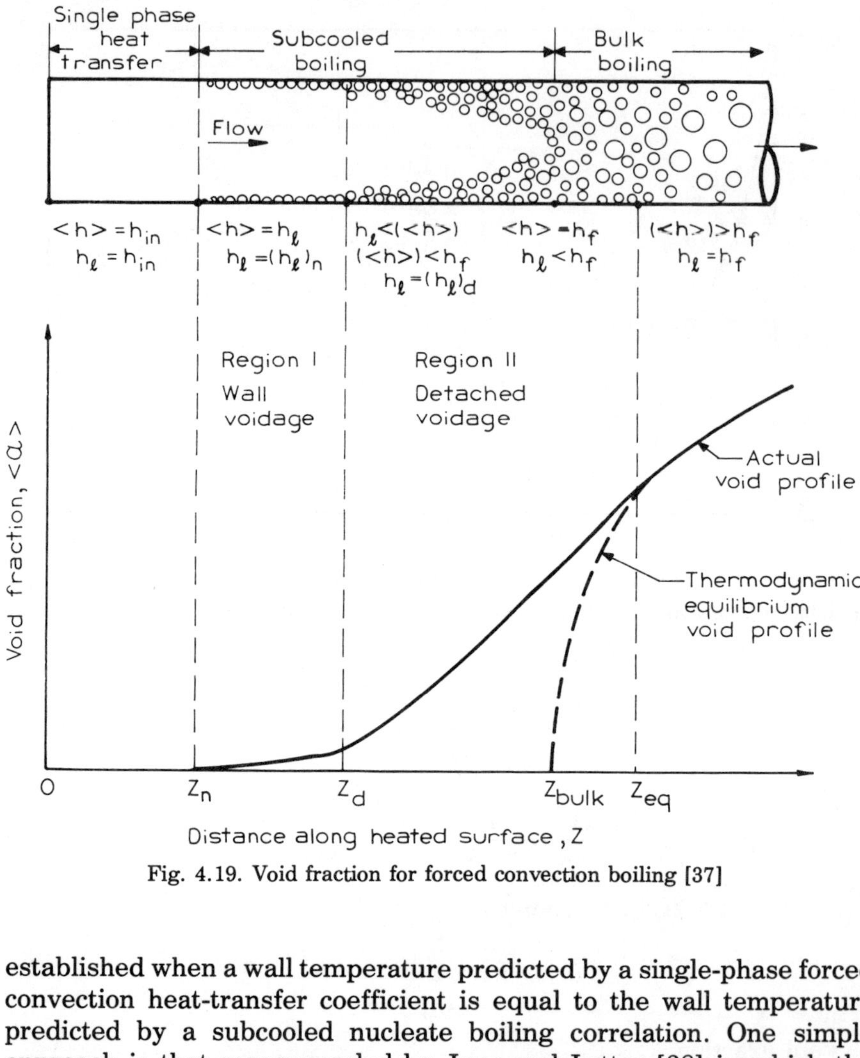

Fig. 4.19. Void fraction for forced convection boiling [37]

established when a wall temperature predicted by a single-phase forced convection heat-transfer coefficient is equal to the wall temperature predicted by a subcooled nucleate boiling correlation. One simple approach is that recommended by Jens and Lottes [38] in which the onset of nucleate boiling occurs when wall temperatures from their nucleate boiling correlation equals wall temperatures calculated by a modified Colburn equation.

$$T_w = T_b + \frac{q''_w}{0.030(k_b/D_e)\,(N_{Re})_b^{0.8}\ (N_{Pr})_b^{1/3}} \tag{4.5.20}$$

$$T_W^* = T_{Sat} + \frac{60(q''_w/10^6)^{1/4}}{e^{p/900}} \tag{4.5.21}$$

where

k_b = liquid thermal conductivity at bulk temperature, Btu/hr–ft–°F
q''_w = wall heat flux, Btu/hr–ft²
p = pressure, psia
$(N_{Re})_b$ = Reynolds number evaluated at bulk fluid temperature
$(N_{Pr})_b$ = Prandtl number evaluated at bulk fluid temperature

If $T_w^* > T_w$ there is no local boiling; when $T_w \geqslant T_w^*$ local boiling exists.

The important location in any subcooled boiling model is to be able to accurately predict the position where significant void fraction appears which is the bubble departure point z_d in Fig. 4.19. Most correlations are given in terms of a critical subcooling for the enthalpy of the liquid phase $(h_\ell)_d$. One correlation proposed by Zuber and Saha is as follows [39]:

$$\text{If} \quad N_{Pe} = \frac{GD_e(c_p)_\ell}{k_\ell} < 70{,}000,$$

$$[h_f - (h_\ell)_d] = 0.0022\,\frac{q''_w D_e (c_p)_\ell}{k_\ell} \tag{4.5.22}$$

$$\text{If} \quad N_{Pe} > 70{,}000,$$

$$[h_f - (h_\ell)_d] = 154\, q''_w/G \tag{4.5.23}$$

Equations (4.5.22) or (4.5.23) give the liquid phase enthalpy at point z_d. This is insufficient to find point z_d because the quality at that point is still unknown. An energy balance on the fluid yields the mixture enthalpy as

$$\dot{m}(h_d - h_{in}) = \int_0^{z_d} q''_w P_h \, dz \tag{4.5.24}$$

The mixture enthalpy is

$$h_d = (1 - x)(h_\ell)_d + x h_g \tag{4.5.25}$$

Techniques for calculating flowing quality are discussed in a later portion of this section. By finding the flowing quality for a known $(h_\ell)_d$ one can then evaluate the mixture enthalpy at z_d by use of Eq. (4.5.25). The position z_d is then found from the energy balance given by Eq. (4.5.24).

The transition from local boiling to bulk boiling is frequently taken as the point where the bulk fluid temperature reaches its saturation temperature. Another approach is to extrapolate the local boiling void fraction or friction multiplier to intersect the bulk boiling curves. In order to prevent discontinuities in axial pressure gradients, Castellana and Bonilla [40] recommend determining the two curves and extrapolating them to a tangential transition.

Many correlations are available for predicting α in the subcooled regions [40]. One correlation by Thom, et al [41] for the mixture density in the detached region (region II) is

$$\rho = \rho_f - \frac{\gamma x_m'(\rho_f - \rho_g)}{1 + x_m(\gamma - 1)} \tag{4.5.26}$$

where

γ = 10 at 1000 psia and 16 at 750 psia

x_m' = reduced mass dryness fraction

The reduced mass dryness fraction is given by

$$x_m' = \frac{h - h_d}{h_g - h_d} \tag{4.5.27}$$

where

h = mixture local enthalpy

h_d = bulk fluid enthalpy at the point of bubble detachment

= $h_f\,[1 - 0.15\, q_w''/G]$

h_f, h_g = saturated liquid and vapor enthalpies

G = mass, velocity,

Tong [42] extended this correlation to 2000 psia by modifying γ

$$\gamma = \exp\left\{4.216\,[(y - 8.353)^2/(8.353)^2 - 1]\right\}^{1/2} \tag{4.5.28}$$

where

$y = \ln\,(p/3206)$; p being pressure in psia.

The treatment of void fraction in the bulk boiling region is more complicated because of its dependence on slip ratio and flow pattern. For high pressure and either very low or very high void fractions, a homogeneous model (S=1) gives satisfactory results for predicting void fractions [34]. Numerous correlations exist for predicting the void fraction for various flow regimes [40]. For bubbly flow, Bankoff proposes [40]

$$a = \frac{K}{1 + [(1 - x)/x]\; v_\ell / v_g} \tag{4.5.29}$$

From experimental data K was found to be

$$K = 0.71 + 0.0001p \tag{4.5.30}$$

where p is in psia.

A widely accepted generalized void-quality model is the drift flux model proposed by Zuber and Findlay [43]. This model proposes the average void fraction is given by

$$a = \frac{1}{C_0[1 + (1 - x/x)(v_\ell / v_g)] + (V_{gj} / x v_g G)} \tag{4.5.31}$$

C_o is called the concentration parameter and a suggested expression by Dix [44] is

$$C_0 = \beta[1 + (1/\beta - 1)^b] \tag{4.5.32}$$

where

$$b = (v_\ell / v_g)^{0.1}$$

and

$$\beta = \frac{1}{1 + [(1 - x)/x]\, v_\ell / v_g} \tag{4.5.33}$$

V_{gj} is the drift velocity which is a void fraction weighted average difference between phase velocities. One expression for this parameter is

$$V_{gj} = 2.9 \left[\frac{(\rho_\ell - \rho_g)\,\sigma g g_c}{\rho_\ell^2}\right]^{0.25} \tag{4.5.34}$$

where σ is the fluid surface tension, lb_f/ft.

Equations (4.5.31) through (4.5.34) are used to predict void fractions in both subcooled and bulk boiling regions. For channel positions which exceed the point where the liquid enthalpy becomes saturated liquid (position z_{eq} on Fig. 4.19), the equilibrium quality (x_e) is used for flow quality in these expressions. For region II on Fig. 4.19 Levy [43] proposes the flowing quality be evaluated from

$$x = x_e - (x_e)_d \exp\left(\frac{x_e}{(x_e)_d} - 1\right) \tag{4.5.35}$$

where x_e is the equilibrium quality at the channel position under consideration given by

$$x_e = \frac{h - h_f}{h_{fg}} \tag{4.5.36}$$

and $(x_e)_d$ is the equilibrium quality at the position of bubble departure z_d given by

$$(x_e)_d = \frac{(h_\ell)_d - h_f}{h_{fg}} \tag{4.5.37}$$

Note, in Eq. (4.5.37) the enthalpy at the point of bubble departure is the liquid phase enthalpy which is determined from Eqs. (4.5.22) or (4.5.23).

Pressure Loss in Two-Phase Flow

In the analysis of boiling channels it is usually necessary to divide channels into the following regions shown in Fig. 4.20: (i) non-boiling region, L', (ii) subcooled boiling region, $L'' - L'$, and (iii) bulk boiling region, $L - L''$. Indentification of the point of transition from region to region is not precise and methods for determining these locations are discussed in the preceding section.

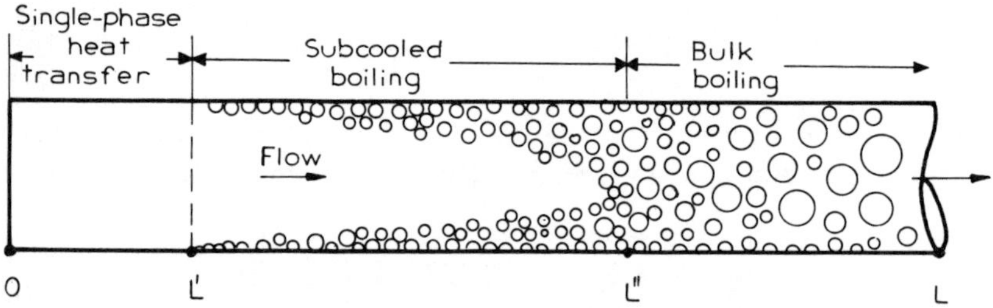

Fig. 4.20. Flow regions in two-phase flow

Pressure loss in the non-boiling region is treated in Sec. 4.2. In boiling regions the mixture density continuously decreases along the channel which results in a significant increase in fluid momentum. For two-phase flow the total pressure change is given by the sum of three components

$$\Delta p_{total} = \Delta p_{friction} + \Delta p_{acceleration} + \Delta p_{elevation} \tag{4.5.38}$$

The elevation pressure difference is given by

$$\Delta p_{elev} = \frac{g}{g_c} K \int_{L''}^{L} [\rho_\ell (1 - a) + \rho_g a] \, dz \tag{4.5.39}$$

One can evaluate the integral over the limits from L'' (corresponding to saturated liquid) to L by numerical or graphical techniques.

For a liquid-phase channel inlet, Martinelli and Nelson assumed the pressure loss due to momentum changes is given by [30]

$$\Delta p_{acc} = (\dot{m}_\ell u_\ell + \dot{m}_g u_g - \dot{m}_T u_0)/(g_c A_c) \tag{4.5.40}$$

where

u_ℓ, u_g = saturated liquid and vapor velocities leaving channel
u_o = liquid velocity at channel inlet

$\dot{m}_\ell, \dot{m}_g$ = mass flow rates of saturated liquid and vapor at channel outlet

$\dot{m}_T$ = mass flow rate at channel inlet

A_c = channel cross section

For saturated liquid at the channel inlet, Eq. (4.5.40) is written

$$\Delta p_{acc} = \frac{G^2}{g_c}\left[\frac{(1-x)^2}{1-\alpha} + \frac{x^2}{\alpha}\left(\frac{v_g}{v_\ell}\right) - 1\right] v_\ell = r\,\frac{G^2}{g_c} \qquad (4.5.41)$$

where quality and void fraction are at the channel exit. For steam-water mixtures Martinelli and Nelson [30] prepared a plot of r as a function of exit quality and pressure. However, one is probably advised to use more recent correlations for α and evaluate pressure loss from Eq. (4.5.41). It should be noted that Eqs. (4.5.40) and (4.5.41) depend only upon channel inlet and exit conditions and, consequently, are independent of the method of heating (sinusoidal, etc.).

Both the elevation and acceleration pressure differences are strongly dependent upon void fraction which in turn depends upon slip ratio. When calculating these components for vertical upflow, errors due to incorrect void fractions or slip ratios tend to compensate for each other.

(1) Local Boiling Pressure Loss

Several empirical equations are proposed for calculating the combined friction and acceleration losses or friction losses alone.

For pressures from 45 to 100 psia, Reynolds [46] recommends for the combined friction and acceleration loss

$$\Delta p_{LB} = \left(\frac{dp}{dL}\right)_o \frac{G D_e c_p (T_{Sat} - T_b)_o}{4 a q''_w} \sinh\left[\frac{4 a q''_w (L'' - L')}{G D_e c_p (T_{Sat} - T_b)_o}\right] \qquad (4.5.42)$$

where $a = 4.6 \times 10^{-6}\, q''_w + 1.2$; $L'' - L'$ is the length over which local boiling occurs; and the subscript o refers to the location where local boiling starts. The wall heat flux is assumed uniform over the channel length. The range of parameters used in the study is as follows: $1.3 \times 10^5 < q''_w < 3.04 \times 10^5$ Btu/hr–ft², $10^6 < G < 2.3 \times 10^6$ lb_m/ft²–hr, and D = 3/8 inch.

Owens and Schrock [47] studied local boiling in tubes and obtained for the local boiling pressure gradient

$$\left(\frac{dp}{dL}\right)_{LB} = \left(\frac{dp}{dL}\right)_o \left\{0.97 + 0.028 \exp[6.13(z - L')/(L'' - L')]\right\} \qquad (4.5.43)$$

where $z - L'$ is the channel length from inception of local boiling. The tests were performed in two tubes with the following range of parameters:

Tubes: 0.1181″ dia. x 15″ and 0.1824″ dia. x 16″

Pressures: 50 to 400 psia
Mass Velocities: 1.1×10^6 to 3.9×10^6 lb_m/ft^2-hr
Heat Flux: 0.27×10^6 to 1.3×10^6 Btu/hr-ft^2

For higher pressures, Mendler, et al [48] recommend for the friction factor

$$\frac{f_{LB}}{f_{iso}} = (1 - 0.0025\Delta T^*)[1 + 0.76(10^6/G)^{2/3}\psi] \qquad (4.5.44)$$

where

$$\psi = 1 - \frac{\Delta T^*}{\Delta T_f}$$

$$\Delta T^* = T_{Sat} + \frac{60(q''_w/10^6)^{1/4}}{e^{-p/900}} - T_b$$

$$\Delta T = \frac{q''_w}{0.030(k_b/D_e)(N_{Re})_b^{0.8}(N_{Pr})_b^{1/3}}$$

(2) Bulk Boiling Friction Loss

Two general approaches are employed in predicting two-phase pressure loss. One approach is to correlate pressure losses in terms of dimensionless groups as is done in single-phase flow. The other approach analyzes the hydrodynamics of individual flow patterns. The former approach suffers because fundamental mechanisms differ over the range of flow conditions, while the latter approach requires a knowledge of the flow pattern for its use.

We will examine a few of the former approaches because the latter approach has not proven satisfactory in nuclear reactor analysis because of uncertainty about the types and locations of flow patterns in heated channels.

Pressure losses due to two-phase friction are usually calculated by applying a multiplier to the pressure gradient evaluated for the liquid phase flowing at the total mixture flow rate. The multiplier is defined as

$$\Phi^2_{\ell o} \equiv \left(\frac{dp}{dz}\right)_{TPF} \Big/ \left(\frac{dp}{dz}\right)_{\ell o} \qquad (4.5.45)$$

where
$(dp/dz)_{TPF}$ = two-phase friction pressure gradient
$(dp/dz)_{\ell o}$ = liquid-phase pressure gradient at total mass flow rate
The liquid-phase pressure gradient is evaluated using saturated liquid properties and a standard expression for the friction factor.

Numerous expressions, both empirical and theoretical, are available for predicting the two phase friction multiplier [40]. The empirical relations are usually obtained from closed channel data and correlated in terms of pressure, quality, and void fraction. Some correlations have additional refinements by incorporating the effects of mass velocity, geometry, and heat flux.

The most widely known correlations for predicting two-phase flow pressure loss are those of Lockhart-Martinelli [49] for adiabatic flows and Martinelli and Nelson [30] for heated flows. These correlations assume that two-phase flows satisfy the following conditions: (1) the static pressure drop for the liquid phase equals the static pressure drop for the gaseous phase regardless of flow pattern, as long as an appreciable radial static pressure difference does not exist; and (2) the volume occupied by liquid plus the volume occupied by gas at any instant equals the total pipe volume. These postulates infer that the flow pattern does not change along pipe length and, consequently, slug flow is eliminated from consideration.

In the analysis of experimental data, flow through horizontal pipes without heat addition is separated into four classifications. The classifications do not correspond directly to any particular flow regime but depend upon whether the flow of either phase separately in the channel is viscous (laminar) or turbulent. Based upon this method of classification, four flow types are possible.

1) Flow of both liquid and gas is turbulent (turbulent-turbulent flow).

2) Flow of liquid is viscous and flow of gas is turbulent (viscous-turbulent flow).

3) Flow of liquid is turbulent and flow of gas is viscous (turbulent-viscous flow).

4) Flow of both liquid and gas is viscous (viscous-viscous flow).

The dividing line between the turbulent and viscous regimes is set as follows:

Table 4.4
Two-phase flow classifications

Flow Mechanism	Symbol	N_{Re} (liquid)	N_{Re} (vapor)
Turbulent-Turbulent	t-t	2000	2000
Viscous-Turbulent	v-t	1000	2000
Turbulent-Viscous	t-v	2000	1000
Viscous-Viscous	v-v	1000	1000

Note: The region $1000 < N_{Re} < 2000$ is a transition region.

In order to calculate pressure loss in a two-phase flowing mixture, a quantity Φ^2 is defined which related two-phase pressure loss to the pressure loss of either component flowing alone at its mass flow rate in the channel. The definitions are

$$\Phi_\ell^2 \equiv (\Delta p/\Delta L)_{TPF} \,/\, (\Delta p/\Delta L)_\ell \tag{4.5.46}$$

and

$$\Phi_g^2 \equiv (\Delta p/\Delta L)_{TPF} \,/\, (\Delta p/\Delta L)_g \tag{4.5.47}$$

where

$(\Delta p/\Delta L)_{TPF}$ = two-phase friction pressure gradient
$(\Delta p/\Delta L)_\ell$ = pressure gradient due to flow of liquid alone in the channel
$(\Delta p/\Delta L)_g$ = pressure gradient due to flow of vapor alone in the channel

On the basis of a simplified analysis, it is postulated that Φ_g^2 and Φ_ℓ^2 are functions of a single variable X^2 which is defined by

$$X^2 \equiv \frac{(\Delta p/\Delta L)_\ell}{(\Delta p/\Delta L)_g} \tag{4.5.48}$$

This ratio is calculated from mass flow rates of the two components and fluid properties. Substituting expressions for the friction pressure loss for each phase into the equation for X^2, we have

$$X^2 = \frac{(N_{Re})_g^m}{(N_{Re})_\ell^n} \frac{C_\ell}{C_g} \left(\frac{\dot{m}_\ell}{\dot{m}_g}\right)^2 \frac{\rho_g}{\rho_\ell} \tag{4.5.49}$$

The exponents m and n and coefficients C_ℓ and C_g are tabulated in Table 4.5. These values are determined from appropriate single-phase expressions for laminar and turbulent flow pressure loss.

Table 4.5
Value of exponents m and n and constants C_ℓ and C_g for various flow types [49]

	t-t	v-t	t-v	v-v
n	0.2	1	0.2	1.0
m	0.2	0.2	1.0	1.0
C_ℓ	0.184*	64	0.184	64
C_g	0.184*	0.184*	64	64

*For smooth pipes.

A good smooth curve correlation is obtained for values of Φ_ℓ^2 and Φ_g^2 as a function of X. Figure 4.21 illustrates values for all flow classifications. Φ is used in the plots because the lines are almost straight and values of the multiplier are not excessively large. It should be kept in mind that the majority of the data used in developing Fig. 4.21 is taken at pressures at or near atmospheric pressure. It turns out that experiments show a more complex pressure dependence than given in X. This feature is discussed later in this section.

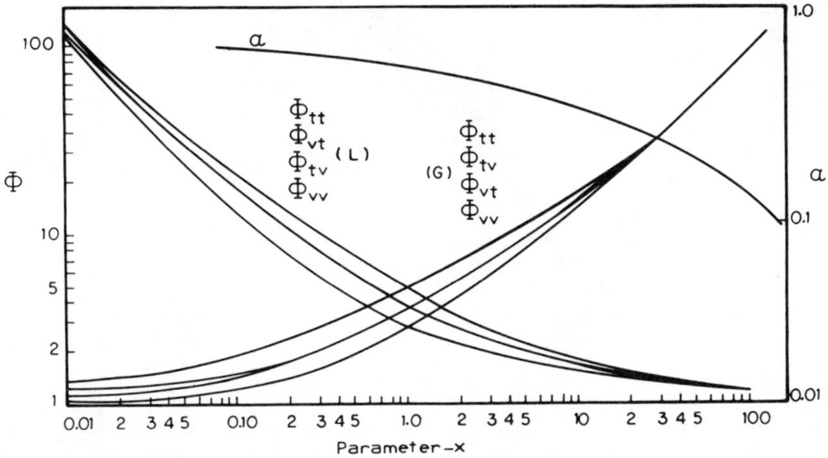

Fig. 4.21. Two-phase flow multiplier and vapor volume fraction for isothermal flow [49]

In addition to the two-phase pressure loss multiplier being a function of X, it is found that void fraction is also a function of X which is shown on Fig. 4.21.

In the treatment of heated horizontal channels, it is assumed the total pressure loss is given by the sum of a pressure loss due to friction and a pressure loss due to increased fluid momentum as vaporization occurs. In order to develop a method for predicting friction pressure losses in boiling water, Martinelli and Nelson used isothermal flow pressure loss data for calculating local friction pressure gradients. It is assumed the only flow classification of practical interest is turbulent-turbulent and, consequently,

$$\left(\frac{dp}{dL}\right)_{TPF} = \left(\frac{dp}{dL}\right)_\ell \Phi_{\ell tt}^2 \tag{4.5.50}$$

The multiplier, $\Phi_{\ell tt}^2$ is correlated in terms of a slightly different parameter than given in Eq. (4.5.49). The correlating parameter is

$$X_{tt} = \left(\frac{\rho_g}{\rho_\ell}\right)^{0.571}\left(\frac{\mu_\ell}{\mu_g}\right)^{0.141}\left(\frac{1}{x} - 1\right) \tag{4.5.51}$$

where x is quality.

In applying Fig. 4.21 to high pressure steam-water mixtures, it is found that pressure losses are over-predicted. Consequently, an empirical dependence of $\Phi_{\ell tt}$ and $(1 - a)$ on pressure is obtained. Figure 4.22 illustrates a plot of $\Phi_{\ell tt}$ and $(1 - a)$ as a function of $\sqrt{X_{tt}}$. Figure 4.22 is used for computing local two-phase friction pressure gradients by use of Eqs. (4.5.50) and (4.5.51).

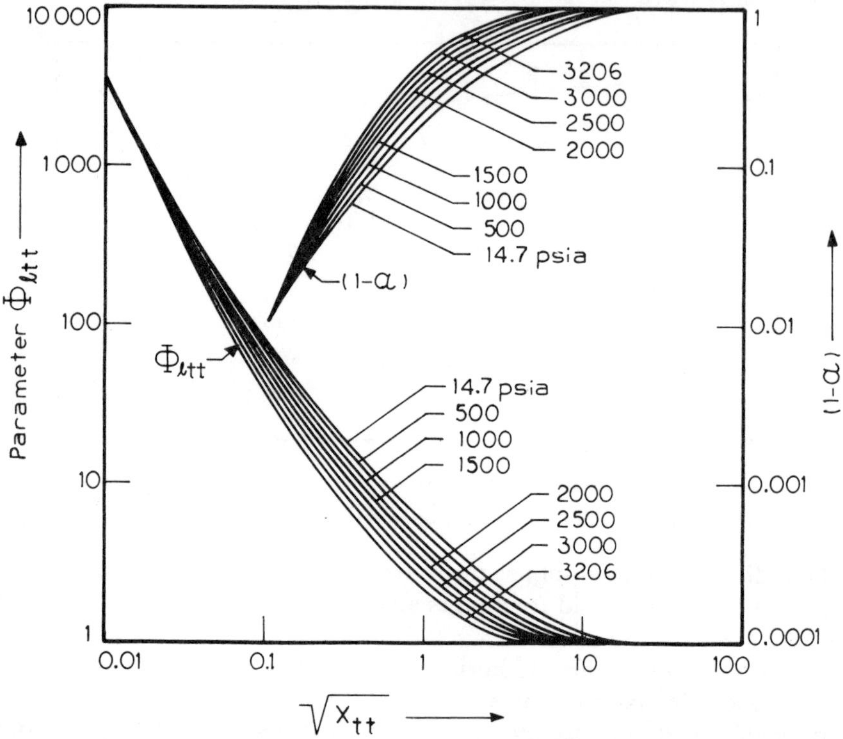

Fig. 4.22. Plot of parameters $\Phi_{\ell tt}$ and $1-a$ versus $\sqrt{X_{tt}}$ at various pressures for steam-water mixtures [30]

The preceding correlations shown in Figs. 4.21 and 4.22 are for flow of two-phase mixtures without heat addition. Consequently, the mixture quality or void fraction remains constant along the channel length. With heat addition, two-phase pressure loss correlations are modified to take into account the change in fluid quality with channel length and to account for acceleration of the fluid because of continually increasing void fraction and, consequently, increase in mixture velocity.

Martinelli and Nelson obtained correlations for two-phase pressure loss with heat addition by making use of experimental data for pressure losses without heat addition. Figure 4.23 illustrates the ratio of two-phase friction pressure gradient to that of pure liquid phase flowing at total mass flow rate, $(dp/dL)_{TPF}/(dp/dL)_0$, as a function of local pressure and quality for steam-water mixtures. The relationship between liquid phase pressure gradients at liquid flow rate and total flow rate is

$$\left(\frac{dp}{dL}\right)_\ell = \left(\frac{dp}{dL}\right)_0 (1 - x)^{1.75} \tag{4.5.52}$$

The ratio of two-phase friction pressure loss to single-phase pressure loss, $\Delta p_{TPF}/\Delta p_0$, is computed by integrating one of the curves in Fig. 4.23 at a constant pressure and heat flux from saturated liquid up to any quality x. The results of the integrations are shown in Fig. 4.24 which illustrates two-phase friction pressure losses as a function of exit quality and system pressure. The friction pressure loss for steam-water mixtures at any pressure and exit quality, where quality varies linearly in the channel, is obtained by multiplying the friction loss for total flow rate as liquid by the ratio obtained from the appropriate curve of Fig. 4.24. Keep in mind this plot is only useful for uniform heating.

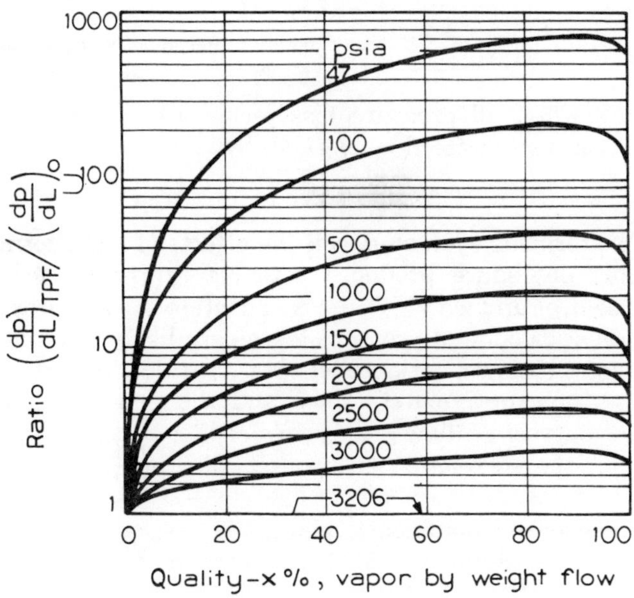

ig. 4.23. Ratio of local two-phase pressure gradient to pressure gradient for 100 percent liquid flow as a function of quality and pressure [30]

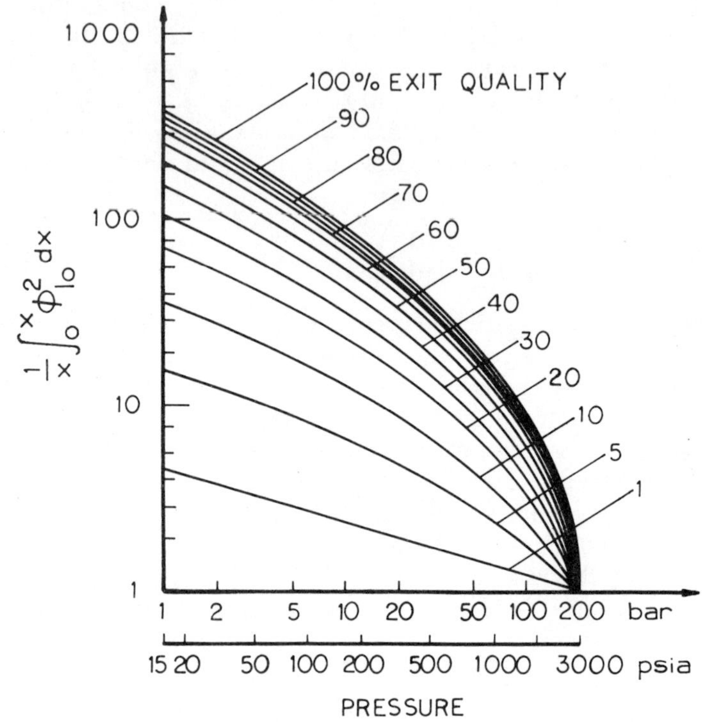

Fig. 4.24. Ratio Δp_{TPF} to Δp_o as a function of exit quality and pressure [30]

In general, when all pressure loss terms are taken into account, pressure loss in two-phase flow is given by

$$\Delta p_{TP} = \Delta p_{TPF} + rG^2/g_c + \Delta p_{elev} \tag{4.5.53}$$

Each of these terms is evaluated as if the effects of other terms are not present. This procedure produces satisfactory results for flows at elevated pressures and with high mass velocities. At low pressures and mass velocities, calculated pressure losses are likely to be in error.

For horizontal flows, calculated pressure losses are generally within $\pm 30\%$ of experimental values and the general shape of the pressure loss curve with flow rate is correct. For vertical channels with heat addition, the correlation is less accurate but improves at high pressures.

For quality varying nonlinearly along the channel, the friction loss term is given by a summation of local pressure gradients where the number of terms is based on the accuracy desired and on the engineer's confidence in the accuracy and/or the applicability of the correlation.

A basic premise of the Martinelli-Nelson correlation is that flow is steady and there are no radial pressure gradients. If this is satisfied, then the correlation is apparently independent of flow pattern. It is noted, however, that these conditions strictly limit the applicability to annular or semi-annular flow patterns, although the correlation is frequently applied to other patterns. Griffith [50] recommends the correlation for annular, wave, and dispersed flow patterns in horizontal flow and for annular and mist flow patterns in vertical flow.

A more general correlation developed by Baroczy [51] accounts for effects of mass velocity and is applied to more than steam-water mixtures. Pressure is eliminated as a parameter by correlating the two-phase multiplier as a function of quality and a property index defined by $(\mu_f/\mu_g)^{0.2}/(\rho_f/\rho_g)$. Figure 4.25 illustrates the multiplier for a mass velocity of 1×10^6 $lb_m/hr\text{-}ft^2$. For other mass velocities, Fig. 4.26 shows a correction factor which is multiplied times $\Phi^2_{\ell o}$ from Fig. 4.25.

The Baroczy correlation is developed from adiabatic flows in circular tubes and rectangular channels. Good agreement is found for heated flows in channels of the same geometry. Pressure loss measurements of adiabatic steam-water flow at 70 atmospheres in a 6 x 6 rod bundle show reasonable agreement with the Baroczy correlation [26]. However, measurements by Tarasova, et al. [52] indicate the two-phase friction multiplier has a heat flux dependence. Consequently, one can probably conclude that there is uncertainty about two-phase pressure loss calculations on reactor rod bundles.

The bulk boiling friction loss is obtained by integrating Eq. (4.5.45) over the bulk boiling channel length.

$$\Delta p_{TPF} = \left(\frac{dp}{dz}\right)_{\ell o} \int_{L''}^{L} \Phi^2_{\ell o}(x)\, dz \qquad (4.5.54)$$

Since $\Phi^2_{\ell o}$ is dependent upon quality it is necessary to determine the mixture quality as a function of axial position. For uniform heating Eq. (4.5.54) is rearranged to yield

$$\Delta p_{TPF} = \Delta p_{\ell o} \frac{1}{x_e} \int_{0}^{x_e} \Phi^2_{\ell o}(x)\, dx \qquad (4.5.55)$$

where x_e is the exit quality. For qualities varying nonlinearly along the channel, Eq. (4.5.54) is solved by performing summations of local pressure gradients along the channel length.

(3) Two-Phase Pressure Losses for Changes in Cross Section

The approach used to evaluate pressure changes from contractions, expansions, and spacer grids for two-phase flow is similar to that employed for single-phase flow. The difference is that a multiplier is employed when evaluating friction loss.

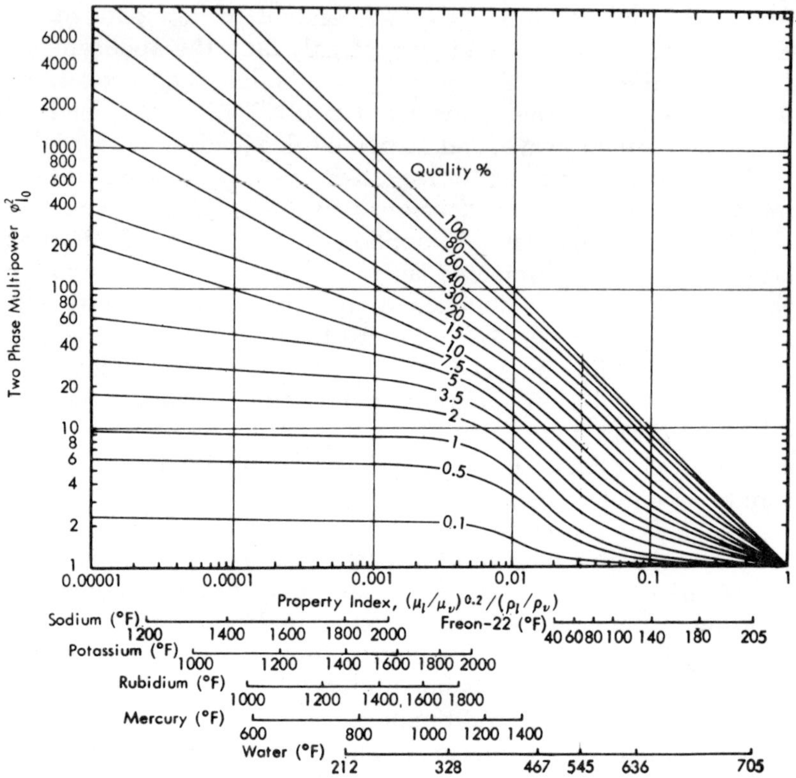

Fig. 4.25. Two-phase friction pressure drop correlation [51]

For sudden contractions, the pressure loss is given by [51]

$$\Delta p_c = \frac{G^2}{2g_c\bar{\rho}}(1 - \sigma^2) + K_c I \frac{G^2}{2g_c\rho_f} \tag{4.5.56}$$

where

G = mass velocity in smaller cross section

$\bar{\rho}$ = $\rho_f(1 - a) + \rho_g a$

σ = ratio of smaller to larger cross section

K_c = contraction pressure-loss coefficient for single-phase flow

I = two-phase friction multiplier

The exit pressure rise for sudden expansions is given by [51]

$$\Delta p_e = \frac{G^2}{2g_c\bar{\rho}}(1 - \sigma^2) - K_e I \frac{G^2}{2g_c\rho_f} \tag{4.5.57}$$

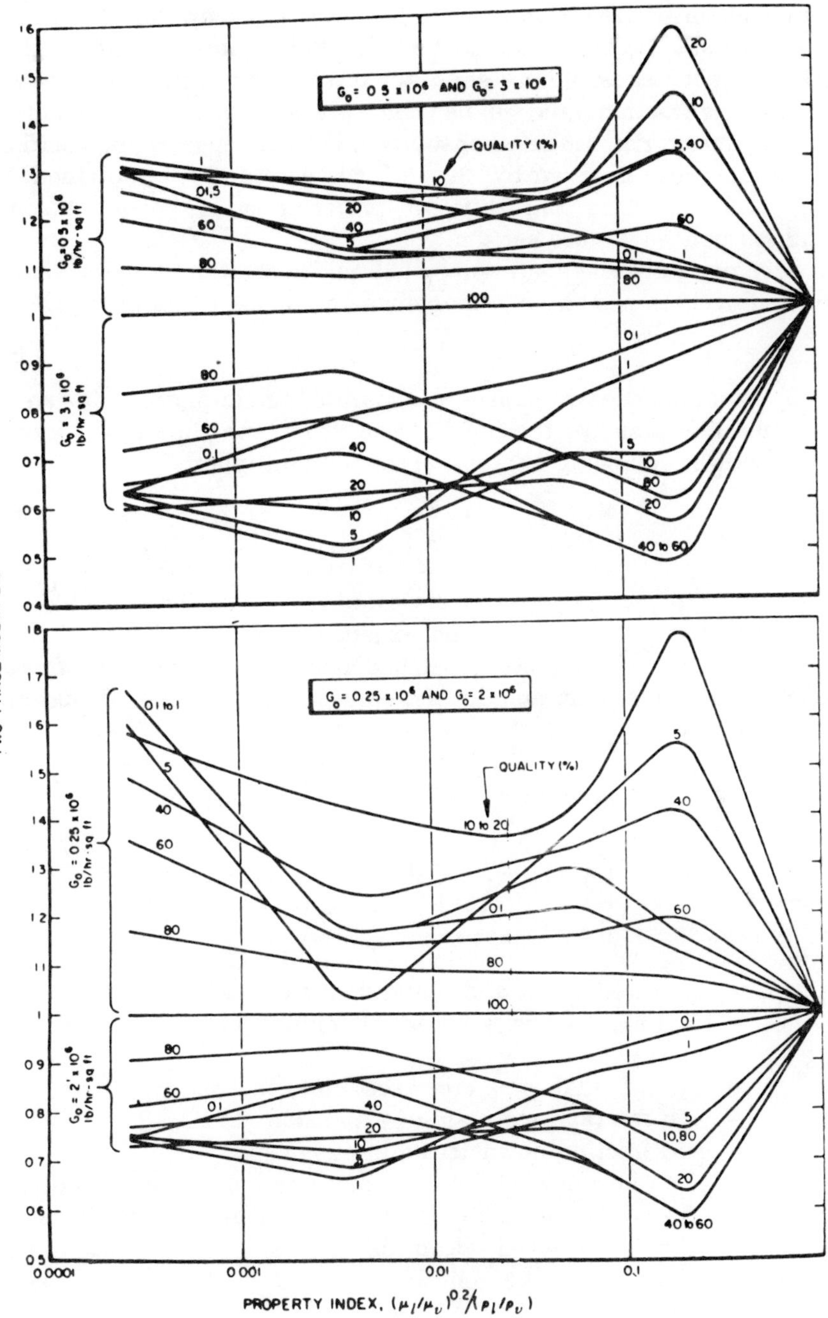

Fig. 4.26. Mass velocity correction vs property index [51]

where all terms are the same as in Eq. (4.5.56) except for K_e which is the expansion pressure-loss coefficient. Both equations imply that void fraction is the same before and after the change in flow cross section.

The contraction and expansion pressure-loss coefficients are obtained from Fig. 4.5 or other sources. For grid spacers the coefficient for friction loss is given by $K = C_v \epsilon^2$ where C_v is obtained from Fig. 4.9 and ϵ is the ratio of projected grid cross section to undisturbed flow cross section.

Several approaches are proposed for evaluating I. The simplest procedure is to assume a homogeneous model and I is then given by

$$I = \frac{\rho_f}{\bar{\rho}} \tag{4.5.58}$$

By assuming the two phases as separate, single-phase fluids and the two-phase friction loss the sum of losses from both phases, Baroczy [51] obtained the following expression for the multiplier:

$$I = \frac{\rho_f}{\rho_g}\frac{x^2}{a} + \frac{(1-x)^2}{1-a} \tag{4.5.59}$$

Hussain and Weisman's [53] review of two-phase data concluded the pressure rise, Δp_e , across an expansion can be computed from a one-dimensional momentum balance based on actual void fractions encountered upstream and downstream the area change. By assuming negligible friction

$$\Delta p_e = \frac{G_T^2 \sigma^2}{\rho_f g_c}\left\{\left[\frac{\rho_f}{\rho_g} x^2 \left(\frac{1}{a_1 \sigma} - \frac{1}{a_2}\right)\right] + \left[(1-x)^2\left(\frac{1}{\sigma(1-a_1)} - \frac{1}{(1-a_2)}\right)\right]\right\} \tag{4.5.60}$$

where

ρ_g ρ_f = saturated vapor and liquid densities, respectively
G_T = total mass velocity based on smaller pipe area, $lb_m/hr\text{-}ft^2$
x = quality
σ = ratio of small pipe cross section to that of large pipe
a_1 = void fraction upstream of expansion
a_2 = void fraction downstream of expansion

Behavior at an abrupt contraction is more complex due to substantial mixing at the vena contracta. However, Hussain and Weisman [53] conclude that pressured drop is closely approximated by a homogeneous model. The contraction pressure drop, Δp_c , is thus approximated by

$$\Delta p_c = \frac{G^2}{2\sigma^2 g_c}\left[\left(\frac{1}{C} - 1\right)^2 + (1-\sigma)^2\right]\left(\frac{x}{\rho_g} + \frac{1-x}{\rho_f}\right) \tag{4.5.61}$$

where G is the total mass velocity based on the large pipe cross section and C is the vena contracta area divided by the smaller size pipe area. Combination of Eqs. (4.5.60) and (4.5.61) yields

$$\Delta p_{loss} = \frac{G^2}{g_c \rho_f}\left\{\frac{\rho_f}{2}\left(\frac{x}{\rho_g} + \frac{1-x}{\rho_f}\right)\left(\frac{1}{\sigma^2}\right)\left[\left(\frac{1}{C} - 1\right)^2 + (1-\sigma)^2\right]\right.$$
$$\left. - \left[\frac{\rho_f}{\rho_g} x^2 \left(\frac{1}{\alpha_1 \sigma} - \frac{1}{\alpha_2}\right)\right] - \left[(1-x)^2\left(\frac{1}{\sigma(1-\alpha_1)} - \frac{1}{(1-\alpha_2)}\right)\right]\right\} \quad (4.5.62)$$

with G based on the large pipe flow area and σ the same for both ends.

4.6 Flow Instability

When a heated fluid undergoes phase changes, the large volume changes that occur in a nonuniform manner can lead to flow instabilities. The term flow instability refers to the condition where a small disturbance produces large changes or oscillations in the coupled system of flow rate, pressure gradient, and void fraction. Flow instabilities can lead to design difficulties in nuclear reactors in three major areas. First, mechanical forces that result from oscillations in flow rate and coolant pressure cause component vibration which eventually leads to fatigue type failures; second, oscillations in coolant flow (also the moderator) cause control problems due to rapidly varying properties of the reactor; and finally, oscillations in flow dramatically reduce the ability of the coolant to remove heat from the fuel. The critical heat flux is reduced by as much as 40% by flow oscillations [54].

Flow instabilities are classified into two broad general categories—static and dynamic.

A static instability arises when a small perturbation from an original steady state flow leads to a new steady state condition which is not in the vicinity of the original state. Static instabilities are analyzed using steady state laws and the threshold for the instability is predicted from these laws. These instabilities lead to either a new steady state condition or a periodic behavior.

Dynamic instabilities exist when inertia and feedback effects are part of the process governing the flow. These instabilities are analyzed using dynamic (unsteady) forms of the conservation equations and applying servo-mechanism theory to explain the phenomena.

Numerous forms of instabilities are known to exist and an excellent review paper is available by Bouré, et al. [55]. We will discuss a few instabilities of interest in reactor design.

Static

- Flow excursion (Ledinegg) instability
- Flow pattern transition instability

Vapor burst (chugging) instability

Dynamic

Density wave oscillations

Pressure drop oscillations

Flow pattern induced instability

Departure from nucleate boiling (DNB) or dryout is also considered a hydrodynamic instability which is of vital interest in reactor design. This topic is covered in detail in Chapter 6.

Flow Excursion

The simplest hydrodynamic instability is the excursive instability in which flow rate undergoes a sudden change (usually to a lower flow rate) of an excursive nature. Excursive flow instabilities can occur when system pressure drops increase with decreasing flow rate. This phenomenon can occur in steam generators or boiling reactors. Boiling increases channel pressure losses and, therefore, for a small decrease in flow rate with a constant heat input, further boiling occurs with a further increase in pressure loss. As a consequence, the flow rate continues to decrease as the pressure loss increases until a region of stable operation or burnout occurs.

Excursive flow instabilities depend upon derivatives of the system static pressure loss (Δp_{system}) and the externally applied pressure difference (Δp_{ext}) with respect to system mass flow rate $(\dot{m})$. The criteria for a stable system is

$$\frac{\partial(\Delta p_{system})}{\partial \dot{m}} > \frac{\partial(\Delta p_{ext})}{\partial \dot{m}} \tag{4.6.1}$$

This criteria is best explained by examining Fig. 4.27 which illustrates a system pressure loss-mass flow curve and four examples of external pressure-flow rate curves.

Case 1 illustrates a situation which would exist if a positive displacement pump is used to provide flow. For the pump

$$\frac{\partial(\Delta p_{ext})}{\partial \dot{m}} = -\infty$$

and, consequently, Eq. (4.6.1) is satisfied and the mass flow rate always stays at position 1.

For Case 2 the external applied pressure is always constant which is similar to what exists for flow through many parallel channels. In this example the flow at point 1 does not satisfy Eq. (4.6.1) and flow is unstable. A slight increase in flow at point 1 causes flow to excursively increase to point 3. As a converse, a slight decrease in flow causes flow to excursively fall to point 2. The decreased flow rate at point 2 could

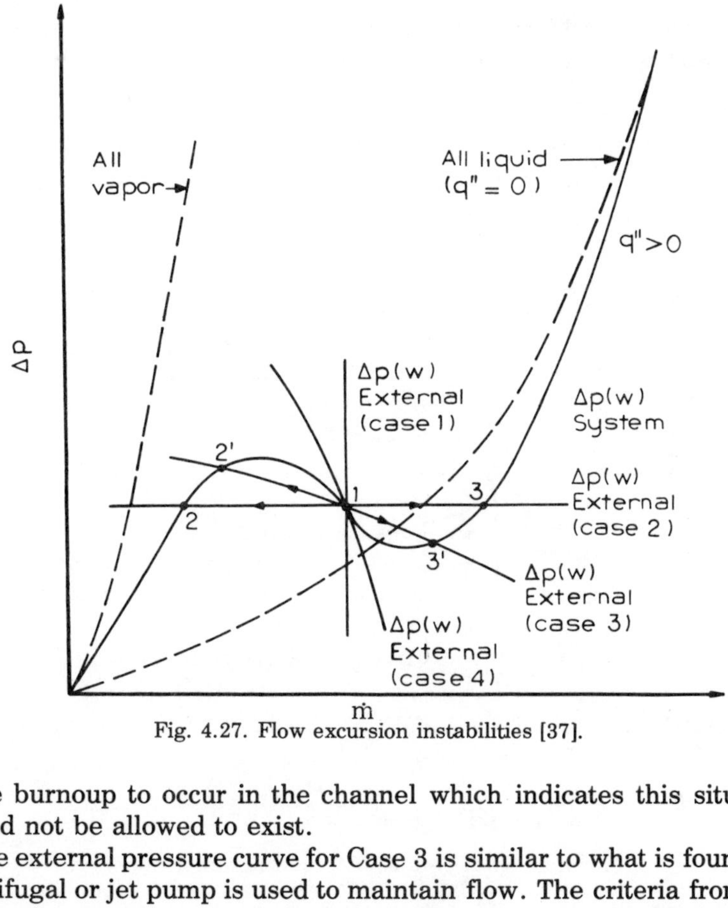

Fig. 4.27. Flow excursion instabilities [37].

cause burnoup to occur in the channel which indicates this situation should not be allowed to exist.

The external pressure curve for Case 3 is similar to what is found if a centrifugal or jet pump is used to maintain flow. The criteria from Eq. (4.6.1) is again not satisfied and the flow is unstable. Small perturbations in flow from operating point 1 cause excursive changes to points $2'$ or $3'$

Case 4 represents a situation in which a centrifugal pump has a non-overloading impellor or the external circuit is well throttled. This example is reported to exist in boiling water reactors [37]. For this situation the criteria from Eq. (4.6.1) is satisfied and flow is stable.

If analysis indicated a possibility of an excursive flow instability in a boiling channel, the condition is remedied by placing friction flow resistances which are proportional to the square of the mass flow rate in series with the reactor channels.

These resistances may be nozzles, orifices, or unheated channel lengths in front of the heated section. Nozzles are ordinarily used because they are usually less subject to erosion and have friction losses which are practically proportional to the square of the mass flow rate.

The increase in stability using additional flow resistances is shown in Fig. 4.28 in which a protective nozzle is added to the channel inlet.

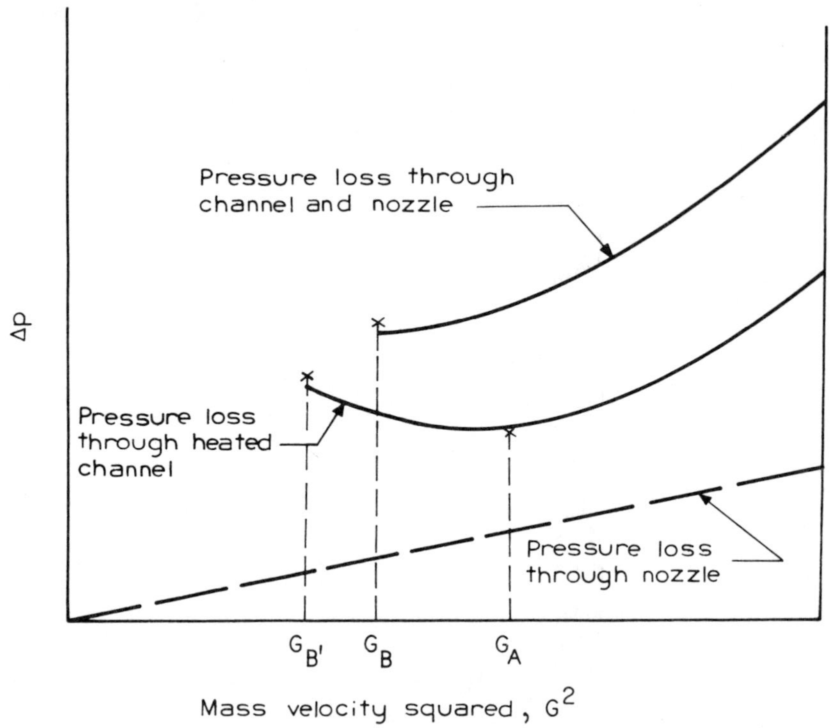

Fig. 4.28. Stability of boiling channel with nozzle

As shown in Fig. 4.28, for an unprotected channel flow instabilities occur at flow rates below G_A. However, by adding a nozzle at the channel inlet, the pressure loss curve is now changed so that it has no portion with a negative slope and, consequently, flow is stable down to G_B where burnout occurs. The burnout mass flow is slightly higher for the protected channel because the fluid boiling point is slightly reduced due to the lower fluid pressure in the heated section of the channel.

The nozzle size required to produce stable flow at a design mass flow, G_0, which is in an unstable region of the pressure loss-mass flow curve, is estimated by a simple analysis. For the unprotected channel, the slope of the pressure loss curve is negative or

$$\left(\frac{d\Delta p}{dG_o^2}\right)_{\text{unprotected channel}} = -K \tag{4.6.2}$$

Consequently, for stability the slope of the pressure loss-mass flow curve for the nozzle must be greater than K or

$$\left(\frac{d\Delta p}{dG_0^2}\right)_{nozzle} > K \tag{4.6.3}$$

For marginal stability, the slopes of the pressure loss curves are equal and consequently, we have

$$\left(\frac{d\Delta p}{dG_0^2}\right)_{nozzle} = -\left(\frac{d\Delta p}{dG_0^2}\right)_{unprotected\ channel} \tag{4.6.4}$$

The pressure loss through a nozzle is approximately given by the friction loss due to a sudden expansion since the losses due to kinetic energy changes across a nozzle cancel each other and the friction loss at the inlet to a well-designed nozzle is negligible. Therefore, the pressure loss in the nozzle is given by [3]

$$(p_1 - p_2)_F = \frac{G_2^2}{\rho 2g_c}\left(\frac{S_2}{S_1} - 1\right)^2 \tag{4.6.5}$$

where S_1 is the cross section area of the nozzle throat, S_2 is the channel cross section area, and G_2 is the mass velocity through the channel. Therefore, for marginal stability, the dimensions of the nozzle are obtained by taking the derivative of the nozzle pressure loss with respect to G^2 and setting this equal to the negative of the slope of the unstable pressure loss versus G^2 curve at the design flow rate G_0 . Therefore, we have

$$-\left(\frac{d\Delta p}{dG_0^2}\right)_{unprotected\ channel} = \frac{1}{\rho 2g_c}\left(\frac{S_2}{S_1} - 1\right)^2 \tag{4.6.6}$$

The effectiveness of a protective nozzle is diminished if it serves two or more parallel tubes, if a tube is subdivided by long axial heat-transfer or centering vanes, or if the channel is an annulus or slot of high perimeter-to-hydraulic radius ratio. This is due to fluid being able to unevenly distribute itself in various channels or parts of a channel if an obstruction or perturbation in flow occurs somewhere in a channel. For effective use of nozzles for more than one channel, the flow resistance of the nozzle must be several times the channel flow resistance and, consequently, pressure losses are too excessive. Therefore, nozzles are not too practical for tube bundle arrangements which are commonly used in high power reactors. One should carefully design reactor fuel assemblies so that flow instabilities are not possible in any of its sub-channels.

Flow Pattern Transition Instability

Flow pattern instabilities arise when two-phase flows are at conditions close to the point of transition from bubbly-slug flows to annular flows. An increase in bubble population for bubbly flow may cause the flow pattern to change to annular flow which characteristically has a lower pressure drop. When a constant pressure difference is applied across the channel, the flow rate increases with a subsequent decrease in vapor generation which causes the flow pattern to revert to a bubble-slug flow. The increased pressure loss from bubble-slug flow decreases the flow rate and the cycle is repeated. This problem is avoided by an appropriate choice of exit quality. PWRs generally operate at exit qualities below this transition; while BWRs operate beyond the transition.

Vapor Burst Instability

Vapor burst instabilities have been referred to as "flashing" instabilities which are caused by a sudden vaporization of the liquid phase with a consequent rapid decrease in mixture density. It is strongly dependent upon fluid properties, system geometry, and conditions of the heated surface. As an example, a very clean and smooth heated surface may require a high wall superheat in order to activate nucleation sites. Fluid adjacent to the surface is highly superheated and bubble growth is extremely rapid once nucleation commences. This in turn ejects liquid from the heated channel. The rapid vaporization cools the adjacent liquid and heater surface which is re-recovered with liquid phase as bubble detachment takes place. Further nucleation is suppressed until the required wall superheat builds up again and the process repeated. This phenomena may occur in liquid metal or low pressure water systems. In most reactor systems the superheats required to initiate boiling are low and vapor burst instability is not a problem.

Vapor burst instabilities are observed in reflood experiments used to test emergency core cooling systems (ECCS) for nuclear power plants [37]. Once water reaches hot fuel rods during reflooding, vapor bursts cause a geysering of a two-phase mixture up along the fuel rods. This effect is beneficial because it contributes to a faster cooldown of fuel rods.

Density-Wave Oscillations

In excursive flow instabilities we saw a small step change in supply pressure lead to a large change in flow. In density-wave instabilities a small oscillation in supply pressure leads to a large oscillation in flow. The mechanism for these oscillations is the transport of density waves.

These oscillations are low frequency in nature (generally less than 1 cps) and are of approximately the same period as the time required for a density wave to travel the channel length. From continuity wave laws this period is approximately 1 to 2 times the time required for a fluid particle to travel through the channel [56].

A physical explanation for density-wave oscillations is as follows: A slight reduction in inlet flow to a heated channel causes an increase in outlet enthalpy and a corresponding decrease in outlet density. This disturbance causes a change in pressure drop because the non-boiling and boiling length in the channel are altered. For certain combinations of flow geometry, operating conditions, and boundary conditions, the perturbations produce a 180° out-of-phase pressure fluctuation at the exit, are immediately transmitted to the inlet flow rate and become self sustained [56]. For parallel channel flows the pressure drop across the channels is constant. An increase in flow across an unstable channel causes an increase in non-boiling length and an increase in single-phase friction loss. Simultaneously the two-phase friction loss decreases because of the decrease in boiling length and decrease in quality. However, the elevation loss also increases because of increase in fluid density due to less void formation. The increase in density eventually causes the flow rate to decrease and the process then repeated.

Density-wave instabilities are analyzed by considering the basic transient equations of fluid flow: (i) continuity, (ii) momentum, (iii) energy, and (iv) equation of state (void-quality model). These equations are solved directly by use of numerical analysis or they are changed to the frequency domain and analyzed using feedback control theory. Reference [37] illustrates the latter form of analysis. Numerous computer programs have been developed for analyzing flow instabilities [55]. In one approach [57] a linearized form of the basic equations for a single, uniformly heated channel is solved. A more general code is HYDNA [58] which solves the nonlinear transient momentum equations for several parallel boiling channels with non-uniform heat flux distributions.

Pressure Drop Oscillations

Pressure drop oscillations occur in systems in which there is an interaction between an upstream compressible volume and a downstream boiling channel near conditions for a flow excursion. The oscillations are at a frequency about one order of magnitude smaller than density-wave oscillations or about 0.1 cycle per second.

A physical explanation for pressure drop oscillations is explained by examining Fig. 4.29 [56]. When the downstream portion CD of a heated channel is at the threshold for a flow excursion instability, a slight decrease in mass velocity in CD causes an increase in pressure loss in CD. When pressures at points D and A are held constant due to external flow conditions, this causes an increase in pressure at point C which in turn causes flow into the compressible volume which acts as a surge tank. At the same time the pressure drop and flow rate in the upstream channel AB begin to decrease. The decreased flow in section CD may cause the pressure loss to decrease in CD. This causes the pressure at C to decrease and fluid to leave the surge tank which in turn increases flow into CD. the interactions between the surge tank and channel then produces sustained oscillations.

Pressure drop instabilities are reported not to be a problem in BWRs [37] although the problem may occur in the design of experimental thermal-hydraulic test loops.

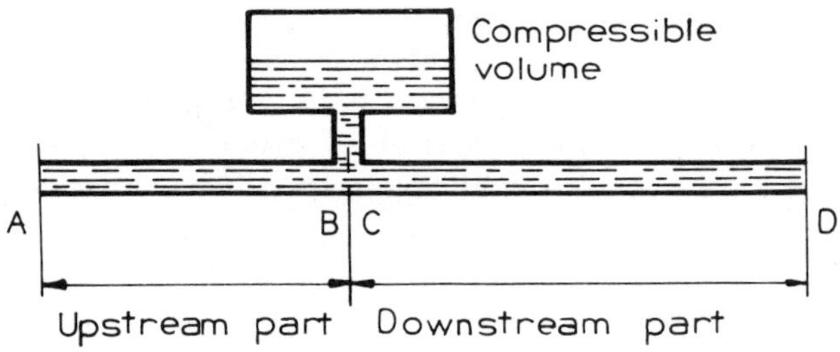

Fig. 4.29. System for pressure drop oscillations [56)

Flow Pattern Induced Instability

The flow pattern induced instability is due to the type of flow pattern present in the channel. Slug flow by its very nature is periodic and this type of flow in an adiabatic channel may produce dynamic oscillations [37]. In the design of BWRs and recirculating boiler steam generators it is necessary to consider this instability in components such as standpipes from which two-phase mixtures enter steam separators. Satisfactory performance of these components is insured by full scale static and dynamic testing [37].

Of the flow instabilities discussed, excursive flow instabilities and density-wave oscillations receive the greatest attention by reactor designers. By proper design these instabilities pose no threat during reactor operation.

4.7 Critical Two-Phase Flow

The flow of a compressible fluid through a pipe or opening cannot exceed a maximum "critical" value which is the speed at which a small downstream disturbance is propagated upstream (speed of sound in the medium). This phenomenon is well understood for gas flows and also occurs in the flow of two-phase mixtures. However, for two-phase flows the sonic velocity is frequently smaller than in the flow of either phase separately. The reason is that the mixture has a density of the order of the heavy liquid phase and a compressibility close to that of the vapor phase.

Under normal operating conditions, flow rates through reactor cores and other power plant components are considerably less than critical flow rates. However, during abnormal transients, such as loss-of-coolant accidents, flow rates may be at critical values. This would certainly be the case for flow through the break caused by a double-ended pipe rupture postulated as the design basis accident for water-cooled reactors. The fact that two-phase critical flows are considerably smaller than that of the liquid phase alone causes an increase in the time to void the reactor vessel which allows time to activate external emergency core cooling systems.

Due to its importance in accident analysis, two-phase critical flows have been studied extensively both theoretically and experimentally. The theoretical treatments are divided into two broad categories: (i) equilibrium models which assume thermodynamic equilibrium between both phases throughout the expansion and (ii) nonequilibrium models which account for nonequilibrium effects between phases. The effect of flow regime is incorporated in these models by either assuming homogeneous flow (slip ratio of unity) or accounting for slip between phases and determining expressions for the slip ratio. Early model development dealt with equilibrium models and later development accounted for non-equilibrium effects as experiments produced disagreement with early models. Nonequilibrium effects appear to be important for critical flows in pipes of short length $(L/D < 6)$ and phase equilibrium may be a reasonable assumption for flows through pipes of greater lengths $(L/D > 12)$.

Reviews of the theory and modeling used in critical two-phase flow are given by Bouré [56,59] and Ardron and Furness [60].

Equilibrium Models

The critical mass velocity (G_c) occurs for compressible fluids when all the energy obtained from a differential pressure loss is transformed into a corresponding increase in kinetic energy. This result follows from an examination of the differential form of the steady-flow energy equation. At a point in a channel at which G_c is reached, one may assume the expansion taking place so fast that neither friction nor heat transfer are important. Consequently, the expansion is assumed isentropic. On further expansion in a channel of constant cross-section area, the fluid velocity would rise even more; however, this is impossible is G_c is reached and, consequently, G_c is reached at the outlet of a channel if at all.

If G is at a maximum, then its differentiation with respect to any variable is zero. Therefore, at G_c

$$\frac{dG}{d\bar{u}} = 0 = \frac{d(\rho\bar{u})}{d\bar{u}} = \rho + \bar{u}\frac{d\rho}{d\bar{u}} \tag{4.7.1}$$

or

$$\frac{d\bar{u}}{\bar{u}} = \frac{-d\rho}{\rho} \tag{4.7.2}$$

From the differential steady flow energy equation, for an isentropic change in energy with $dz = 0$

$$\frac{dp}{\rho} + \frac{\bar{u}d\bar{u}}{g_c} = 0 \tag{4.7.3}$$

By eliminating $d\bar{u}$ from Eq. (4.7.3) through Eq. (4.7.2) the critical velocity is given by

$$\bar{u}_c^2 = g_c\left(\frac{dp}{d\rho}\right)_s \tag{4.7.4}$$

or

$$G_c^2 = \bar{u}_c^2/v^2 = \frac{g_c}{v^2}\left(\frac{dp}{d(1/v)}\right)_s$$

$$= -g_c\left(\frac{dp}{dv}\right)_s \tag{4.7.5}$$

where v is the mixture specific volume and the subscript s refers to differentiation taken at constant entropy. Equation (4.7.5) is the familiar equation for calculating critical mass velocities for homogeneous, compressible fluids.

The simplest way to apply Eq. (4.7.5) is to assume a one-dimensional homogeneous flow. The mixture specific volume is then given by

$$v = (1 - x)v_f + xv_g \tag{4.7.6}$$

For a given quality and pressure just inside the channel exit, the differential is evaluated using steam tables. This is done by assuming a small Δp at the exit, such a 1 psi, and then finding the change in quality holding s constant. The new mixture specific volume is then found and the differential is evaluated from $\Delta p/\Delta v$. Table 4.6 shows calculated critical mass velocities for steam-water mixtures using this approach. This approach yields correct trends of G_c with respect to system pressure and quality; however, it usually underestimates G_c. Consequently, this model is not conservative in accident analysis because it overestimates time required to void a pressure vessel.

Table 4.6
Calculated critical mass velocities for steam-water mixtures

Quality	G_c, lb_m/ft^2-sec				
	Pressure, psia				
	50	200	500	1000	2000
0.995	177	674	1660	3340	6700
0.75	201	760	1850	3720	7380
0.50	240	912	2160	4270	8390
0.25	317	1140	2620	5160	9960
0.10	425	1510	3250	6080	11500
0.005	607	2140	3890	7000	12900

Fauske [61] improved upon the homogeneous, equilibrium model by making the following additional assumptions:

(i) Annular flow pattern with liquid along the pipe walls at a uniform velocity u_ℓ and vapor flowing in the pipe core at a uniform velocity u_g.

(ii) Critical flow is attained when flow rate no longer increases with decreasing downstream static pressure. Further, the pressure gradient attains a finite maximum value for a given flow rate and quality. The maximization of the pressure gradient is achieved by varying the slip ratio, $S = u_g/u_\ell$.

For isentropic annular flow the critical mass velocity is still given by Eq. (4.7.5). However, the mean specific volume is given by the expression for annular (separated) flow

$$v = \frac{x^2 v_g}{a} + \frac{(1 - x)^2 v_f}{1 - a} \tag{4.7.7}$$

By introducing the slip ratio S given by

$$S = \frac{u_g}{u_\ell} = \frac{x}{1 - x}\,\frac{1 - a}{a}\,\frac{v_f}{v_g} \tag{4.7.8}$$

the void fraction a is eliminated from Eq. (4.7.7) to give

$$v = \frac{[(1 - x)v_f S + x v_g]\,[1 + x(S - 1)]}{S} \tag{4.7.9}$$

The pressure gradient at the point of critical flow is given by an expression in the form

$$\frac{dp}{dz} = f_W \frac{vG_c^2}{2g_c} \frac{1}{D_e} \tag{4.7.10}$$

At critical flow all terms in Eq. (4.7.10) are constant with the exception of v which is maximized according to assumption (ii). Since for given exit conditions all terms in Eq. (4.7.9) are constant except S, the maximum value of v is obtained by setting dv/dS = 0. Therefore,

$$\frac{dv}{dS} = (x - x^2)(v_f - v_g/S^2) = 0 \tag{4.7.11}$$

Consequently, at critical flow the slip ratio is

$$S = u_g/u_\ell = \sqrt{v_g/v_f} \tag{4.7.12}$$

for $0 < x < 1.0$.

The critical mass velocity is given by substituting Eq. (4.7.9) into Eq. (4.7.5)

$$G_c^2 = \frac{-g_c}{\frac{d}{dp}\left\{[(1 - x)v_\ell S + xv_g][1 + x(S - 1)]/S\right\}} \tag{4.7.13}$$

or, neglecting the insignificant term, dv_ℓ/dp ,

$$G_c^2 = \frac{-g_c S}{[(1 - x + Sx)x]\frac{dv_g}{dp} + [v_g(1 + 2Sx - 2x) + v_f(2xS - 2S - 2xS^2 + S^2)]\frac{dx}{dp}} \tag{4.7.14}$$

Isenthalpic conditions are used for evaluating dv_g/dp at the outlet pressure where conditions are known. The reason for this assumption is that a slip ratio is assumed and a real flow cannot be isentropic. The term dx/dp is also evaluated from the constant mixture enthalpy since

$$\frac{dh}{dp} = \frac{dh_f}{dp} + x\frac{dh_{fg}}{dp} + h_{fg}\frac{dx}{dp} = 0 \tag{4.7.15}$$

or

$$\frac{dx}{dp} = -\frac{1}{h_{fg}}\left[\frac{dh_f}{dp} + x\frac{dh_{fg}}{dp}\right] \tag{4.7.16)}$$

For a given exit quality and pressure, the critical mass velocity is evaluated using Eqs. (4.7.12), (4.7.14), and (4.7.16) and steam tables. Figure 4.30 illustrates calculated critical mass velocities using Fauske's method and a comparison with Table 4.6 shows that the slip (separated) flow model yields critical mass velocities from 1.1 to as high as 4 times greater than the homogeneous flow model. At high pressures and quality approaching unity the models have their closest agreement.

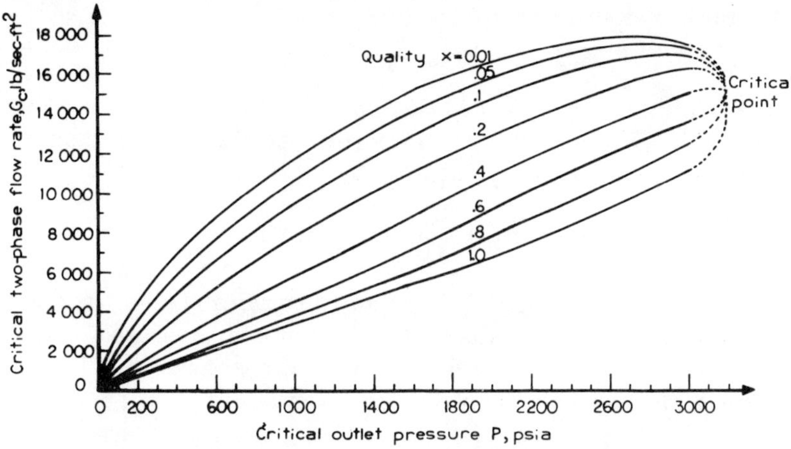

Fig. 4.30. Theoretical predictions of critical flow rates by Fauske's model [61]

The equilibrium homogeneous and separated flow models presented have a serious disadvantage in that critical mass velocities are determined in terms of exit quality and pressure which are parameters not immediately known in a pipe rupture situation. Consequently, an iterative procedure is required to analyze blowdown of a vessel of known internal pressure and enthalpy. This problem is eliminated by the well-known Moody [62] critical flow model which determined critical mass velocities in terms of upstream vessel stagnation properties. This model is a follows:

For critical two-phase flow with no energy transfer

$$h_0 = h + x\frac{u_g^2}{2g_c J} + (1 - x)\frac{u_\ell^2}{2g_c J} = \text{constant} \qquad (4.7.17)$$

where h_0 is the vessel stagnation or total enthalpy. Applying the continuity equation to each phase and making use of the definition of slip ratio, Eq. (4.7.17) is rearranged to give

$$h_0 = h + \frac{G^2}{2g_c J}[S(1 - x)v_f + xv_g]^2 \; [x + (1 - x)/S^2] \qquad (4.7.18)$$

The mass velocity is then given by

$$G = \frac{\sqrt{2g_c J(h_0 - h)}}{[S(1 - x)v_f + xv_g]\ [x + (1 - x)/S^2]^{1/2}} \tag{4.7.19}$$

The enthalpy of the saturated mixture is given by

$$h = h_f + xh_{fg}\,. \tag{4.7.20}$$

By assuming flow is an isentropic process

$$s_f + xs_{fg} = s_0 = \text{constant} \tag{4.7.21}$$

Quality is eliminated from Eq. (4.7.20) by use of Eq. (4.7.21) to give the mixture enthalpy in terms of properties at the vessel stagnation pressure p_0 and the exit pressure p_e.

$$h = h_f + \frac{h_{fg}}{s_{fg}}(s_o - s_f). \tag{4.7.22}$$

Since s_0 can be determined knowing pressure vessel stagnation pressure and enthalpy, h is then a function of exit pressure p_e and s_0. Examination of Eqs. (4.7.19) and (4.7.22) shows the functional dependence of mass velocity at the exit is given by

$$G = G(p_e, S; p_o, h_o). \tag{4.7.23}$$

For given upstream stagnation conditions, G is a function of p_e and S. From an analogy with single-phase critical flow, Moody reasoned that at critical flow the following conditions must be satisfied:

$$(\partial G/\partial S)_{p_e} = 0; \quad (\partial^2 G/\partial S^2)_{p_e} < 0 \tag{4.7.24}$$

and

$$(\partial G/\partial p)_S = 0; \quad (\partial^2 G/\partial p^2)_S < 0 \tag{4.7.25}$$

Substituting Eq. (4.7.19) into Eq. (4.7.24) yields at critical flow

$$S = S(p_e) = (v_g/v_f)^{1/3}. \tag{4.7.26}$$

Equation (4.7.26) shows the slip ratio at critical flow is a function of exit pressure only.

Substituting Eqs. (4.7.19) and (4.7.26) into Eq. (4.7.25) and performing the differentiation with the use of property tables, the critical mass velocity G_c is found as a function of p_o and h_o. Moody performed these calculations for steam-water mixtures which is illustrated in Fig. 4.31. For the same exit conditions, the Fauske and Moody models agree with each other fairly well. However, as noted, the Moody model is more convenient because it required knowledge of only upstream conditions.

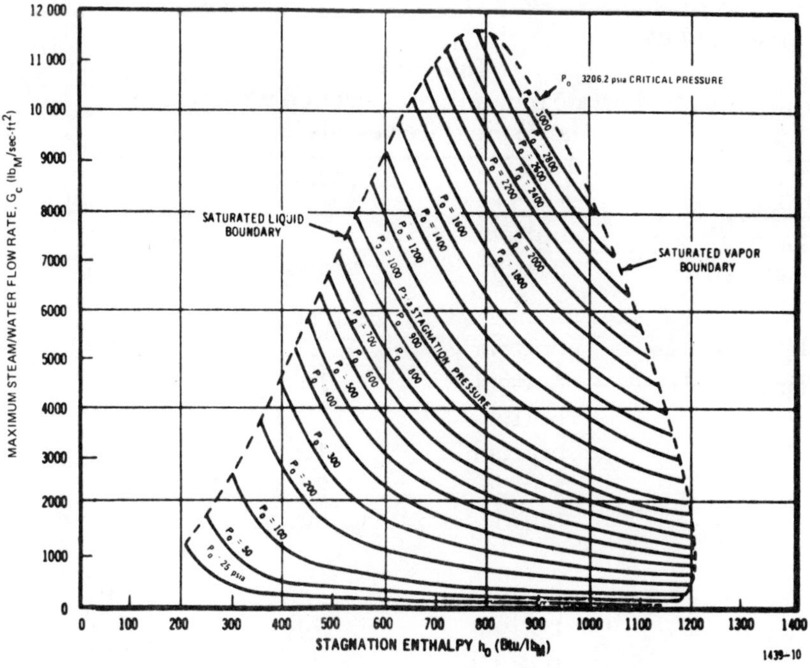

Fig. 4.31. Maximum steam/water flow rate and local stagnation properties (Moody model) [62].

Moody extended his work to accommodate critical flow through long pipes where wall friction effects can be of significance [63]. This work is published in the form of graphs showing steam-water critical flow rates as a function of pipe fL/D and upstream stagnation properties.

Experimental data at low qualities disagree with a slip ratio of either $(v_g/v_f)^{1/2}$ or $(v_g/v_f)^{1/3}$. This is probably due to a metastable liquid phase. The correct slip ratio is not constant but a function of quality and pressure. Fauske [64] obtained better agreement for low qualities using Armand's slip ratio [65] which gives for the void fraction

$$a = \frac{(0.833 + 0.167x)xv_g}{(1 - x)v_f + xv_g} \tag{4.7.27}$$

This correlation, which is used in Eq. (4.7.7), is recommended for calculation of a during low-pressure critical flow from pipes [42].

For reactor safety studies it is common practice to use the Moody model because it tends to overpredict critical flows when $L/D > 12$ and upstream stagnation qualities are above 0.01 [60]. One method for improving calculations is to use the Moody multiplier defined as the time-averaged ratio of instantaneous blowdown flow rate to that predicted by the Moody theory. Table 4.7 lists the multiplier for various rupture situations.

Table 4.7
Moody flowrate multipliers calculated in blowdown of large pressure vessels [60]

Initial Conditions	Break location	$\frac{\text{Orifice dia.}}{\text{Pipe dia.}}$ = % area	Moody multiplier
P_o = 39 bar	bottom	4	0.85
(saturated)	bottom	8.16	0.75
P_o = 52 bar	bottom	7.5	0.8
(saturated)	bottom	10	0.8
	bottom	12.5	0.7
	bottom	15	0.7
	bottom	25	0.65
	bottom	50	0.60
P_o = 67 bar	bottom	2.04	0.80
(saturated)	bottom	4.0	0.73
	bottom	8.16	0.63
	bottom	12.8	0.6
P_o = 98 bar	bottom	4	0.73
(saturated)	bottom	8.16	0.63
	bottom	12.8	0.60
P_o = 39 bar	top	0.51	0.75
(saturated)	top	4	0.6
P_o = 67 bar	top	0.51	0.8
(saturated)	top	2.04	0.65
P_o = 155 bar	top	10	0.6
(subcooled)	bottom	2	0.8

Nonequilibrium Flow

A finite amount of time is required for bubble growth resulting from a rapid change in fluid pressure from a high value to a low value. This time is of the order of 0.5 to 1 msec [37]. Consequently, if the length of time for a fluid particle to be expelled through an opening is less than 1 msec, one should expect nonequilibrium effects during critical flow.

Henry and Fauske [66] developed a nonequilibrium model which enables predicting critical flows in terms of stagnation properties. This model shows good agreement for flow through short pipes ($L/D \approx 3$).

Zaloudek [67] studied critical steam-water flow in a series of pipes in which $L/D < 6$. He observed two types of choking mechanisms which are illustrated in Fig. 4.32. The first critical flow rate occurs at point B in which the upstream portion of the vena contracta approached the water saturation pressure. Visual observations show flashing between the wall and vena contracta. As the pressure differential is increased to point C, the flow pattern changes to the free discharge of a metastable liquid. In this region the pressure throughout the pipe is nearly equal to the downstream pressure and the flow rate again increases with increasing pressure differential. Once point D is reached the fluid flashes violently at the discharge and the second (downstream) choking occurs. This flow rate is not exceeded with a further differential pressure increase and this type of choking is considered the type which occurs with single-phase, compressible fluid flows.

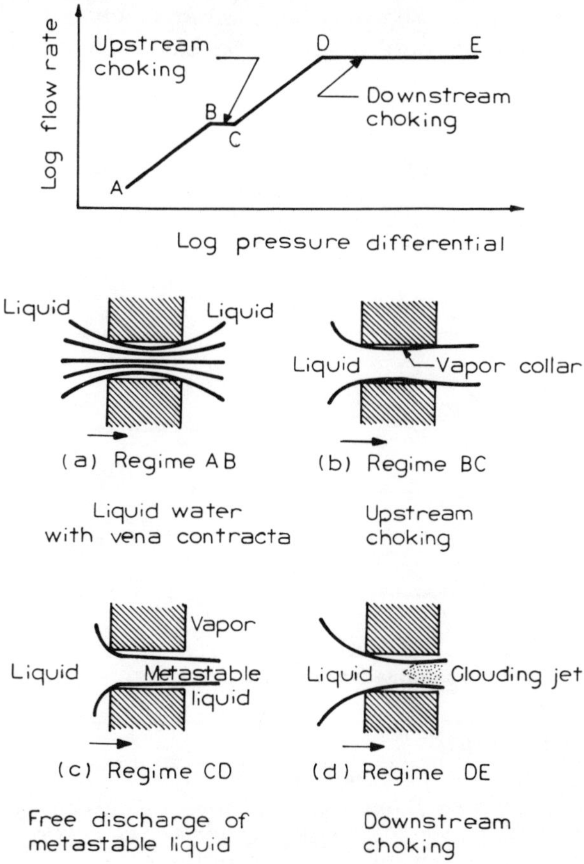

Fig. 4.32. Short pipe critical flows [67].

The first critical mass velocity (region BC) is given by [67]

$$G_c = C_1\sqrt{2g_c\rho_f\,(p_{upstream} - p_{sat}\,)} \tag{4.7.28}$$

where C_1 is an empirical constant found to range from 0.61 to 0.64. Measured second critical flow rates agree well with the correlation of Burnell [68] which is given by

$$G_c = \sqrt{2g_c\rho_f\,[p_{upstream} - (1 - C_2)p_{sat}\,]} \tag{4.7.29}$$

where C_2 is a constant dependent upon water surface tension σ

$$C_2 = 0.294\,\frac{\sigma \text{ for } p_{sat}}{\sigma \text{ for } p_{sat} = 175 \text{ psia}} \tag{4.7.30}$$

Most critical flow models are able to find experimental data which agree with the model. However, no model is developed at this writing which is accurate over a range of L/D ratios from very small to large or over pressure vessel fluids ranging from subcooled water to steam-water mixtures. Figure 4.33 compares some models with a variety of experimental data. As seen, no model fits all data.

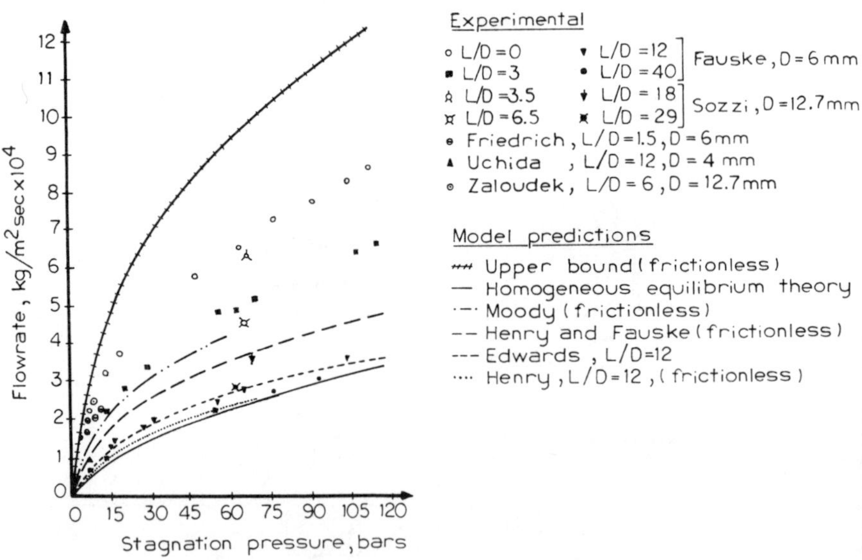

Fig. 4.33. Discharge of saturated water through orifices, nozzles and pipes [60].

4.8 Pump Performance

With the information provided in previous sections it is now possible to make estimates of pressure losses through reactor cores and assorted ducts, pipes, valves and other fixtures which are contained in nuclear reactor flow loops. It is apparent that pressure loss calculations are only approximate and unless the majority of the flow path is through simple channels of sufficient length that entrance effects are small, errors in calculations can be considerable. Depending upon the use being made of pressure loss data, it may be desirable to adjust calculations so as to provide conservatism to the analysis.

One obvious use of pressure loss calculations is for specifying heads, capacities, and power requirements for pumps or compressors used to maintain fluid circulation through reactor system flow loops. However, before one can reasonably interpret the significance of pressure loss calculations, it is necessary to understand in some detail the characteristics of pumps or compressors that are employed.

Pumping Power Requirements

Pumping power requirements are calculated by applying the steady-flow energy equation across a pump inlet and outlet and multiplying the evaluated energy requirement by the fluid mass flow rate. The pump is represented by the following diagram:

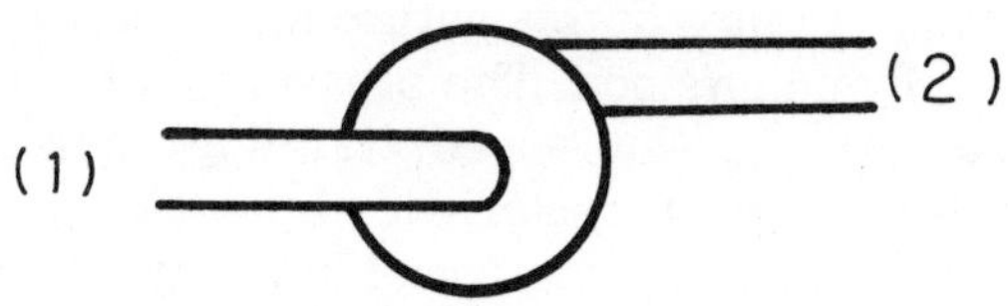

Consequently, the ideal pumping work W_p, or total head is given by

$$W_p = \frac{p_2 - p_1}{\rho} + a\left(\frac{u_2^2 - u_1^2}{2g_c}\right) + (z_2 - z_1)\frac{g}{g_c} \tag{4.8.1}$$

where W_p is expressed in $ft\text{-}lb_f/lb_m$. Ordinarily, one can neglect the kinetic and potential energy terms and

$$W_p = \frac{p_2 - p_1}{\rho}, \frac{ft\text{-}lb_f}{lb_m} \tag{4.8.2}$$

The pressure change across the pump, $p_2 - p_1$, is calculated by analyzing all pressure losses in flow system components and performing appropriate summations. One then determines pump power requirements from

$$P = \dot{m}_T \frac{\Delta p}{\rho}, \ ft\text{-}lb_f/sec \tag{4.8.3}$$

where $\dot{m}_T$ is mass flow rate through the pump in lb_m/sec.

In more standard engineering terms

$$P = \frac{\dot{m}_T \Delta p}{550\rho}, \text{ horsepower}$$

$$= \frac{\dot{m}_T \Delta p}{737\rho}, \text{ Kilowatts} \tag{4.8.4}$$

Pumps are unable to convert 100% of prime mover power into a pressure rise in the flowing fluid. Consequently, a pump efficiency is defined by

$$\eta_p \equiv \frac{\text{hydraulic power}}{\text{Inlet power from prime mower}} \tag{4.8.5}$$

Therefore, the external power source to sustain flow is given by

$$P = \frac{\dot{m}_T}{\eta_p} \frac{\Delta p}{550\rho}, \text{ horsepower} \tag{4.8.6}$$

Variables in Pump Performance

Circulation of liquids in reactor systems is normally achieved by use of centrifugal pumps. Pumps of this nature have certain characteristics which should be fully understood. The pressure output of these pumps are a function of (1) impellor speed, (2) fluid flow rate, (3) fluid properties, and (4) pump characteristic dimensions. For any given pump, the pressure is normally expressed in terms of total head, W_p . The total head and efficiency for a given pump operating at constant impellor speed is then a function of volumetric flow rate through the pump. The variables used in discussing pump performance are as follows:

D = rotor diameter, ft.
N = speed of rotation (rpm)
Δp = psf or W_p = ft-lb_f/lb_m
ρ = lb_m/ft^3
η = viscosity = lb_m/ft-sec
Q_T = gpm or ft^3/sec

Pump performance is given in terms of these variables and vendors normally provide specifications using water as the fluid. A few important parameters used in discussing pumps are as follows:

(1) Pump characteristic curves

Centrifugal pump performance is given by a plot of total head, power requirements, and efficiency as a function of volumetric flow rate. Figure 4.34 illustrates a typical set of curves for a pump operating at a fixed impellor speed of rotation. Increasing impellor speed increases the total head output and power requirements.

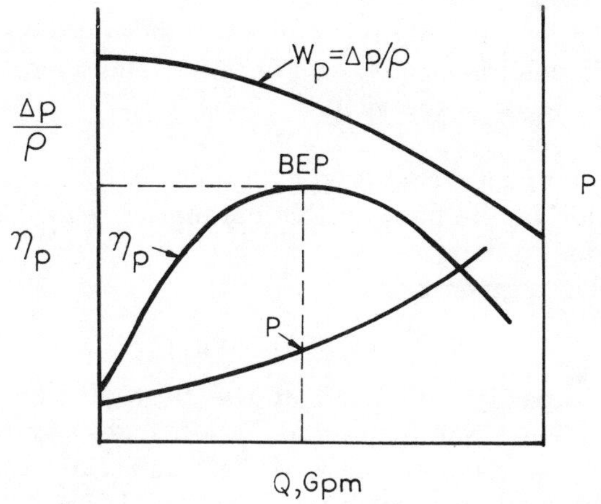

Fig. 4.34. Performance characteristic of a centrifugal pump operating at constant speed

The ratings of given prime movers are added to Fig. 4.34 to give Fig. 4.35 which shows regions in which satisfactory pump operation are achieved for a given motor attached to the pump. Areas to the left of a given motor rating are operating regions in which that motor provides adequate power.

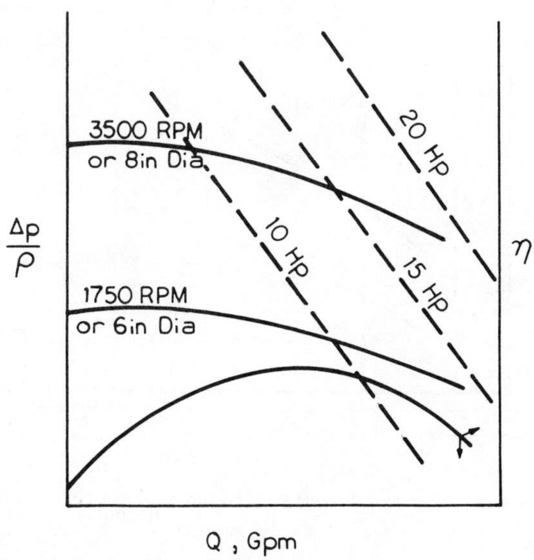

Fig. 4.35. Prime mover power requirements plus pump characteristic curve

(2) B.E.P.—Best efficiency point

The B.E.P. is the point of maximum efficiency on the efficiency-flow rate curve. This point is located on Fig. 4.34. The operating point of a pump should be at or in the vicinity of the B.E.P.

(3) Specific speed, N_s

Centrifugal pumps are classified hydraulically by specific speed and may be compared on the basis of this parameter regardless of relative sizes of individual pumps.

Specific speed is given by

$$N_s = NQ_T^{1/2} (\rho/\Delta p g_c)^{3/4} \tag{4.8.7}$$

which relates capacity and head at the best efficiency point. The specific speed gives a rough guide to the efficiency of pumps as shown in Fig. 4.36. For a given capacity, the efficiency of pumps increases with increasing specific speeds up to a range of 2500 to 3000 [69]. At higher specific speeds the efficiency begins to fall off.

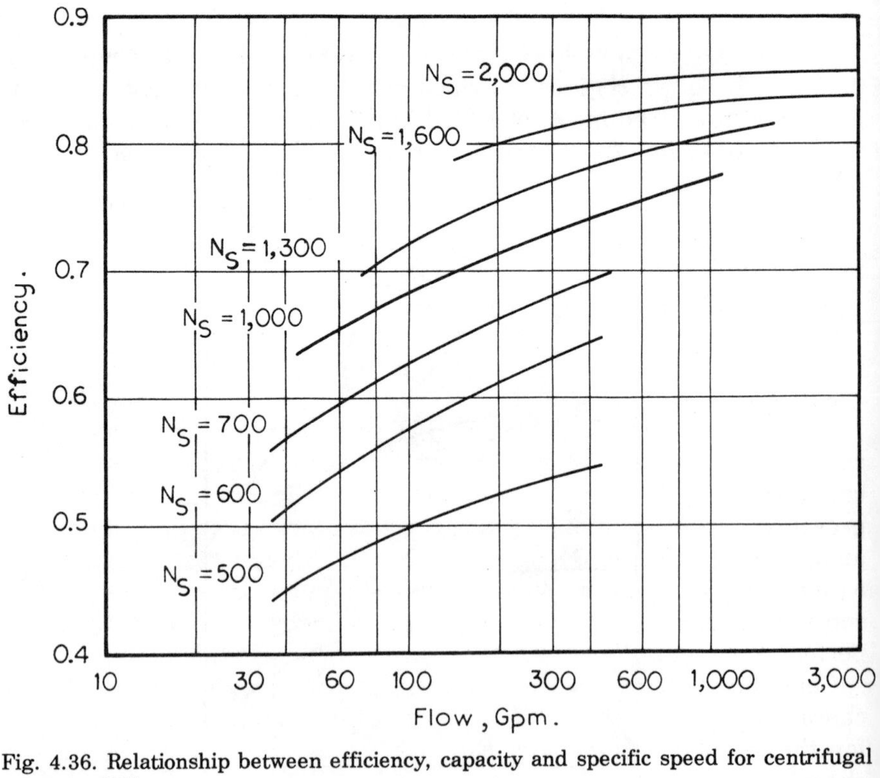

Fig. 4.36. Relationship between efficiency, capacity and specific speed for centrifugal pumps [69]

(4) NPSH—Net positive suction head

The NPSH is the difference in absolute pressure measured at the center line of the suction nozzle and the lowest pressure that exists in the liquid passage within the pump before the liquid is acted upon by the pump. This parameter is important in order to prevent cavitation (formation of vapor bubbles) within the pump during operation. If p_1 is the inlet pressure to the pump, cavitation is prevented when

$$p_1 - \text{NPSH} > p_v \tag{4.8.9}$$

where p_v is the liquid vapor pressure at the pumping temperature. NPSH is usually given for the B.E.P. by pump manufacturers. This parameter depends upon impellor speed of rotation and pump capacity as shown in Fig. 4.37.

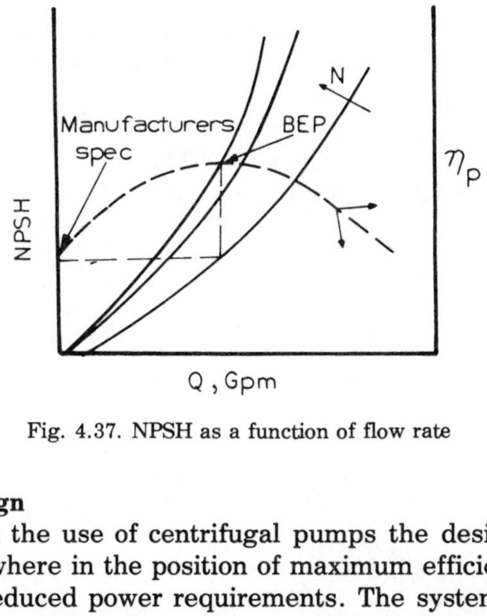

Fig. 4.37. NPSH as a function of flow rate

System Design

Ordinarily in the use of centrifugal pumps the design flow rate is selected somewhere in the position of maximum efficiency in order to capitalize on reduced power requirements. The system flow rate and pressure loss is established from a thermodynamic and fluid flow analysis of the entire powerplant. Consequently, an operating point on a pump characteristic curve is established and one must, therefore, use some judgement in selection of pumps in order to insure satisfactory performance and efficiency is achieved.

Centrifugal pumps only produce the performance indicated by their characteristic curves. One can compute a flow rate-pressure loss curve for the reactor system and plot it simultaneously with the pump characteristic curve as shown in Fig. 4.38. For a given pump, the

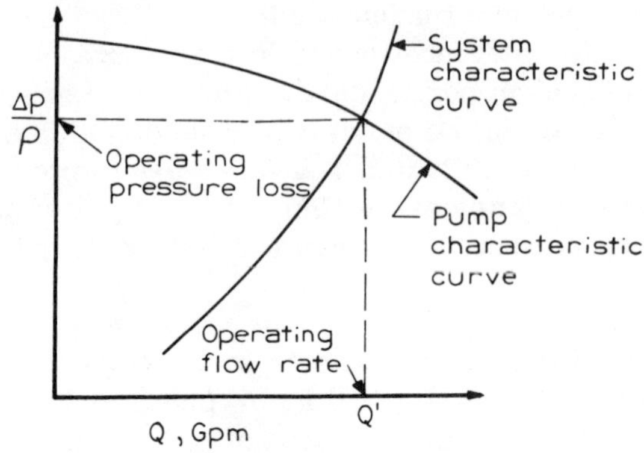

Fig. 4.38. Pump and reactor system pressure loss vs. flow rate characteristic curves.

system flow rate is fixed because, by Fig. 4.38, the operating flow rate (Q') only occurs at the intersection of the pump and system characteristic curves.

Can adding a safety factor to calculated system response curves cause any problems in pump selection? The answer is yes as shown in Fig. 4.39. The actual system curve is given by line OA_1, and the

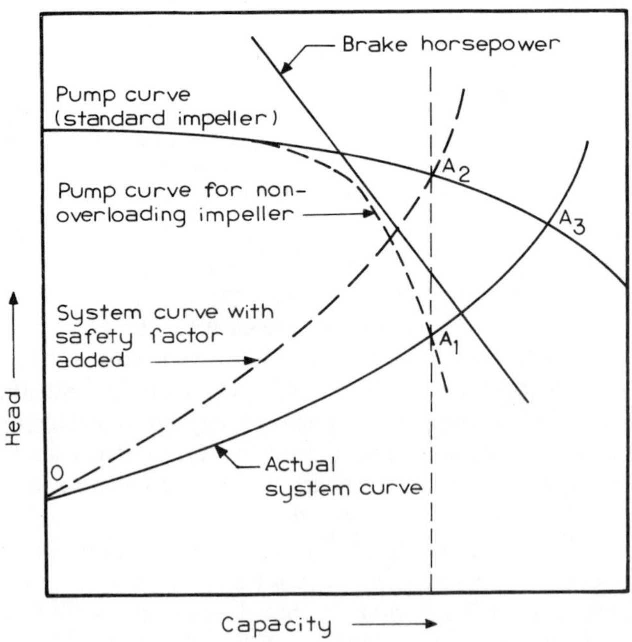

Fig. 4.39. Effect of using safety factor on pump performance [70]

pump is required to operate at point A_1 . Adding a safety factor gives a system curve OA_2 . On the basis of this curve a pump manufacturer supplies a pump to operate at point A_2 . However, since the total required head at this capacity is A_1 and not A_2 , the pump operates at point A_3 which is a capacity greater than required by the system. At this higher capacity the pump motor may be overloaded.

Here we see a conservative calculation of system response curve may result in specifying a motor too small to handle the pump. Non-overloading pump impellors can solve the problem; however, their efficiency is poor and thus operating costs are increased.

Pumps contribute to system stability because of their characteristic curves. Because the slope of their pressure rise-flow rate curve is usually negative, a slight decrease in flow-rate results in an increase in output pressure which tends to increase flow rate. Similarily, a slight increase in flow rate produces a decrease in pressure output and the usual system response of increased pressure loss tends to reduce flow rate. Consequently, barring the occurrence of some unexpected flow resistance, a steady flow output is produced by centrifugal pumps.

Some judgement must be exercised in selecting the location of centrifugal pumps in a flow loop. Cavitation can occur due to formation of air or vapor bubbles inside the pump chamber. This is prevented by using deaerated water and locating pumps in positions in the flow loop where the inlet pressure to the pump is considerably above the fluid vapor pressure corresponding to its temperature. This is a consideration in providing sub-cooling for liquids exiting from condensers.

Cavitation is not permissible in pumps because it drastically reduces pump efficiency and, because of its hammering effects, causes erosion and possible mechanical breakdown.

Another type of pump may be used in reactor system cooled by liquid metals. This is the electromagnetic pump which operates on a principle similar to an electric motor; however, the moving conductor is now the conducting fluid. These pumps have a strong advantage in that they are completely self-containing and, consequently, eliminate the problem of working fluid leakage which occurs with ordinary centrifugal pumps. However, E. M. pumps have a disadvantage in that their overall efficiencies are quite low (ranging from 10 to 40 percent) and their current requirements are very high (thousands of amperes). Present LMFBR designs call for the use of centrifugal pumps for primary coolant circulation.

REFERENCES

1. El-Wakil, M. M., **Nuclear Heat Transport**, International Textbook, Scranton, Penn., 1971.
2. Bonilla, C. F., ed., **Nuclear Engineering**, McGraw-Hill, New York, 1957.
3. Perry, J. P., ed., **Chemical Engineer's Handbook**, 3rd ed., McGraw-Hill, New York, 1950.
4. Knudsen, J. G. and Katz, D. L., **Fluid Dynamics and Heat Transfer**, McGraw-Hill, New York, 1958.
5. von Karman, T., NACA Techincal Memorandum No. 611 (1931).
6. McAdams, W. A., **Heat Transmission,** 3rd ed., McGraw-Hill, New York, 1954.
7. Colebrook, C. F., Turbulent flow in pipes with particular reference to the transition region between the smooth and rough pipe laws, **Proc. Inst. Civil Eng.** 11, 133 (1939).
8. Mark, L. S., ed., **Mechanical Engineer's Handbook**, 5th ed., McGraw-Hill, New York, 1951.
9. Moody, L. F., Friction factors for pipe flows, **Trans. of the ASME,** 671 (November 1944).
10. Esselman, W. H., et al., Thermal and hydraulic experiments for pressurized water reactors, **Proc. of the 2nd United National Int. Conference on the Peaceful Uses of Atomic Energy** 7, 748, United Nations, Geneva (1958)
11. Gambill, W. R., and Bundy, R. D., HFIR heat-transfer studies of turbulent water flow in thin rectangular channels, ORNL-3079 (June 1961).
12. Levy, S., Turbulent flow in an annulus, **Trans. of the ASME, Series C** 89, 25-31 (1967).
13. Crandall, J. R., et. al., The Savannah River high flux demonstration, DP-999 (June 1965).
14. Rotta, J. C., Recent developments in calculation methods for turbulent boundary layers with pressure gradients and heat transfer, **ASME J. of Applied Mechanics** 33, 429-437 (June 1966).
15. Rohsenow, W. M. and Clark, J. A., Heat transfer and pressure drop data for high heat flux densities to water at high sub-critical pressures, **1951 Heat Transfer and Fluid Mechanics Institute**, Stanford University Press, Stanford, California, June 1951.
16. LeTourneau, B. W., Grimble, R. E., and Zerbe, J. E., Pressure drop for parallel flow through rod bundles, **Trans. ASME** 79, 483 (1957).
17. Taylor, M. F., Correlation of friction coefficients for laminar and

turbulent flow with ratios of surface to bulk temperature from 0.35 to 7.35, NASA TR R-267 (September 1967).

18. Humble, L. V., Loudermilk, W. H., and Desmon, L. G., Measurements of average heat-transfer and friction coefficients for subsonic flow of air in smooth tubes at high surface and fluid temperatures, NACA TR 1020 (1951).
19. Deissler, R. G. and Eian, C. S., Analytical and experimental investigation of fully developed turbulent flow of air in a smooth tube with heat transfer with variable fluid properties, NACA TN 2629 (1952).
20. Magee, P. M. and McEligot, D. M., Effect of property variation on the turbulent flow of gases in tubes: The thermal entry, **Nuclear Science and Engineering** 31, 337-341 (February 1968).
21. Kays, W. M. and London, A. L., **Compact Heat Exchangers**, 2nd ed., McGraw-Hill, New York, 1964.
22. Kays, W. M., **Convective Heat and Mass Transfer**, McGraw-Hill, New York, 1966.
23. Subbotin, V. I., Heat removal from the reactor fuel elements cooled by liquid metals, **Proc. 3rd. United Nations Int. Conference on Peaceful Uses of Atomic Energy**, P/328 (1964).
24. Deissler, R. G. and Taylor, M. F., Analysis of axial turbulent flow and heat transfer through banks of rods or tubes, TID-7529, Reactor Heat Transfer Confernece, Part 1, Book 2, USAEC (1957).
25. Miller, P., Byrnes, J. J., and Benforado, D. M., Heat transfer to water flowing parallel to a rod bundle, **A.I.Ch.E.J.** 2, 226 (1956).
26. Grillo, P. and Mazzone, G., Single- and two-phase pressure drops on a 6 x 6 rod bundle at 70 atm, **Nuclear Technology** 15, 25-35- (July 1972).
27. de Stordeur, A. M., Drag coefficients for fuel elements spacers, **Nucleonics** 19, No. 6, 74 (1961).
28. Brindley, J. H., Fuel-spacers drag coefficients in sodium, **Nucleonics** 23, No. 8, 94 (1965).
29. Rehme, K., Pressure drop correlations for fuel element spacers, **Nuclear Technology** 17, 15-23 (January 1973).
30. Martinelli, R. C. and D. B. Nelson, Prediction of pressure drop during forced circulation boiling of water, **Trans. ASME** 70, 695-702 (August 1948).
31. Anderson, R. J. and T. W. F. Russell, Designing for two-phase flow—part I, **Chemical Engineering**, 139-144 (December 6, 1965).
32. Goldman, K., Firstenberg, H. and C. Lombardi, Burnout in turbulent flow—A droplet diffusion model, **Trans. ASME J. Heat Transfer** 83, 158-162 (1961).

33. Griffith, P. and G. B. Wallis, Two-phase slug flow, **Trans. ASME J. Heat Transfer** 83, 307-320 (1961).
34. Tong, L. S., **Boiling Heat Transfer and Two-Phase Flow**, John Wiley & Sons, Inc., New York, 1965.
35. Griffith, P., The slug-annular flow regime transition at elevated pressure, ANL-6796 (November 1965).
36. Bergles, A. E., Ross, J. P. and J. G. Bourne, Investigation of boiling flow regimes and critical heat flux, Dynatech Final Summary Report 797, NYO-3304-13 (July 1968)
37. Lahey, R. T. and F. J. Moody, **The Thermal-Hydraulics of a Boiling Water Nuclear Reactor**, American Nuclear Society, La Grange Park, Ill., 1977.
38. Jens, W. H. and P. A. Lottes, Analysis of heat transfer, burnout, pressure drop, and density data for high pressure water, ANL-4627 (1951).
39. Saha, P. and N. Zuber, Point of net vapor generations and vapor void fraction in subcooled boiling, **Proc. 5th Int. Heat Transfer Conf.** IV (1974).
40. Castellana, F. S. and C. F. Bonilla, Two-phase pressure drop and heat transfer in rod bundles, **Two-Phase Flow and Heat Transfer in Rod Bundles**, ASME, New York, 15-30, 1969.
41. Thom, J. R. S., Walker, W. M., Fallow, T. A., and G. F. S. Reising, Boiling in subcooled water during flow up heated tubes or annuli, **Proc. Inst. M. E.**, 1965-1966, Part 3C, 226-246 (1966).
42. Tong, L. S. and J. Weisman, **Thermal Analysis Of Pressurized Water Reactors**, American Nuclear Society, La Grange Park, Illinois, 1970.
43. Zuber, N., and J. A. Findlay, Average volumetric concentration in two-phase flow systems, **J. Heat Trans.** 88, 453 (1965).
44. Dix, G. E., Vapor void fractions for forced convection with subcooled boiling at low flow rates, NEDO-10491, General Electric Co (1971).
45. Levy, S., Forced convection subcooled boiling—prediction of vapor volumetric fraction, GEAP-5157, General Electric Company (1966).
46. Reynolds, J. B., Local boiling pressure drop, ANL-5178, Argonne National Lab. (1954).
47. Owens, W. L. and V. E. Schrock, Local pressure gradients for subcooled boiling of water in vertical tubes; ASME 60-WA-249 (1960).
48. Mendler, O. J., et al., Natural-circulation tests with water at 800 to 2000 psia under nonboiling, local boiling, and bulk boiling conditions, **J. of Heat Transfer** 83, 261-273 (1961).

49. Lockhart, R. W. and R. C. Martinelli, Proposed correlation of data for isothermal two-phase two-component flow in pipes, **Chem. Engrg. Progr.** 45, 39-48 (1949).
50 Griffith, P., Two phase flow in pipes, **Development in Heat Transfer**, M.I.T. Press, Cambridge, Massachusetts, 1964.
51. Baroczy, C. J., A systematic correlation for two-phase pressure drop, A.I.Ch.E. paper 37, Proc. 8th Int. Heat Transfer Conf., Los Angeles (August 1965).
52. Tarasova, N. V., et al., Pressure drop of boiling subcooled water and steam-water mixture flowing in heated channels, **Proc. of the 3rd Int. Heat Transfer Conf. iv**, 178-183 (August 1966).
53. Hussain, A., and J. Weisman, Two-phase pressure drop across abrupt area changes, COO 2152-15, NRC Report (1975).
54. Lowdermilk, W. H., Lanzo, D. C., and B. L. Siegal, Investigation of boiling burnout and flow stability for water flowing in tubes, NACA-TN-4382 (1958).
55. Bouré, J. A., Bergles, A. E., and Tong, L. S., Review of two-phase flow instability, **Nuc. Eng. Des.** 25, 165-192 (1973).
56. Ginoux, J. J. (ed). **Two-Phase Flows and Heat Transfer with Application to Nuclear Reactor Design Problems**, Hemisphere Publishing Corporation, Washington, D. C. (1978).
57. Jones, A. B. and A. G. Dight, Hydrodynamic stability of a boiling channel, Part II, KAPL-2208 (1962).
58. Currin, H. B., et al., HYDNA—Digital computer program for hydrodynamic transients in a pressure tube reactor or a closed channel core, CVNA-77 (1961).
59. Bankoff, S. G. and Jones, O. C., **Thermal and Hydraulic Aspects of Nuclear Reactor Safety, Vol. I: Light Water Reactors**, ASME, New York (1977).
60. Ardron, K. H. and Furness, K. H., A study of critical flow models used in reactor blowdown analysis, **Nucl Eng. Des.** 39, 257 (1976).
61. Fauske, H. K., Contribution to the theory of two-phase, one-component critical flow, ANL-6633 (1962).
62. Moody, F. J., Maximum flow rate of a single-component, two-phase mixture, **Trans. ASME, J. Heat Transfer, Series C** 87, 134 (1965).
63. Moody, F. J., Maximum two-phase vessel blowdown from pipes, **Trans. ASME, J. Heat Transfer, Series C** 88, 285 (1966).
64. Fauske, H. K., What's new in two-phase flow, **Power Reactor Technology** 9, No. 1, 35 (1966).
65. Armand, A. A., The resistance during the movement of a two-phase system in horizontal pipes, AERE-Trans. 828 (1959).
66. Henry, R. E. and H. K. Fauske, The two-phase critical flow of

one-component mixtures in nozzles, orifices, and short tubes, **Trans. ASME, J. Heat Transfer, Series C** 93, 179 (1971).
67. Zaloudek, F. R., The critical flow of hot water through short tubes, HW-77594 (May 1, 1963).
68. Burnell, J. G., Flow of boiling water through nozzles, orifices, and pipes, **Engineering** 164, 572 (1947).
69. Pollak, H. M., How to select centrifugal pumps, **Chemical Engr.** 70, 81-96 (February 4, 1963).
70. Holland, F. A. and Chapman, F. S., Centrifugal pumps—types—characteristics—system design, **Chemical Engr.** 73, 91-112 (July 4, 1966).

PROBLEMS

4.1 Prove for fully-developed laminar flow in a pipe that the kinetic energy flow can be expressed as u_m^2/g_c , ft–lb$_f$/lb$_m$

4.2 A large water tank has a 2 in. diameter cast iron drain pipe attached to its bottom. The drain pipe is 60 ft. long and inclined at an angle of $30°$ from the horizontal. If the water level in the tank is 20 ft., what is the flow out the drain pipe. Neglect inlet friction loss to the drain pipe.

4.3 A proposed fuel element contains 225 rods of 0.400 in. diameter, each being 12 ft. long and arranged in a square lattice. The rods are held in position by 8 dimpled honeycomb grids of 0.05 in. equivalent thickness. The rods are contained in a fuel element shroud of 7.44 in. on a side inside dimensions. Find the pressure loss when the flow through the fuel element is 2000 G.P.M. of water at $600°F$. Ignore entrance and exit losses.

4.4 A proposed gas-cooled reactor has an annular flow passage with a 1/4 in. inside dia. and 1/2 in. outside dia. The heat flux on the inside portion of the flow channel has an average value of 150,000 Btu/hr-ft^2 . The channel length is 12 feet and the coolant is helium with an inlet pressure of 600 psia and an inlet temperature of $1000°F$. For downflow calculate the pressure loss across the channel. Neglect inlet and exit losses and assume mass flow rates of 500, 1000, 2000, and 4000 lb$_m$/hr. . The specific heat of helium is 1.25 Btu/lb$_m$-°F and its viscosity is given by $0.0653\ \sqrt{T/960}$, lb$_m$/ft–hr. Assume the wall temperature on the hot side of the channel is $500°F$ higher than the bulk fluid temperature. In order to obtain the most instructional value from this problem it is recommended that each contribution to the pressure loss be calculated separately—friction, momentum, and elevation.

4.5 A 10 ft. long fuel element channel has the mass flow varied in such a fashion so as to maintain constant inlet and outlet temperatures during operation $T_{in} = 540°F$ and $T_{out} = 580°F$. The fuel elements are 0.4 in. dia. rods with a pitch of 0.55 in. Examine the stability of this system for power variations from 1 to 100% full power. At full power, the average heat flux is 1×10^5 Btu/hr–ft². Downflow is incorporated and the system pressure is 2000 psia.

4.6 Water enters a 6 ft. long uniformly heated 1 in. dia. tube at 460°F and 600 psia. The entering velocity is 2.5 ft/sec. Neglecting pressure losses in the channel, determine the maximum heat input, in Btu/hr-ft of channel length, if the exit void fraction is not to exceed 60 percent.

4.7 A boiling water reactor operates at a pressure of 1000 psia. In one of its fuel assemblies the inlet velocity is 10 ft/sec, inlet temperature is 520°F, and coolant mass flow rate is 200,000 lb_m/hr. If the heat generated in the fuel assembly is 2×10^7 Btu/hr, what is the fuel assembly exit void fraction?

4.8 A boiling water reactor has fuel rods of 0.45 in. dia., 12 feet long arranged in a square lattice with a pitch of 0.65 in. The pressure is 1000 psia and the water inlet temperature is 520°F. Assume a fluid outlet quality of 0.12 and a uniform heat flux along the rod length of 200,000 Btu/hr–ft². Calculate the following

a. Water inlet velocity

b. Location for the onset of bulk boiling

c. Location of the onset of nucleate boiling (assume the single-phase heat-transfer coefficient is 6000 Btu/hr–ft²–°F and the heat capacity of water is 1.23Btu/lb–°F).

e. Exit void fraction.

4.9 A two-phase mixture flows upward through a vertical, 1 in. dia. pipe. The mixture quality is 0.1 and the mixture has a mass flow rate of 3.03 lb/sec. The fluid pressure is 1000 psia at the bottom of the pipe. For a 10 foot pipe length, calculate the following for adiabatic flow:

a. Mixture density accounting for slip

b. Elevation pressure difference

c. Two-phase friction pressure loss

d. What is the momentum pressure loss?

4.10 A nominal subchannel of a BWR has 0.400 in. dia. rods located in a square array with a pitch of 0.600 in. The inlet coolant is at 1000 psia, has a velocity of 12 ft/sec, and is at 520°F. The exit quality is 12 percent. Voids at the top of the core cause a strong neutron flux depression so the axial heat flux profile is represented by

$$q''_w = q''_o \exp(-\pi z/H) \sin(\pi z/H)$$

where q''_o is a constant, z = 0 is the channel entrance, and H is the channel height of 12 feet. Find the pressure loss in the channel.

4.11 A large pressure vessel contains saturated water at 1600 psia. Determine the initial mass flow rate exiting from a long 36 inch diameter pipe attached to the pressure vessel. Assume the ambient pressure is atmospheric.

4.12 A backup safety system for a nuclear power plant requires pumping water from a deep well into a pressurized water tank. Air pressure in the tank is 50 psig (64.7 psia). The pipe diameter is 6 inches and it is made of drawn tubing. The system must be capable of delivering 2000 gallons/minute of water at 60°F. If the pump efficiency is 85%, what is the minimum size motor required on the pump?

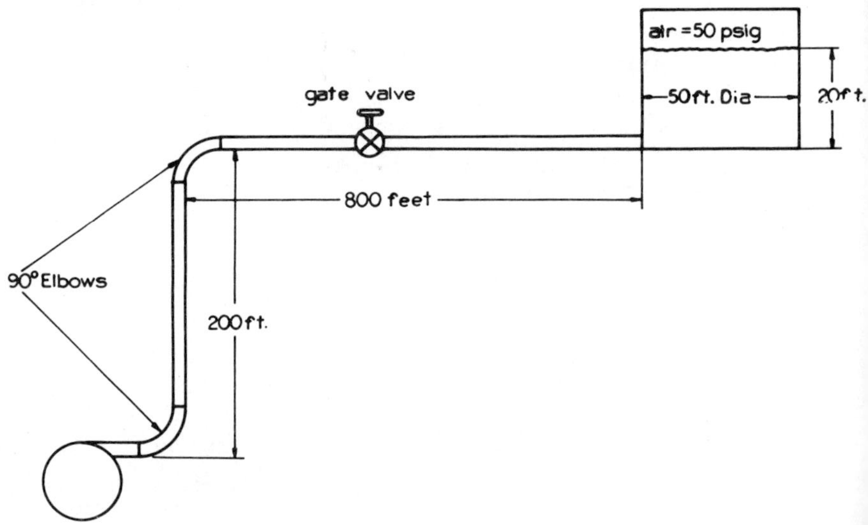

CHAPTER 5

HEAT GENERATION IN NUCLEAR REACTORS

5.1 Introduction

All the heat generated in reactor systems is directly or indirectly due to the fission process. Table 5.1 illustrates various sources of energy in a reactor due to fission of U^{235}. The energy from fission is essentially the same for all fissionable materials and Table 5.1 can be used for estimating the energy from fission for plutonium and other isotopes of uranium. All fission fragment kinetic energy and β energy from decay of fission products is essentially dissipated in the reactor fuel. A substantial amount of gamma energy from fission, fission product decay, and neutron reactions is dissipated in the reactor fuel. Heat generation also takes place in the cladding, reactor structure, coolant, moderator, control rods, reflector, thermal shields, pressure vessel, and biological shield. The heat sources in these materials is primarily due to slowing down of fission neutrons by elastic scattering and absorption of gamma rays resulting from fission, fission product decay, non-fission neutron capture, and fast neutron inelastic scattering.

In thermal reactors the majority of fission neutron kinetic energy is dissipated in the moderator by way of elastic scattering collisions. Gamma heating takes place in all reactor materials. To a good approximation it may be assumed that all fission energy, with the exception of neutrino energy, is retained within the reactor core and recovered by the reactor coolant.

5.2 Steady-State Heat Generation in the Reactor Fuel

Since the majority of fission energy is due to fission fragment kinetic energy, it is reasonable to assume that most of the recoverable fission energy is dissipated in the fuel when analyzing heterogeneous reactors. This assumption is somewhat conservative and results in slightly overestimating fuel temperatures. The volumetric heat generation rate in reactor fuels is determined by a core physics analysis and is given by the following equation:

$$q''' = 176 \sum_i N_i \int_0^\infty \phi(E)\, \sigma_{f,i}(E)\, dE + E_c \qquad (5.2.1)$$

TABLE 5.1
Approximate distribution of energy in a thermal reactor [1]
(MeV per Atom of U^{235})

	Energy	Range
Prompt energy of fission		
1. Fission fragment kinetic energy	168	Short
2. Kinetic energy of fission neutrons	5	Medium
3. Energy of Prompt γ	7	Long
Radioactive decay of fission products		
4. β	8	Short
5. γ	7	Long
6. Neutrino	12 (Non-recoverable)	
Recoverable energy directly from fission process	195	
Non-fission reactions with neutrons*		
7. $\beta + \gamma$	~7-10	Short and Long
Total energy recovered by coolant	~202-205	

*In a critical reactor, ν-1 neutrons (1.4-2.0 neutrons/fission) undergo absorption reactions without fission. These neutrons are absorbed in the fuel, fertile material, moderator, coolant, structure, and shields. Immediately upon capture the neutron binding energy, which ranges from 2.2 MeV in hydrogen to 6-8 MeV in heavy materials, is released in the form of γ's. Many of the products are unstable and decay to a stable state at later times by beta and gamma ray emission.

where

q''' = volumetric heat generation rate, MeV/cm^3-sec

N_i = atom density of ith fuel material, atoms/cm^3

$\phi(E)$ = neutron flux per unit energy, neutrons/(cm)(sec)(MeV)

$\sigma_{f,i}(E)$ = energy dependent fission cross section of ith fuel material, cm^2

E_c = gamma volumetric heat generation rate (from items 3, 5, and 7 Table 5.1) MeV/cm^3-sec

Equation (5.2.1) is converted to fuel specific power by dividing by the fuel density. It is written in a simpler form by using appropriately averaged fission cross sections and the total neutron flux .

$$q''' = 176\,\Phi \sum_i N_i \overline{\sigma}_{f,i} + E_c\text{, MeV/cm}^3\text{- sec} \qquad (5.2.2)$$

where $\overline{\sigma}_{f,i}$ is the spectrum-averaged fission cross section of the fuel material. $q'''$is spatially and time dependent because of the space-time dependence of the neutron flux. In addition, q'''may have a spatial dependence because of non-homogeneity of reactor fuel and a time dependence because atom densities change as fuel is consumed and fertile material converted to fissile material.

In many reactors the mass of fuel is large compared to other materials and the majority of the gamma energy is dissipated in the fuel. Consequently, E_c will contain a substantial amount of the 20-25 MeV of gamma energy per fission. The spatial dependence of the gamma energy is approximately the same as the neutron flux and Eq. (5.2.2) is simplified

$$q''' = 195\,\Phi \sum_i N_i\,\bar{\sigma}_{f,i}\,,\ \mathrm{MeV/cm^3-sec}$$

$$= 2.96 \times 10^{-14}\,\Phi \sum_i N_i\,\bar{\sigma}_{f,i}\,,\ \mathrm{Btu/cm^3-sec} \qquad (5.2.3)$$

In order to use Eq. (5.2.3) it is necessary to have a detailed map of the neutron flux distribution throughout all fuel bearing regions of the reactor core. Obtaining accurate neutron flux or power distributions throughout a reactor core is a very complicated process which usually requires numerical solution of multi-group diffusion or transport theory equations. These techniques are discussed in standard books on nuclear reactor theory.

For homogeneous nuclear reactors, the core neutron flux distribution is approximated by solutions to the wave equation [1]

$$\nabla^2\Phi + B^2\Phi = 0 \qquad (5.2.4)$$

where B^2 is the geometric buckling. Nuclear reactors with a uniform fuel enrichment which have fuel elements uniformly dispersed throughout the reactor core may be treated as homogeneous reactors for determing gross features of the neutron flux distribution. Uniformly loaded pressurized-water reactors fall in this category. Non-homogeneous poisons such as control rods invalidate use of Eq. (5.2.4) for finding neutron flux distributions. However, frequently in pressurized-water reactors boric acid is dissolved in the moderator to provide reactivity control. These reactors can operate at full power with essentially all control rods (non-homogeneous poisons) withdrawn from the core.

Solutions are readily found to Eq. (5.2.4) for bare reactors of simple geometry. The boundary conditions are the neutron flux is finite within the reactor core and the neutron flux goes to zero at the extrapolation distance from the core boundary. For practical purposes the extrapolation distance is taken as $0.71\lambda_{tr}$, where λ_{tr} is the transport mean free path of neutrons in the reactor core. Therefore, for a cylindrical reactor of actual length L and radius R the important length parameters in solutions to Eq. (5.2.4) are the extrapolated length given by

$$L_e = L + 2(0.71\lambda_{tr}) \qquad (5.2.5)$$

and the extrapolated radius given by

$$R_e = R + 0.71\lambda_{tr} \qquad (5.2.6)$$

Since λ_{tr} is usually of the order of a few centimeters, the extrapolation distance can be ignored for large reactors.

Solutions to Eq.(5.2.4) for several simple reactor geometries are given in Table 5.2. These neutron flux distributions are used as approximations for flux shapes in unreflected, homogeneous or quasi-homogeneous reactors which contain no non-homogeneous poisons or control rods. These flux distributions are probably more representative of actual flux distributions at the start of life for a reactor core because after significant operating times flux shapes are altered due to the accumulation of fission product poisons and fuel burnup. Figure 5.1(a) illustrates a typical flux distribution in a bare, homogeneous reactor.

Frequently in a reactor thermal analysis the core size and reactor power level are specified. Under these conditions it is not necessary to use Eq. (5.2.3) to determine heat source distributions. For reactors with uniform fuel enrichment, one may determine the average volumetric heat generation rate in the fuel $\overline{q'''}$ by dividing the reactor thermal power level Q by the fuel volume V_f. The local fuel volumetric heat generation rate is then given by

$$q''' = (\Phi/\overline{\Phi})\,\overline{q'''} \qquad (5.2.7)$$

where $\Phi/\overline{\Phi}$ is the ratio of local neutron flux to average neutron flux. If fuel enrichment is varied in the core, Eq. (5.2.7) is multiplied by the ratio of local enrichment to core average enrichment. Table 5.2 gives expressions for $\Phi/\overline{\Phi}$ for bare homogeneous reactors which have extrapolation distances negligible in comparison to actual core dimensions.

Unreflected reactors are generally not considered desirable for power production because of inefficient use of fuel in regions adjacent to core boundaries. Heat transfer rates are low in these regions and fuel burnup is considerably lower than in central regions of the reactor core. Consequently, in order to improve upon fuel utilization, most power reactors use radial and axial reflectors to produce a more uniform neutron flux distribution within the core. Equation (5.2.4) is still valid for determining neutron flux distributions in homogeneous or quasi-homogeneous reactors with reflectors. The solutions are altered because of different boundary conditions at the core-reflector interface.

TABLE 5.2

Neutron flux distributions in bare homogeneous reactors

Geometry	Coordinate	Flux Distributions Φ/Φ_o	$\overline{\Phi}/\Phi_o$	$\Phi/\overline{\Phi}$
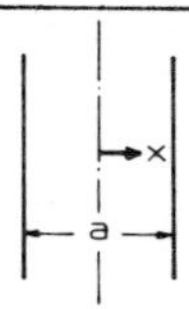 Infinite slab	x	$\cos\left(\frac{\pi x}{a_e}\right)$	$2/\pi$	$\frac{\pi}{2}\cos\left(\frac{\pi x}{a_e}\right)$
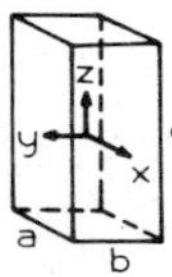 Parallel-epiped	x,y,z	$\cos\left(\frac{\pi x}{a_e}\right)\cos\left(\frac{\pi y}{b_e}\right)\cos\left(\frac{\pi z}{c_e}\right)$	$8/\pi^3$	$\frac{\pi^3}{8}\cos\left(\frac{\pi x}{a_e}\right)\cos\left(\frac{\pi y}{b_e}\right)\cos\left(\frac{\pi z}{c_e}\right)$
 Sphere	r	$\frac{\sin(\pi r/R_e)}{\pi r/R_e}$	$3/\pi^2$	$\frac{\pi^2}{3}\frac{\sin(\pi r/R_e)}{\pi r/R_e}$
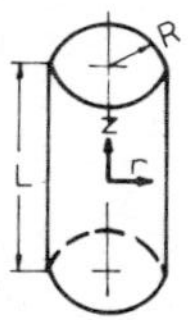 Finite cylinder	r,z	$J_o\left(\frac{2.405r}{R_e}\right)\cos\left(\frac{\pi z}{L_e}\right)$	$\frac{1}{2.32}\ \frac{2}{\pi}$	$2.32\ J_o\left(\frac{2.405r}{R_e}\right)\frac{\pi}{2}\cos\left(\frac{\pi z}{L_e}\right)$

Φ/Φ_o = ratio of local flux to center-line flux
$\overline{\Phi}/\Phi_o$ = ratio of average flux to center-line flux
$\Phi/\overline{\Phi}$ = ratio of local flux to average flux

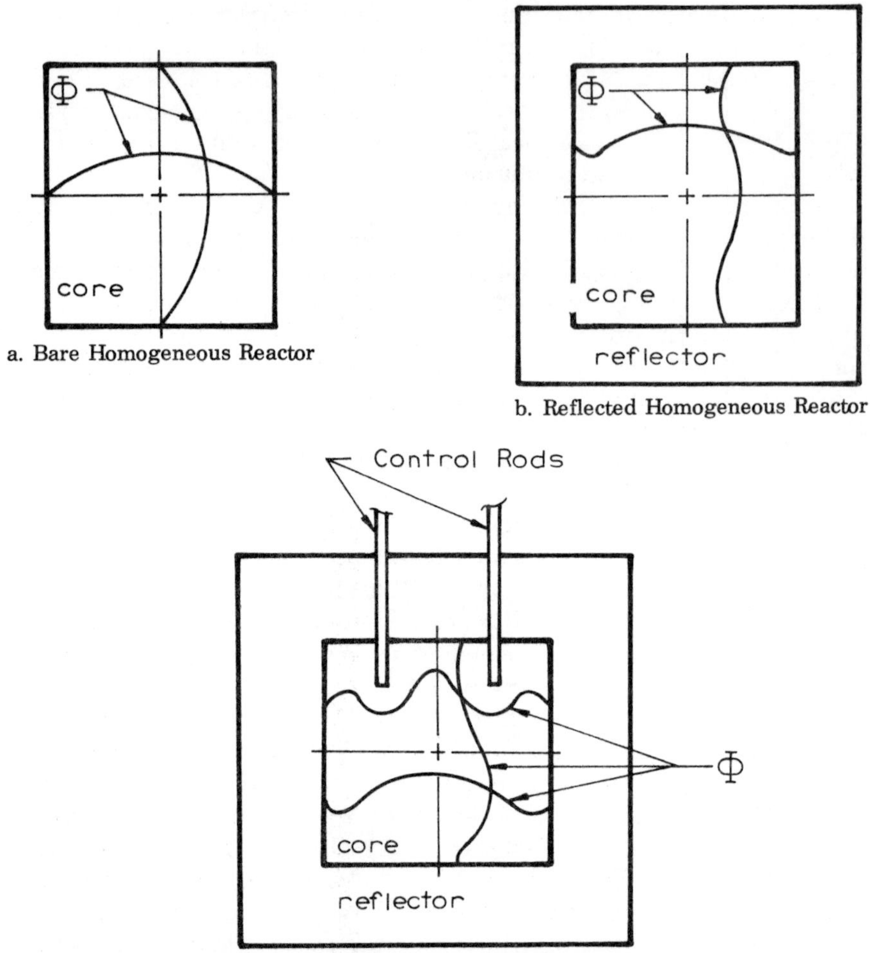

a. Bare Homogeneous Reactor

b. Reflected Homogeneous Reactor

c. Reflected Reactor With Control Rods

Fig. 5.1. Flux distributions in reactor systems

For reflected homogeneous reactors the solutions to Eq.(5.2.4) are the same as solutions given in Table 5.2; however, the length parameters are now modified to include the reflector savings(δ). The reflector savings is the decrease in critical dimensions of the reactor core due to the presence of the reflector. For a cylindrical reactor of radius R and length L the important radial parameter in the flux distribution is

$$R' = R + \delta \tag{5.2.8}$$

and the length parameter is

$$L' = L + 2\delta \tag{5.2.9}$$

Table 5.3 lists approximate core flux distributions for several homogeneous reactors with reflectors. Also included in Table 5.3 are expressions for ratios of average-to-maximum flux which are used in Eq. (5.2.7) for calculating local fuel volumetric heat generation rates. Figure 5.1(b) illustrates typical flux distributions found in reactor cores with a reflector around all surfaces.

Techniques for calculating reflector savings for the one-group diffusion theory model are described in standard books on reactor theory [1,2]. For reactors with similar diffusion coefficients in core and reflector and infinitely thick reflectors, the reflector savings is approximately the diffusion length (L) of the reflector. Reflectors with a thickness of 1.5 to 2 migration lengths (square root of the sum of the Fermi age and square of the diffusion length) are assumed equivalent to an infinite reflector [3].

The expressions given in Tables 5.2 and 5.3 are only suitable for approximating neutron flux distributions in homogeneous or quasi-homogeneous cores. Inhomogeneities are introduced into most reactor cores by the large number of control rods used to control excess reactivity built into the system. Control rods produce large perturbations in neutron flux distributions. Since from a safety standpoint control rods are usually withdrawn in an upward direction out of cores,* flux depressions are usually produced at the core top in power reactors. Additional flux perturbations occur in boiling water reactors since vaporization essentially displaces moderator (water) with void (steam) which results in a lesser degree of neutron thermalization and a reduced power level at the core top. An additional complication is that power distributions change with time in reactor cores because of greater fuel depletion in regions of higher neutron flux. The effects of control rods, fuel depletion, and void produce difficult problems in the theoretical physics analysis of power reactors. Figure 5.1(c) illustrates an approximate flux distribution expected in poison-controlled reactors.

In heterogeneous reactors, neutron flux distributions are usually assumed uniform across fuel rods or plates. This assumption is reasonable for thin plates or small diameter rods if fuel enrichment is low or fuel is dilutely dispersed in a weak absorbing material. In fuel regions the neutron flux is at a maximum at the fuel surface and decreases due to neutron absorption toward the fuel center. For some reactor fuel elements it may be desirable to account for neutron flux depressions when calculating fuel temperature distributions.

---

*Exceptions are some boiling water reactors which have control rod drives which withdraw rods through the core bottom.

TABLE 5.3

Core neutron flux distributions in reflected homogeneous reactors

Geometry	Coordinate	Flux Distributions	
		Φ/Φ_o	$\bar{\Phi}/\Phi_o$
Infinite Slab	x	$\cos\left(\frac{\pi x}{a'}\right)$	$\frac{2a'}{\pi a}\sin\left(\frac{\pi a}{2a'}\right)$
Parallelepiped	x,y,z	$\cos\left(\frac{\pi x}{a'}\right)\cos\left(\frac{\pi y}{b'}\right)\cos\left(\frac{\pi z}{c'}\right)$	$\frac{2a'}{\pi a}\sin\left(\frac{\pi a}{2a'}\right)\frac{2b'}{\pi b}\sin\left(\frac{\pi b}{2b'}\right)\frac{2c'}{\pi c}\sin\left(\frac{\pi c}{2c'}\right)$
Sphere	r	$\frac{\sin(\pi r/R')}{\pi r/R'}$	$\frac{3\left[\sin\left(\frac{\pi R}{R'}\right)-\frac{\pi R}{R'}\cos\left(\frac{\pi R}{R'}\right)\right]}{(\pi R/R')^3}$
Finite Cylinder	r,z	$J_o\left(\frac{2.405r}{R'}\right)\cos\left(\frac{\pi z}{L'}\right)$	$\frac{2R'}{2.405R}J_1\left(\frac{2.405R}{R'}\right)\frac{2L'}{\pi L}\sin\left(\frac{\pi L}{2L'}\right)$

Φ/Φ_o = ratio of local flux to center-line flux
$\bar{\Phi}/\Phi_o$ = ratio of average flux to center-line flux

By assuming neutron slowing down occurs only in the moderator, the one-group thermal neutron diffusion equation for a homogeneous fuel region in heterogeneous reactors is given by

$$\nabla^2\Phi - \kappa^2\Phi = 0 \tag{5.2.10}$$

where κ is the reciprocal of the thermal diffusion length of the fuel material. Boundary conditions for solutions to Eq.(5.2.10) are the flux is finite and non-negative within the fuel material and the flux is given by Φ_o on the fuel material surface. Table 5.4 lists solutions to Eq. (5.2.10) for several fuel geometries. Also included in Table 5.4 are ratios of average-to-maximum fluxes $(\overline{\Phi}/\Phi_o)$ which are used with Eq. (5.2.7) when the average volumetric heat generation rate in the fuel is known. Diffusion theory underestimates the flux depression in the fuel center which makes these solutions conservative when calculating fuel temperature distributions.

TABLE 5.4

Neutron flux distributions in homogeneous fuel elements

Geometry	Coordinate	Flux Distributions	
		Φ/Φ_o	$\overline{\Phi}/\Phi_o$
Infinite Slab (half thickness a/2)	x	$\dfrac{\cosh(\kappa x)}{\cosh(\kappa a/2)}$	$\dfrac{\tanh(\kappa a/2)}{\kappa a/2}$
Infinite Cylinder (radius r_o)	r	$\dfrac{I_0(\kappa r)}{I_0(\kappa R_0)}$	$\dfrac{2 I_1(\kappa r_0)}{\kappa r_0 I_0(\kappa r_0)}$
Sphere (radius r_o)	r	$\dfrac{\kappa r_0}{\kappa r}\,\dfrac{\sinh(\kappa r)}{\sinh(\kappa r_0)}$	$\dfrac{3[\kappa r_0 - \tanh(\kappa r_0)]}{\kappa^2 r_0^2 \tanh(\kappa r_0)}$

Φ/Φ_o = ratio of local flux to flux at surface of the fuel
$\overline{\Phi}/\Phi_o$ = ratio of average flux to surface flux

5.3 Heat Generation in the Fuel During Transients

Since approximately 7 percent of the fission energy is due to fission product decay, heat generation in the fuel does not exactly follow changes in neutron flux during transient reactor operation. However, from a practical point of view, one can assume the heat generation rate is proportional to the neutron flux for power level increases and decreases in power down to about 50 percent of the steady operating level. For quick reductions in power level from a high level to a low level, Eq. (5.2.3) may seriously underestimate the volumetric heat generation rate. The heat release from fission product decay due to high-power operation may temporarily be higher than the heat generated from low-power operation fission processes. Under these circumstances one must consider individual time-dependent energy sources in order to formulate q'''.

An additional complication arises in determining heat source distributions in large power reactors which have undergone power level changes due to rapid motion of control rods. Changes occur in the spatial dependence of fission product poisons of which Xe^{135} is of greatest importance. The change in xenon distribution can produce substantial power shifts after control rod movement. Figure 5.2(a) illustrates a typical shift of neutron flux along the core axis due to withdrawal of a control rod. Immediately upon control rod removal, the peak neutron flux shifts in the direction of the control rod withdrawal; later the peak neutron flux shifts back toward its original position [4]. The time dependence of the peak neutron flux at the original steady state position A and transient peak at position B are shown in Fig. 5.2(b). As illustrated, the neutron flux during a transient may be larger than the final steady state value. Consequently, in power reactors which have temperature or heat flux limitations, heat source distributions during transient and steady state operation following control rod movement must be carefully evaluated.

Heat source distributions are evaluated during reactor shutdown in order to determine fuel element cooling requirements during normal reactor shutdown procedures and consequences of accident situations such as loss-of-coolant or loss-of-flow accidents. Heat generation in a reactor after shutdown is due to fissions caused by delayed neutrons or photo-neutrons and the decay of fission products, fertile materials (U^{239}, etc.), and other activation products from neutron capture. In most reactors the important heat sources after reactor shutdown are due to fissions resulting from delayed neutrons, the decay of fission products, and for low enriched fuels the decay of fertile materials. Fission is important as a heat source for only the first 5-10 minutes after a reactor scram.

After a step insertion of a large negative reactivity, conditions corresponding to a reactor scram, the neutron flux rapidly decreases and thereafter decreases exponentially with a period approximately equal to the mean life of the longest-lived delayed neutron precursor.

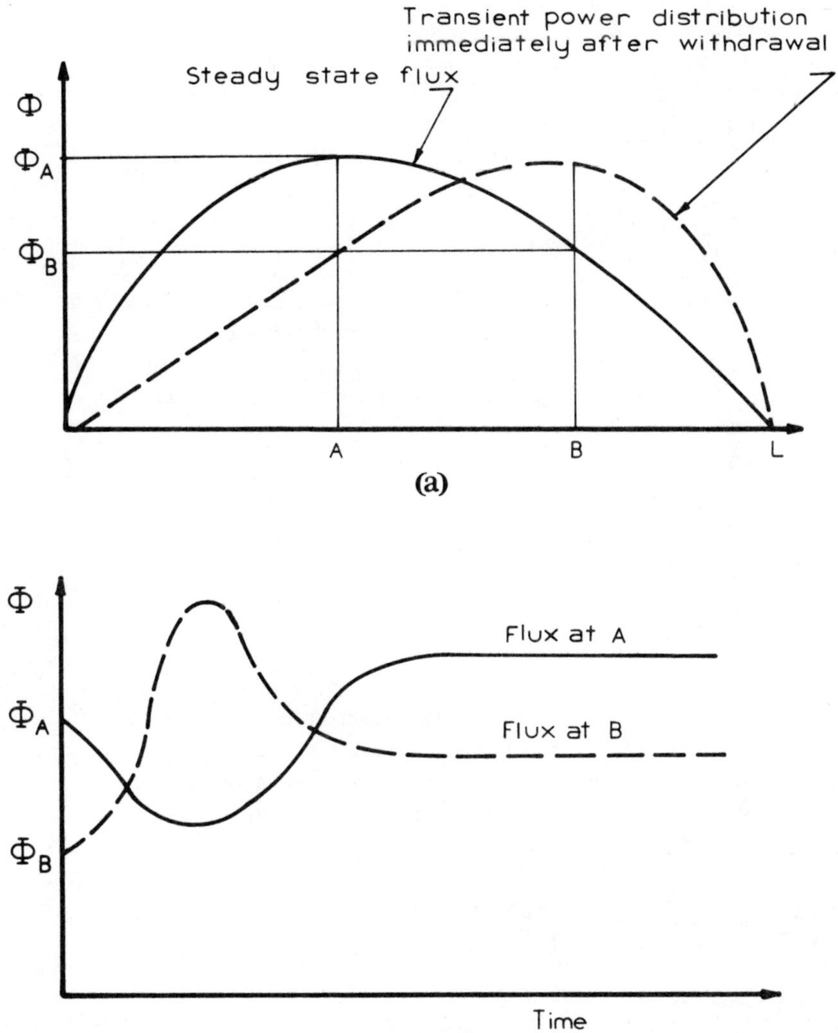

Fig. 5.2. Axial neutron flux shift after control rod withdrawal [4]

By employing solutions for the stable reactor period for large negative reactivities ($|\rho| > 0.01$) and delayed neutrons approximated by a single group, the time-dependent neutron flux for a step reactivity change is approximated by

$$\Phi\,(t) = \Phi_0 \left[\frac{\beta}{\beta-\rho}\, e^{-\gamma_1 t} - \frac{\rho}{\beta-\rho}\, e^{\frac{-(\beta-\rho)\,t}{\ell}} \right] \tag{5.3.1}$$

where

Φ_0 = steady state neutron flux before shutdown, n/cm^2 - sec
β = total delayed neutron fraction
ρ = step reactivity change
ℓ = prompt neutron lifetime, sec
γ_1 = decay constant for longest-lived delayed neutron precursor, sec^{-1}

Equation (5.3.1) overestimates neutron flux after a scram. The coefficient of the first term is eventually determined by the delayed neutron fraction of the longest-lived delayed neutron precursor which is considerably smaller than the total delayed neutron fraction. A more accurate representation of the time-dependent neutron flux is obtained by solving the reactor kinetics equations and employing the six or more groups of delayed neutron sources.

For a U^{235} fueled, water-moderated reactor, typical values for the parameters in Eq. (5.3.1) are γ_1 = 0.0124 sec^{-1}, β = 0.0065, and ℓ = 10^{-4}sec. Therefore, for a reactor scram the reactivity insertion may be of the order of -0.09 and the reactor power level due to fission decreases in the following manner:

$$P/P_0 = 0.067\, e^{-0.0124t} + 0.933\, e^{-965t} \qquad (5.3.2)$$

The second term in Eq. (5.3.2) decays to insignificance in less than 0.01 second and the reactor power then decreases exponentially with a period of about 80 seconds. For heat transfer analysis one can assume all fission energy (195 MeV/fission) is deposited in the fuel material.

Heat generation from fission product decay results from the beta and gamma emissions from isotopes which decay with half lives ranging from millionths of a second to millions of years. A simple empirical formula, accurate within a factor of two, for the rate of energy release from fission products is [3]

$$\begin{aligned} \text{Beta energy release} &= 1.40\, t^{-1.2}, \text{ MeV/fission/sec} \\ \text{Gamma energy release} &= 1.26\, t^{-1.2}, \text{ MeV/fission/sec} \end{aligned} \qquad (5.3.3)$$

where t is in seconds and the range of validity is for 10 sec $< t <$ 100 days. In order to determine the rate of energy release from fission products after a reactor is shutdown, it is necessary to integrate Eq. (5.3.3) over the time of reactor operation. By assuming a total energy release rate of 202 MeV/fission, it requires 3.1 x 10^{10} fissions/sec to generate a power level of one watt. Therefore, the number of fissions/cc produced in the reactor fuel for a volumetric heat generation rate of q_0''' watts/cc over a time increment dT seconds is 3.1 x $10^{10} q_0'''$ dT. The energy release rate from fission product decay at some later time τ due to fissions in the time increment dT is given by

$$\text{Beta energy} = 1.4(\tau-T)^{-1.2} \times 3.1 \times 10^{10} q_0''' \, dT, \text{ MeV/cc-sec}$$
$$\text{Gamma energy} = 1.26(\tau-T)^{-1.2} \times 3.1 \times 10^{10} q_0''' \, dT, \text{ MeV/cc-sec} \tag{5.3.4}$$

The fission product energy release rate at some later time τ due to reactor operation over T seconds is given by

$$\text{Beta energy} = 4.35 \times 10^{10} \int_0^{T_0} q_0''' (T) \times (\tau-T)^{-1.2} dT, \text{ MeV/cc-sec}$$
$$\text{Gamma energy} = 3.9 \times 10^{10} \int_0^{T_0} q_0''' (T) \times (\tau-T)^{-1.2} dT, \text{ MeV/cc-sec} \tag{5.3.5}$$

for 10 sec $< (\tau-T_0) <$ 100 days. Some error is incorporated in Eq. (5.3.5) because Eq. (5.3.3) is not recommended for times after fission in excess of 100 days. For reactors operating at constant power levels over the time interval T_0 , Eq. (5.3.5) is integrated directly.

$$\text{Beta energy} = 2.18 \times 10^{11} q_0''' \left[(\tau-T_0)^{-0.2} - \tau^{-0.2}\right], \text{ MeV/cc-sec}$$
$$\text{Gamma energy} = 1.95 \times 10^{11} q_0''' \left[(\tau-T_0)^{-0.2} - \tau^{-0.2}\right], \text{ MeV/cc-sec} \tag{5.3.6}$$

By converting units on Eq. (5.3.6) to watts, the fraction of reactor steady state volumetric heat generation rate q_0''' due to fission product beta decay and gamma emission is given by

$$q_\beta''' / q_0''' = 0.035 \left[(\tau-T_0)^{-0.2} - \tau^{-0.2}\right]$$
$$q_\gamma''' / q_0''' = 0.031 \left[(\tau-T_0)^{-0.2} - \tau^{-0.2}\right] \tag{5.3.7}$$

Equation (5.3.7) is used for estimating decay heat generation rates in reactor fuel elements after shutdown. Usually all energy from beta particles is deposited in the fuel material. Depending upon the reactor configuration somewhere from 10 to 70 percent of the gamma energy is deposited in the fuel material. For conservative calculations one can assume that all gamma energy is deposited in fuel elements when determining cooling requirements for shutdown reactors or spent fuel elements in storage. Figure 5.3 illustrates the sum of the decay power due to beta and gamma radiation as a function of time after reactor shutdown for different times of reactor operation. As seen, for long operating times the decay power immediately after shutdown is practically independent of operating time.

For low enriched fuels the decay of U^{239} and Np^{239} in the decay chain for plutonium production is a significant contributor to the decay after heat for several days after reactor shutdown. An experimental equation for the decay heat production in a one inch diameter uranium

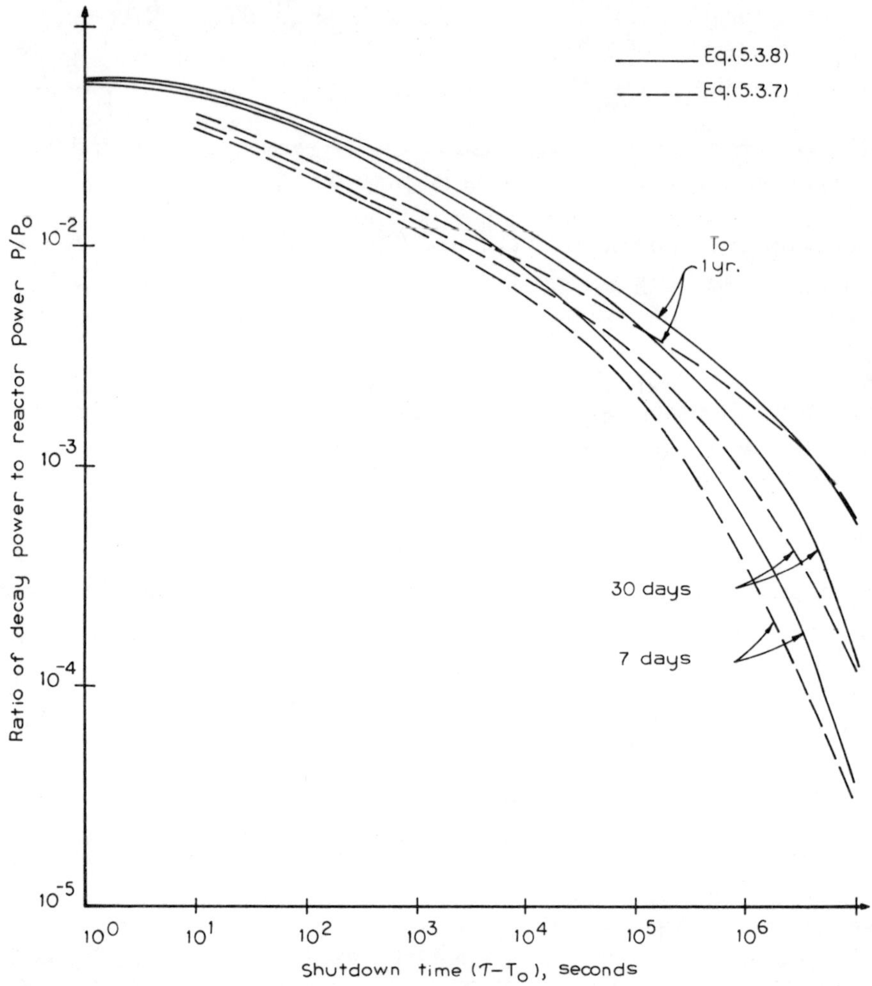

Fig. 5.3. Decay power from U^{235} fission products as a function of shutdown time.

rod which accounts for U^{235} fission product decay and U^{239} and $N_p{}^{239}$ decay is given by [3]

$$P/P_0 = 0.1 \left[(\tau - T_0 + 10)^{-0.2} - (\tau + 10)^{-0.2} + 0.87(\tau + 2 \times 10^7)^{-0.2} - 0.87(\tau - T_0 + 2 \times 10^7)^{-0.2}\right] \quad (5.3.8)$$

where P is the rate of heat production $\tau - T_0$ seconds after shutdown and P_0 is the steady state rate of heat production during the reactor operating time of T_0 seconds. An estimate of the error in Eq. (5.3.8) is as follows: under 1 sec, large error; 1-10^2 sec, ± 50%; 10^2 –10^4 sec, ± 30%; 10^4 -10^6 sec, ± 10%; and 10^6 -10^8 sec, ± 50%.

Equation (5.3.8) is illustrated in Fig. 5.3 for several reactor operating times. As shown, decay heat production rates are considerably larger than predicted by Eq. (5.3.7). For low-enriched fuels it is probably advisable to use Eq. (5.3.8) for estimating decay heat production rates because of its incorporating the effects of U^{239} and Np^{239} decay.

The total heat generation rate in fuel materials after shutdown is given by the sum of the heat generation rate due to fission and the heat generation rate due to fission product decay. By examining Eqs. (5.3.1) or (5.3.2) and comparing with Fig. 5.3, it is apparent that heat generation due to fission is of little significance 5 to 10 minutes after shutdown.

5.4 Heat Generation in the Cladding, Coolant, Moderator, Core Structure, and Control Rods

The importance of heat generation in non-fuel materials is dependent upon the reactor type. Generally heat generation in the cladding is neglected because of large heat fluxes originating from the fuel. Heat generation in the coolant is usually of no significance when compared to heat transfer rates to the coolant. Heat generation in the moderator is of little significance when the moderator is used as the coolant such as in present light water-moderated and -cooled reactors. In graphite-moderated, gas-cooled reactors which have fuel uniformly dispersed throughout the graphite, the volumetric heat generation rate in the moderator can be assumed the same as that due to a fission source homogeneously dispersed in the moderator.

In heterogeneous graphite-moderated reactors which have fuel slugs dispersed in a graphite matrix, further thought is needed in arriving at expressions for the volumetric heat generation rate. It is reasonable to assume that all fission neutron kinetic energy is dissipated in the moderator which amounts to 5 MeV per fission. For reactors operating over times sufficiently long to insure an equilibrium concentration of fission products the gamma energy produced per fission ranges from 18 to 22 MeV. The fractional amount of gamma energy absorbed in the moderator is approximated by the ratio of the mass of moderator to the mass of all materials in the core. Consequently, the fractional amount of reactor power dissipated in the moderator is approximately given by

$$Q_m / Q = \left[5 + (18\text{–}22)\,(w_m / w_T)\right] /202 \qquad (5.4.1)$$

where w_m / w_T is the ratio of moderator mass to total core mass. The spatial distribution of the volumetric heat generation rate in the moderator is assumed similar to that of the neutron flux. Therefore, the moderator volumetric heat generation rate for a heterogeneous reactor is approximated by

$$q'''_m = \frac{[5 + (18\text{–}22)(w_m/w_T)]\,Q}{202V_m}\frac{\Phi}{\overline{\Phi}} \tag{5.4.2}$$

where Q is the reactor thermal power, V_m is the moderator volume, and $\Phi/\overline{\Phi}$ is the ratio of local neutron flux to average neutron flux given in Tables 5.2 and 5.3.

Heat generation in the reactor structure is generally due to the absorption of gamma rays. Obtaining an accurate profile of heat generation in the structure requires a detailed analysis of the production and dissipation of gamma rays in the reactor system. For an engineering approximation as to the magnitude of the heat generation in the reactor structure, one may assume the fraction of gamma rays absorbed in the structure is given by the ratio of the mass of structure to the total core mass. Since structural materials are usually relatively thin, the volumetric heat source may be assumed uniform across the thickness. In the analysis of structural components one can initially make liberal overestimates of volumetric heat generation rates to determine whether any heating or thermal stress problems arise. If a potential problem arises a more accurate description of the volumetric heat generation rate must be made in order to determine whether design changes are in order.

Heat generation in control rods is due to absorption of gamma rays originating from the fission process and absorption of radiation resulting from neutron capture in the control rod material. Neutron capture may produce capture gamma rays, alpha particles, beta particles, and gamma rays from decay of daughter products. In particular, large volumetric heat sources can occur in control rods using boron-10 because the capture reaction produces an alpha particle. The energy released from the reaction is 2.31 MeV in reaction product kinetic energy and 0.48 MeV in gamma energy. The volumetric heat generation rate in a boron containing control rod is given by

$$q'''_c = 2.31\,\Phi\Sigma_a + E_c\,,\ \text{MeV/cm}^3\text{–sec} \tag{5.4.3}$$

where

Φ = local average neutron flux, neutrons/cm^2-sec

Σ_a = boron-10 average absorption cross section, cm^{-1}

E_c = volumetric heat generation rate due to gamma absorption, MeV/cm^3–sec

Depending upon the neutron capture reaction, expressions similar to Eq. (5.4.3) are formulated for estimating volumetric heat generation rates in control rods.

5.5 Heat Generation in Reflectors, Thermal Shields, Pressure Vessel, and Biological Shield

Heat generation in the out-of-core materials is due to leakage of gamma rays and neutrons from the core. Since about one-third of the gamma rays produced from fission are due to fission product decay, volumetric heat generation rates in out-of-core materials are not proportional to reactor power level during reactor transients or shutdown. As an example, one hour after reactor shutdown the heat generation in the core is about one percent of operating level while in the reflector and shields the heat generation is about 10 percent of operating level [3].

Depending upon the nature of reflector materials, heat generation may be due to elastic neutron scattering and gamma heating from core gammas, capture and activation gammas, and inelastic neutron scattering. In some reactor designs the reflector contains fertile materials such as thorium or depleted uranium and fast neutron fission is an additional heat source.

In order to reduce the amount of heat deposition in reactor pressure vessels and biological shields in power reactors, thermal shields consisting of successive layers of a dense metal are placed between the reactor core or reflector and the pressure vessel. Heat generation in thermal shields is due to gamma heating from core gammas, inelastic neutron scattering, and thermal neutron capture and activation gammas. Due to thermal stress limitations, the thickness of thermal shields is usually quite small in comparison to their diameter. Consequently, thermal shields can be assumed to have a slab geometry when developing expressions for volumetric heat generation rates.

The attenuation of core gammas in thermal shields is exponential and variation of the uncollided gamma flux of energy E_γ through a slab with a source on one side is given by [3]

$$\phi_\gamma(x) = \phi_\gamma(0)\, e^{-\mu x} \tag{5.5.1}$$

where

$\phi_\gamma(0)$ = gamma flux at shield surface, photons/cm^2–sec

μ = linear attenuation coefficient for gammas of energy E_γ, cm^{-1}

x = distance from shield surface, cm

The volumetric heat generation rate due to gamma flux $\phi_\gamma(x)$ of energy E_γ is given by

$$q'''_\gamma(x) = B_e \mu_e E_\gamma \phi_\gamma(0)\, e^{-\mu x}, \ \text{MeV/cm}^3\text{–sec} \tag{5.5.2}$$

where

B_e = energy absorption buildup factor

μ_e = energy absorption coefficient, cm^{-1}

The energy absorption buildup factor accounts for absorption of secondary gammas resulting from Compton scattering of the uncollided gamma flux. Buildup factors are dependent upon gamma ray energy, type of material, and depth of penetration into the material . Standard textbooks on shielding give tables or graphs of linear attenuation and absorption coefficients and buildup factors [5,6].

Since the gamma flux striking the surface of thermal shields has a wide range of energies, the volumetric heat generation rate is represented as the sum of heat generation rates due to gamma rays in different energy increments. Therefore, the volumetric heat generation rate in thermal shields due to external gamma ray sources is given by

$$q_\gamma''' = \sum_i \left[B_e \mu_e E_\gamma \phi_\gamma(0)\, e^{-\mu x} \right]_i \qquad (5.5.3)$$

where the subscript "i" is used to represent the average gamma ray energy over a specified energy increment.

For thin thermal shields ($\mu x < 1.0$), the energy absorption buildup factor is approximately unity and the volumetric heat generation rate due to an external gamma source is represented by a sum of exponentials. Consequently, the major portion of the heat generation occurs on the inner surface of thermal shields.

Expressions for heat generation due to gamma rays resulting from inelastic neutron scattering and neutron capture are quite complicated. Techniques for evaluating heat generation rates are discussed in books on reactor shielding or physics. A simplified approach for estimating the volumetric heat generation rate due to inelastic neutron scattering is to assume the heat generation due to gamma ray absorption occurs at the point of neutron collision. Consequently, the volumetric heat generation rate due to inelastic scattering of a fast neutron flux $\Phi_f(x)$ of average energy E_n is given by [3]

$$q_i'''(x) \approx \Sigma_i \Phi_f(x)\, f(E_n), \ \text{MeV/cm}^3\text{–sec} \qquad (5.5.4)$$

where Σ_i is the macroscopic inelastic scattering cross section and $f(E_n)$ is the fraction of neutron energy lost in the collision. A similar expression is used for estimating the gamma heating due to neutron capture.

$$q_\gamma'''(x) \approx \Sigma_a \Phi(x) N_\gamma, \ \text{MeV/cm}^3\text{–sec} \qquad (5.5.5)$$

where Σ_a is the average macroscopic absorption cross, $\Phi(x)$ is the local average neutron flux, and $N\gamma$ is total gamma energy liberated per neutron capture. Equations (5.5.4) and (5.5.5) are approximate and if predicted heat generation rates become excessive, more sophisticated techniques should be employed.

Heat generation in pressure vessel walls is due to gamma heating from gamma rays originating inside the pressure vessel and gamma rays resulting from neutron inelastic scattering and neutron capture within the pressure vessel. Equations (5.5.3), (5.5.4), and (5.5.5) are used for estimating volumetric heat generation rates. For pressurized-water and boiling-water reactors, pressure vessel walls are quite thick and gamma ray buildup factors cannot be assumed unity.

Heat generation in biological shields is due to fast neutron elastic scattering and gamma heating due to gamma rays penetrating the shield and gamma rays resulting from inelastic scattering and neutron capture in the shield material. Additional heat sources in shield materials is due to non-capture neutron reactions such as (n,α) reactions with Li^6 and B^{10}. Equations (5.5.3), (5.5.4), and (5.5.5) are used for estimating volumetric heat generation rates from gamma heating. Similar expressions can be developed for estimating heat generation rates from other heat sources. Volumetric heat generation rates in biological shields are usually not very intense because of strong attenuation of gamma ray and neutron fluxes in thermal shields and pressure vessel walls.

REFERENCES

1. Lamarsh, J. R., **Introduction to Nuclear Reactor Theory**, Addison-Wesley, Reading, Massachusetts, 1966.
2. Henry, A. F., **Nuclear Reactor Analysis**, MIT Press, Cambridge, Massachusetts, 1975.
3. Glasstone, S. and Sesonske, A., **Nuclear Reactor Engineering**, D. Van Nostrand Co., Princeton, New Jersey, 1967.
4. Tong, L. S. and Weisman, J., **Thermal Analysis of Pressurized Water Reactors**, ANS, Lagrange, Illinois, 1970.
5. Schaeffer, N. M., Reactor shielding for nuclear engineers, TID-25951, U.S.A.E.C. (1973).
6. Rockwell, T., Reactor shielding design manual, TID-7004, U.S.A.E.C. (March 1956).
7. El-Wakil, M. M., **Nuclear Heat Transport**, International Textbook Co., Scranton, Pennsylvania, 1971.

PROBLEMS

5.1 A pressurized water reactor has an active height of 12 ft. and an equivalent active diameter of 11.5 ft. The core contains 54,000 fuel rods of 2.9 percent enriched UO_2 pellets of 0.324 in. diameter. The UO_2 density is 10 g/cc. The water average temperature is 600°F and the reactor power level is 3800 MW(t). Assuming the neutron flux distribution is that of a bare, homogeneous reactor, determine the maximum thermal neutron flux in the core. The 2200 m/sec fission cross section for U^{235} is 582 barns.

5.2 For the reactor in problem 5.1 determine the average and maximum volumetric heat generation rates in the UO_2 fuel.

5.3 If the maximum thermal neutron flux is 5×10^{13} n/cm^2-sec for the reactor described in problem 5.1, what is the reactor power output in MW(t)?

5.4 An HTGR is designed as a parallelepiped with core dimensions of 21 ft. height and transverse dimensions of 25 ft. on each side. The core is surrounded on all sides by a graphite reflector of thickness sufficient to give a reflector savings of 2 ft. in all directions. What is the ratio of the maximum-to-average neutron flux in this reactor? Compare this result with a bare parallelepiped reactor with negligible extrapolation distance.

5.5 The reactor in problem 5.4 contains 2000 kg U^{235} which may be assumed dispersed uniformly throughout the core. If the reactor power level is 2900 MW(t) and the average moderator temperature is 1340°F, what is the average and maximum thermal neutron flux in the core?

5.6 Unit 2 of the Three Mile Island nuclear power station was operating at 2700 MW(t) prior to its loss of pressure accident. The plant had been operating for 90 days at full power. Determine the decay heat removal rate one hour and 24 hours after shutdown.

5.7 The decay heat removal rate for a shut down reactor that had operated for a long period of time ($T_0 = \infty$) is well-approximated by $P/P_0 = 0.095\,(\tau - T_0)^{-0.26}$, where $\tau - T_0$ is time after shutdown in seconds [7]. For a reactor that had initially operated at 2700 MW(t), what would be the amount of heat removed in Btu during the second hour after shutdown.

CHAPTER 6

HEAT TRANSFER IN NUCLEAR POWER SYSTEMS

6.1 Introduction

Removing heat from reactor cores is one of the major problems encountered in power reactor design. This is primarily due to power densities (expressed in kW/liter or MW/ft^3) in power reactor cores being orders of magnitude higher than other conventional heat sources. Producing heat in reactor cores presents no difficulty, the problem is removing heat without having excessively high temperatures inside fuel elements or on their surfaces.

A major portion of reactor system thermal analysis is determining fuel element surface and interior temperature distributions and coolant heat removal rates. An additional consideration with reactors employing large fuel element surface heat fluxes is thermal stability or prevention of fuel element melting due to vapor blanketing of heat transfer surfaces. Determining fuel element temperature distributions usually involves solutions to the heat conduction equation; however, for rigorous analysis, the presence of spatially dependent heat generation terms and temperature dependent material properties usually makes it impossible to find closed-form solutions to problems. An analysis of heat removal rates usually involves convection heat transfer because reactor cores are normally cooled by circulating fluids. Because of unusual geometries and high heat fluxes encountered in reactor systems, the analysis of convection heat transfer is most difficult and requires heavy reliance upon experimental data.

Due to high power densities in many nuclear reactors, the heat removal rate from any given coolant channel is usually quite large. Consequently, some of the problems involved in reactor thermal analysis are seen by examining a simplified equation for heat removal rate from a coolant channel.

$$Q = \bar{h}A(T_w - T_b)_{avg} \tag{6.1.1}$$

where

Q = heat removed, Btu/hr
$\bar{h}$ = average heat-transfer coefficient, Btu/(hr)(ft^2)(°F)
A = channel heat-transfer surface area, ft^2
$(T_w - T_b)_{avg}$ = average channel wall and fluid temperature difference, °F

Inspection of Eq. (6.1.1) shows the heat removal rate is increased by increasing either $\bar{h}$, A, or $(T_w - T_b)_{avg}$. In order to achieve maximum capability of nuclear reactors, it is desirable to maximize each of these quantities.

Large heat transfer surface areas are desired in reactor cores. Consequently, due to ease in fabrication and good neutron economy, most power reactors use small diameter, rod-type fuel elements in order to obtain large surface areas for a given core volume. A minimum fuel rod diameter exists due to the necessity of each rod containing enough fuel with respect to its cladding to enable reactor criticality. In addition, practical design limitations and the need for rod structural strength also limit minimum fuel rod diameters.

Heat-transfer coefficients are increased by using high flow rates and fluids with good heat transfer properties. Fuel element vibration and pumping power requirements place upper limits on coolant velocities. Large thermal conductivities and heat capacities are desirable properties for reactor coolants. Some liquid metals possess these properties; however, nuclear considerations make the use of these materials undesirable for some reactor types. Heavy water or gases are the most suitable coolants for natural-uranium fueled reactors because of their low neutron absorption properties.

The last factor to be considered is the average temperature difference between the fuel element surface and reactor coolant. The fuel element surface temperature is limited by the cladding melting point, although from a practical standpoint the cladding temperature is limited to a lower value because of the reduction in cladding strength with high temperatures. Other temperature limitations may be imposed by the fuel melting point or phase changes. Uranium metal undergoes a phase change at about1235°F which results in a decrease in density. Ceramic fuels (UO_2, UC, PuO_2, etc.) are more desirable for use in power reactors, since they have no temperature restrictions due to phase changes. However, carbide fuels have not seen much use to date because of their high rate of swelling with heating and irradiation and chemical reactions with water. The burnup lifetime of fuels also limits cladding temperatures, since the higher the temperatures, the lower the fuel element lifetime. From the standpoint of removing reactor heat, low coolant temperatures are desirable. However, this imposes a conflict with the thermodynamic requirements needed to produce reasonable energy conversion efficiencies. Consequently, a compromise in fuel element surface and coolant temperatures is necessary in order to have optimum system performance.

From the discussion of the quantities in Eq. (6.1.1) it is seen that many facets are involved in core design. Practical design, thermodynamics, fluid flow, metallurgy, nucleonics, and stress limitations all place restrictions upon the maximum value of these quantities. As a consequence, the core designer must be familiar with all these fields of study so that sensible compromises are made in design.

In the thermal analysis of nuclear reactors, several parameters are used to describe reactor performance. These parameters are the **fuel specific power, core power density, linear heat rate,** and **cladding or fuel surface heat flux**. However, constraints are placed upon their maximum values due to materials failure or economic considerations. The fuel specific power is expressed in kW/kgU and large values are desirable to reduce fuel inventory and its associated cost. However, a conflict arises with increased fuel specific powers due to burnup considerations and the need for more frequent reactor shutdowns due to refueling. Large core power densities (kW/liter) are desirable due to economic benefits from reduced core and pressure vessel sizes. Linear heat rates are given in kW per foot of fuel element length and are normally used as a performance index for rod-type fuel elements. For fuel rods, maximum fuel temperatures are proportional to the linear heat rate and, consequently, any limits on fuel temperatures impose corresponding limits on linear heat rates. Large cladding or fuel surface heat fluxes (Btu/hr–ft^2) are desirable in order to achieve high core power densities. However, this parameter determines the temperature difference between the bulk coolant and fluid-solid interface. Temperature limitations on solid surfaces or interior fuel regions limit surface heat fluxes.

6.2 Heat Transfer to the Coolant

Heat transfer to reactor coolants is frequently expressed in terms of

$$q''_w = h\Delta T \tag{6.2.1}$$

where

q''_w = wall heat flux, Btu/hr–ft^2

h = heat-transfer coefficient, Btu/hr–ft^2–°F

$\Delta T = T_w - T_{ref}$, °F

The term $T_w - T_{ref}$ is the difference between the wall surface temperature and some fluid reference temperature which is usually the bulk fluid temperature T_b. For channel flows, T_b is obtained from an energy balance on the fluid.

The major problem in using Eq. (6.2.1) is finding proper expressions for h. Methods for evaluating heat-transfer coefficients usually classify fluids into three categories—(i) ordinary fluids which have Prandtl numbers greater than 0.6, (ii) liquid metals which are fluids with Prandtl numbers less than 0.1, and (iii) fluids with subcooled or bulk boiling. Heat-transfer coefficients are somewhat analogous to friction factors in that they are larger in entrance regions of heated channels than in downstream regions. This is due to the buildup of thermal boundary layers in the channel which are similar to velocity boundary layers at channel inlets.

Heat Transfer to Ordinary Fluids

For turbulent flow, heat-transfer coefficients are usually obtained from experiments with uniform wall heat flux in circular tubes. For fluids with Prandtl numbers greater than 0.6, one can use the equivalent diameter concept for evaluating Nusselt numbers for non-circular tube geometries by using circular tube correlations. This is due to the fact that the majority of the resistance to heat transfer is adjacent to the channel wall and, as was the case in the study of hydrodynamics, a small region of the channel wall is not strongly influenced by its surroundings. An exception may be regions like the apex to triangular shaped channels in which there may be stagnant flow zones. A necessary requirement is that the heat flux or wall temperature be uniform around the channel periphery.

(1) Dittus-Boelter Equations for Pipe Flow

The most famous expression for turbulent flow is the Dittus-Boelter equation given by [1].

$$N_{Nu} = \left(\frac{\bar{h}D_e}{k}\right)_{mb} = 0.023\ (N_{Re})_{mb}^{0.8}\ (N_{Pr})_{mb}^{n} \tag{6.2.2}$$

where

N_{Nu} = Nusselt number, dimensionless
$\bar{h}$ = heat-transfer coefficient, Btu/hr-ft^2-°F
D_e = equivalent diameter, ft
k = fluid thermal conductivity, Btu/hr-ft-°F
N_{Re} = Reynolds number, $\rho u D_e/\mu$
N_{Pr} = fluid Prandtl number, $c_p \mu/k$

The conditions on Eq. (6.2.2) are

1) Fluid properties evaluated at the arithmetic mean bulk temperature

$$T_{mb} = \frac{T_{b1} + T_{b2}}{2}$$

2) $N_{Re} > 10{,}000$
3) $0.7 < N_{Pr} < 100$
4) n = 0.4 for heating, 0.3 for cooling
5) L/D > 60 (entrance effects negligible)

Although Eq. (6.2.2) is obtained from experiments in circular tubes it is frequently used for other flow geometries. For high heat fluxes in which there may be substantial variation between the fluid viscosity at the wall and bulk fluid temperatures, Eq. (6.2.2) may be modified by using the Sieder-Tate [2] correction which multiplies Eq. (6.2.2) by $(\mu_{mb}/\mu_w)^{0.14}$.

(2) Colburn Analogy

Another correlation is the Colburn analogy which relates heat-transfer coefficients to friction factors [3]

$$N_{St} = \frac{h}{\rho_f u c_{pmb}} = \frac{f_W}{8}(N_{Pr})_f^{-2/3} \qquad (6.2.3)$$

where

N_{St} = Stanton number, dimensionless
f_W = Darcy-Weisbach friction factor

The subscripts mb and f denote mean bulk and film temperatures, respectively. The film temperature is given by $(T_w + T_b)/2$. Equation (6.2.3) then includes roughness, entrance, and curvature effects in heat-transfer coefficients to the same extent as included in friction factors. Because the wall temperature is needed to evaluate fluid properties in Eq. (6.2.3), an iteration procedure is required in its use. This equation is frequently used in the analysis of HTGRs.

(3) Flow Through an Annulus

Several equations are available for flow through an annulus. For the inside surface of the annulus or outside surface of the inside tube, the following relationship is obtained [4]:

$$N_{Nu} = \left(\frac{h_{1m} D_e}{k}\right)_{mb} = 0.02 \left(\frac{D_e \rho u_b}{\mu}\right)_{mb}^{0.8} (N_{Pr})_{mb}^{1/3} (D_2/D_1)^{0.53} \qquad (6.2.4)$$

The conditions are

1) Properties evaluated at the arithmetic mean bulk temperature

2) $D_e = D_2 - D_1 = \dfrac{4 \text{ area}}{\text{wetted perimeter}}$

3) $N_{Pr} > 0.6$

4) $N_{Re} > 10^4$

5) $L/D_e > 20$

For the outside surface of the annulus or inner surface of the outer tube, Eq. (6.2.2) is used providing the equivalent diameter is taken as the characteristic length.

(4) Thin Rectangular Channels

The use of the standard pipe equation has proven satisfactory for rectangular channel flow providing the equivalent diameter is used. An empirical equation proposed by Hausen for computing local heat-transfer coefficients in rectangular channels is given by [5]

$$(N_{Nu})_z = 0.116\left[(N_{Re})_{bz}^{2/3} - 125\right](N_{Pr})_b^{1/3} \times \left[1 + 1/3\left(\frac{D_e}{z}\right)^{2/3}\right]\left(\frac{\mu_b}{\mu_w}\right)_z^{0.14} \tag{6.2.5}$$

where z is position along the axis. This equation is restricted to film temperatures $\geqslant 15°C$ and $z/D_e \geqslant 30$. The mean heat-transfer coefficient for a channel is obtained by integrating Eq. (6.2.5 to give

$$(N_{Nu})_m = \frac{1}{L}\int_o^L (N_{Nu})_z\, dz = 0.116\left[(N_{Re})_b^{2/3} - 125\right]\left[1 + \left(\frac{D_e}{L}\right)^{2/3}\right](N_{Pr})_b^{1/3}\left(\frac{\mu_b}{\mu_w}\right)^{0.14} \tag{6.2.6}$$

Equation (6.2.5) was recommended for the thermal analysis of the High Flux Isotope Reactor (HFIR).

(5) Rod Bundles

Weisman [6] correlated rod bundle heat transfer with

$$N_{Nu} = C(N_{Re})_b^{0.8}(N_{Pr})_b^{1/3} \tag{6.2.7}$$

where for a triangular array with $1.1 \leqslant P/D \leqslant 1.5$

$$C = 0.026\,(P/D) - 0.006$$

and for a square array with $1.1 \leqslant P/D \leqslant 1.3$

$$C = 0.042\,(P/D) - 0.024$$

P/D is the rod pitch-to-diameter ratio. Rod bundle data collected by Subbotin [7] using water agree with Eq. (6.2.7); however, there is wide discrepancy in data for air.

Experiments indicate that rod bundle heat-transfer coefficients increase with increasing rod spacing and are above circular tube correlations for $P/D > 1.10$. In addition, circumferential wall temperature variations are not significant for $P/D > 1.10$; however, they are of concern for closely spaced rods at low Reynolds numbers. There is little transverse mixing between adjacent subchannels unless spacers designed to promote mixing are used [8].

The effects of rod spacers and surrounding channel walls on local heat-transfer coefficients have not been resolved. Different fuel assemblies have their own peculiar flow characteristics and prototype testing is necessary in order to confirm an analytical method. In general, grid spacers trip the boundary layer and increase heat-transfer coefficients by starting another thermal entry length. However, it has been reported that local hot spots occur immediately downstream from grid spacers due to local flow retardation over the heated surface [9]. Spacers which produce gross rod bundle mixing such as wire wraps alter flow patterns to such an extent that basic heat transfer correlations are no longer valid [8].

(6) Peripheral Heat Flux Variation

In some applications of nuclear fuel element analysis, the wall heat flux may vary around a channel periphery. This may exist in pressure tube reactor fuel elements or in regions adjacent to control rods.

For the case of a flux tilt, the heat flux is approximated by

$$q''_w(\theta) = \overline{q''_w}\,(1 + \epsilon \cos\theta) \tag{6.2.8}$$

where $\overline{q''_w}$ is the average heat flux around the channel periphery. By assuming the eddy diffusivity for heat transfer in the circumferential direction is the same as in the radial direction, a theoretical study for fully-developed turbulent flow in a circular tube yielded for the Nusselt number [10]

$$N_{Nu}(\theta) = \frac{1 + \epsilon\cos\theta}{(1/N_{Nu}) + (S_1\epsilon/2)\cos\theta} \tag{6.2.9}$$

where N_{Nu} is the fully-developed Nusselt number for constant heat flux. The function S_1 is tabulated in Table 6.1 for a wide range of Reynolds and Prandtl numbers. The effect of peripheral heat flux variations is more pronounced with decreasing Prandtl number as should be expected.

As an example, let N_{Pr} = 0.7 and N_{Re} = 100,000. Then N_{Nu} = 199 and from Table 6.1, S_1 = 0.018 . Substitution into Eq. (6.2.9) for an ϵ = 0.20 gives local variations in the Nusselt number ranging from 176 to 245. This prediction is rather pronounced for a 20 percent variation in heat flux from its average value.

It should be noted that Eq. (6.2.9) predicts high heat-transfer coefficients in regions of high heat flux and tends to predict uniform temperatures around a channel periphery for fluids with $N_{Pr} > 1.0$. Thus in an analysis one could use uniform heat flux heat-transfer coefficients and calculate wall temperatures based upon $\overline{q_w''}$.

TABLE 6.1
Circumferential heat flux functions S_1 for fully developed turbulent flow in a circular tube with constant heat rate [10]

Pr \ Re	Laminar	10^4	3×10^4	10^5	3×10^5	10^6
0	1.000	1.000	1.000	1.000	1.000	1.000
0.001		1.000	1.000	0.999	0.974	0.901
0.003		0.999	0.994	0.957	0.831	0.473
0.01		0.991	0.952	0.733	0.409	0.161
0.03		0.923	0.699	0.348	0.145	0.0535
0.7		0.121	0.0490	0.0180	0.00721	0.00275
3		0.0448	0.0178	0.00629	0.00246	0.000902
10		0.0239	0.00931	0.00322	0.00123	0.000438
30		0.0151	0.00582	0.00199	0.00751	0.000166
100		0.00994	0.00383	0.00130	0.000486	0.000166
1000		0.00513	0.00198	0.000667	0.000248	0.0000841

(7) Transition Flow

Heat transfer in the range of Reynolds numbers from 2000 to 10,000 is poorly understood. This is because the change from laminar to fully-developed turbulent flow is not abrupt. None of the equations developed so far is applicable in this region. Figure 6.1 illustrates a graphical correlation of the Stanton number as a function of Reynolds number in this region. This correlation is only valid for $N_{Pr} > 0.7$.

Flows in a transition Reynolds number range are not encountered in power reactors; however, such flows might exist in low powered research reactors.

(8) Transient Convection Heat Transfer

Part of reactor analysis deals with transients which may be startups or shutdowns, loss-of-flow accidents, or loss-of-coolant accidents. Common practice is to use heat-transfer coefficient data from steady-state experiments. However, there is some evidence that

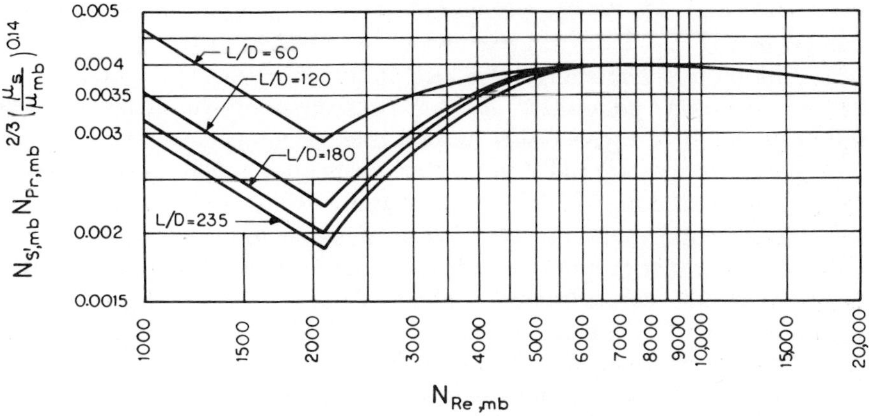

Fig. 6.1. Forced convection transition flow [11].

steady-state heat-transfer coefficients are not suitable for analyzing transients. From analysis of axially varying heat flux, heat-transfer coefficients are higher in regions of increasing heat flux and lower in regions of decreasing heat flux. One can probably infer from these results that, for transients with increasing heat flux, heat-transfer coefficients are higher and, for transients with decreasing heat flux, heat-transfer coefficients are lower.

The theoretical analysis of transient heat convection is quite complicated because one must simultaneously solve the energy equation for the fluid and heat conduction equation for the heated solid. The heat capacity of the solid has a strong influence on temperature distributions in both materials during a transient.

Some theoretical studies for laminar, slug flow between parallel plates indicate that the use of steady-state heat-transfer coefficients improves as the ratio of the solid-to-fluid heat capacities increases [12]. Further improvements would exist for turbulent flow of ordinary fluids.

Heat Transfer to Liquid Metals

In the early 1950's to the mid 1960's liquid metals were used as coolants for a variety of nuclear reactors. Such reactors were the Experimental Breeder Reactors I and II (EBR-I and EBR-II), Sodium Reactor Experiment (SRE), Enrico Fermi Fast Breeder Reactor, and Sodium Graphite Moderated Reactor (SGR). All of these reactors employed sodium as a coolant. Unfortunate accidents during the operation of several sodium-cooled reactors, none of which were attributable directly to the use of sodium, caused a considerable disinterest in the use of liquid metal coolants. As a consequence, no new construction of liquid-metal-cooled power reactors took place for over a decade. However, we should see construction of liquid-metal-cooled fast reactors toward the end of the 1970's.

Liquid metals considered for use in nuclear reactor systems are lithium, sodium, potassium, sodium-potassium alloys, bismuth, and mercury. Each of these fluids has its own peculiar set of advantages and disadvantages for nuclear reactor applications. The most common fluid used to date is sodium, although there is considerable interest in the sodium-potassium eutectic alloy because its melting point is 53°F. Two advantages that liquid metals have over other reactor coolants are large thermal conductivities which enable larger heat-transfer coefficients and relatively high boiling temperatures at atmospheric pressure which eliminates the need for highly pressurized systems.

With heat transfer in liquid metals, molecular conduction is an important heat transfer mechanism throughout the entire flow channel. Consequently, Nusselt numbers are strongly a function of channel geometry and boundary conditions on heated surfaces. This is somewhat analogous to the situation with laminar flow through channels; friction factors are strongly dependent upon channel geometry.

Furthermore, conduction reduces the effect of turbulent convection as a mode of heat transfer because eddies lose heat by conduction during the course of their movement in a flow field. As a consequence, the eddy diffusivity for heat transfer (ϵ_H) is smaller than the eddy diffusivity for momentum transfer (ϵ_M).

Because of the importance of liquid-metal heat transfer and its complexity, it is worthwhile to study the subject in detail. A general equation for predicting liquid-metal heat-transfer coefficients is derived by R. N. Lyon [13]. The following assumptions are make in developing this equation:

1. Steady State
2. Constant fluid physical properties
3. Axial conduction is negligible (this is valid for $N_{Pe} > 100$) .
4. Uniform heat flux
5. Fully developed flow and heat transfer through a circular tube of radius r_w .

Under these assumptions, temperatures increase linearly with distance down the tube axis. Consequently, the axial temperature gradient, dT/dz, is independent of radial or axial position.

By definition, the heat-transfer coefficient is given by

$$h \equiv \frac{q''_w}{T_w - T_b} \tag{6.2.10}$$

for heat transfer across a fluid-solid interface.

By considering an energy balance on a unit tube length, the heat flux, q''_w , is given by

$$2\pi r_w q_w = \pi r_w^2 \rho c_p u_m dT/dz \tag{6.2.11}$$

At any axial position, the fluid mixing cup or bulk temperature is given by (assuming ρc_p = const)

$$\begin{aligned} T_b &= \frac{\int_o^{r_w} 2\pi r u T \, dr}{\int_o^{r_w} 2\pi r u \, dr} \\ &= \frac{2\int_o^{r_w} r u T \, dr}{r_w^2 u_m} \end{aligned} \tag{6.2.12}$$

Therefore, since the axial temperature gradient is constant

$$T_w - T_b = (T_w - T)_b \tag{6.2.13}$$

and we have

$$T_w - T_b = \frac{2\int_o^{r_w} r u (T_w - T) \, dr}{r_w^2 u_m} \tag{6.2.14}$$

For a given axial position, the radial temperature distribution is given from

$$T_w - T(r) = \int_r^{r_w} \frac{\partial T}{\partial r} dr \tag{6.2.15}$$

In addition, the radial heat flux in the fluid at any radial position is given by

$$q'' = K \frac{\partial T}{\partial r} \tag{6.2.16}$$

where K is the total conductivity which is given by the sum of the thermal conductivity (k) and the heat transfer eddy diffusivity ($\rho c_p \epsilon_H$). Therefore, Eq. (6.2.15) is written

$$T_w - T(r) = \int_r^{r_w} \frac{q''}{K} dr \tag{6.2.17}$$

An energy balance around a cylinder of unit length and radius r_q in the fluid gives

$$q'' 2\pi r_q = \int_o^{r_q} 2\pi r u \rho c_p \frac{\partial T}{\partial z} dr \tag{6.2.18}$$

By substituting Eq. (6.2.18) into Eq. (6.2.17) and Eq. (6.2.17) into Eq. (6.2.14) and Eqs. (6.2.14) and (6.2.11) into Eq. (6.2.10), we have the following expression for the heat-transfer coefficient:

$$h = \frac{r_w^3 u_m^2}{4\int_o^{r_w} ru \left[\int_r^{r_w} \frac{\int_o^{r_q} ru\, dr}{r_q K} dr_q\right] dr} \tag{6.2.19}$$

By defining dimensionless velocity and radial parameters by

$$V = u/u_m$$
$$S = r/r_w \tag{6.2.20}$$

and substituting into Eq. (6.2.19), the heat-transfer coefficient is given by

$$\frac{1}{h} = 4r_w \int_o^1 \int_{S_t}^1 \int_o^{S_q} \frac{S_t V_t SV}{S_q K} dS_t\, dS_q\, dS \tag{6.2.21}$$

Since the Nusselt number is given by

$$N_{Nu} = \frac{h 2r_w}{k} \tag{6.2.22}$$

Eq. (6.2.21) is rearranged to give

$$\frac{1}{N_{Nu}} = 2\int_o^1 \frac{\left[\int_o^{S_q} VSdS\right]^2}{S_q K/k} dS_q \tag{6.2.23}$$

The ratio of total conductivity to thermal conductivity is written

$$K/k = 1 + \psi N_{Pr} \epsilon_M/\nu \tag{6.2.24}$$

where ψ is the ratio of eddy diffusivities, $\psi = \epsilon_H/\epsilon_M$. Substituting Eq. (6.2.24) into Eq. (6.2.23) gives

$$\frac{1}{N_{Nu}} = 2\int_o^1 \frac{\left[\int_o^{S_q} VSdS\right]^2}{S_q(1 + \psi N_{Pr}\epsilon_M/\nu)} dS_q \tag{6.2.25}$$

Equation (6.2.25) is solved using appropriate expressions for tube velocity distribution, eddy diffusivity for momentum transfer, and eddy diffusivity ratio. Lyon [13] solved Eq. (6.2.25) for a wide range of

Reynolds numbers and a range of Prandtl numbers from 0 to 0.1 using the circular tube velocity data of Nikuradse. It was assumed that the eddy diffusivity ratio was unity. The results of the numerical solution are given with a reasonable accuracy by a simple equation [13]

$$N_{Nu} = 7.0 + 0.025(N_{Re} \times N_{Pr})^{0.8} \tag{6.2.26}$$

Equation (6.2.26) usually overestimates liquid metal heat-transfer coefficients. This may be attributed to assuming the eddy diffusivity

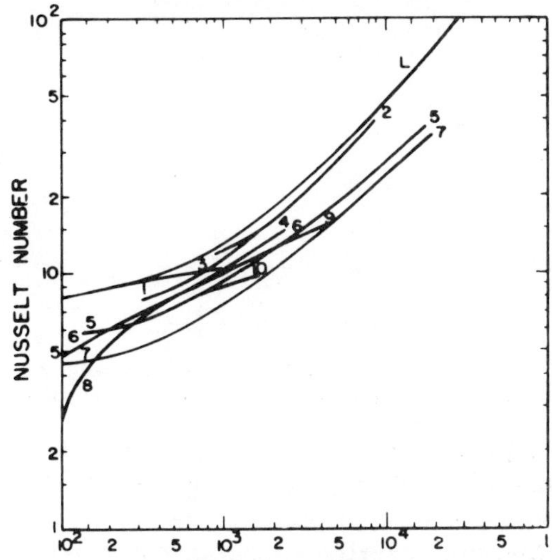

Fig. 6.2. Experimental values for the fully developed Nusselt number; L = Lyon; 1 = Baker and Sesonske; 2 = Isakoff and Drew; 3 = Untermeyer (magnesium added); 4 = Styrikovick and Semenovker; 5 = Johnson, Clabaugh, and Hartnett; 6 = Trefethen; 7 = Stromquist; 8 = English and Barrett; 9 = Johnson, Hartnett, and Clabaugh; 10 = Seban [14].

ratio being unity. Usually alkali liquid metals have better agreement than heavy liquid metals such as mercury. It should be noted that Eq. (6.2.26) does not yield the laminar flow Nusselt number for uniform heat flux for low Reynolds number flows. For this case, the energy equation is solved exactly and $N_{Nu} = 4.36$. Consequently, it is probably not advisable to use Eq. (6.2.26) for Reynolds numbers less than 10,000.

Figure 6.2 illustrates curves which best illustrate liquid metal heat-transfer coefficients from several investigations [14]. The data shown is from measurements in which velocity and temperature profiles are fully developed. As seen, Lyon's equation (Eq. 6.2.26) overestimates heat-transfer coefficients for all investigations illustrated.

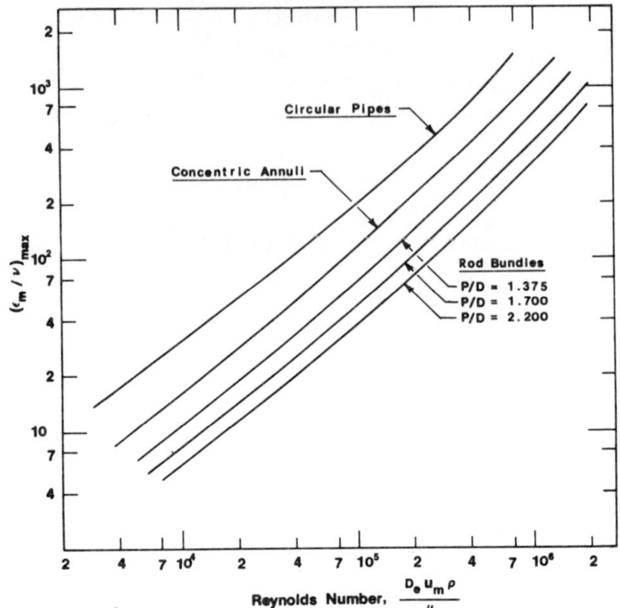

Fig. 6.3. Values of $(\epsilon_M/\upsilon)_{max}$ for fully-developed turbulent flow of liquid metals through circular tubes, annuli, and rod bundles with equilateral triangular spacing [16].

A parametric analysis of Eq. (6.2.25) indicates that calculated Nusselt numbers are not strongly dependent upon choice of expressions for velocity distributions or eddy diffusivities for momentum transfer [14]. Since the appearance of Lyon's paper in 1951, numerous expressions were developed for the eddy diffusivity ratio ψ . A parametric substitution of eight different expressions for ψ into Eq. (6.2.25) showed significant differences in calculated Nusselt numbers [14]. The majority of the Nusselt numbers are lower than predicted by Eq. (6.2.26).

In a more recent analysis of equations similar to Eq. (6.2.25), Lyon and Dwyer [15] propose that liquid-metal Nusselt numbers be calculated with the following equation:

$$N_{Nu} = a + \beta(\overline{\psi} N_{Pe})^{\gamma} \tag{6.2.27}$$

for $10^2 \leqslant N_{Pe} \leqslant 10^4$. Both N_{Nu} and N_{Pe} are based upon channel equivalent diameter. The average ratio of eddy diffusivities is calculated with

$$\overline{\psi} = \frac{\overline{\epsilon_H}}{\epsilon_M} = 1 - \frac{1.82}{N_{Pr}(\epsilon_M/\nu)^{1.4}_{Max}} \tag{6.2.28}$$

Table 6.2
Summary of empirical constants for use in modified Lyon equation [15].

Flow through	Heat transfer conditions		Y	α	β	γ
1. Pipes			—	7.0	0.025	0.8
2. Annuli	(a) Heat transfer from inner wall		1—7	$4.82 + 0.697y$	0.022	$0.758y^{0.053}$
	(b) Heat transfer from outer wall		1—7	$5.54 - 0.023y$	$0.0189 + 0.00316y + 0.000087y^2$	$0.758y^{-0.0204}$
	Heat flux from both walls	Inner	1.5	11.80	0.0726	0.701
			2.0	15.30	0.0855	0.704
			3.0	27.0	0.1095	0.707
		Outer	1.5	8.23	0.0418	0.733
			2.0	7.55	0.0415	0.736
			3.0	7.08	0.0414	0.739
	Heat transfer from both walls with $T_1 = T_2$ at any axial position	Inner	1.5	10.40	0.0603	0.708
			2.0	11.05	0.0613	0.718
			3.0	12.52	0.0647	0.732
		Outer	1.5	8.92	0.0462	0.721
			2.0	8.58	0.0432	0.731
			3.0	8.16	0.0403	0.741
3. Parallel plates	Heat transfer from one plate		—	5.60	0.01905	0.775
	Equal heat flux from both plates		—	9.49	0.0596	0.688
4. In-line flow between rod bundles triangular spacing			$P/D = 1.3$—3.0	$6.66 + 3.126(P/D) + 1.184(P/D)^2$	0.0155	0.86
5. In line flow between rod bundles sq. spacing			$P/D = 1.7$—10.0	$7.0 + 4.24 (P/D)^{1.52}$	$0.0275 \times (P/D)^{0.27}$	0.8

Equation (6.2.27) is recommended for flow through pipes, parallel plates, annuli, and rod bundles. Figure 6.3 illustrates values of $(\epsilon_M/\nu)_{Max}$ for pipes, annuli, and rod bundles with equilateral triangular spacing as a function of channel Reynolds number. Values of a, β, and γ are listed in Table 6.2 for liquid-metal heat transfer with uniform heat flux and fully-developed velocity and temperature profiles. For heat transfer in pipes and parallel plates, a, β, and γ are constants; for flow through annuli a, β, and γ are a function of the ratio of outer-to-inner radii, Y, and for in-line flow through rod bundles a, β, and γ are dependent upon pitch-to-diameter ratio P/D.

Inspection of Eq. (6.2.27) and Fig. 6.3 shows that for a given Prandtl number there are lower values of Reynolds numbers in which $\overline{\psi}$ is zero or negative. Negative values of $\overline{\psi}$ have no physical meaning when used in Eq. (6.2.27) and, consequently, $\overline{\psi}$ should be considered zero when negative values are computed. Peclet numbers at which $\overline{\psi}$ is zero in Eq. (6.2.28) are called critical Peclet numbers. At or below critical Peclet numbers the contribution of turbulence or eddy convection to the heat transfer process is negligible. Under these circumstances, Nusselt numbers are constant or

$$N_{Nu} = a \tag{6.2.29}$$

for

$$N_{Pe} \leqslant N_{Pe}\ (\text{critical})$$

Table 6.3 gives critical Peclet numbers for circular tubes, annuli, and equilateral spacing rod bundles. As seen, critical Peclet numbers are considerably higher for rod bundles than for either tubes or annuli. An interesting facet exists in the design of heat exchangers using liquid metals. For reasonably high Reynolds numbers, heat-transfer coefficients may be no higher than for flow conditions close to laminar flow. As an example, for a fluid with $N_{Pr} = 0.02$, the critical Peclet number is 800 for a rod bundle with $P/D = 1.700$. Consequently, Reynolds numbers must be in excess of 40,000 in order to have an increase in heat-transfer coefficients due to turbulent convection. From a practical point of view Reynolds numbers must exceed 100,000 in order to have any significant increase in heat-transfer coefficients.

Equations of the form of Eq. (6.2.25), which are used to determine Eq. (6.2.27), neglect circumferential heat flow. Dwyer [17] has shown that circumferential heat flow may be neglected in triangular-spaced rod bundles only when $P/D \geqslant 1.35$. Consequently, in the application to rod bundles, Eq. (6.2.27) is restricted to $P/D \geqslant 1.3$ for a triangular spacing and $P/D \geqslant 1.7$ for a square spacing. LMFBRs will most likely be fueled with close-packed rod bundles with $P/D < 1.3$ and circumferential variations in heat-transfer coefficients will be appreciable. Therefore, Eq. (6.2.27) is not suitable for analyzing heat transfer in LMFBR fuel assemblies.

The analysis of close-packed rod bundles is more complicated than for widely-spaced rod bundles because circumferential heat transfer is taken into account. Dwyer presents the results of his calculations and those of Borishansky, et al. in Ref. 17. Both investigations analyzed the cross-hatched area shown in Fig. 6.4 for fully developed flow and heat

Table 6.3

Critical Peclet numbers [16]

N_{Pr}	Tubes	Annuli*	Δ Rod Bundles		
			P/D=1.375	1.700	2.200
0.005	117	270	460	622	770
0.01	131	300	530	720	890
0.02	144	330	582	800	1000
0.03	150	345	603	840	1056

* For heat transfer through either inner or outer wall

transfer with a uniform heat flux. Local and average heat-transfer coefficients are defined by the following equations:

$$h \equiv \frac{q''_w}{T_{w,\theta} - T_b} \tag{6.2.30}$$

and

$$\bar{h} \equiv \frac{q''_w}{\bar{T}_{w,\theta} - T_b} \tag{6.2.31}$$

where $\bar{T}_{w,\theta}$ is averaged over the 30-degree arc.

Figure 6.4 shows calculated average Nusselt numbers as a function of $\bar{\psi} N_{Pe}$ and P/D. Table 6.4 presents Dwyer's calculated average and local heat-transfer coefficients for three different pitch-to-diameter ratios. As seen, there is considerable variation in local heat-transfer coefficients for small P/D ratios.

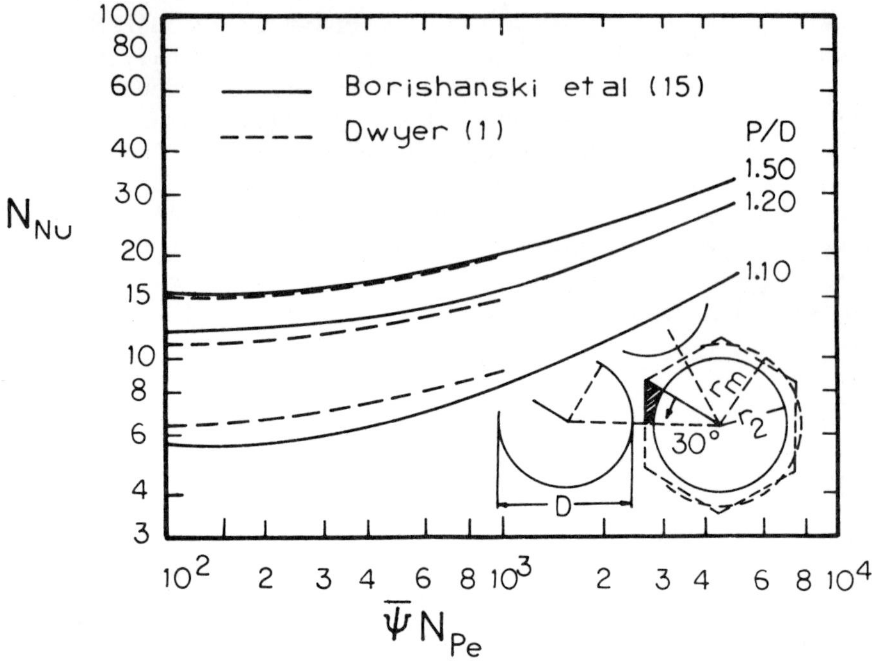

Fig. 6.4. Heat transfer to liquid metals flowing in-line through unbaffled rod bundles under conditions of uniform heat flux and fully-developed velocity and temperature profiles [17].

Table 6.4

Predicted average and local heat-transfer coefficients for sodium flowing in-line through closely-packed rod bundles [17].

	$h/\overline{h}$				
θ	P/D = 1.100 $\overline{\psi}N_{Pe}$ = 206 $\overline{h}$ = 20,520	P/D = 1.100 $\overline{\psi}N_{Pe}$ = 1015 $\overline{h}$ = 32,550	P/D = 1.200 $\overline{\psi}N_{Pe}$ = 98.2 $\overline{h}$ = 20,160	P/D = 1.200 $\overline{\psi}N_{Pe}$ = 885 $\overline{h}$ = 29,340	P/D = 1.375 $\overline{\psi}N_{Pe}$ = 998 $\overline{h}$ = 18,840
0.00	0.494	0.477	0.794	0.782	0.925
1.88	0.497	0.480	0.799	0.786	0.942
5.62	0.520	0.511	0.821	0.806	0.948
9.38	0.590	0.598	0.876	0.852	0.968
13.12	0.804	0.844	0.959	0.926	0.988
16.88	1.431	1.437	1.070	1.052	1.011
20.62	3.295	3.034	1.182	1.192	1.036
24.38	9.71	11.7	1.261	1.311	1.059
28.12	32	27	1.314	1.386	1.082
30.00	48		1.326	1.369	1.087

A comparison of some experimental close-packed rod bundle data with Dwyer's calculations is shown in Fig. 6.5. The curve A-A is from Fig. 6.4 with $\overline{\psi} = 1$. The curve B-B is from Fig. 6.4 with $\overline{\psi}$ evaluated from Eq. (6.2.28) and C-C assumes eddy convection is negligible. The agreement between calculations and experiments is not good which may be explained by the liquid metal in the experiments not wetting the surface which added an interface resistance or the measurements not being very accurate.

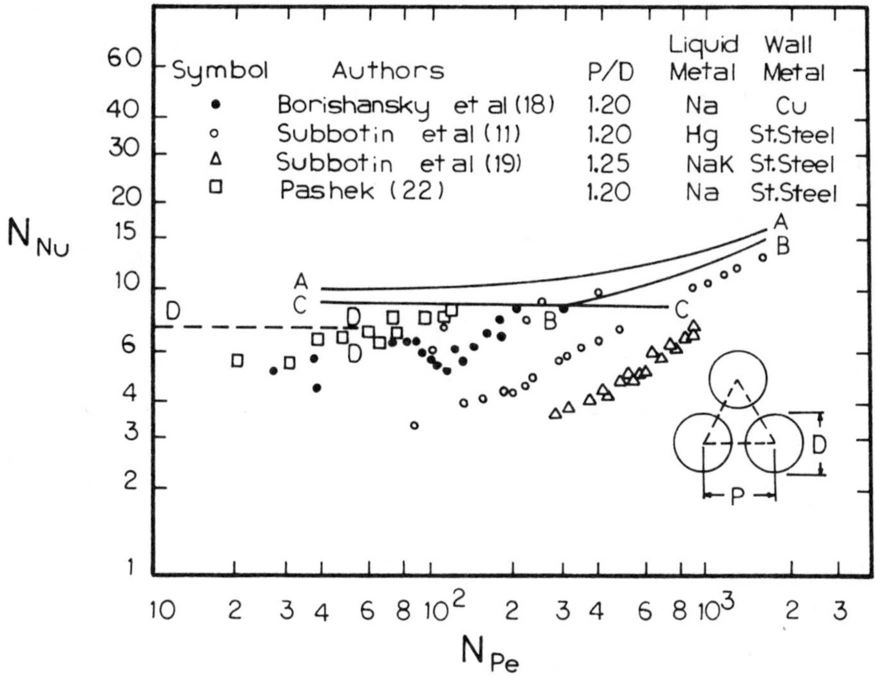

Fig. 6.5. Comparison of analytical predictions with experimental results for in-line flow of liquid metals through unbaffled rod bundles under conditions of nominally uniform wall heat flux and fully-developed velocity and temperature profiles [17].

Subbotin [7] and co-workers analyzed theoretically and investigated experimentally the flow of liquid metal coolants through rod bundles. For a close-packed triangular array, in which the circumferential non-uniformity in wall temperature influences heat-transfer coefficients, they offer the following equation:

$$N_{Nu} = 0.15\left(\frac{k_w}{k_f}\right)^{0.02} \times (N_{Pe})^{0.3 + 0.04\sqrt{k_w/k_f}} \tag{6.2.32}$$

where k_w and k_f are the thermal conductivities of the wall and fluid and Nusselt and Peclet numbers are based on the equivalent diameter. The range of validity of Eq. (6.2.32) is

$$N_{Pr} << 1\,,\ 80 \leqslant N_{Pe} \leqslant 600\,,\ 0.69 \leqslant k_w/k_f \leqslant 16.3\,,\ \text{and}\ 1.0 \leqslant P/D \leqslant 1.1$$

For the non-close packed array, experiments indicate that Nusselt numbers are given by

$$N_{Nu} = 0.58\left(\frac{D_e}{D}\right)^{0.55} (N_{Pe})^{0.45} \tag{6.2.33}$$

for $80 \leqslant N_{Pe} \leqslant 4000$ and $1.1 \leqslant P/D \leqslant 1.5$, where D_e is the equivalent diameter and D is the tube diameter. Most Russian data fell within $\pm 15\%$ of these proposed correlations.

Rod bundle heat-transfer coefficients for Eqs. (6.2.32) and (6.2.33) are defined by

$$h = \frac{q''_w}{T_w - T_b}$$

where T_b is the fluid temperature in subchannels associated with the central rod.

A correlation for Nusselt numbers for liquid metals flowing in the shell side of a shell-and-tube heat exchanger is also proposed by Subbotin [7]. Their experiment used sodium flowing in the tubes at high temperatures and NaK flowing in the shell side. The range of geometry and flow variables studied in the shell side are

$$1.1 \leqslant P/D \leqslant 1.4\,,\ 60 \leqslant L/D \leqslant 260\,,\ \text{and}\ 200 \leqslant N_{Pe} \leqslant 1200$$

The data for the mean shell-side heat-transfer coefficient correlated to $\pm 10\%$ by the following equation:

$$N_{Nu} = 8\,[(D_e/L) + 0.027\,(P/D - 1.1)^{0.46}](N_{Pe})^{0.6} \tag{6.2.34}$$

The characteristic dimension of the Nusselt number and Peclet number is the equivalent diameter, D_e, based upon the entire shell side flow area.

Equation (6.2.34) differs considerably from Eq. (6.2.33) for in-line flow through rod bundles with uniform wall heat flux. The difference is thought to be due to the influence of the starting length, non-uniformly distributed heat flux, and presence of the non-heated shell surface.

There is some experimental verification for use of Eqs. (6.2.27) and (6.2.28) for computing liquid-metal Nusselt numbers with uniform heat flux and fully-developed temperature and velocity profiles for tubes, annuli, and rod bundles. In addition, there is a considerable body of data which indicates Nusselt numbers are lower than theoretical predictions. Frequent arguments for the discrepancy are the liquid metal did not wet the heat transfer surface, the liquid metal was contaminated by oxides or other particulates, or inert gases were entrained in the liquid metal. A combination of several of these factors appears to produce a contact thermal resistance at the interface [17].

Another factor leading to smaller heat-transfer coefficients with liquid metals has to do with non-ideal geometry in annuli or rod bundles. For annuli, it is experimentally verified that eccentricity substantially reduces heat-transfer coefficients. Bowing of rods in rod bundles is shown to reduce heat-transfer coefficients by as much as 50% below those of straight rods. Consequently, asymmetry in geometry can produce great problems in the design of heat transfer equipment using liquid metals. The problem of bowing is not considered as serious in the design of baffled heat exchangers because very little flow is parallel to the tubes.

The majority of the theoretical work done in liquid-metal heat transfer assumes uniform heat flux, fully-developed temperature and velocity profiles, and constant physical properties. Since temperature distributions in turbulent flowing liquid metals are somewhat similar to temperature distributions in laminar flow, heat-transfer coefficients are dependent upon boundary conditions, i.e., uniform-wall temperature, unsteady state, or variable heat flux. This situation normally does not exist in heat transfer to turbulent flowing ordinary fluids. Some work has been done on variable boundary conditions; however, there is still room for considerable more research, both experimental and theoretical, in liquid-metal heat transfer.

LMFBR fuel assemblies will probably contain wire-wrapped fuel rods and the analytical approaches presented are not accurate for analysis. If an accurate prediction of wall temperatures is necessary, then prototype testing is required. Since heat-transfer coefficients are large, the wall-bulk fluid temperature difference is small and sizable errors in predicting heat-transfer coefficients should not lead to large errors in determining cladding or fuel temperatures.

Boiling Heat Transfer

Heat transfer with changes from liquid to vapor phases is very complex and not as well understood as single-phase heat transfer. Numerous types of boiling heat transfer are possible depending upon fluid bulk temperature, T_b ; fluid saturation temperature, T_{sat} ; temperature of the wall at the fluid-solid interface, T_w ; and nature of the flow system—natural or forced convection.

Boiling cannot take place unless temperatures some place in the fluid exceed T_{sat} . Ordinarily in boiling problems, some surface temperature, T_w , is greater than T_{sat} . The fluid bulk temperature also has an influence upon the character of boiling regimes. Two types of boiling are characterized by bulk temperatures:

1. When $T_b < T_{sat}$, one has **subcooled boiling** at the fluid-solid interface.

2. When $T_b = T_{sat}$, one has **bulk boiling** or **saturated boiling.**

The nature of the flow system also produces four subdivisions in boiling. This is dependent upon whether the system is bounded or unbounded, and whether the fluid motion is forced or natural convection.

A final type of boiling is due to internal heat generation within the fluid. This type of boiling is somestimes called **homogeneous boiling** and boiling takes place throughout the fluid.

Experiments show that $T_w - T_{sat}$ is the important temperature parameter in boiling. Boiling does not take place when the liquid is at T_{sat} because a certain amount of superheat is necessary for forming a bubble. For highly subcooled liquids, boiling can occur at the liquid-solid interface because the liquid is superheated in the region adjacent to the walls. With subcooled fluids bubbles may not travel far from heated surfaces because they condense in the subcooled liquid.

Depending upon the magnitude of $T_w - T_{sat}$, three boiling regions are identified—(i) **nucleate boiling** in which bubbles form on the heated surface, (ii) **transition** (unstable-film) **boiling** in which unstable regions of nucleate boiling and vapor patches form on the heated surface, and (iii) **stable-film boiling** in which vapor covers the heated surface. These regions are shown in Fig. 6.6 as a plot of wall heat flux as a function of $T_w - T_{sat}$

The theoretical treatment of boiling heat transfer presents great difficulty because of poor understanding of the variables involved. At present there is no single correlation for heat-transfer coefficients or relationships between q''_w and ΔT which is applicable for all regions shown on Fig. 6.6. The approach is to provide empirical or semi-empirical correlations over distinct regions.

In region I the mode of heat transfer is by natural or forced convection since no boiling takes place. The effect of increasing mass velocity (G) is shown by an arrow. In this region the relationships for

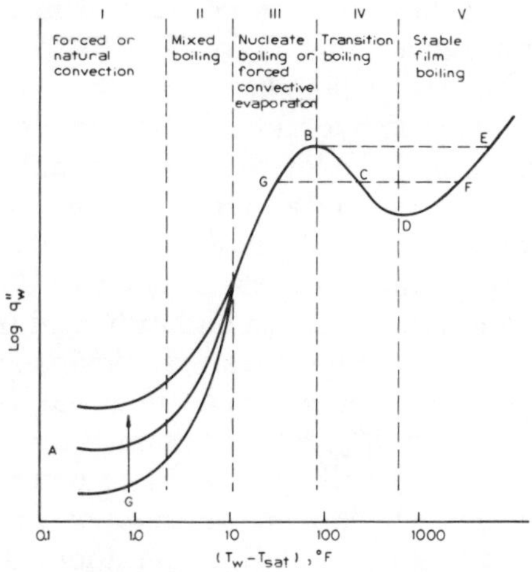

Fig. 6.6. Typical boiling curve

single-phase natural or forced convection are applicable for heat transfer analysis.

Region II describes a regime in which both single-phase convection and nucleate boiling are important mechanisms for heat transfer. In region III nucleate boiling is the dominant heat transfer mechanism and nucleate boiling correlations are used in analysis. As discussed later, the transition boiling region IV is unstable and no highly reliable method is available at this time for describing that portion of the boiling curve [18]. For steady state operation of nuclear reactors, there is little interest in region IV because no boiling reactor could operate in this region. However, in accident analysis region IV is important because fuel rods pass through this region in both loss-of-coolant (blowdown) portion and reflood portion of severe reactor accidents.

The film boiling region V is also important in accident analysis and important heat transfer mechanisms are convection to the vapor film and radiation through the vapor film.

By applying physical reasoning to the boiling curve shown in Fig. 6.6, it is apparent that an instability exists over a portion of the boiling curve labeled B-D. Increasing heat generation rates in materials by increasing an electric current or fission rate cause heat fluxes to increase. In the region from A to B increasing heat fluxes produce an increase in $T_w - T_{sat}$. However, beyond point B, an increase in wall temperature is unable to generate an increase in heat flux. As shown in Fig. 6.6, wall temperatures higher than point B produce lower heat fluxes than at point B. If the heat generation rate in a solid is sufficient to require heat fluxes higher than at point B in order to maintain steady state, the situation arises in which heat is accumulated in the material with a resulting increase in wall temperatures. The wall temperature increases to point E or higher at which the heat flux is sufficient to sustain the heat generation rate. Unfortunately, for most materials temperatures at point E are in excess of the melting temperature.

The boiling instability due to the q''_w vs. ΔT relationship is analogous to the hydraulic instability also found in two-phase systems. The instability is understood by examining point C on the boiling curve. A slight increase in wall temperature at point C reduces the heat flux. The reduction in heat flux causes further rise in wall temperatures and wall temperatures increase to point F. Conversely, a slight decrease in wall temperature enables higher heat fluxes. This further reduces wall temperatures and temperatures decrease to point G.

As previously mentioned, material failure or burnout is usually achieved when q''_w reaches point B for most materials. Consequently, in the design of boiling equipment one must be careful not to achieve heat fluxes which lead to the possibility of reaching heat fluxes corresponding to point B. In addition, design should not lead to the possibility of wall temperatures being in the transition boiling region because an increase in wall temperature might result in material failure. In some boiling situations, wall temperatures may be the controlled variable and it may be impossible to achieve temperatures which result in material failure. Such a situation might be boiling a fluid by exchanging heat with a condensing vapor. However, in processes in which heat generation is the controlling variable, as in nuclear reactors, burnout is always a distinct possibility.

In nuclear reactors, heat flux is the independent variable since reactor power level is the quantity under control by the operator. Therefore, for proper orientation of the boiling curve it is more reasonable to represent q''_w as the absicca. Figure 6.7 illustrates the characteristic boiling curve with the change in variable orientation.

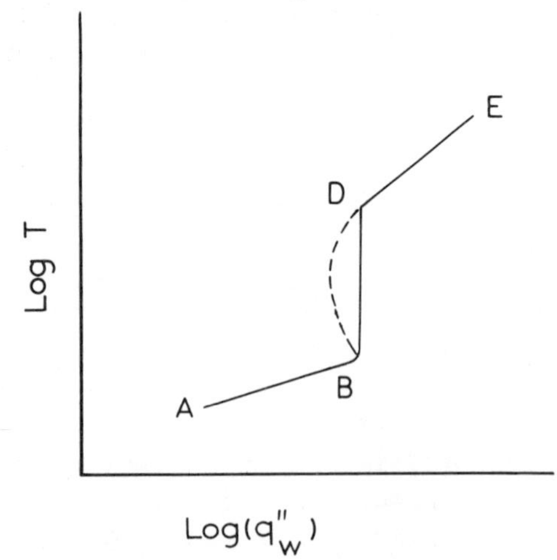

Fig. 6.7. Characteristic boiling curve

The heat flux at point B is given a variety of names. Because early tests with electrically heated test elements detected the limiting heat flux by test section failure the word **burnout heat flux** is frequently used. Other names are the **boiling crisis**, **departure from nucleate boiling** (DNB) in subcooled or low quality regions, **dryout** in high quality regions, and **critical heat flux** (CHF).

In nuclear reactor analysis, the principal areas of interest in boiling heat transfer are predicting the onset of nucleate boiling, wall temperatures in the nucleate boiling regime, maximum heat flux (burnout heat flux) just beyond the nucleate boiling regime, and wall temperatures in the transition and film boiling regions for purposes of accident analysis. In the next few sections, various correlations are presented for predicting quantities of interest in boiling heat transfer analysis. A wealth of information has been published in this area over the last two decades. The amount of material presented is an infinitesimal amount of this information and interested readers are urged to consult the literature on boiling heat transfer. A good survey of some of the boiling heat transfer literature is contained in Refs. [18], [19], and [20].

(1) Onset of Nucleate Boiling

Predicting the onset of nucleate boiling is important in reactor analysis because at this location in heated channels pressure losses are no longer calculated using single-phase correlations developed in Chapter 4. Once boiling commences, pressure losses are larger than in single-phase flows. An additional factor is that in some reactors boiling is not permitted or considered safe and the heat flux at the onset of nucleate boiling establishes an upper limit in reactor power level.

The onset of nucleate boiling is predicted using the Jens and Lottes approach given by Eqs. (4.5.20) and (4.5.21) in Chapter 4. Another approach is to use the Bergles and Rohsenow [21] correlation for calculating the incipient boiling heat flux in water for a range of pressures from 15 to 2000 psia.

$$(q''_w)_i = 15.60\, p^{1.156} (T_w - T_{sat})^{2.30 p^{0.0234}} \qquad (6.2.35)$$

where

$(q''_w)_i$ = incipient boiling heat flux, Btu/hr–ft^2

p = pressure, psia

Wall temperatures or heat fluxes at the onset of nucleate boiling are calculated from the intersection of Eq. (6.2.35) and appropriate forced convection heat transfer expressions (Eqs. (6.2.2 to 6.2.7)) relating heat flux to wall temperature. Figure 6.8 illustrates graphically the method for calculating heat flux and wall temperature at the onset of nucleate boiling.

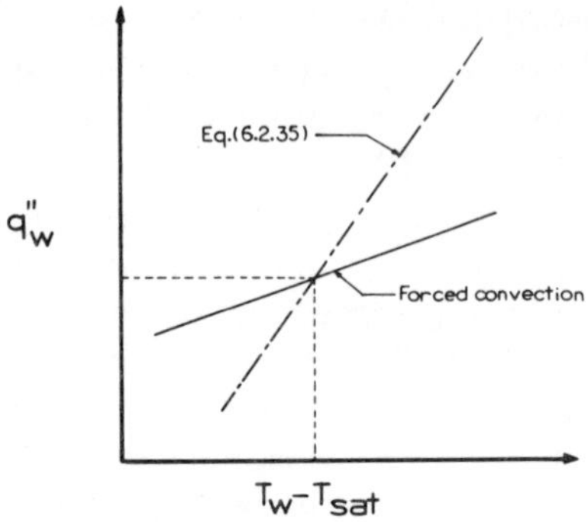

Fig. 6.8. Prediction of the onset of nucleate boiling.

A comparison of the Bergles and Rohsenow technique for predicting onset of nucleate boiling with experimental data from the Oak Ridge Research Reactor (ORR) is shown in Fig. 6.9. The ORR has plate-type fuel elements and Eq. (6.2.5) is used for calculating forced convection heat-transfer coefficients. As seen, agreement with experimental data is quite good.

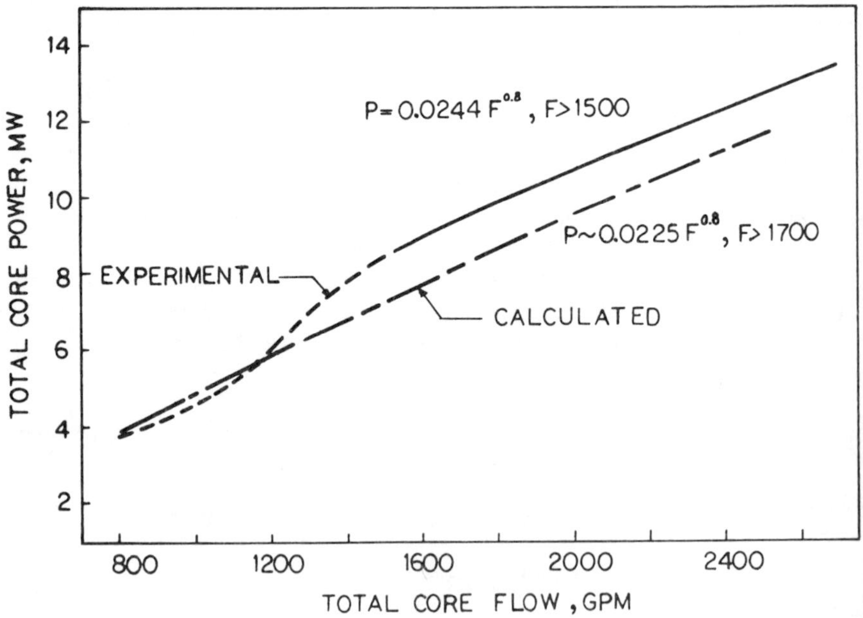

Fig. 6.9. Calculated vs. experimental incipient boiling in the Oak Ridge Research Reactor [22].

(2) Nucleate Boiling with Convection

Nucleate boiling with forced convection is represented by regions II and III in Fig. 6.6. Region II is normally associated with subcooled nucleate boiling and is a complicated mixture of single-phase forced convection and nucleate boiling modes of heat transfer. Both increasing flow velocity or subcooling increase heat transfer rates for a given wall superheat. A qualitative effect of increasing flow velocity is shown in Fig. 6.6. Wall superheats in region II are not very high and, consequently, it is frequently deemed adequate to extend the heat transfer correlation from region III down into region II which is conservative in that it slightly overestimates wall temperatures.

Numerous correlations exist for calculating the heat flux-wall temperature relationship in region II. A well-accepted, and somewhat complicated, model is the Chen correlation discussed in Ref. [18]. Another model is the Bergles-Rohsenow [21] correlation given by

$$q''_w = (q''_w)_{FC}\left[1 + \frac{(q''_w)_B}{(q''_w)_{FC}}\left\{1 - \frac{(q''_w)_{Bi}}{(q''_w)_B}\right\}^2\right]^{1/2} \tag{6.2.36}$$

where $(q''_w)_{FC}$ is the heat flux associated with single-phase forced convection, $(q''_w)_B$ the fully-developed boiling heat flux, and $(q''_w)_{Bi}$ is the fully-developed boiling heat flux at incipient boiling. The forced convection heat flux is computed using heat-transfer coefficient correlations, Eqs. (6.2.2-6.2.7), previously discussed.

$$(q''_w)_{FC} = h(T_w - T_b) \tag{6.2.37}$$

The fully-developed boiling heat flux represents a heat flux due to boiling in which effects due to velocity or subcooling are small. It is obtained from experimental data at high heat flux and frequently is correlated in the following manner:

$$(q''_w)_B = C(T_w - T_{sat})^n \tag{6.2.38}$$

The Jens-Lottes equation given by Eq. (4.5.20) in Chapter 4 is rearranged into the form of Eq. (6.2.38) to yield

$$(q''_w)_B/10^6 = \frac{\exp(4p/900)}{(60)^4}(T_w - T_{sat})^4 \tag{6.2.39}$$

The fully-developed heat flux at incipient boiling represents the value obtained for $(q''_w)_B$ for the wall temperature difference $(T_w - T_{sat})$ at the onset of nucleate boiling. The value of $(T_w - T_{sat})$ at the onset of nucleate boiling is obtained from the intersection of the single-phase forced convection heat flux curve with the equation for the incipient boiling heat flux $(q''_w)_i$. For water, a correlation for the incipient boiling heat flux is given by Eq. (6.2.35).

The design procedure for calculating the heat flux in region II is shown in Fig. 6.10. From the value of $(T_w - T_{sat})$ obtained from the intersection of $(q''_w)_i$ and $(q''_w)_{FC}$, the heat flux at fully-developed incipient boiling $(q''_w)_{Bi}$ is evaluated and substituted into Eq. (6.2.36). By taking a series of wall superheats, the heat flux in region II is evaluated through the use of Eqs. (6.2.37) and (6.2.38).

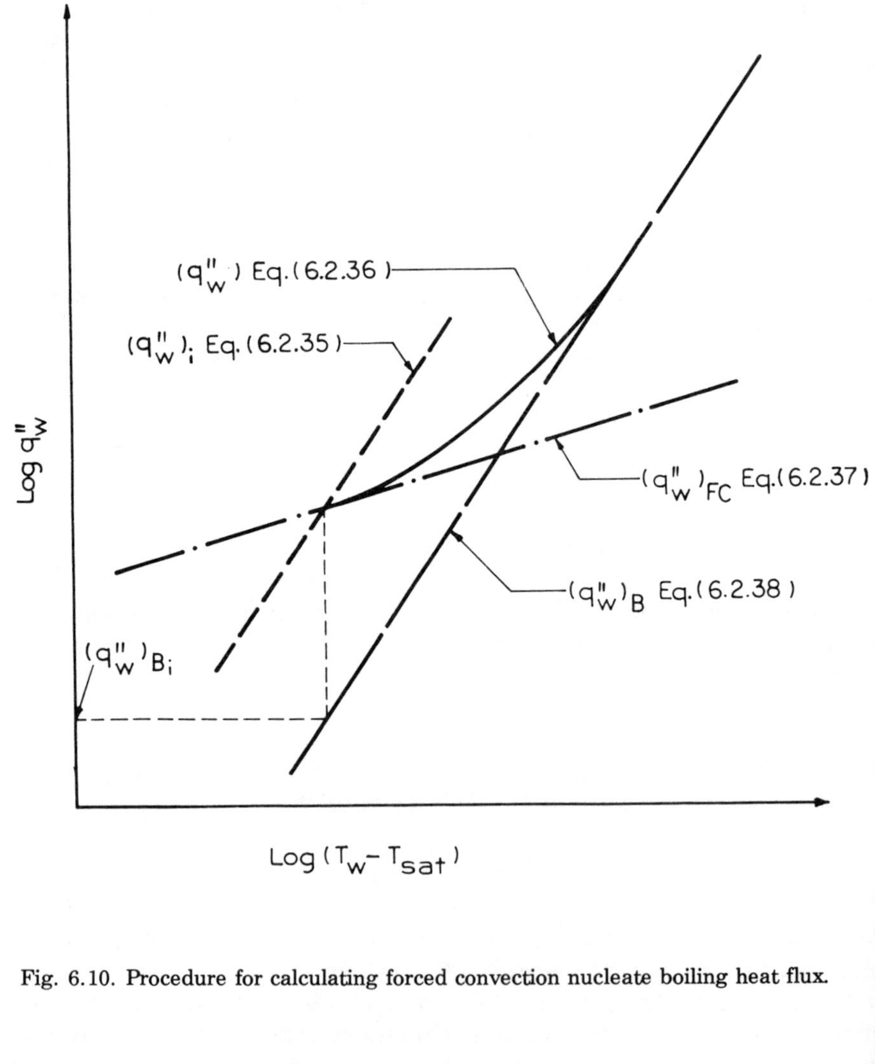

Fig. 6.10. Procedure for calculating forced convection nucleate boiling heat flux.

Both the Chen correlation described in Ref. [18] and the Bergles-Rohsenow correlation just presented compute heat fluxes based upon known wall temperatures. If heat flux is the controlling variable, as is usually encountered in steady state reactor analysis, then wall temperatures are determined by an iterative approach.

Several heat transfer mechanisms are possible for describing region III. For subcooled or low quality boiling the mechanism is nucleate boiling under conditions in which the amount of liquid phase subcooling or flow velocity is unimportant. These conditions probably occur in the analysis of PWRs and can be described by equations of the form given by Eqs. (6.2.38) or (6.2.39). It should be noted that Eq. (6.2.39) is independent of flow quality and strongly dependent upon fluid pressure.

For high quality annular flow the heat transfer mechanism in region III is **forced convective evaporation**. This regime is also called **quality boiling** and is important in the analysis of BWRs and in postulated power surges in PWRs. In this regime the vapor velocity and interfacial turbulence levels are high and the liquid film layer at the wall thin. Consequently, nucleation at the wall is suppressed and heat is transferred by conduction through the thin liquid film and evaporation at the liquid-vapor interface with a flow effect similar to single-phase forced convection. Correlations for forced convection evaporation are frequently presented in the following form:

$$h_{TP}/h_{LP} = A(X_{tt})^{-n} \tag{6.2.40}$$

where

h_{TP} = forced convective evaporation heat-transfer coefficient

h_{LP} = single-phase liquid heat-transfer coefficient based upon total flow or actual flow of saturated liquid, Equations(6.2.2–6.2.7)

A = constant

$$X_{tt} = \left(\frac{1-x}{x}\right)^{0.9}\left(\frac{\rho_g}{\rho_f}\right)^{0.5}\left(\frac{\mu_f}{\mu_g}\right)^{0.1}$$

Reference [22] contains a review of several experimental correlations for predicting forced convective evaporation heat-transfer coefficients for circular tubes and rod bundles with helical spacers. The rod bundle heat-transfer coefficients are higher than those for circular tubes because of the swirling effect on the flow due to the helical spacers. For forced convective evaporation in rod bundles without helical spacers, the authors of Ref. [22] recommend using the correlation of Schrock and Grossman obtained from circular tubes which is given by

$$h_{TP}/h_{LP} = 2.5(X_{tt})^{-0.75} \tag{6.2.41}$$

where h_{LP} is evaluated based upon flow of the actual saturated liquid phase alone. Equation (6.2.41) is probably conservative when applied to rod bundles because it should underestimate heat-transfer coefficients.

One of the difficulties in applying forced convective evaporation heat-transfer correlations, such as Eq. (6.2.41), is to know the range of qualities and flow velocities where they are applicable. This problem is eliminated by the Chen correlation because it is deemed suitable for both fully-developed nucleate boiling and forced convective evaporation [20].

(3) Burnout Heat Flux Correlations

A parameter of great importance in the analysis of nuclear reactors is determining the limiting heat flux at which values in excess of this limit will probably result in fuel element cladding failure. For liquid cooled reactors this limiting heat flux lies between regions III and IV described on the boiling curve. For heat fluxes above the limiting value the degradation of the heat transfer mechanism is so fast that cladding melting may occur in a short period of time. The burnout phenomenon occurs with all liquids whether they be water, organic fluids, or liquid metals.

The importance of predicting burnout heat fluxes in nuclear reactors prompted a considerable number of worldwide theoretical and experimental investigations. Because theoretical investigations have not met with much success, it is necessary to rely on experimental burnout data in nuclear reactor design. The most desirable design data are that which simulate most closely the geometry and the range of operating conditions of the actual reactor core.

Experiments show that burnout is a function of local coolant enthalpy, inlet subcooling, mass velocity, pressure, geometry, and axial heat generation distribution along the coolant channel. For nuclear reactors which frequently employ rod bundle geometry with parallel coolant flow, some of the important geometry factors are (i) number of rods, (ii) length, (iii) diameter, (iv) spacing, (v) heat flux variations between rods, (vi) rod array (square or triangular, etc.), (vii) type of grid spacers, and (viii) unique position in a rod bundle (such as a corner or side rod in a fuel element which faces an unheated surface). An additional geometric consideration is the orientation of the rod bundle. This orientation may be vertical or horizontal which means gravitational forces are either parallel or transverse to the direction of mean coolant flow. A further complicating factor is that burnout also depends on the nature of the test loop and coolant compressibility upstream from the heated channel. Upstream void or large volumes of slightly compressible liquid enables flow oscillations to occur in heated channels which may decrease burnout heat fluxes.

Since burnout is affected by a large number of variables, it is difficult to obtain reliable burnout data which covers the range of all variables considered in a reactor design. Common practice in the burnout analysis of nuclear reactors employing rod bundle geometries is to use empirical correlations or lower limit curves based on burnout data from tube, annular, rectangular, and rod bundle geometries. New data for specific reactor geometries are continually forthcoming and reactor designers are urged to consult the current literature before making a decision on using burnout correlation.

For the range of interest for most pressurized water reactors, Tong [23] developed the following correlation (W-3) for uniformly heated channels:

$$\frac{q''_{DNB,EU}}{10^6} = \{(2.022 - 0.0004302p) + (0.1722 - 0.0000984p) \times \exp[(18.77 - 0.004129p)X]\}$$
$$\times [(0.1484 - 1.596X + 0.1729X\,|X|)\,G/10^6 + 1.037]$$
$$\times [1.157 - 0.869X] \times [0.2664 + 0.8357\exp(-3.15De)]$$
$$\times [0.8258 + 0.000794(h_f - h_{in})], \quad (6.2.42)$$

where

$q''_{DNB,EU}$ = equivalent uniform channel DNB heat flux, Btu/hr-ft^2

p = pressure, psia

X = equilibrium quality = $(h - h_f)/h_{fg}$

G = mass velocity, lb$_m$/ft^2-hr

h = enthalpy, Btu/lb$_m$

D_e = channel equivalent diameter, in

The ranges of parameters for the data used in developing Eq. (6.2.42)

(i) Geometries: circular tube and rectangular channel
(ii) Mass velocity: 0.5 x 10^6-5.0 x 10^6 lb_m/hr-ft^2
(iii) Pressure: 1000 – 2300 psia
(iv) Length: 10–144 in.
(v) Inlet enthalpy: $h_{inlet} \geqslant 400$ Btu/lb_m
(vi) Exit quality: -0.15 to 0.15
(vii) Equivalent diameter: 0.2 to 0.7 in.
(viii) Heated-to-wetted perimeter ratio: 0.88–1.0

Tong [23] extended his correlation for use with nonuniform axial heat flux distributions by employing a correction factor

$$q''_{DNB,N} = q''_{DNB,EU}/F \tag{6.2.43}$$

where

$q''_{DNB,N}$ = DNB heat flux for nonuniformly heated channels
$q''_{DNB,EU}$ = equivalent uniform DNB flux from Eq. (6.2.42)

and

$$F = \frac{C\int_o^{\ell_{DNB}} q''(z) \exp[-C(\ell_{DNB} - z)]\, dz}{q''_{local}\,[1 - \exp(-C\ell_{DNB})]} \tag{6.2.44}$$

where

$$C = 0.44\,\frac{(1 - X_{DNB})^{7.9}}{(G/10^6)^{1.72}}\ \text{in.}^{-1}$$

q''_{local} = local heat flux at ℓ_{DNB}
ℓ_{DNB} = axial location at which DNB occurs, in.

The presence of spacers in rod bundles increases burnout heat fluxes because the boundary layer is tripped by the spacer and the spacer causes increased mixing between subchannels. The Westinghouse approach is to multiply Eq. (6.2.43) by a spacer factor F_s which is given by [23]

$$F_s = 1.0 + 0.03(G/10^6)\left(\frac{TDC}{0.019}\right)^{0.35} \tag{6.2.45}$$

where TDC is the thermal diffusion coefficient which is experimentally determined for a given rod bundle-spacer design. Reference 23 quotes a value of 0.019 for TDC in a simple rod bundle grid spacer and higher values for grid spacers with mixing vanes.

In applying the W-3 correlation (Eq. (6.2.42)) to burnout analysis of rod bundles it is necessary to use subchannel values for quality and mass velocity. Consequently, a subchannel analysis instead of a bundle-average analysis is required for each fuel assembly. The subchannel analysis evaluates mass velocity and local enthalpy of each subchannel of interest. This evaluation should include cross-flow redistribution and mixing between adjacent channels. The importance of subchannel analysis, especially for close-packed rod bundles with nonuniform lattices, is shown in Knudsen's analysis of rod bundle mixing discussed in Ref. [24]. Failures to use this approach may result in serious underestimates of burnout.

For the treatment of boiling water reactors at 1000 psia, Hench and Levy [25] propose the lower limit curves shown in Fig. 6.11 for predicting critical heat fluxes. Lower limit curves are obtained by drawing lines below the available data for a given flow rate and quality. Figure 6.11 is based upon data from 4- and 9-rod test section. At system pressures other than 1000 psia, the following correction factor is recommended [20]:

$$q_c''(p) = q_c''(1000)\left\{1.1 - 0.1\,[(p - 600)/400]^{1.25}\right\} \tag{6.2.46}$$

Note that Eq. (6.2.46) is only valid over a range of pressures from 600 psia to 1450 psia.

(4) Film Boiling

As shown on the boiling curve, if the heat flux exceeds the burnout heat flux the heat transfer regime passes into the film boiling region. Wall temperatures during film boiling at full power reactor operation would probably exceed the cladding melting point. Consequently, the design conditions for reactors at full power operation are such that heat fluxes are considerably below burnout heat fluxes and film boiling is a highly improbable event.

In the analysis of reactors during loss-of-flow or -coolant accidents, film boiling might occur at lower reactor power levels in which cladding melting might not occur. Consequently, it is necessary to evaluate film

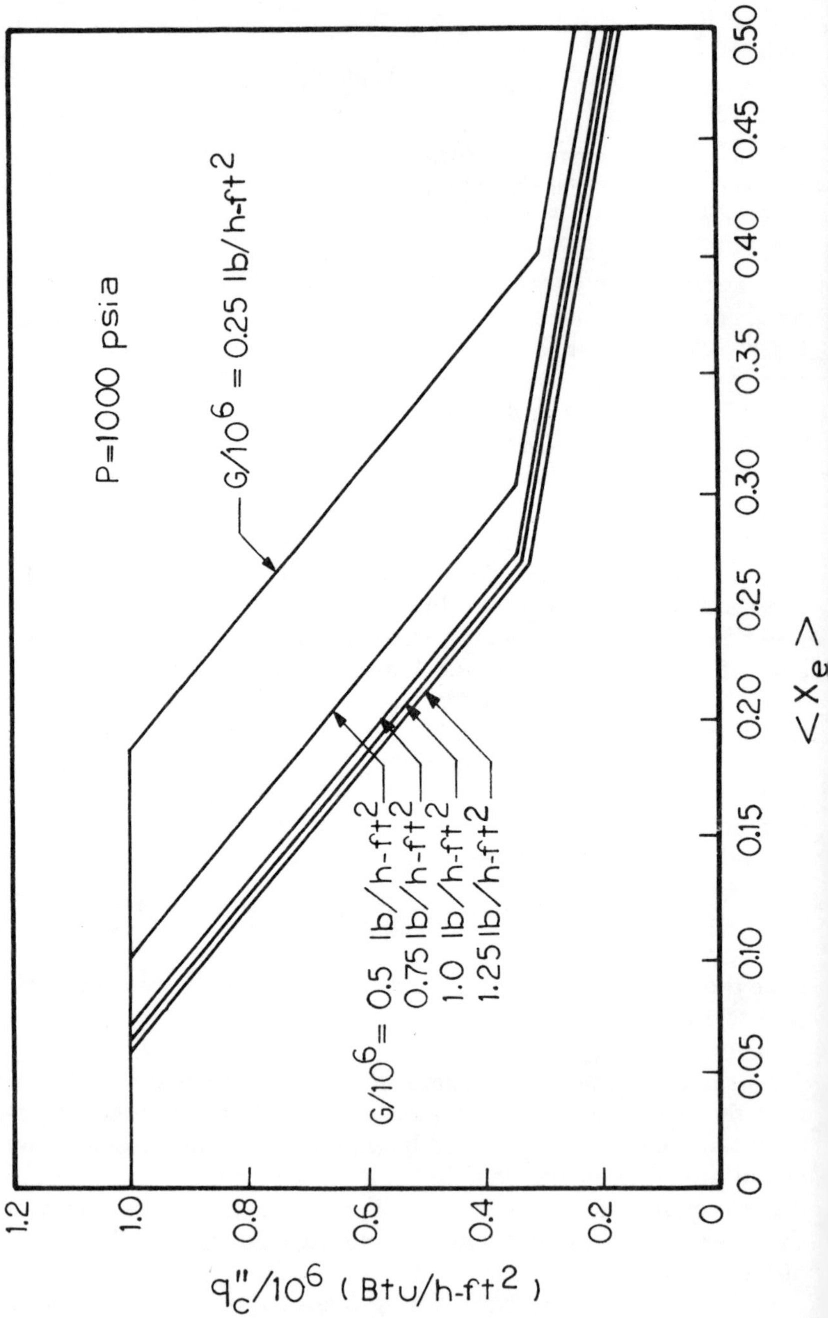

Fig. 6.11. Hench-Levy limit lines [20].

boiling heat-transfer coefficients so as to make estimates of cladding wall temperatures during the accident.

Figure 6.12 gives a qualitative picture of the type of heat transfer regimes expected during the blowdown portion of a loss-of-coolant accident [18]. For a given pressure and flow rate, the illustration shows a map of heat transfer regimes and their boundaries as a function of wall superheat and quality coordinates. The dotted line is a hypothetical trajectory at a point in the core during the course of the blowdown illustrating expected heat transfer regimes and their order of occurrence.

As mentioned previously, transition boiling cannot exist during steady reactor operation and will only exist during short intervals of loss-of-coolant accidents. Groeneveld's review of methods for predicting transition boiling indicates no satisfactory method is

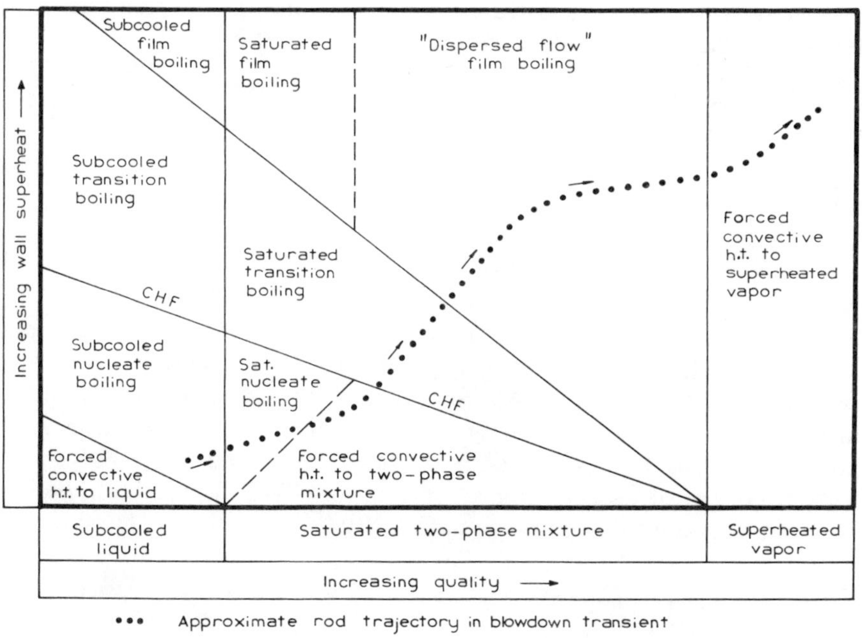

Fig. 6.12. Simplified boiling map for high flow region [18].

available [26]. Reference [18] outlines a simple method which is probably adequate for accident analysis.

When heated surfaces are entirely covered by a stable vapor film the heat transfer region is called stable film boiling. For sufficiently low wall superheats, liquid droplets may strike heated surfaces and greatly enhance the heat transfer process. When wall temperatures reach the **Leidenfrost temperature** the wall is sufficiently hot that a vapor layer occurs between the liquid droplet and heated surface which greatly diminishes the cooling effect. For steam-water mixtures at 30 psia and wall superheats less than 50°F, heat-transfer coefficients are observed to be three to six times greater than for flow of dry steam. For wall superheats exceeding 50°F, the heat-transfer coefficients are almost identical to those for dry steam. Additional data indicates that wall superheats necessary for reaching the Leidenfrost temperature increases with increasing pressure. For steam-water mixtures at 250 psia the wall superheat required is about 160°F and 250°F at 1000 psia [9].

A widely recognized correlation for predicting film boiling heat transfer is the Groeneveld correlation given by [27]

$$\frac{hD_e}{k_g} = 0.052\left[\frac{GD_e}{\mu_g}\left(x_e + (\rho_g/\rho_f)(1 - x_e)\right)\right]^{0.688} (N_{Pr})_{wall}^{1.26}\, Y^{-1.06} \tag{6.2.47}$$

where x_e is the equilibrium quality and

$$Y = 1 - 0.1\left(\frac{\rho_f}{\rho_g} - 1\right)^{0.4}(1 - x_e)^{0.4}$$

Equation (6.2.47) is accepted by the U.S.N.R.C. for the design and safety analysis of BWRs. For a quality of zero, Y goes to zero at a pressure of 78 psia and goes to zero at lower pressures when qualities are above zero. This singularity causes problems in using Eq. (6.2.47) and its use may be questionable at pressures in this range [20].

For system pressures more typical of PWR operation, Bishop recommends the following correlation [9]:

$$\left(\frac{hD_e}{k_g}\right)_f = 0.0193\left(\frac{D_e G}{\mu}\right)_f^{0.8}\left(\frac{c_p \mu}{k}\right)_f^{1.23}\left(\frac{\rho_g}{\rho_{bulk}}\right)^{0.68}\left(\frac{\rho_g}{\rho_f}\right)^{0.068} \tag{6.2.48}$$

where the subscript f outside the parenthesis indicates properties are evaluated at the film temperature $(T_w + T_b)/2$. Equation (6.2.48) is developed from data covering the following ranges:

$$q''_w = 0.11 \times 10^6 \text{ to } 0.61 \times 10^6 \text{ Btu/hr-ft}^2$$
$$G = 0.88 \times 10^6 \text{ to } 2.5 \times 10^6 \text{ lb}_m\text{/hr-ft}^2$$

$$p = 580 \text{ to } 3190 \text{ psia}$$
$$D_e = 0.10 \text{ to } 0.32$$
$$T_b = 483 \text{ to } 705°F$$
$$T_w = 658 \text{ to } 1100°F$$

Equations (6.2.47) and (6.2.48) are empirical equations arbitrarily referenced to the saturation temperature by

$$h \equiv \frac{q''_w}{T_w - T_{sat}} \qquad (6.2.49)$$

These equations are highly conservative in that they usually underestimate film boiling heat-transfer coefficients. In addition, they do not provide great insight into the heat transfer mechanisms because they do not account for such effects as vapor superheat [20].

For lower mass velocities and pressures, Refs. [9] and [18] list additional correlations for predicting film boiling heat-transfer coefficients. When wall temperatures exceed the Leidenfrost temperature, heat-transfer coefficients can be evaluated using correlations for steam. McEligot, et al. [28] recommend

$$\frac{hD_e}{k} = 0.020(N_{Re})_b^{0.8} \ (N_{Pr})_b^{0.4} \ (T_b/T_w)^{0.5} \qquad (6.2.50)$$

where subscript b indicates bulk temperature and all fluid properties are for steam only.

6.3 Heat Transfer Within Fuels

The heat conduction equation for solids is used in the analysis of reactor fuel element temperature distributions.

$$\nabla \cdot k(\vec{r},T) \, \nabla \cdot T(\vec{r},t) + q'''(\vec{r},t) = \rho c_p(\vec{r},t) \frac{\partial T(\vec{r},t)}{\partial t} \qquad (6.3.1)$$

where

$k(\vec{r},T)$ = spatial and temperature dependent thermal conductivity, Btu/hr–ft–°F

$T(\vec{r},t)$ = spatial and time dependent temperature, °F

$q'''(\vec{r},t)$ = spatial and time dependent volumetric heat source (Chapter 5), Btu/ft^3-hr

$\rho c_p(\vec{r},T)$ = spatial and temperature dependent volumetric heat capacity, Btu/ft^3–°F

t = time, hr

The first term in Eq. (6.3.1) represents the net leakage of heat from a differential volume element, the second term the heat generated in the element, and the third term the rate of heat accumulation within the element.

Great difficulty is experienced in solving Eq. (6.3.1) for a nuclear reactor environment because large temperature gradients in the fuel may induce significant variations in thermophysical properties. An additional complicating factor is that thermophysical properties may change with time during steady power operation of a reactor because of irradiation damage to materials and buildup of fission products. Physical property data as a function of temperature are given in Appendix D for some common materials used in nuclear reactor systems.

Handbook values for thermophysical property data are readily found for most of materials used in nuclear reactor systems. Some uncertainty exists in using property data for reactor analysis because the effects of irradiation on property values are not completely resolved.

In nuclear reactor analysis, Eq. (6.3.1) is usually solved numerically because of its extreme complexity. However, in some simplified cases its complexity can be reduced so that it is easy to obtain closed-form solutions for temperature distributions.

Temperature Distributions in Plate Fuel Elements

Figure 6.13 illustrates a typical fuel plate used in plate fuel elements. Because plate widths are small, the neutron flux or heat generation in individual plates does not undergo substantial variations across the plate width. In addition, over small plate lengths the neutron flux or

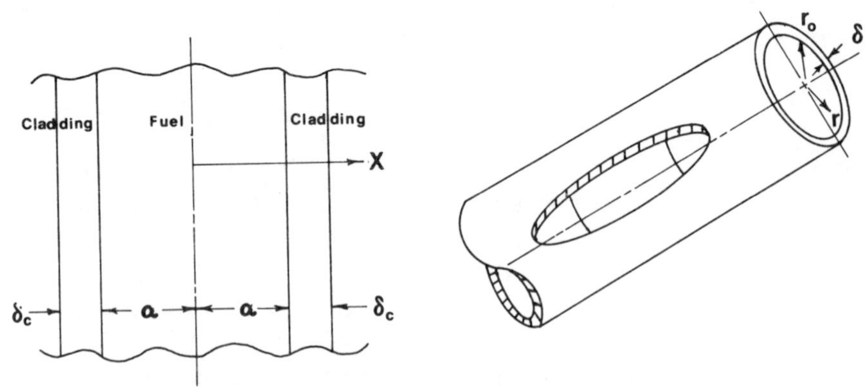

Fig. 6.13. Plate and rod fuel elements.

heat generation is not strongly dependent upon axial position. Consequently, for these reasons and also because of small plate thicknesses compared to dimensions in other directions, the most severe temperature gradients in fuel plates are across the plate thickness. It may be assumed, without encountering great error, that heat conduction in thin fuel plates is one-dimensional and occurs in a direction along the plate thickness. This assumption is conservative, since it tends to slightly overestimate temperatures in fuel plates due to neglecting heat conduction along the plate width and length. The errors are less than a few percent based upon an analysis by Thorpe [29].

By assuming steady state and a one-dimensional temperature dependence, Eq. (6.3.1) is written for the fuel region

$$\frac{d}{dx}\left(k_f \frac{dT}{dx}\right) + q'''(x) = 0 \tag{6.3.2}$$

Frequently, the coolant temperature at a given axial position is the same on both sides of the fuel plate and the volumetric heat generation rate is symmetric with respect to the fuel midplane. Under these conditions the plate temperature distribution in symmetric and typical boundary conditions are at $x = 0$, $dT/dx = 0$, and at $x = \pm a$, $T = T_s$ Direct integration of Eq. (6.3.2) gives

$$\int_{T_s}^{T} k_f \, dT = -\int_a^x \int_o^x q'''(x') \, dx' \, dx \tag{6.3.3}$$

In plate fuel elements the fuel is usually dilutely dispersed in a weak absorbing material. As a consequence, because fuel thicknesses are small the volumetric heat generation rate is essentially uniform. If the fuel thermal conductivity is constant or a power series expansion for the thermal conductivity as a function of a temperature is known, the left hand side of Eq. (6.3.3) is directly integrated. Usually the temperature rise across the fuel is not large and an average value is assumed for the fuel thermal conductivity. Consequently, integration of Eq. (6.3.3) gives

$$T - T_s = \frac{q_o'''(a^2 - x^2)}{2k_f} \tag{6.3.4}$$

Because of symmetry about $x = 0$, the heat flux at the fuel surface is given by

$$q_s'' = \int_o^a q''' \, dx = q_o''' a \tag{6.3.5}$$

Equation (6.3.4) is also written in terms of the fuel surface heat flux as

$$T - T_s = \frac{q_s''a}{2k_f}[1 - (x/a)^2] \tag{6.3.6}$$

Equation (6.3.2) is also used for evaluating the cladding temperature distribution. The volumetric heat generation rate is assumed negligible because of the large heat flow through the cladding originating from the fuel region. Equation (6.3.5) also gives the cladding heat flux which is used as a boundary condition on either cladding surface. Therefore, if T_w is the wall temperature at the outer cladding surface and k_c is the cladding thermal conductivity, integration of Eq. (6.3.2) for the coordinate system shown in Fig. 6.13 gives

$$T_{\delta c} - T_w = \frac{q_c''\delta_c}{k_c} \tag{6.3.7}$$

where $T_{\delta c}$ is the temperature at the cladding inner surface.

It is assumed there is no thermal resistance between the fuel surface and cladding inner surface. Consequently, the temperature at the fuel surface T_s equals the temperature at the cladding inner surface $T_{\delta c}$. This assumption is quite good if a good metallurgical bond exists between the fuel and cladding. The fuel surface temperature is eliminated by adding Eqs. (6.3.6) and (6.3.7)

$$T - T_w = q_s''\left[\frac{a^2 - x^2}{2k_f a} + \frac{\delta_c}{k_c}\right] \tag{6.3.8}$$

for $-a \leqslant x \leqslant a$.

The rate of heat transfer at any position on the cladding surface to an adjacent flowing fluid is given by

$$q_w'' = h(T_w - T_b) \tag{6.3.9}$$

where

q_w'' = local wall heat flux, Btu/hr–ft^2
h = local convection heat-transfer coefficient, Btu/hr–ft^2-°F
T_b = local bulk coolant temperature, °F

Because of the plate geometry $q_s'' = q_w''$. Therefore, adding Eqs. (6.3.8) and (6.3.9) eliminates T_w and gives

$$T - T_b = q_w\left[\frac{a^2 - x^2}{2k_f\, a} + \frac{\delta_c}{k_c} + \frac{1}{h}\right] \tag{6.3.10}$$

for $-a \leqslant x \leqslant a$.

Temperature Distribution is Solid Cylindrical Fuel Elements

The majority of nuclear reactors used in power production employ rod bundle fuel elements. Present fuel pins consist of small UO_2 or PuO_2 pellets encapsulated in a zirconium or stainless steel tube, as shown in Fig. 6.13. If a fuel rod is noτ in the vicinity of a control rod, the neutron flux distribution in a fuel pellet is fairly independent of angular position. In addition, for a fuel rod being cooled with highly turbulent flow, the cladding temperature for a given axial location is reasonably independent of angular position.* Consequently, by taking into account these factors, both T and q''' can be assumed independent of angular position. Over small rod lengths the neutron flux will not change significantly and, consequently, conduction along the rod length is assumed negligible. Therefore, the temperature distribution for a given axial position is considered one-dimensional and Eq. (6.3.1) is written

$$\frac{1}{r}\frac{d}{dr}\left(k_f\, r\,\frac{dT}{dr}\right) + q'''(r) = 0 \tag{6.3.11}$$

Since $dT/dr = 0$ at $r = 0$,direct integration of Eq. (6.3.11) for a fuel rod of radius r_o and surface temperature T_s yields

$$\int_{T_s}^{T} k_f\, dT = -\int_{r_o}^{r}\frac{1}{r}\int_{o}^{r} r'q'''(r')\, dr'\, dr \tag{6.3.12}$$

By assuming the thermal conductivity is independent of temperature and the volumetric heat source replaced by an appropriately averaged value, $\bar{q}'''$, Eq. (6.3.12) is written

$$T - T_s = \frac{\bar{q}'''r_o^2}{4k_f}\left[1 - (r/r_o)^2\right] \tag{6.3.13}$$

It is readily proved that Eq. (6.3.13) is valid when the thermal conductivity is a linear function of temperature if k_f is evaluated at $(T + T_s)/2$.

*An exception may be liquid-metal-cooled reactors in which cladding wall temperatures in a rod bundle can display a considerable amount of angular dependence.

Since the heat flux at the fuel surface is given by

$$q_s'' = \frac{\int_0^{r_o} q'''(r)2\pi r \, dr}{2\pi r_o} = \frac{\bar{q}'''r_o}{2} \tag{6.3.14}$$

Equation (6.3.13) is written

$$T - T_s = \frac{q_s''r_o}{2k_f}\left[1 - (r/r_o)^2\right] \tag{6.3.15}$$

which is the same form of solution found for the flat plate, Eq. (6.3.7).

The linear heat rate q', which is the rate of heat generation per unit length of fuel rod, is given by

$$q' = \int_0^{r_o} q'''(r)2\pi r \, dr$$

$$= \pi r_o^2\bar{q}''' = 2\pi r_o q_s'' \tag{6.3.16}$$

Linear heat rates are frequently used as an index of performance for fuel rods. Typical values for BWR's and PWR's range from about 6 to 15kW/ft. In terms of linear heat rate, Eq. (6.3.13) is written

$$T - T_s = \frac{q'}{4\pi k_f}\left[1 - (r/r_o)^2\right] \tag{6.3.17}$$

The maximum temperature difference across the fuel pellet occurs when $r = 0$ or

$$T_{max} - T_s = \frac{q'}{4\pi k_f} \tag{6.3.18}$$

Equation (6.3.18) shows the interesting result that the maximum temperature difference in a fuel pellet expressed in terms of linear heat rate is independent of pellet radius. Consequently, by specifying a linear heat rate for a fuel rod, the maximum temperature difference in the fuel is fixed.

Equations (6.3.13) and (6.3.18) are derived on the assumption of a uniform radial heat generation rate. Actually, self shielding effects within the fuel produce a thermal neutron flux depression and corresponding heat generation depression at the fuel center. For low enriched fuels($<3\%$)in thermal reactors, this variation will not significantly affect temperature distributions. However, in more highly enriched fuels this effect should be considered. It is noted that, since neutron flux depressions causes a reduction in central temperature for the same total heat generation rate, the assumption of constant heat generation overestimates central temperatures and is thus conservative. In addition, it is assumed that an appropriately averaged thermal conductivity is used in the heat conduction equation. In power reactors, temperature gradients in fuel materials are quite large and, consequently, substantial variations in thermal conductivity occur over small distances. This is particularly true for oxide-fueled reactors in which temperature differences of 2000-3000°F occur over a distance of 0.2 inch.

Referring back to Eq. (6.3.12) and keeping a temperature dependent thermal conductivity, fuel temperatures for uniform heat generation in terms of linear heat rates are given by

$$\int_{T_s}^{T} k_f(T')\, dT' = \frac{q'}{4\pi}\left[1 - (r/r_o)^2\right] \qquad (6.3.19)$$

and

$$\int_{T_s}^{T_{max}} k_f(T')\, dT' = q'/4\pi \qquad (6.3.20)$$

The left-hand side of Eqs. (6.3.19) and (6.3.20) is called the integral conductivity of the fuel material. This quantity is easy to evaluate if one knows the temperature dependence of the fuel material thermal conductivity. Figure 6.14 illustrates a typical integral conductivity curve for UO_2 of 95% theoretical density.

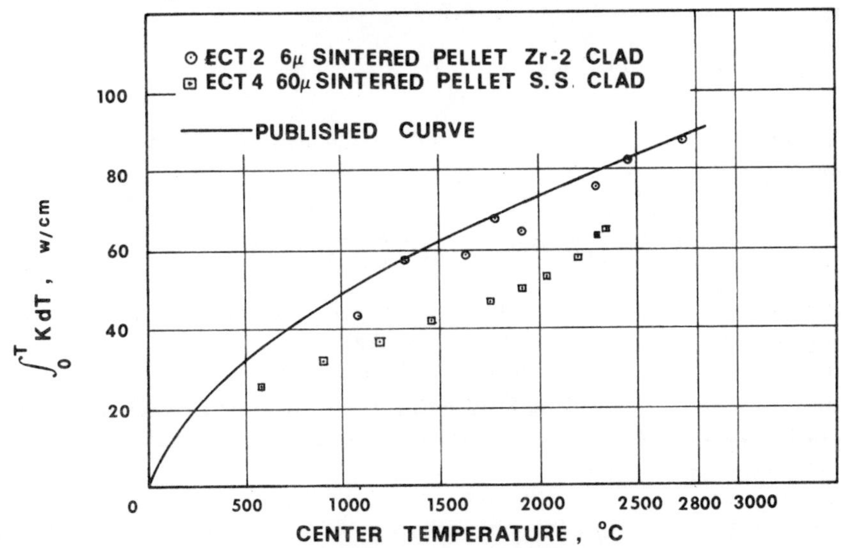

Fig. 6.14. $\int k(T)\,dT$ for UO_2 fuel pellets, 95% theoretical density [30].

Oxide fuel pellets are usually fabricated by sintering a powder to achieve desired fuel density. The density obtained is less than the theoretical density; typically, 93-95% of the theoretical density. The thermal conductivity is reduced by the presence of these voids. The Maxwell-Euken correlation is commonly used to account for this [9]

$$k_p = \frac{1 - \epsilon}{1 + \beta\epsilon} k_{100\%} \tag{6.3.21}$$

where

k_p = conductivity of porous fuel
ϵ = porosity
$k_{100\%}$ = conductivity of fuel of 100% theoretical density
β = pore-shape factor (β = 0.5 for fuel of 90% theoretical density and above, β = 0.7 for fuels of lower density)

The value of $\int k_f \, dt$ is, of course, similarly reduced and Eq. (6.3.21) must be applied to Fig. 6.14 when UO_2 density differs from 95% theoretical.

The use of Eq. (6.3.20) and Fig. 6.14 to determine maximum fuel temperatures in UO_2 pellets is illustrated by the following example:

Example 6.1

It is desired to find the maximum fuel temperature of a 95% density sintered UO_2 pellet operating at a linear heat rate of 22 kW/ft with a fuel surface temperature of 1130°F. From Eq. (6.320)

$$\int_0^{T_{max}} k_f(T')\, dT' - \int_0^{T_s} k_f(T')\, dT' = q'/4\pi \tag{6.3.22}$$

From Fig. 6.14, $\int_0^{610} k(T')\, dT' = 27$ watts/cm. Therefore, from Eq. (6.3.22) we have

$$\int_0^{T_{max}} k_f(T')\, dT' = \frac{22{,}000}{30.5 \times 4\pi} + 27 = 84.4$$

By inspection of Fig. 6.14, it is seen that the maximum fuel temperature is T_{max} = 2600°C or 4700°F.

Robertson of Chalk River evaluated the effect of nonuniform heat generation in solid and hollow cylindrical fuels [31]. For the solid cylinder he assumed a thermal neutron flux and heat generation distribution of the form

$$q'''(r) = q_o''' I_o(\kappa r) \tag{6.3.23}$$

where q_o''' is a constant, I_o is the modified Bessel function, and κ is the fuel inverse diffusion length. The temperature distribution is then found by

$$\int_{T_s}^{T} k_f(T')\,dT' = \frac{q_0'}{4\pi}\left[\frac{I_0(\kappa a) - I_0(\kappa r)}{\frac{1}{2}\kappa a\, I_1(\kappa a)}\right] \qquad (6.3.24)$$

and the fuel centerline temperature is

$$\int_{T_s}^{T_{max}} k_f(T')\,dT' = \frac{q_0'}{4\pi}\left[\frac{I_0(\kappa a) - 1}{\frac{1}{2}\kappa a\, I_1(\kappa a)}\right] = \frac{q_0' f}{4\pi} \qquad (6.3.25)$$

where "a" is the fuel radius. Since κ is a function of enrichment, the function in brackets (called the flux depression factor f) may be evaluated for various combinations of fuel radii and enrichments. Based on the Robertson data, the value of

$$f = \left[\frac{I_0(\kappa a) - 1}{\frac{1}{2}\kappa a\, I_1(\kappa a)}\right]$$

for a typical PWR pellet size (~ 0.37 inch diameter) is given in Table 6.5 as a function of enrichment.

Table 6.5

Values of $\dfrac{I_o(\kappa a) - 1}{\frac{1}{2}\kappa a\, I_1(\kappa a)}$ versus enrichment [31]

Enrichment, %	$\dfrac{I_o(\kappa a) - 1}{\frac{1}{2}\kappa a\, I_1(\kappa a)}$ *
0	1.0
5	0.94
10	0.87
15	0.795
20	0.72

*For 95% dense UO_2 solid cylinders (0.37 inch diameter)

Using Table 6.5 we return to the preceding example problem and evaluate the effect of using 5 percent enriched fuel on the central temperature.

Example 6.2

$$\int_0^{T_{max}} k_f(T')\, dT' = 27 + \frac{22{,}000}{4\pi\ 30.49}(0.94) = 81 \qquad (6.3.26)$$

From Fig. 6.14, T_{max} = 4350°F which is 350°F lower than occurs when assuming uniform heat generation.

Although the Robertson method is useful for approximating the effects of self shielding, the diffusion theory solution for the neutron flux distribution, Eq. (6.3.23), underestimates neutron flux depression and, consequently, overestimates fuel temperatures.

Due to the process used in fabricating rod-type fuel elements, a resistance to heat transfer exists at the interface between the fuel pellet surface and the cladding inside surface. This resistance produces a substantial temperature difference at the interface which is calculated using a contact heat-transfer coefficient

$$T_s - T_{sc} = \frac{q''_s}{h_c} \qquad (6.3.27)$$

or

$$T_s - T_{sc} = \frac{q'}{2\pi r_0 h_c} \qquad (6.3.28)$$

Methods for estimating h_c follow this section.

For the cladding, the internal heat generation is usually negligible compared to the heat flow through it from the fuel. Consequently, the cladding temperature difference is obtained by solving Eq. (6.3.11) without a volumetric heat source. Direct integration yields

$$T_{sc} - T_w = \frac{q''_s r_0}{k_c} \ln\left(\frac{r_0 + \delta_c}{r_0}\right) \qquad (6.3.29)$$

where T_w is the outside wall temperature and k_c is the cladding mean thermal conductivity.

The heat flux at the cladding outer surface is less than q''_s because the surface area on the outside of a hollow cylinder is larger than on the inside. In terms of the outer wall heat flux q''_w , Eq. (6.3.29) is written

$$T_{sc} - T_w = \frac{q''_w (r_0 + \delta_0)}{k_c} \ln\left(\frac{r_0 + \delta_c}{r_0}\right) \qquad (6.3.30)$$

The wall temperature T_w in Eq. (6.3.30) is replaced by a fluid bulk temperature by introducing the heat-transfer coefficient

$$T_{sc} - T_b = q''_w \left[\frac{(r_0 + \delta_c)}{k_c} \ln \left(\frac{r_0 + \delta_c}{r_0} \right) + \frac{1}{h} \right] \tag{6.3.31}$$

For rod bundle geometries, T_b is the fluid bulk temperature at a given axial location for a rod bundle subchannel. Naturally, employing Eq. (6.3.31) for calculating fuel temperatures assumes T_b is the same for all subchannels surrounding the fuel pin.

Contact Heat-Transfer Coefficients

For most rod-type fuel elements, a gap or at least a poor bonding exists between the fuel and cladding. The temperature drop between the fuel surface and cladding inner surface is defined by

$$\frac{1}{h_c} \equiv \frac{T_s - T_{sc}}{q''_w} \tag{6.3.32}$$

The contact resistance is difficult to evaluate. Some fuel elements use helium or other gases to fill the gap between fuel and cladding. Consequently, the modes of heat transfer across gas-filled gaps are conduction through the gas and thermal radiation between the two surfaces. With heating and irradiation, swelling occurs in reactor fuels and physical contact takes place between fuel and cladding. However, even for materials in contact under pressure, a finite contact heat-transfer coefficient exists which experiments show is roughly proportional to the applied pressure. Since the problem of determining h_c is difficult to resolve, one should use conservative answers. Normal design use is for h_c to vary between 750-2000 Btu/hr- ft^2 -°F.

For gas-filled gaps in which mean free paths of gas molecules and dimensions of heat transfer surfaces are large compared to gap widths, the contact heat-transfer coefficient is approximated by

$$h_c \quad \left(\frac{k}{x}\right)_g + \sigma \left(\frac{1}{\epsilon_s} + \frac{1}{\epsilon_{sc}} - 1 \right) \frac{T_s^4 - T_{sc}^4}{T_s - T_{sc}} \tag{6.3.33}$$

where

$(k/x)_g$ = gas thermal conductivity in the gap divided by gap width, Btu/hr- ft^2 -°F

σ = Stephan-Boltzmann constant, 0.173 x 10^{-8} Btu/hr–ft^2-.°R^4

ϵ_s = emissivity of fuel surface, dimensionless

ϵ_{sc} = emissivity of cladding inner surface, dimensionless
T_s = fuel surface absolute temperature, °R
T_{sc} = cladding inner surface absolute temperature, °R

Fuel swelling or cladding shrinkage may cause contact between fuel and cladding. However, due to surface roughness on both materials, surfaces are not completely in contact with one another but meet only at "higher points." In many cases that exist in reactor fuel elements, the actual amount of surface contact may be only about 10 percent. Consequently, heat transfer across gas-filled gaps is an important contribution to the heat transfer process. Ross and Stoute propose a model for evaluating contact heat-transfer coefficients which consider separately the fuel-to-cladding contact and gas-gap heat transfer [32].

$$h_c = h_{solid} + h_{gap} \tag{6.3.34}$$

where h_{solid} and h_{gap} are both defined for the total interface area.

The solid-to-solid conductance is given by

$$h_{solid} = \frac{k_m p}{a_o R^{1/2} H} \tag{6.3.35}$$

where
k_m = harmonic mean conductivity of interface materials $2k_1k_2/(k_1 + k_2)$, Btu/hr-ft-°F
p = contact pressure, psia
a_0 = empirical constant, 0.0905 $ft^{1/2}$
R = root-mean-square of contact materials surface roughness $[(R_1^2 + R_2^2)/2]^{1/2}$, ft.
H = Meyer hardness of softer material, psi

Typical values of the Meyer hardness for stainless steel is 13×10^4 psi and 14.2×10^4 psi for zircaloy.

The second portion of the heat-transfer correlation pertains to heat transfer across the gas medium. It is concentrated solely on the conduction mechanism since radiation is assumed negligible and the gap is too small to allow free convection. The gas medium is assumed to lie between two plane parallel surfaces with a separation distance which is a direct function of the surface roughness. Since this separation distance is usually the same order of magnitude as the mean free path of gas molecules, the accomodation effect causes the temperature difference imposed upon the gas to be less than that existing between the interfaces. This reduces the amount of heat transferred by conduction through the gas. The reduction in heat transferred may be considered the result of an artificial increase in gap thickness (see Fig. 6.15). Thus, the heat-transfer coefficient for conduction is given by

$$h_{gap} = \frac{k_{gas}}{t + (g_1 + g_2)} \qquad (6.3.36)$$

where
k_{gas} = gas thermal conductivity
t = actual gap thickness
g_1, g_2 = "increase" in gap at each surface due to accomodation effect
The "g" terms are called temperature jump distances and are defined by the relation

$$\Delta T = -g \frac{dT}{dx}$$

where ΔT is the temperature discontinuity at any one surface and dT/dx is the temperature gradient normal to the surface.

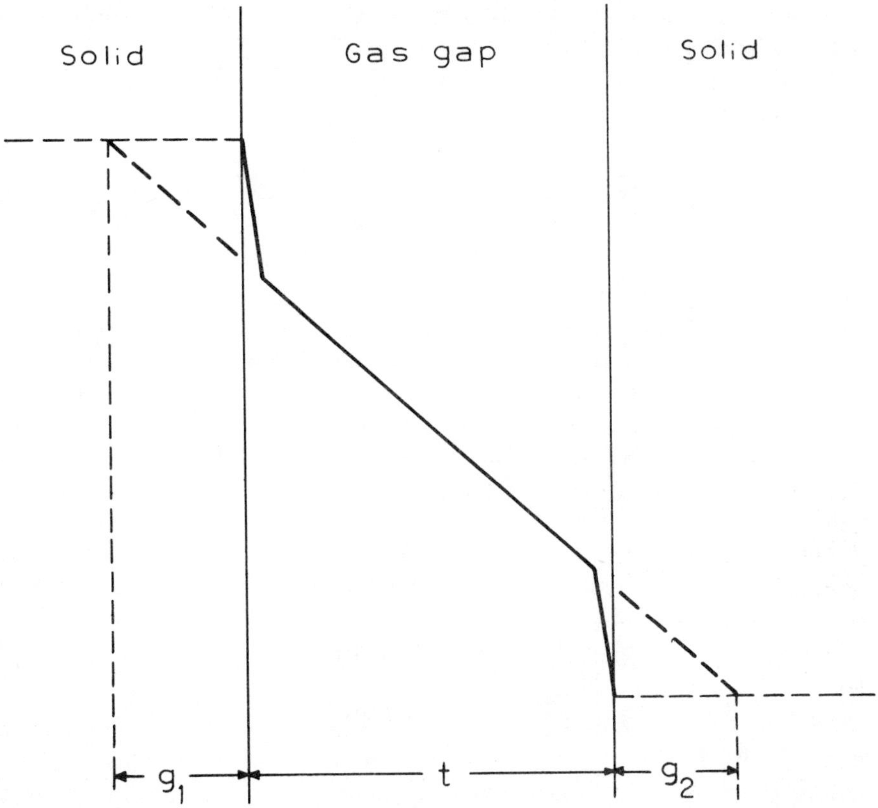

Fig. 6.15. Schematic diagram of gas-gap temperature distribution with accommodation effect.

The value of t in Eq. (6.3.36) is found to be a direct function of surface roughness given by

$$t = c(R_1 + R_2) \tag{6.3.37}$$

where R_1 and R_2 are surface roughness of interface materials. Ross and Stoute found the value of "c" to be dependent upon interface pressure. At pressures about 1420 psi the value was 2.5, decreasing to 1.5 at higher pressures, ~ 7000 psi. Using a linear relationship the value of "c" is correlated by

$$c = 2.75 - 0.17 \times 10^{-3} p \tag{6.3.38}$$

where p is interface pressure in psi.

Ross and Stoute calculated values of $(g_1 + g_2)$ for various gases by comparing analysis with experimental data. Their values are summarized in Table 6.5.

Combining Eqs. (6.3.35), (6.3.36), and (6.3.38) gives the following equation for the contact heat-transfer coefficient:

$$h = \frac{k_m p}{0.0905\, R^{1/2} H} + \frac{k_{gas}}{(2.75 - 1.7 \times 10^{-4} p)(R_1 + R_2) + (g_1 + g_2)} \tag{6.3.39}$$

Table 6.5

Values of temperature jump distances for various gases [32]

Gas	$(g_1 + g_2)$, cm	Temperature Range, °C
Helium	10×10^{-4}	150-250
Argon	5×10^{-4}	180-320
Krypton	1×10^{-4}	180-330
Xenon	1×10^{-4}	180-330

The gas conductivity k_{gas} poses a problem because it changes with time during reactor operation. At initial reactor startup the gas is the fill gas which is usually helium. During reactor operation the helium is diluted with fission product gases which considerably reduce mixture conductivity. Mixture conductivities should not be calculated by simple mole fraction averaging because this places too great an emphasis on low molecular weight, high conductivity gases and overestimates mixture conductivities. One method for estimating mixture conductivities is the Lindsay-Bromley [33] approach given for an N component mixture

$$k_{gas} = \sum_{i=1}^{N} \frac{k_i}{1 + \sum_{\substack{j=1 \\ j \neq i}}^{N} A_{ij}(x_j / x_i)} \tag{6.3.40}$$

where k_i = component conductivities
x_i , x_j = mole fractions
A_{ij} = weighting factor

The weighting factor is a complex function of mixture absolute temperature, pure component viscosities and molecular weights, and Sutherland constants. H. von Ubisch, et al., [34] report thermal conductivities for helium and fission product mixtures.

References [9], [20], and [35] describe other models used for estimating contact heat-transfer coefficients. All contact heat transfer models are dependent upon interface pressure which is directly related to relative expansion of fuel and cladding. Since expansions are functions of temperature, an iterative procedure is necessary. However, this is not easily performed by hand calculations.

The contact pressure is estimated by computing dimesions of fuel pellet and cladding based only upon thermal expansion; thus ignoring elastic deformation of cladding due to pressure differential between cladding inside and outside surfaces [9]. The internal pressure required to produce an elastic deformation in the cladding that gives a zero gap width is obtained from Eq. (6.3.41) for the elastic deformation of the cladding inside diameter (ΔD_i) due to a pressure differential [9]

$$\Delta D_i = \frac{D_i}{E(K^2 - 1)} \left\{ p_i [(1 - \nu) + (1 + \nu) K^2] - 2K^2 p_o \right\} \tag{6.3.41}$$

where D_i = internal diameter
D_o = external diameter
p_i = internal pressure
p_o = external pressure
E = Young's modulus for the cladding
K = D_o/D_i
ν = Poisson's ratio for the cladding

Since the cladding and pellet are in contact over only a fraction of their surface areas the internal pressure is given by

$$p_i = p_{gas} + p_{contact} \tag{6.3.42}$$

Values of contact heat-transfer coefficients are dynamic in the sense that gas thermal conductivities decrease with time due to dilution of the fill gas, such as helium, with fission products during reactor operation. Another factor is that gap widths or contact pressures also change with time due to fuel expansion with irradiation, buildup of fission product gases, and cladding creep. Figure 6.16 illustrates expected behavior of contact heat-transfer coefficients for zircaloy clad UO_2 (95% density) as a function of linear heat rate at beginning-of-life (BOL) and end-of-life (EOL).

Temperature Distributions in Oxide Fuels

Calculation of temperature distribution in oxide fuels (UO_2) is more complex than heretofore presented because of observed effects due to irradiation such as decreases in contact heat-transfer coefficients due to dilution of fill gases with released fission product gases, fuel densification due to grain growth, formation of a central void in cylindrical pellets, fuel swelling due to heating and burnup, decreases in fuel thermal conductivities with fuel burnup, and changes in conductivity due to changes in UO_2 stoichiometry.

Oxide fuel pellets are usually fabricated by sintering a powder until the desired fuel density is obtained. The pellet density is less than theoretical density because of the need to provide void space to accomodate fission product swelling. Typical densities for UO_2 fuels used in water-cooled reactors are 93-95% of theoretical. The thermal conductivity of fuels containing voids is less than that of 100% dense fuels and a typical correction is to use the Maxwell-Euken factor given by Eq. (6.3.21).

Numerous correlations are available for the thermal conductivity of UO_2 as a function of temperature and a few of these also account for fuel burnup. A survey by Belle, et al. [36] yielded the following correlation which accounts for porosity, temperature, and burnup:

$$k(UO_2) = \frac{(1 - \epsilon)/(1 + \beta\epsilon)}{0.116 + 1.88 \times 10^{-4}T + (56.05/T) + (19.3F/T)} \quad (6.3.43)$$

where

$k(UO_2)$ = thermal conductivity, Btu/hr-ft-°F

T = temperature, °R

F = burnup (fissions/cm³) x 10^{-20}

ϵ = porosity

β = values used in Equation (6.3.21)

For unirradiated fuels, the terms containing 56.05 and F are omitted in the use of Eq. (6.3.43). In the application of Eq. (6.3.43), the effect of burnup is minor for fuel temperatures in excess of 1500°F.

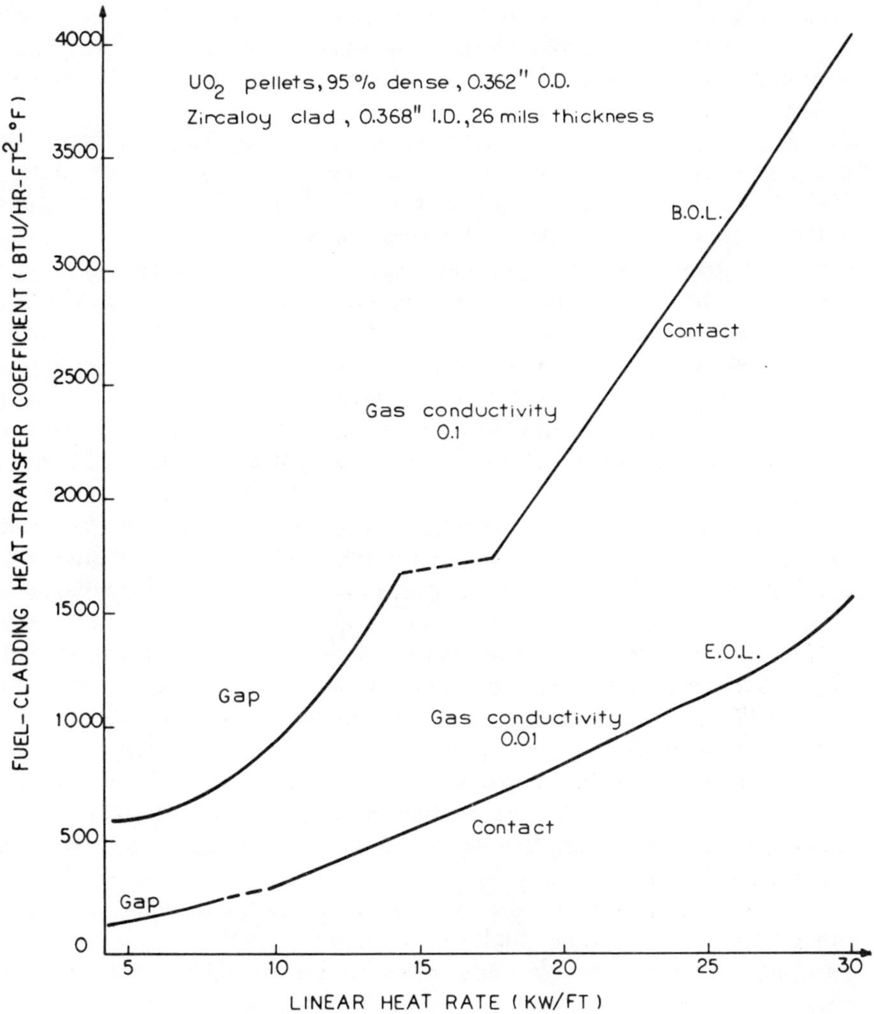

Fig. 6.16. Fuel-cladding heat-transfer coefficient

Figure 6.17 illustrates experimental measurements of UO_2 thermal conductivity and a correlation reported by Lyons and an International Panel of Experts in 1965 [20]. Equation (6.3.43) agrees almost perfectly with this correlation over a range of temperatures from 500°F to 3000°F for unirradiated UO_2 at 100% theoretical density. At higher temperatures the predictions are lower which makes Eq. (6.3.43) conservative for design analysis.

With irradiation and at high temperatures, mass transfer effects take place in oxide fuels which complicates thermal analysis. This results in fuel densification due to grain growth mechanisms. For regions in UO_2 above about 2700°F, the average grain size increases due to equiaxed grain growth. This is a predictable phenomenon and it is observed that for average grain diameters above 20 microns the fuel porosity is independent of grain size and is approximately 3 percent [37]. For temperatures above about 3100°F and in the presence of large temperature gradients, columnar grains are formed in a radial orientation within the fuel [38]. By a process thought to be due to evaporation and condensation on pore boundaries, the pores move up the thermal gradient and eventually leave behind columnar grains of 100% theoretical density. The migration of pores toward the center of fuel pellets causes a central void which increases in radius with irradiation time. Figure 6.18 illustrates the various grain boundaries in irradiated UO_2 .

Nichols [39] developed a theoretical model for predicting the velocity of pores moving up the thermal gradient. With this model it is possible to predict central void radius (r_v) as a function of irradiation time and the radial boundary of the columnar grain growth region (r_{cg}) . Nichols takes the region in which all pores have had sufficient time to migrate to the central void as the columnar grain growth region. Consequently, the fuel density in this region is taken as 100% of theoretical density.

Figure 6.19 illustrates the results of Nichols' calculations for a typical UO_2 pellet. The fuel temperature distribution is assumed parabolic with a pellet surface temperature of 980°F. The primary vapor constituent is assumed helium. Temperatures shown on the ordinate are UO_2 temperatures at the central void boundary or maximum fuel temperatures. For high temperatures the columnar grain growth radius can be a significant fraction of the fuel pellet radius (r_0) . The radius of the central void (r_v) is given by

$$r_v = r_0(r_{cg}/r_0)\sqrt{\epsilon} \qquad (6.3.44)$$

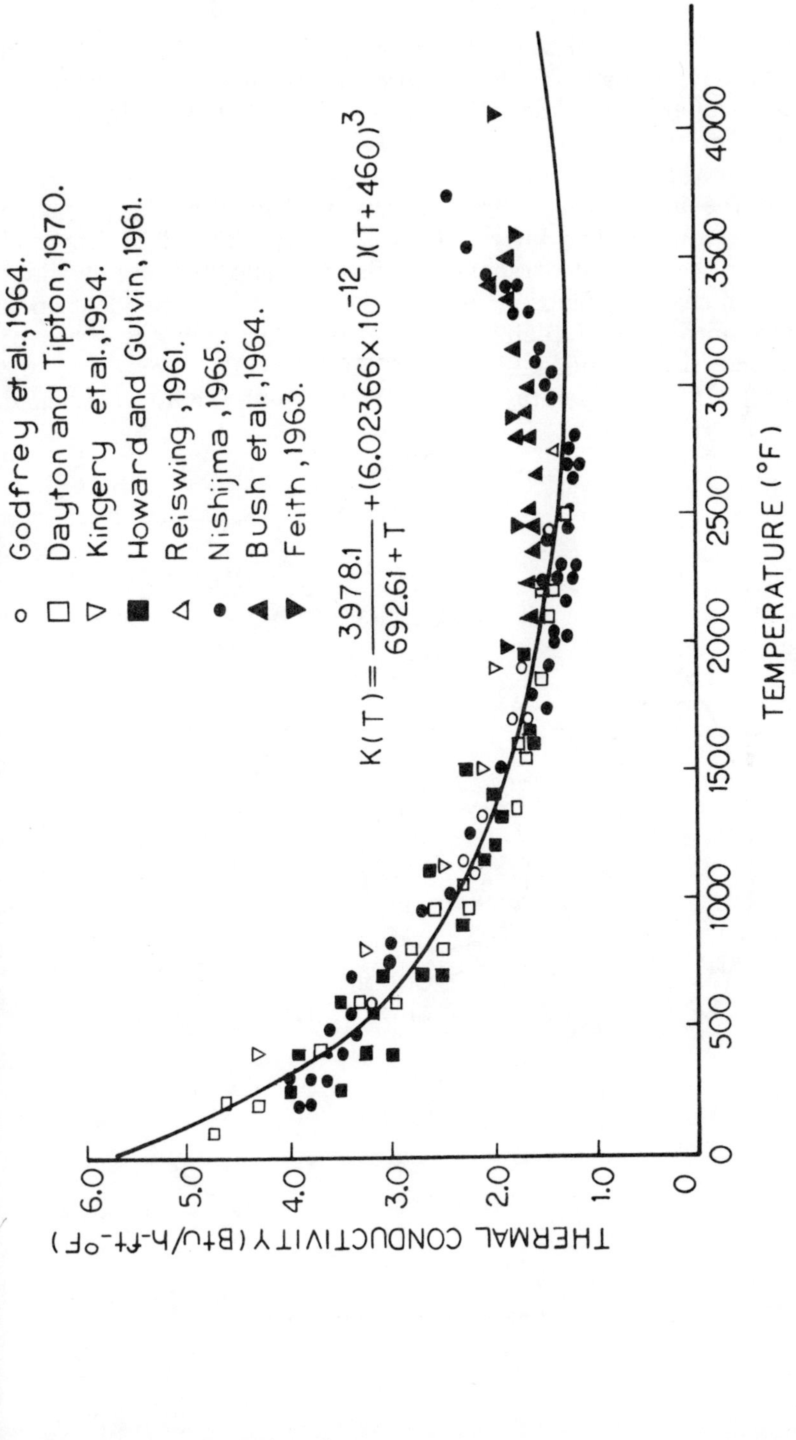

Fig. 6.17. The thermal conductivity of UO_2 [20].

Upon initiation of a central void the solution to Eq. (6.3.11) must be altered by incorporating the boundary condition $dT/dr = 0$ at $r = r_v$.

Because of effects of void migration, burnup, fuel swelling, and cladding creep, it is impossible to find closed-form analytical solutions for fuel temperatures and, therefore, numerical techniques are required. The usual procedure is to divide the fuel pellet, gap, and cladding into a series of annular rings and determine average temperatures in the rings using iterative procedures. Appropriate models are used to determine void migration, fuel thermal conductivity, contact-heat-transfer coefficients, and fuel and cladding dimensional changes. Since many of these models are dependent upon time and burnup, fuel temperatures are evaluated for a series of burnup and time periods. Typical computer codes developed for determining fuel temperatures are CYGRO [40], FIGRO [41], and FMODEL [42].

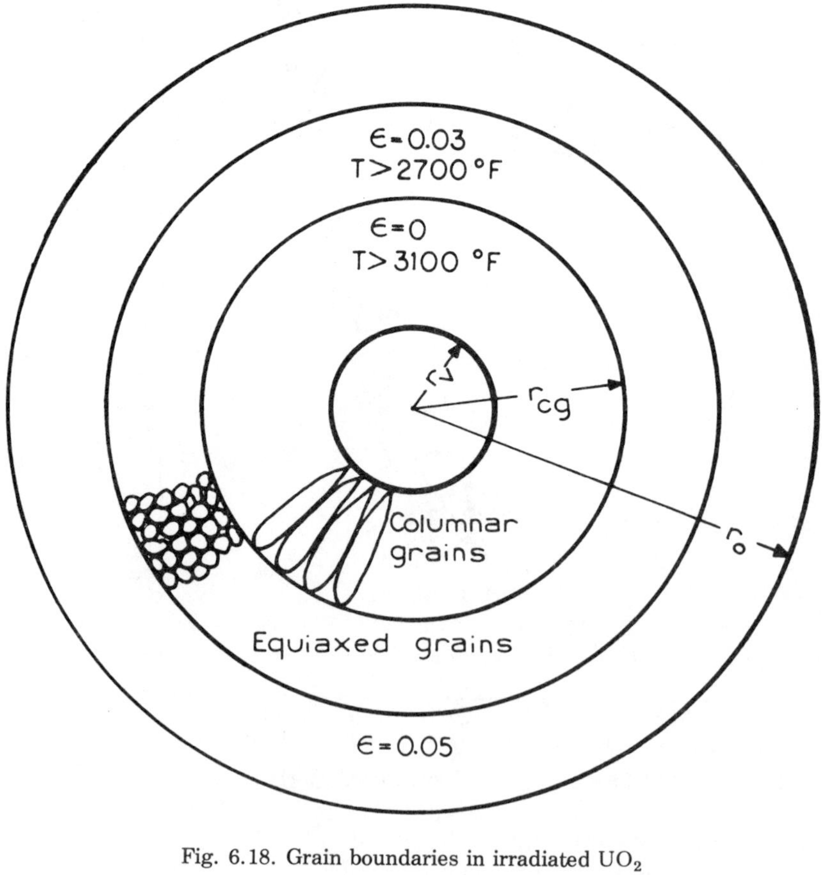

Fig. 6.18. Grain boundaries in irradiated UO_2

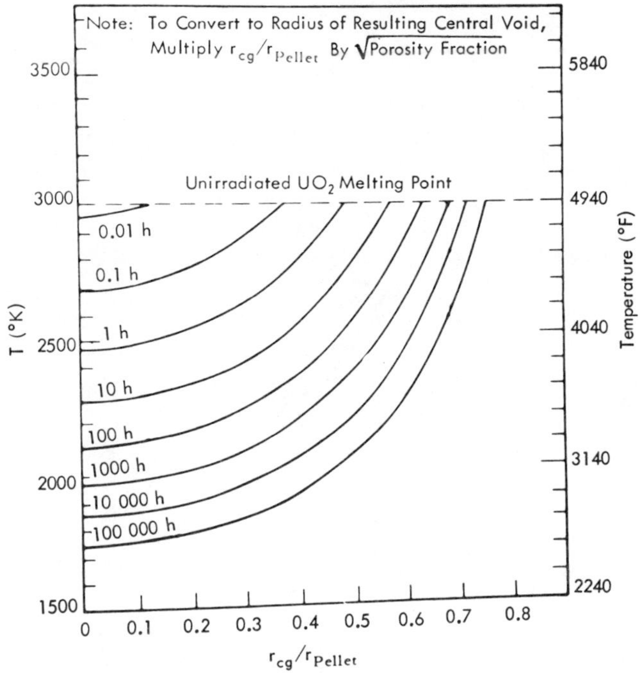

Fig. 6.19. Columnar grain growth radius in UO_2 rods [39].

Because of the dynamic nature of oxide fuel behavior, temperature distributions need to be evaluated at beginning of life (B.O.L) and end of life (E.O.L) in order to determine if design basis limitations are exceeded. Figure 6.20 illustrates typical temperature distributions that would be expected in a 0.37 in. diameter UO_2 fuel pellet with a linear heat rate of 26 kW/ft. (26 kW/ft is a linear heat rate about 60% higher than the maximum expected in water-cooled power reactors). At the end of life a central void is created due to columnar grain growth and the temperature discontinuity at the fuel-cladding interface increases due to the decrease in contact heat-transfer coefficient from dilution of the helium fill gas with fission product gases.

Most codes used to predict in-core fuel behavior assume temperature distributions depend only upon the radial coordinate. However, if significant fuel densification occurs so that gas-filled gaps occur at the fuel-cladding interface, the pellets may be stacked eccentrically within the cladding. With these conditions the contact heat-transfer coefficient is angularly dependent. A parametric study by Goldberg [44] shows that fuel temperatures are asymmetric with these conditions; however, maximum fuel temperatures are unchanged or smaller when comparing eccentric fuel pellets with concentric fuel pellets.

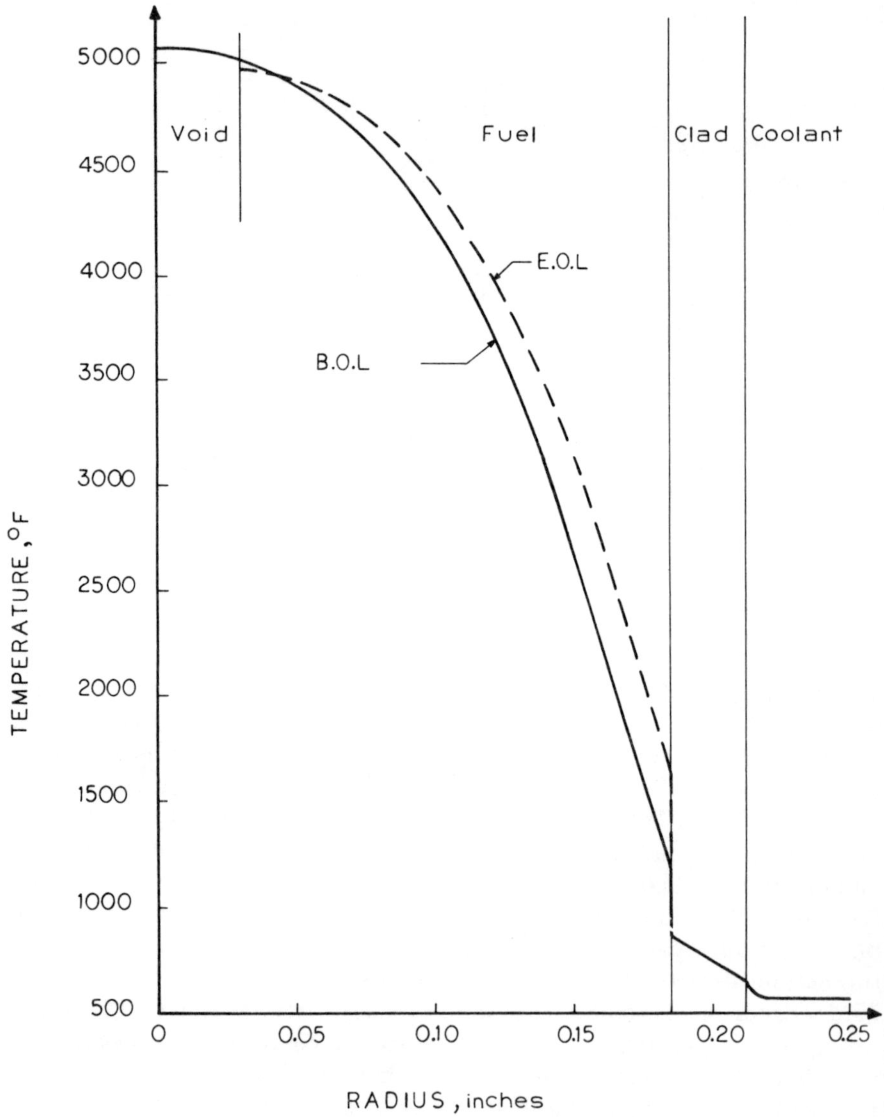

Fig. 6.20. Typical pin temperature profile (Linear heat rate 26 kW/ft) [43]

REFERENCES

1. Dittus, F. W. and L. M. K. Boelter, **Univ. Calif. Pub. Eng.** 2, 443 (1930).
2. Seider, E. M. and G. E. Tate, **Ind Eng. Chem.** 28, 1429-35 (1936).
3. Colburn, A. P., A method of correlating forced convection heat transfer data and a comparison with fluid friction, **Trans. AIChE** 29, 174 (1933).
4. Monrad, C. C. and J. F. Pelton, **Trans. ASME** 38, 593 (1942).
5. Eckert, E. R. and R. M. Drake, Jr., **Heat and Mass Transfer**, McGraw-Hill, New York, 1964.
6. Weisman, J., Heat transfer to water flowing parallel to tube bundles, **Nucl. Sci. Engr.** 6, 79 (1959).
7. Subbotin, V. I., et al., Heat removal from the reactor fuel elements cooled by liquid metals, **Proc. 3rd U. N. Intern. Conf. Peaceful Uses of Atomic Energy, September 1964** 8, United Nations, New York, 192-200 (1965).
8. Sutherland, W. A., Experimental heat transfer in rod bundles, **Heat Transfer in Rod Bundles**, ASME, New York, 104-138, 1968.
9. Tong, L. S. and J. Weisman, **Thermal Analysis of Pressurized Water Reactors**, American Nuclear Society, Hinsdale, Illinois, 1970.
10. Kays, W. M., **Convective Heat and Mass Transfer**, McGraw-Hill, New York, 1966.
11. Hsu, S. T., **Engineering Heat Transfer**, D. Van Nostrand, New York, 1963.
12. Siegal, R., Forced convection in a channel with wall heat capacity and with wall heating variable with axial position and time, **Int. J. Heat Mass Transfer** 6, 607-620 (1963).
13. Lyon, R. M., Liquid metal heat-transfer coefficients, **Chem. Eng. Prog.** 47, 75 (1951).
14. Rust, J. H., A parametric analysis of Lyon's integral equation for liquid-metal heat transfer coefficients, **Nucl. Eng. and Design** 16, 223-236 (1971).
15. Dwyer, O. E. and R. N. Lyon, Liquid metal heat transfer, **Proc. 3rd U. N. Intern. Conf. Peaceful Uses of Atomic Energy, September 1964** 8, United Nations, New York, 182-189 (1965).
16. Dwyer, O. E., Eddy transport in liquid metal heat transfer, **AIChE J.** 9, 261-268 (March, 1963).
17. Dwyer, O. E. Heat transfer to liquid metals flowing in-line through unbaffled rod bundles: a review, **Heat Transfer in Rod Bundles**, ASME, New York, 139-168, 1968.
18. Bjornard, T. A. and P. Griffith, PWR blowdown heat transfer,

Symp. on the Thermal and Hydraulic Aspects of Nuclear Reactor Safety, Vol 1: Light Water Reactors, ASME, New York, 17-41, 1977.
19. Tong, L. S., **Boiling Heat Transfer and Two-Phase Flow**, John Wiley, New York, 1965.
20. Lahey, R. T. and F. J. Moody, **The Thermal-Hydraulics of a Boiling Water Nuclear Reactor**, American Nuclear Society, LaGrange Park, Illinois, 1977.
21. Bergles, A. and W. Rohsenow, The determination of forced-convection surface boiling heat transfer, **Trans. ASME, J. Heat Transfer** 86, 365-372 (August, 1963).
22. Castellana, F. S. and C. F. Bonilla, Two-phase pressure drop and heat transfer in rod bundles, **Two-Phase Flow and Heat Transfer in Rod Bundles**, ASME, New York, 15-30, 1969.
23. Tong, L. S. Critical heat fluxes in rod bundles, **Two-Phase Flow and Heat Transfer in Rod Bundles**, ASME, New York, 31-46, 1969.
24. Knudsen, J. G., The effect of mixing on burnout in multirod bundles, HW-84525 (December 1964).
25. Healzer, J. M., J. E. Hench, E. Janssen, and S. Levy, Design basis for critical heat flux condition in boiling water reactors, APED-5286, General Electric Company (1966).
26. Groeneveld, D. C. and K. K. Fung, Forced convective transition boiling—review of literature and comparison of prediction methods, AECL-5543, Chalk River Nuclear Laboratories (June 1976).
27. Groeneveld, D. C., Post-dryout heat transfer at operating conditions, **Proc. Topl. Mtg. Water Reactor Safety**, CONF-730304, U.S.A.E.C. (1973).
28. McEligot, D. M., Magee, P. M., and G. Leppert, Effect of large temperature gradients on convective heat transfer in the downstream region, **Trans. ASME, J. Heat Transfer** 87,67 (1965).
29. Thorpe, J. F., A method for calculation of oxide-fuel elements, **Nucl. Sci. Engr.** 23, 329-334 (1965).
30. Lyons, M. F., et al., UO_2 powder and pellet thermal conductivity during irradiation, GEAP-5100-1 (March, 1966).
31. Robertson, J. A. L., $\int kd\theta$ in fuel irradiations, AECL-807 (1959).
32. Ross, A. M. and R. L. Stoute, Heat transfer coefficient between uranium dioxide and zircaloy 2, CFRD-1075 (1962).
33. Lindsay, A. L. and L. A. Bromley, **Ind. Eng. Chem.** 42, 1, 508 (1950).
34. Von Ubisch, H., et al., Thermal conductivities of mixtures of fission product gases with helium and with argon, **Proc. 2nd U. N.**

Intern. Conf. Peaceful Uses Atomic Energy 7, United Nations, New York, 697 (1958).

35. Lustman, B., Irradiation effects in UO_2 fuel elements, UO_2 : **Properties and Nuclear Applications**, U. S. Government Printing Office, 1961.
36. Belle, J., et al., Thermal conductivity of bulk oxide fuels, WAPD-TM-586, Bettis Atomic Power Laboratory (February 1967).
37. MacEwan, J. R., Grain growth in sintered uranium dioxide: I, equiaxed grain growth, **J. Amer. Cer. Soc.** 45 (1962).
38. MacEwan, J. and V. B. Lawson, Grain growth in sintered uranium dioxide: II columnar grain growth, **J. Amer. Cer. Soc.** 45, 42 (1962).
39. Nichols, F. H., Theory of columnar grain growth and central void formation in oxide fuel rods, **J. Nucl. Mater.** 22, 214 (1967).
40. Duncomb, E., et al., CYGRO-3—a computer program to determine temperature, stresses, and deformations in oxide fuel rods, WAPD-TM-961, Bettis Atomic Power Laboratory (1970).
41. FIGRO—FORTRAN IV digital computer program for the analysis of fuel swelling and calculation of the temperature in bulk-oxide cylindrical fuel elements, WAPD-TM-618, Bettis Atomic Power Laboratory (1966).
42. Homan, F. J., et al., FMODEL—a FORTRAN IV computer code to predict in-reactor behavior of LMFBR fuel pins, ORNL-4835, Oak Ridge National Laboratory (1973).
43. Pierce, W. R., **Temperature Distributions in Oxide Fuel Pins**, Master's thesis, The University of Virginia, Charlottesville, Virginia, 1967.
44. Goldberg, M. D. and J. H. Rust, Parametric study of heat transfer from UO_2 fuel pellets located eccentrically within the cladding, **Trans. Amer. Nucl. Soc.** 19, 318-19 (1974).

PROBLEMS

6.1 The Unit 1 pressurized-water reactor of the Oconee Nuclear Power Station has a thermal output of 2568 MW. The core equivalent diameter is 128 inches and core height is 144 inches. The core contains 177 fuel assemblies, each containing on the average 203 fuel rods of 0.430 in. diameter. The weight of the UO_2 fuel is 207,485 pounds. Calculate the average fuel specific power, core power density, linear heat rate, and cladding surface heat flux. Assume 97.4% of the core power is dissipated in the fuel.

6.2 Starting with the energy equation, derive an expression for the Nusselt number for laminar flow with a uniform velocity distribution between 2 infinite parallel plates with spacing 2a. Assume a uniform heat flux, q''_w , on each plate surface and constant fluid properties.

6.3 By examining the energy equation, show that the ratio of axial convection to axial conduction in channel flows is given by $N_{Pe}/4$.

6.4 A research reactor core is cubical, 20 ft. on the side. The fuel elements are made of 1 in. diameter solid natural-uranium metal rods. They are placed horizontally in the center of 3 in. diameter holes. Air is used as the coolant at atmospheric pressure and an initial velocity of 15 fps. For the centermost fuel element, air enters at 80°F and leaves at 190°F. Calculate (a) the heat-transfer coefficient, and (b) the maximum neutron flux in the core, neglecting cladding and extrapolation lengths.

6.5 A 0.5 in. inside diameter tube 10 ft. long is electrically heated so the wall heat flux is constant. Sodium with an inlet velocity of 25 ft/sec and an inlet temperature of 700°F flows down this tube. If the maximum wall temperature of the tube (T_w) is to be 1100°F, what is the wall heat flux?

6.6 A PWR operating at 2300 psia contains 0.43 in. diameter fuel rods arranged on a square lattice with a pitch of 0.55 in. At a particular point in the reactor the fluid bulk temperature is 580°F and the heat flux is 300,000 Btu/hr–ft^2 . If it is desired to prevent subcooled nucleate boiling at that point, what is the minimum water velocity in the subchannel?

6.7 A BWR operating at 1000 psia has a coolant inlet enthalpy of 510 Btu/lb_m. The fuel rods are 0.5 in. in diameter and 12 ft. long. In one of the subchannels the coolant mass velocity is $10^6\ lb_m/ft^2$-hr and the flow cross section is 0.002 sq. ft. If the fuel rods for that subchannel have a uniform heat flux of 200,000 Btu/hr–ft^2, determine the critical heat flux at the core midplane and reactor exit.

6.8 A boiling water reactor operates at a pressure of 1000 psia. Fuel rods are 14 ft. long and 1/2 in. in diameter. Assume heating is uniform along the rod length with a linear heat rate of 7kW/ft. The fuel rod pitch is 0.7 in. and the coolant velocity at the inlet is 10 ft/sec. The coolant inlet temperature is 520°F. You may assume the heat capacity of water at this pressure and temperatures is 1.23 Btu/lb_m–°F. Calculate the position along the rod length where nucleate boiling commences.

6.9 For a fuel material whose thermal conductivity is given by $k = k_0[1 + bT]$, show the maximum temperature difference in the fuel is evaluated assuming a constant thermal conductivity if the conductivity is evaluated at the average fuel temperature.

6.10 Derive an expression for the temperature distributions in an unclad annular fuel element with uniform volumetric heat generation, q''', and constant thermal conductivity. The inner and outer radii of the fuel material are r_i and r_o, respectively. Assume the fuel inside wall temperature is T_{wi} and the outside wall temperature is T_{wo}.

6.11 A 95% theoretical density UO_2 fuel pellet is 0.37 in. in diameter and clad with zircaloy of thickness 0.026 inches. The average volumetric heat generation rate is 6×10^7 Btu/hr–ft^3. If the cladding outside wall temperature is 630°F, calculate the maximum fuel temperature assuming constant thermal conductivity and uniform heat generation.

6.12 Rework problem 6.11 assuming 90% theoretical density UO_2 enriched to 3.2% in U^{235}. Account for a temperature dependent thermal conductivity and neutron flux depression.

6.13 With irradiation, a central void is created in UO_2 fuels. Solve Eq. (6.3.11) for an annular fuel pellet with outside surface temperature T_s , outside radius r_0 , and inner radius r_i for uniform heat generation and constant thermal conductivity.

6.14 Using the result from problem 6.13, calculate the maximum fuel temperature for a 0.37 in. outside diameter, 95% theoretical density UO_2 fuel pellet with a surface temperature of 980°F. The volumetric heat generation rate is 6×10^7 Btu/hr-ft^3 and the thermal conductivity is 2 Btu/hr-ft-°F. The fuel is irradiated for 10,000 hrs.

CHAPTER 7

REACTOR THERMAL-HYDRAULIC ANALYSIS

7.1 Introduction

The reactor thermal analysis usually determines upper limits on thermal output of fuel elements which in turn places upper limits on reactor power levels. These limitations on thermal performance usually have to do with maintaining fuel element integrity. Additional limitations exist from the standpoint of economical operation of the power plant. Several common design constraints on fuel element thermal outputs for steady-state and normal transient reactor operation are as follows:

(i) Thermal outputs of fuel elements are limited because excessive fuel temperatures may cause undesirable phase changes or melting in the fuel. Significant center-line fuel melting is presently not allowed in oxide fuel pellets used in power reactors because of adverse effects that may result from displacement of molten oxide. Such effects may be penetration of molten fuel to the cladding with cladding failure or a higher local concentration of fuel resulting in excessive cladding strain and volumetric heat generation rates. Since maximum fuel temperatures in fuel pellets are proportional to linear heat rates, center-line melting may limit linear heat rates in pressurized-water and liquid-metal-cooled power reactors employing oxide fuel pellets.

(ii) Thermal outputs of fuel elements may be limited by maximum allowable heat fluxes at the fuel element-coolant interface. Because cladding strength usually decreases with increasing temperatures, maximum cladding temperatures may be far lower than maximum fuel temperatures. Since cladding temperatures are related to surface heat fluxes, limits on cladding temperature may establish maximum values for surface heat fluxes. Another factor which limits heat fluxes for liquid-cooled reactors is the possibility of burnout (see Sec. 7.4) which may lead to cladding melting and hydraulic instabilities (see Sec. 4.6) which may lead to burnout.

(iii) Thermal outputs of fuel elements may be limited due to fuel burnup and, consequently, release of fission product gases causing excessive internal pressure and cladding creep or embrittlement.

(iv) Thermal outputs of fuel elements may have upper and lower limits on fuel specific powers (kW/kg of U) from an economic point of view so that refueling times and fuel fabrication costs are optimized.

The first two limitations involve knowing the magnitude and location of the most adverse heat flux conditions within the reactor core. Core physics studies determine the location of the highest volumetric heat generation within the fuel and the fuel rod or coolant subchannel which has the highest power input along its axial length. The heat fluxes and coolant enthalpy rises may be higher than those obtained from nuclear considerations because of deviations of conditions from nominal design. These deviations are accounted for by use of the concept of hot channel and hot spot factors discussed in Sec. 7.3. The hot channel concept lumps all adverse engineering effects into a single flow channel which has the highest power output integrated along the channel length. The hot spot concept lumps all adverse engineering effects into a local hot spot which has the highest local power output. The thermal design limitations are applied to the hot spot and hot channel which should insure that the entire core is operating safely under steady state and normal transient operation.

Additional thermal design limitations arise when considering abnormal transients such as the loss-of-flow or loss-of-coolant accidents examined in reactor safety analysis. These limitations may be more restrictive than those imposed by steady-state operation. As an example, for pressurized water reactors maximum linear heat rates imposed by a loss-of-coolant accident are lower than that required to prevent fuel center-line melting. Limitations due to abnormal transients are not easily dealt with because they depend upon specific reactor types and engineered safeguards employed to reduce the severity of accidents. An effort is now underway to develop a set of general criteria which may be applied to a given reactor type. These criteria can be expected to change as our understanding of reactor behavior improves.

For light-water-cooled power reactors using zircaloy-clad oxide fuel pellets, a set of criteria based upon calculated cladding damage has been established to judge the acceptability of the combination of design and operating limits in the event of a loss-of-coolant accident. The acceptance criteria for emergency core cooling systems is as follows [1]:

(i) **Peak Cladding Temperature**. The calculated maximum fuel element cladding temperature shall not exceed 2,200°F.

(ii) **Maximum Cladding Oxidation**. The calculated total oxidation of the cladding shall nowhere exceed 0.17 times the total cladding thickness before oxidation. As used in this subparagraph, total oxidation means the total thickness of cladding metal that would be locally converted to oxide if all the oxygen absorbed by and reacted with the cladding locally were converted to stoichiometric zirconium dioxide. If cladding rupture is calculated to occur, the inside surfaces of the cladding shall be included in the oxidation, beginning at the calculated time of rupture. Cladding thickness before oxidation means the radial distance from inside to outside the cladding, after any calculated rupture or swelling has occurred but before significant oxidation. Where the calculated conditions of transient pressure and temperature lead to a prediction of cladding swelling, with or without cladding rupture, the unoxidized cladding thickness shall be defined as the cladding cross-sectional area, taken at a horizontal plane at the elevation of the rupture, if it occurs, or at the elevation of the highest cladding temperature if no rupture is calculated to occur, divided by the average circumference at that elevation. For ruptured cladding the circumference does not include the rupture opening.

(iii) **Maximum Hydrogen Generation**. The calculated total amount of hydrogen generated from the chemical reaction of the cladding with water or steam shall not exceed 0.01 times the hypothetical amount that would be generated if all of the metal in the cladding cylinders surrounding the fuel, excluding the cladding surrounding the plenum volume, were to react.

(iv) **Coolable Geometry**. Calculated changes in core geometry shall be such that the core remains amenable to cooling.

(v) **Long-Term Cooling**. After any calculated successful initial operation of the ECCS, the calculated core temperature shall be maintained at an acceptably low value and decay heat shall be removed for the extended period of time required by the long-lived radioactivity remaining in the core.

Generally, nuclear reactor system thermal analysis is divided into the following steps:

(i). Determine the types of heat sources and their time and spatial dependence throughout the reactor system. This is predominantly a problem in reactor core physics and shielding.

(ii) Determine temperature distributions throughout the solid (fuel elements, structure, etc.) parts of the reactor system by a heat conduction analysis.

(iii) Analysis of convection heat transfer to the fluids flowing through the reactor system by use of appropriate energy balances. This enables calculating surface temperatures of solid components.

(iv) Analysis of heat transfer in out-of-core components. This is usually convection heat transfer between the core coolant and secondary coolants and/or steam generators. All of these steps are interconnected because heat source distributions are temperature dependent. The neutron spectrum depends upon the moderator temperature and fuel absorption cross sections may depend upon fuel temperatures because of Doppler effects. In addition, heating causes expansion of most materials and, consequently, displacements of core materials may occur with increasing reactor power.

The reactor system thermal analysis is further subdivided into two general areas.

(i) Steady-state analysis, which is the mapping of temperature distributions and heat transfer rates during normal full power operation.

(ii) Unsteady-state analysis, which deals with mapping of temperature distributions during various reactor system transients. Transients occur during reactor startup and shutdown and when coolant flow rates are altered. For most reactor systems, the most severe transients occur as a result of a loss-of-coolant accident.

7.2 Core Flow Distribution

The approaches developed so far are for analysis of single channels or individual fuel assemblies in which it is assumed that all subchannels have the same velocity. This is inadequate for precise reactor thermal analysis because the flow is not uniformly distributed through the reactor core and mixing takes place between subchannels in open channel flows.

Core Inlet Flow

Maldistribution of flow occurs at the core inlet because of the method of coolant discharge into the inlet plenum region. Additional effects are caused by nonuniform heating in the reactor core, orificing the inlet nozzles to fuel assemblies, different subchannel geometries, and the presence of grid spacers, mixing vanes, or wire wraps.

When boiling occurs, increased vapor generation occurs in high-power channels. The increased resistance encountered in these channels leads to lower inlet flows unless steps are taken to remedy this. In boiling-water reactors, where fuel assemblies are enclosed by cans, inlet orifices are generally provided so that approximately uniform exit qualities are obtained. Lower inlet flows are than provided to low-power assemblies. In pressurized-water reactors, where fuel assembly cans are not used, no inlet orificing is attempted. Here the high-power assemblies have inlet flows which are lower than average. This effect must be superimposed upon any maldistribution due to inlet effects.

For many power reactors the inlet core flow is complicated because inlet nozzles are located at the top of the pressure vessel and flow proceeds downward alongside the core barrel and thermal shields before entering the lower plenum region. One can calculate inlet flow distribution to fuel assemblies by use of potential flow theory; however, a more reliable approach is to measure the flow distribution by use of hydraulic models of the entire flow system. By use of dimensional analysis, Hetsroni [2] showed that component Euler numbers given by $\Delta p/(\rho u^2/2g_c)$, should be similar in the model and prototype.

Figure 7.1 illustrates the cross section of a PWR hydraulic model used to study the effects of lower plenum length and flow skirt taper on core flow distributions.

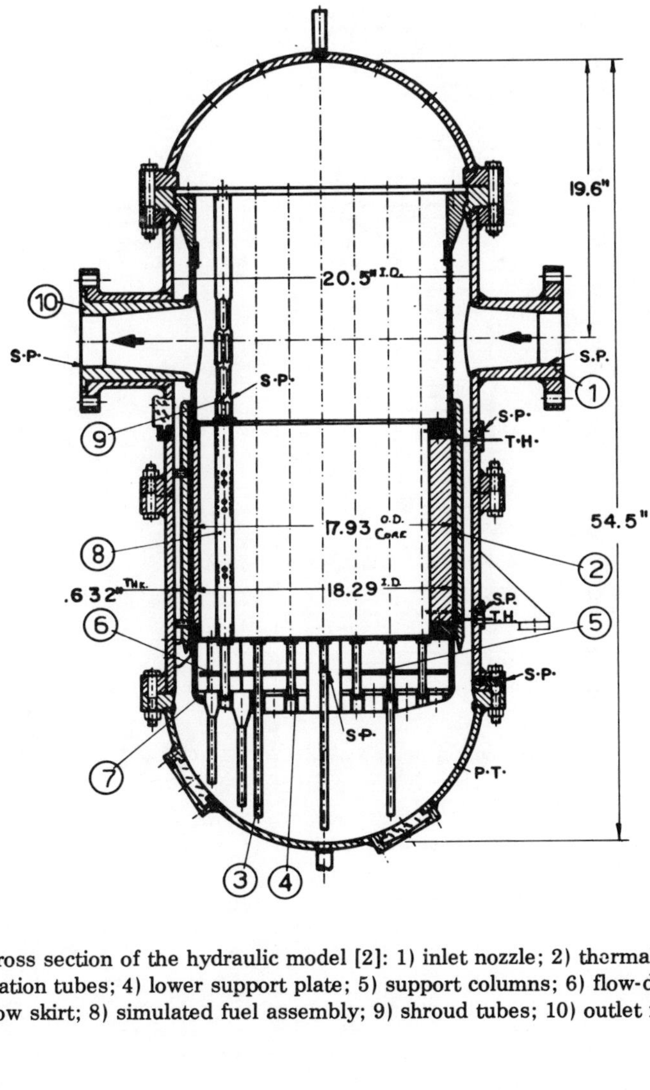

Fig. 7.1. Cross section of the hydraulic model [2]: 1) inlet nozzle; 2) thermal shield; 3) instrumentation tubes; 4) lower support plate; 5) support columns; 6) flow-distribution plate; 7) flow skirt; 8) simulated fuel assembly; 9) shroud tubes; 10) outlet nozzle.

Table 7.1 illustrates component configurations and Euler numbers

Table 7.1

Summary of Euler numbers for various flow configurations [2]

Component	Flow Area (Model) ft^2	Velocity Head (Model) psi	Long Vessel Tapered Skirt	Long Vessel Straight Skirt	Short Vessel Straight Skirt	Short Vessel Tapered Skirt	Intermediate Vessel Straight Skirt
			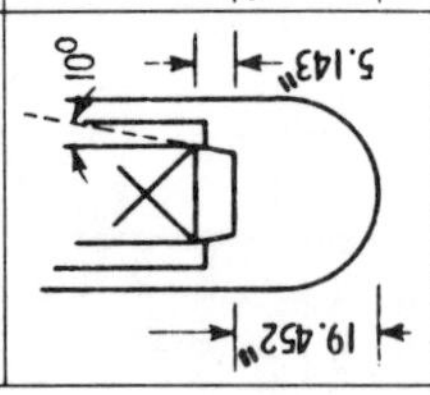	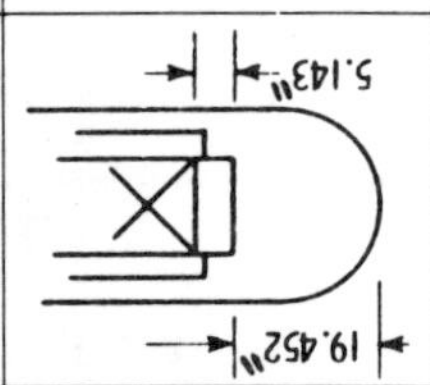	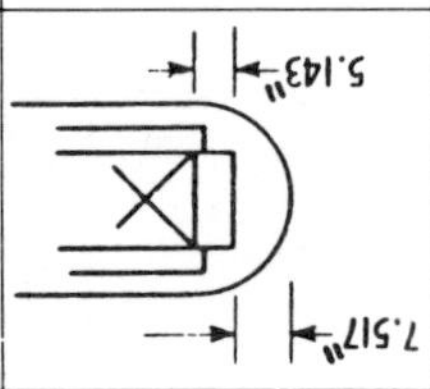	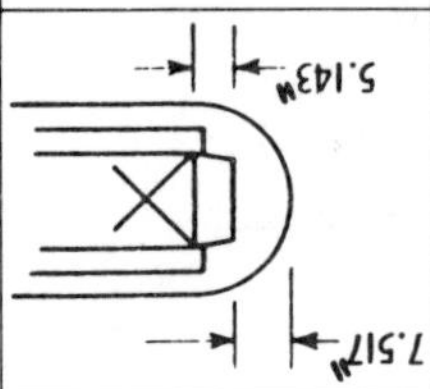	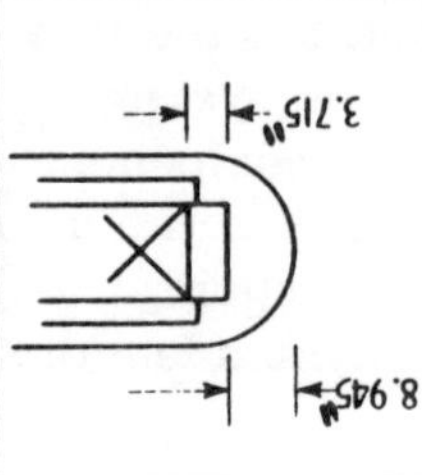
Inlet Nozzles	0.2532	5.14	0.76	0.76	0.73	0.73	0.74
Thermal Shield	0.3449	2.77	0.38	0.38	0.78	0.78	0.43
Lower Plenum	1.5045	0.15	21.37	14.79	22.12	35.62	21.00
Core	0.6556	0.77	24.52	24.52	24.52	24.52	24.55
Outlet Nozzles	0.2532	5.14	0.12	0.12	0.11	0.11	0.15

for the various geometries studied. The most uniform core flow distribution was obtained with a long vessel with a straight skirt and Fig. 7.2 illustrates normalized flow distributions through each fuel assembly.

						1.05	0.98	0.88						
				1.09	1.01	1.00	0.99	1.01	0.94	0.94				
			1.05	1.00	1.09	1.01	1.02	0.97	0.99	0.97	0.97			
		1.00	1.02	1.09	0.96	1.02	1.01	1.00	1.05	0.99	0.95	1.10		
	0.98	0.98	1.03	1.00	0.97	0.98	1.05	1.04	1.08	0.91	1.07	1.02	0.96	
	0.98	1.12	1.07	1.11	1.01	1.01	1.08	1.00	0.93	1.08	1.00	1.02	0.96	
0.88	1.00	0.96	1.04	0.93	1.03	1.00	1.03	0.98	1.04	0.98	1.05	1.01	0.98	0.97
0.88	0.92	0.95	1.00	0.95	1.05	1.04	0.98	1.09	1.03	1.03	1.01	1.00	1.04	1.01
0.93	0.96	1.00	0.99	1.00	0.97	1.01	0.99	0.98	1.05	1.09	1.12	1.01	0.96	1.00
	0.98	0.99	1.00	1.00	1.00	0.98	0.96	1.07	1.06	0.98	1.08	1.04	1.01	
	0.96	0.99	0.97	0.95	0.98	1.00	1.02	1.01	1.03	0.91	0.99	0.99	0.95	
		0.97	0.95	0.97	0.92	0.96	1.01	0.97	1.01	1.06	1.06	1.09		
			1.03	1.00	0.97	0.97	0.99	0.96	0.99	1.00	1.00			
				0.92	1.00	0.96	0.98	0.91	0.98	1.01				
						0.92	1.05	0.96						

Fig. 7.2. Normalized flow distribution with long vessel and straight skirt [2].

Core Flow Distribution

It is important to be able to predict mixing between subchannels in reactors employing rod bundles so as to accurately evaluate coolant properties. This in turn allows improvements in reactor performance and safety for the following reasons:

(i) More accurate predictions of critical heat fluxes (or DNB) in water-cooled reactors.

(ii) More accurate calculations of fuel and clad hot spot temperatures for gas-cooled and sodium-cooled reactors.

(iii) More accurate prediction of effects of other fuel bundle conditions such as limits due to fuel rod swelling and bowing, subchannel flow blockages, and onset of flow instabilities.

Coolant mixing between subchannels in rod bundles is classified into four areas — (i) turbulent interchange, (ii) diversion cross-flow, (iii) flow scattering, and (iv) flow sweeping [3]. The first two areas are natural mixing effects which always occur in rod bundles and the last two areas are forced mixing effects due to mechanical means.

(1) Turbulent Interchange

Turbulent interchange is natural eddy diffusion between subchannels which can be characterized by eddy diffusivities, ϵ_M or ϵ_H . For computational purposes turbulent interchange is frequently represented by a hypothetical equivalent transverse flow rate per unit channel length, w'_{km} , given in lb_m/hr–ft . At a subchannel boundary the time average of w'_{km} is zero, or between subchannels k and m, $w'_{km} = w'_{mk}$. As an example, for the subchannel configurations shown in Fig. 7.3 the shear stress per unit axial length due to turbulent exchange at the boundary between subchannels k and m is represented by

$$\frac{w'_{km}}{g_c}(u_k - u_m) = \frac{\rho d_{mk}\epsilon_{M,km}}{g_c}\left(\frac{d\bar{u}_z}{dn}\right)_{km} \quad (7.2.1)$$

where $(d\bar{u}_z/dn)_{km}$ is the gradient of the axial velocity component in a direction normal to the km interface. The velocity gradient may be approximated by

$$\left(\frac{d\bar{u}_z}{dn}\right)_{km} = \frac{u_k - u_m}{\delta_{mk}} \quad (7.2.2)$$

and we have

$$w'_{km} = \frac{\rho d_{mk}\epsilon_{M,km}}{\delta_{mk}} \quad (7.2.3)$$

Dividing Eq. (7.2.3) by $G_k d_{mk}$ ($G_k = \rho u_k$) gives

$$M_{km} = \frac{w'_{km}}{G_k d_{km}} = \frac{\rho \epsilon_{M,km}}{G_k \delta_{mk}} \quad (7.2.4)$$

where M_{km} is the ratio of the mass velocity due to turbulent interchange transverse to the rods to the mass velocity in the subchannel. For friction factors given in the form $f_w = a(N_{Re})^{-n}$, one expression for M_{km} is [3]

$$M_{km} = \frac{\sqrt{a/2}}{40}\left(\frac{D_{e,k}}{\delta_{mk}}\right)(N_{Re})^{-n/2} \quad (7.2.5)$$

For single-phase mixing between channels having the same D_e the turbulent interchange may be estimated from [4]

$$w'_{km} = 0.0005\, d_{mk} G \left(\frac{D_e}{D}\right) N_{Re}^{-0.1}\left(\frac{d_{mk}}{D}\right)^{-0.894} \quad (7.2.6)$$

where

G = mass velocity, $lb_m/hr\text{-}ft^2$
d_{mk} = gap between rods, ft
D = rod diameter, ft
D_e = equivalent diameter of subchannel, ft

Under two-phase conditions, mixing may be substantially enhanced. Beus [5] has correlated the data at low and moderate voids by

$$w'_I = w'_{km} + B_1\left(\frac{AG}{D_e}\right)\left(\frac{\rho_f}{\rho_g}\right)\left(\frac{S-1}{S}\right) x \quad (7.2.7)$$

where

w'_I = exchange flow per unit length for two-phase mixture
w'_{km} = exchange flow per unit length for pure liquid
A = channel area
S = slip ratio
x = quality
ρ_f, ρ_g = liquid and vapor densities, respectively
B_1 = empirical constant.

The peak mixing w'_c is assumed to take place at the slug-annular transition at a quality, x_c, given by the transition relationship as

$$x_c = \frac{\frac{0.4}{G}[g\rho_f D_e(\rho_f - \rho_g)]^{1/2} + 0.6}{0.6 + (\rho_f/\rho_g)^{1/2}} \quad (7.2.8)$$

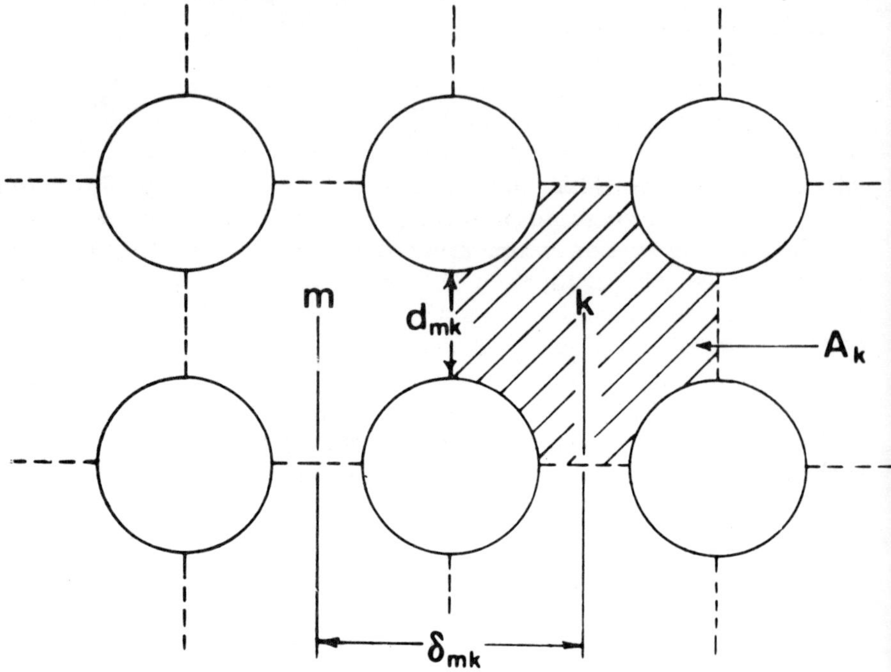

Fig. 7.3. Rod bundle subchannel configurations.

The peak mixing rate is determined by solving Eq. (7.2.7) when $x = x_c$.

The transition from peak mixing to the mixing, w'_g, obtained in the presence of pure vapor, is given by an empirical equation that has the form

$$w'_{II} = w'_g + (w'_c - w'_g)\left(\frac{1 - k}{x/x_c - k}\right) \tag{7.2.9}$$

where w'_{II} is the exchange flow rate per unit length for two-phase mixtures after peak mixing and k is an empirical constant. Beus found that he could reasonably fit the available air-water and steam-water mixing data by setting

$$B_1 = 0.04\left(\frac{d_{mk}}{D_e}\right)^{1.5} \tag{7.2.10}$$

and

$$k = 0.57\,(N_{Re})^{0.0417} \tag{7.2.11}$$

(2) Diversion Cross-Flow

Diversion cross-flow is a directed flow between subchannels caused by radial pressure gradients between adjacent subchannels. Radial pressure gradients may be caused by heat flux differences in adjacent subchannels, different subchannel equivalent diameters, rod eccentricity or bowing, changes in equivalent diameter in the flow duct, and rod bundle eccentricity within a flow duct. Regions where radial pressure gradients are most significant are in the entrance length because of flow redistribution, locations of abrupt changes in flow cross section (i.e., grid spacers), and locations where heat flux or flow asymmetries cause one subchannel to boil sooner than others.

With radial pressure gradients there is a net flow from one subchannel to an adjacent subchannel. This is given by a transverse flow rate per unit channel length, w_{km} , given in $lb_m/ft\text{-}hr$. For computation purposes, diversion cross-flow is calculated by

$$\Delta p_{km} = K_m \frac{(w_{km}/d_{mk})^2}{2g_c \rho} \tag{7.2.12}$$

where K_m is the transverse resistance coefficient. Very little information is known about K_m and Ref.[3] gives a few expressions that are used in thermal-hydraulic codes. In comparison to other transverse mixing effects in rod bundles it is thought that diversion cross-flow is of little importance when evaluating subchannel flow distributions and pressure losses.

(3) Flow Scattering

Flow scattering is a non-directional mixing associated with the presence of grid spacers, wart-type spacers, axial or circumferential fins, and end plates on short fuel bundles stacked in columns. The analysis of flow scattering is analogous to that of turbulent interchange and several analytical approaches simply increase the turbulent interchange downstream from disturbances.

Little is known about flow scattering though it appears that its effects increase with decreasing ratio of spacer flow area to open bundle flow area and decreasing axial pitch between spacers. Flow scattering may increase mixing flow rates about a factor of 2 over that of bare rods; however, this contribution might be unimportant when flow sweeping is employed.

(4) Flow Sweeping

Flow sweeping is a directed cross flow due to mixing vanes on grid spacers, contoured grids, wire wrap spacers, and helical fins. Because flow sweeping is due to a directed transverse driving force rather than random flow perturbations and results in a net flow between adjacent subchannels it is analogous to diversion cross-flow. For fully-developed flow, the mass flow rate out of a subchannel equals the mass flow rate into the subchannel or

$$\sum_m w_{km} = - \sum_m w_{mk} \tag{7.2.13}$$

Numerous models have been proposed for estimating transverse flow rates due to flow sweeping which are well summarized in Ref. [3]. One of the simpler models for wire-wrapped rods is [3]

$$\frac{w_{km} H}{\dot{m}_k} = c_w \tag{7.2.14}$$

where H is the axial pitch of the wire wrap (ft), and c_w is a constant dependent upon rod diameter and pitch. Reported values of c_w range from 0.23 to 0.35; however, the agreement with experiment is poor. c_w represents the fraction of the subchannel axial flow which is exchanged with an adjacent subchannel per wire wrap revolution.

7.3 Hot-Channel and Hot-Spot Factors

Power reactors contain up to 50,000 or more parallel coolant channels and, consequently, it is convenient to define a "hot channel" as the flow channel which has the maximum enthalpy rise and the "hot spot" as the location in the core where the heat flux is at a maximum. Frequently, the hot spot will be located in the hot channel. In addition, due to power flattening techniques used in reactor design there may be many hot channels. Since the thermal design constraints discussed in Sec. 7.1 will be most closely approached in the hot channel, a considerable portion of the core thermal-hydraulic analysis is directed toward this channel.

By use of theoretical or empirical equations for conduction and convection heat transfer, it is possible to calculate temperatures at any point in a fuel element. However, uncertainty exists in these calculations because of possible deviations of parameters used in the calculations from parameter values specified in the calculations. These uncertainties may be due to poorly known property values such as the thermal conductivity of the fuel, changes in coolant flow rates, dimensional changes because of fabrication tolerances or thermal stresses, inaccuracies in computed power distributions within the core, and many other factors, some of which are interrelated with each other. These uncertainties may produce an underprediction or an overprediction of temperatures or other design constraints within fuel elements. The overpredictions may result in poorer reactor performance while an underprediction may have disastrous consequences.

Therefore, one must guard against underpredicting fuel element design limits and this is accomplished by the of **hot-channel** or **hot-spot factors**.

The subchannel enthalpy rise, heat flux, and coolant film temperature difference vary considerably in reactor cores. Consequently, it is customary to define hot-channel factors in the manner described by LeTourneau and Grimble [6] by

$$F^T_{\Delta h} \equiv \frac{\Delta h_{max}}{\Delta h_{nom\ av}} = \text{hot-channel factor for enthalpy rise} \quad (7.3.1)$$

$$F^T_Q \equiv \frac{q''_{max}}{q''_{nom\ av}} = \text{hot-channel factor for heat flux} \quad (7.3.2)$$

$$F^T_{\Delta t} \equiv \frac{\Delta t_{max}}{\Delta t_{nom\ av}} = \text{hot-channel factor for film temperature difference} \quad (7.3.3)$$

The symbol F is used to indicate hot-channel factor and the subscripts Δh , Q, and Δt are used to denote the thermal parameter under consideration. The superscript T means this is the total hot-channel factor including both nuclear and engineering deviations from the nominal or average parameters.

The factor $F^T_{\Delta h}$ is defined as the ratio of the maximum channel enthalpy rise in the reactor to the enthalpy rise based upon average channel core flow through a channel of nominal dimensions with average heat flux at the heated surface. The factor $F^T_{\Delta t}$ is the ratio of the maximum coolant film temperature rise to the temperature rise that would be calculated if the core is composed entirely of nominal channels. $F^T_{\Delta t}$ is used for calculating the cladding surface temperature at the hot spot. The factor F^T_Q is the ratio of the maximum surface heat flux in the core to the average surface heat flux computed for a core consisting entirely of nominal channels.

The total hot-channel or hot-spot factors are composed of contributions due to nuclear and engineering considerations which are assumed to be separative. Consequently,

$$F^T_{\Delta h} = F^N_{\Delta h} F^E_{\Delta h} \quad (7.3.4)$$

$$F^T_Q = F^N_Q F^E_Q \quad (7.3.5)$$

$$F^T_{\Delta t} = F^N_{\Delta t}\ F^E_{\Delta t} \quad (7.3.6)$$

where the superscripts N and E refer to nuclear and engineering factors, respectively.

Nuclear Hot-Channel Factors

Nuclear hot-channel factors account for variations in core power density due to variations in neutron flux. These variations may be due to fuel burnup, control rods, nonuniform fuel loading, voids, water gaps, etc.

Nuclear hot-channel factors are best illustrated by examining the nuclear hot-channel factor for heat flux, F_Q^N [7].

$$F_Q^N \equiv \frac{\text{Est. Maximum}^* \text{ Heat Flux in Core}}{\text{Mean Heat Flux in Core}} \tag{7.3.7}$$

$$= \text{Nuclear Heat Flux Factor}$$

$$F_Z^N \equiv \frac{\text{Est. Maximum}^* \text{ Heat Flux in Hot-Channel}}{\text{Mean Heat Flux in Hot-Channel}} \tag{7.3.8}$$

$$= \text{Axial Nuclear Factor}$$

$$F_R^N \equiv \frac{\text{Mean Heat Flux in Hot-Channel}}{\text{Mean Heat Flux in Core Average Channel}} \tag{7.3.9}$$

$$= \text{Radial Nuclear Factor}$$

From these definitions it follows that

$$F_Q^N = F_Z^N F_R^N \tag{7.3.10}$$

The nuclear hot-channel factor for enthalpy rise is then given by

$$F_{\Delta h}^N = F_R^N \tag{7.3.11}$$

and the nuclear factor for film temperature difference is given by

$$F_{\Delta t}^N = F_Z^N F_R^N \tag{7.3.12}$$

Frequently a nuclear hot-channel factor for local neutron flux peaking on a fuel rod, F_P^N , is also applied to F_Q^N , and $F_{\Delta t}^N$ to account for flux peaks due to moderator concentration.

Nuclear hot-channel factors are determined from core power

* Maximum flux estimated on basis of nominal fuel enrichment, density, and dimensions.

distributions obtained from reactor physics analyses. For bare, homogeneous thermal reactors, Table 5.2 can be used to give approximate values for nuclear hot-channel factors. As an example, for a cylinder $F_Z^N = \pi/2$ and $F_R^N = 2.32$. For reflected, homogeneous reactors, Table 5.3 can be used to obtain estimated nuclear hot-channel factors.

Modern thermal-hydraulic codes do not use radial or axial nuclear hot channel factors because their input data consists of actual power distributions for the core. Nuclear peaking factors may be used to account for local flux peaks.

Engineering Hot-Channel Factors

Engineering hot-channel factors deal with effects of departure from nominal design specifications on enthalpy rise, heat flux, and film temperature difference. These hot-channel factors are usually smaller than nuclear hot-channel factors, however, they are still of great importance in assessing whether thermal design limits are exceeded. Engineering hot-channel factors are frequently subdivided into **mechanical subfactors** and **flow distribution subfactors.**

(1) Mechanical Subfactors

Mechanical subfactors are numerous. Some examples would be increased heat flux due to fuel density, enrichment, or pellet diameter being larger than nominal. Other examples are higher temperatures due to fuel and cladding thermal conductivities being lower than nominal. Further effects are due to cladding thicknesses and diameters varying from manufacturers nominal specifications.

Mechanical subfactors are frequently evaluated from knowledge about deviations of a large sample of a statistically varying quantity about its nominal value. Such items as fuel pellet enrichment, density, and diameter would be randomly distributed in accordance with the normal distribution function and various numbers of standard deviations could be used to describe the range of values of these quantities within a certain confidence level. As an example, if fuel pellet diameter is the quantity being considered, one could fairly well expect that 99.87 percent of all fuel pellets would have diameters within ± three standard deviations from the nominal (design) value. References [8] and [9] go into some detail describing statistical variations in mechanical subfactors.

From sample analysis the nuclear power industry has a considerable store of data on variations of such items as fuel pellet and cladding characteristics from nominal values.

(2) Flow Distribution Subfactors

Flow distribution subfactors are also due to a variety of effects. Some of the primary system total flow may not pass through fuel assemblies because of coolant by-pass between fuel assemblies, the gap in control rod slots, the gap between the core barrel and baffle, and the gap between the baffle and the outlet nozzles. If total primary flow is used to evaluate the average coolant enthalpy rise, then a core by-pass flow hot-channel factor is used to account for the reduced total core flow.

Other flow subfactors are due to the inlet flow to the core from the lower plenum, redistribution of flow in fuel assemblies, and interchannel mixing between subchannels. Inlet flows to fuel assemblies from the lower plenum region are variable as discussed in Sec. 7.2. Flows within the core are redistributed due to deviations from nominal design of subchannel dimensions due to manufacturing tolerances, thermal expansion, and creep from pressure differentials and irradiation. An additional factor affecting flow distribution is the presence of subcooled or bulk boiling in the hot channel. Interchannel mixing produces a flow distribution subfactor less than one because it accounts for mixing of high enthalpy fluid from the hot channel with lower enthalpy fluid from adjacent subchannels.

Flow distribution subfactors cannot be statistically analyzed because they are not random variables as was the case of mechanical subfactors. These subfactors are evaluated from prototype testing or thermal-hydraulic computations using interchannel mixing correlations discussed in Sec. 7.2.

Application of Hot-Channel Factors

Using the example from Fenech and Gueron [10], the internal cladding temperature at the hot spot would be given by

$$T_{sc} = T_{in} + \Delta T_1 + \Delta T_2 + \Delta T_3 \tag{7.3.13}$$

where T_{in} is the coolant inlet temperature to the hot-channel, ΔT_1 is the coolant temperature rise to the point of maximum cladding temperature, ΔT_2 is the film temperature difference at the hot-spot, and ΔT_3 is the temperature drop across the cladding at the hot-spot.

In estimating T_{sc}, hot-channel factors may be applied in the following manner:

(i) ΔT_1 would incorporate $F^T_{\Delta h}$ with its appropriate nuclear and engineering hot-channel factors.

(ii) ΔT_2 would incorporate $F^T_{\Delta t}$ with its appropriate nuclear and engineering hot-channel factors.

(iii) ΔT_3 would incorporate F^T_Q with its appropriate nuclear and engineering hot-channel factors.

In determining ΔT_1, ΔT_2, and ΔT_3 it is assumed that the worst possible conditions exist among the numerous variables required to make the calculations. This approach is called the "deterministic method" which is unduly pessimistic in its assumption that worst possible values of variables behaving in a stochastic manner all appear simultaneously in the hot-channel.

Tables 7.2 and 7.3 list hot-channel factors used in the analysis of the seed and blanket for the first core of the Shippingport PWR, respectivly. This reactor was the first power reactor to come into commercial operation in the United States. An examination of these tables shows that the nuclear and engineering hot-channel factors were obtained by taking the product of all the individual subfactors. This approach is the deterministic method described in the preceding paragraph which resulted in a highly conservative design. The power level for the first core was 231 MW(t).

Because of the stochastic nature of the mechanical subfactors in the engineering hot-channel factor, it is unreasonable to assume that the worst possible adverse conditions will all simultaneously occur at the hot spot and in the hot channel. LeTourneau and Grimble [6] recommended that statistical approaches be used. References [7] and [9] describe statistical approaches based upon random subfactors being described by normal distributions. Antognetti, et al., [12] describe the "spot method" where Monte Carlo techniques are used to determine maximum values of design constraints. Fenech and Gueron [10]

Table 7.2
Hot-channel factors for the seed at the beginning of Core I life [11].

Type of factor	Coolant temp. rise, $F_{\Delta h}$	Fluid-film temp. rise, $F_{\Delta t}$	Heat flux, F_Q
Engineering			
Flow distribution	1.07	1.05	1.00
Meat thickness variation	1.05	1.13	1.13
Fuel concentration	1.02	1.04	1.04
Meat eccentricity	1.05	1.14	1.14
Channel thickness	1.10	1.11	1.00
Heat-transfer correlation	1.00	1.25	1.00
End conduction from fuel	0.95	0.95	0.95
End coolant channel	0.97	0.95	0.95
Product	1.22	1.77	1.21
Nuclear			
Maximum to average, radial	1.25	1.25	1.25
Maximum to average, axial		2.00	2.00
Maximum to average, water hole	1.29	1.29	1.29
Blanket end-cap effect	—	1.09	1.09
Product	1.61	3.51	3.51

Table 7.3

Hot-channel factors for the blanket at the beginning of Core I life [11].

Type of factor	Coolant temp. rise, $F_{\Delta h}$	Fluid-film temp. rise, $F_{\Delta t}$	Heat flux, F_Q
Engineering			
Flow distribution	1.07	1.05	1.00
Fuel diameter	1.01	1.01	1.01
Fuel concentration (1.01 density, variation, 1.002 concentration variation)	1.01	1.01	1.01
Meat eccentricity	1.02	1.03	1.03
Heat-transfer correlation	1.00	1.25	1.00
Pitch, bowing, and rod diameter	1.11	1.14	1.00
Absence of one fuel rod (for FEDAL system)	1.01	1.01	1.01
Cooling effectiveness	1.05	1.05	1.00
Differential orificing	0.87	0.89	1.00
Product	1.15	1.49	1.05
Nuclear			
Maximum to average,*radial			
Region 1	1.24	1.24	1.24
Region 2	1.94	1.94	1.94
Region 3	1.82	1.82	1.82
Region 4	0.90	0.90	0.90
Maximum to average,*axial			
Region 1		1.77	1.77
Region 2		2.00	2.00
Region 3		2.00	2.00
Region 4		1.77	1.77
Maximum to average,* bundle depression all regions	1.24	1.28	1.28
End-cap peaking, all regions		1.18	1.18
Product			
Region 1	1.54	3.33	3.33
Region 2	2.40	5.85	5.85
Region 3	2.25	5.49	5.49
Region 4	1.11	2.41	2.41
Overall Factors			
Region 1	1.78	4.95	3.52
Region 2	2.77	8.71	6.19
Region 3	2.60	8.17	5.80
Region 4	1.29	3.58	2.54

*Average of the entire blanket region.

describe the "synthesis method" which is a combination of the former methods.

Due to increased understanding of boiling heat transfer, improved analytical methods which enabled eliminating certain hot-channel factors, and the use of statistical methods of using engineering hot-channel factors, it was possible to raise the power level of the Shippingport PWR from 231 MW(t) to 505 (MW(t) when core 2 was placed in operation in 1964 [11].

The early statistical approach is fairly easy to understand and suitable for hand calculations. Weisman [7] describes a method for evaluating the heat flux engineering hot-channel factor F_Q^E. The heat flux at the hot-spot is subject to tolerances on fuel enrichment, density, and pellet diameter. Consequently, we may define engineering subfactors F_e^E, F_ρ^E, and F_d^E to account for variations from nominal of enrichment, density, and diameter, respectively. By use of the deterministic method F_Q^E would be given by

$$F_Q^E = F_e^E \, F_\rho^E \, F_d^E \tag{7.3.14}$$

If each of the subfactors had a value of 1.03, then $F_Q^E = 1.093$.

The heat flux from the fuel pellet is proportional to the product of (enrichment)x(density)x(pellet diameter). Therefore, for a fuel pellet we may describe deviations due to tolerances by

$$\frac{q''_{local}}{q''_{nominal}} = \left(\frac{e_{local}}{e_{nominal}}\right)\left(\frac{\rho_{local}}{\rho_{nominal}}\right)\left(\frac{d_{local}}{d_{nominal}}\right) \tag{7.3.15}$$

where

e = fuel enrichment
ρ = fuel density
d = pellet diameter

Each of the ratios in Eq. (7.3.15) vary randomly and should have a mean value of one because the mean of the local value is the nominal value. Because of close tolerances on manufacturing, each of the ratios will not vary much from one even under the most extreme circumstances. Consequently, Eq. (7.3.15) is approximated by

$$\frac{q''_{local}}{q''_{nominal}} = 1 + \left(\frac{e_{local}}{e_{nominal}} - 1\right) + \left(\frac{\rho_{local}}{\rho_{nominal}} - 1\right) + \left(\frac{d_{local}}{d_{nominal}} - 1\right) \tag{7.3.16}$$

The ratios in Eq. (7.3.16) are normally distributed and, consequently, can be described in terms of their variances σ_e^2, σ_ρ^2, and σ_d^2. Since $q''_{local}/q''_{nominal}$ is the sum of normally distributed quantities, it will be normally distributed and can be described by its variance, $\sigma_{q''}^2$, which is given by

$$\sigma_{q''}^2 = \sigma_e^2 + \sigma_\rho^2 + \sigma_d^2 \tag{7.3.17}$$

The mean value of $q''_{local}/q''_{nominal}$ is one, consequently, this ratio is described by its standard deviation in the form

$$\frac{q''_{local}}{q''_{nominal}} = 1 + n\sigma_{q''} \tag{7.3.18}$$

where n is a factor which determines the degree of confidence that $q''_{local}/q''_{nominal}$ will have a probability of not exceeding a given value. Table 8.4 lists percentages that a normally distributed random variable will not exceed a specified number of standard deviations from its mean. For a normally distributed random variable, 99.865% of its value lies within 3 standard deviations from its mean. Consequently, if n = 3, then the probability of $q''_{local}/q''_{nominal}$ not exceeding 1 + 3 $\sigma_{q''}$ is 0.99865. It is apparent that F_Q^E corresponds to an acceptable maximum value of $q''_{local}/q''_{nominal}$. Therefore, assigning

$$F_Q^E = 1 + 3\,\sigma_{q''} \tag{7.3.19}$$

gives us the confidence that only 0.13% of the time would F_Q^E exceed this value. By use of the statistical method it has been possible to reduce F_Q^E from a range of 1.08 - 1.1 down to the order of 1.04 for PWRs [7].

Example 7.1

Assume the standard deviations for $e_{local}/e_{nominal}$ = 0.01, $\rho_{local}/\rho_{nominal}$ = 0.02 and $d_{local}/d_{nominal}$ = 0.03. Calculate F_Q^E by (a) the statistical method with a confidence level of 99.379% and (b) deterministic method with a confidence level of 99.379% for each subfactor.

Statistical Method

From Eq. (7.3.17) the variance for F_Q^E is given by

$$\sigma_{q''}^2 = (0.01)^2 + (0.02)^2 + (0.03)^2 = 14 \times 10^{-4}$$

Consequently, the standard deviation $\sigma_{q''}$ = 0.0374. In order to have a confidence level of 99.379%, Table 7.4 shows that n = 2.5 and

Table 7.4

Probability distributions for a normal distribution [9]

Number of Standard Deviations, A	Probability $P(Q > Q_h)$	Probable Failures in 1,000	Confidence Limit percent
0.	.500000	500.000	50.0000
0.5	.308538	308.538	69.1462
1.0	.158655	158.655	84.1345
1.5	.066807	66.807	93.3193
2.0	.022750	22.750	97.7250
2.5	.006210	6.210	99.3790
3.0	.001350	1.350	99.8650
3.2	.000687	.687	99.9313
3.4	.000337	.337	99.9663
3.6	.000159	.159	99.9841
3.8	.000072	.072	99.9928
4.0	.000032	.032	99.9968
4.2	.000013	.013	99.9987
4.4	.000005	.005	99.9995
4.6	.000002	.002	99.9998
4.8	7.93×10^{-7}	.001	99.9999
5.0	2.87×10^{-7}	~ .000	~100.0000

$$F_Q^E = 1 + 2.5\,\sigma_q = 1.0935$$

Deterministic Method

In order to have a confidence level of 99.379%, each subfactor is given by $F = 1 + 2.5\,\sigma$. Therefore, $F_e^E = 1.025$, $F_\rho^E = 1.05$, and $F_d^E = 1.075$. From the product method

$$F_Q^E = (1.025)(1.05)(1.075) = 1.157$$

For this value of F_Q^E, n = 4.2 which means that the probability that F_Q^E is never exceeded is 99.9987% which is unreasonably high.

The statistical method of hot-channel analysis was used in the design of the Enrico Fermi fast-breeder reactor. Table 7.5 summarizes engineering hot-channel factors and their application to various temperature rises used to determine the maximum fuel centerline temperature.

Table 7.5

Hot-channel factors used in the Enrico Fermi fast-breeder reactor [9].

Physical Uncertainty	Factor for Coolant Temperature Rise $\Delta t_c = 285°F$		Factor for Temperature Rise Through Film $\Delta t_f = 41.5°F$		Factor for Temperature Rise Through Oxide Layer $\Delta t_o = 15.5°F$		Factor for Temperature Rise Through Clad $\Delta t_{Zr} = 39.8°F$		Factor for Temperature Rise Through Fuel $\Delta t_U = 196°F$		Summary	
	f	3σ	f	3σ	f	3σ	f	3σ	f	3σ	$\Sigma(3\sigma)$	$[\Sigma(3\sigma)]^2$
Maldistribution of coolant (a) To subassemblies (b) Within subassemblies	1.15	42.8									42.8	1,831.8
Deviation from nominal dimensions	1.03	8.6									8.6	74.0
			1.04	1.7							1.7	2.9
					2.00	15.5					15.5	240.3
							1.02	0.8			0.8	0.6
									1.02	3.9	3.9	15.2
Maldistribution of U-235	1.02	5.7	1.02	0.8	1.02	0.3	1.02	0.8	1.02	3.9	11.5	132.3
Maldistribution of flux	1.05	14.3	1.05	2.1	1.05	0.8	1.05	2.0	1.05	9.8	29.0	841.0
Burnup in core fuel pin	1.10	28.5	1.10	4.2	1.10	1.6	1.10	4.0	1.10	19.6	57.9	3,352.4
Power measurement and control	1.08	22.8	1.08	3.3	1.08	1.2	1.08	3.2	1.08	15.7	46.2	2,134.4
Film heat-transfer coefficient			1.30	12.5							12.5	156.3
Thermal conductivity of zirconium oxide					1.20	3.1					3.1	9.6
Thermal conductivity of zirconium							1.10	4.0			4.0	16.0
Thermal conductivity of fuel alloy									1.20	39.2	39.2	1,536.6

Totals: 277.0

$\Sigma[\Sigma(3\sigma)]^2$ = 10,343.4

Square root of sum of squares = 101°F

Uranium hot spot located at fuel alloy length fraction, $z/L = 0.70$.
Maximum uranium temperature without hot-channel factors = 1134°F (50% confidence)
Maximum uranium temperature with hot-channel factors: 1168°F (1σ, 84.135% confidence), 1202°F (2σ, 97.725% confidence), 1235°F (3σ, 99.865% confidence).

A 99.865% confidence limit (3σ) is chosen for selecting individual subfactors represented by f. The hot spot is located at position z/L = 0.7 in the channel. After applying nuclear hot-channel factors, the nominal maximum fuel temperature is

$$T_{max,\ nominal} = T_i + \Delta t_c + \Delta t_f + \Delta t_o + \Delta t_{Zr} + \Delta t_U \quad (7.3.20)$$

where

T_{in} = coolant inlet temperature, 577.8° F
Δt_c = coolant temperature rise to position z/L = 0.7, 285° F
Δt_f = film temperature rise, 41.5° F
Δt_o = temperature rise through oxide layer on cladding, 15.5° F
Δt_{Zr} = temperature rise through zircaloy cladding, 39.8° F
Δt_U = temperature rise in metallic uranium fuel, 196° F

From nominal values at the hot spot

$$T_{max,\ nominal} = 577.8 + 285 + 41.5 + 15.5 + 39.8 + 196 = 1134°F$$

The increase in various Δt's due to the engineering hot-channel subfactors are calculated by multiplying (f – 1) time the Δt it would affect. These temperature increases are listed next to its corresponding f.

Using the deterministic method one would add all of the temperature increases together which is shown under $\Sigma(3\sigma)$ column in the summary portion of Table 7.5. For this example the engineering hot-channel subfactors would cause an additional temperature rise of 277°F which would result in a maximum fuel temperature of 1134 + 277 = 1411°F.

In applying the statistical method one has to determine whether temperature increases due to individual subfactors are statistically dependent or independent before calculating variances of the temperatures. As an example, the hot-channel subfactor due to power measurement and control would affect all five Δt's in the same manner and, therefore, these temperature increases are statistically dependent. Consequently, the five temperature increases are summed for inclusion in the summary column. The temperature increases due to deviation from nominal dimensions are statistically independent because the dimensions of concern are different for each Δt. The temperatures recorded in the $\Sigma(3\sigma)$ column are all statistically independent and are combined statistically by taking the square root of the sum of their squares. This yields a temperature rise of 101°F due to engineering uncertainties. Therefore, the maximum fuel temperature is 1134 + 101 = 1235°F. Because the engineering subfactors are given by $1 + 3\sigma$, it is expected that 99.865% of the time this temperature is not exceeded.

As this example shows, the statistical method produces significant decreases in predicted maximum fuel temperature over the deterministic approach while still giving a high degree of confidence that a design limit is not exceeded.

As previously discussed, flow distribution subfactors are not random quantities so they are not statistically combined in the process of calculating engineering hot-channel factors. The engineering hot-channel factors for enthalpy rise are subdivided into three major components—subfactor accounting for fabrication tolerances, subfactor accounting for deviations in hot-channel flow, and an interchannel mixing subfactor [7].

Fabrication factors affecting enthalpy rise are the weight of fuel providing energy to the subchannel may be greater than nominal, the fuel enrichment may be greater than nominal, and the fuel assembly may be put together in such a manner that the flow in the hot channel is reduced. These effects are used to define a fabrication subfactor in the following form [7]:

$$F^E_{\Delta h, fab} = \left(\frac{G_{nominal}}{G_{hot\text{-}channel}}\right) \left(\frac{W_{hot\text{-}channel}}{W_{nominal}}\right) \left(\frac{\bar{e}_{hot\text{-}channel}}{\bar{e}_{nominal}}\right) \quad (7.3.21)$$

where

G = channel mass velocity

W = average fuel weight in channel

$\bar{e}$ = average fuel enrichment in channel

Each of these effects are due to random events, so they are described by their variances in the same manner as done with F^E_Q. Consequently, the variance for $F^E_{\Delta h, fab}$ is given by

$$\sigma^2_{\Delta h, fab} = \sigma^2_G + \sigma^2_W + \sigma^2_e \quad (7.3.22)$$

and the fabrication subfactor on enthalpy rise written as

$$F^E_{\Delta h, fab} = 1 + 3\sigma_{\Delta h, fab} \quad (7.3.23)$$

The subfactor on hot-channel flow accounts for core bypass previously discussed and reduced flow from the lower plenum at the subchannel inlet even if the hydraulic resistance of the hot channel was the same as nominal channels.

The interchannel mixing subfactor helps counteract the effects of the other subfactors because this tends to reduce hot-channel enthalpy.

Combining all of these effects gives for the enthalpy rise hot-channel factor

$$F^E_{\triangle h} = (1 + 3\sigma_{\triangle h,fab})(F^E_{\triangle h,flow})(F^E_{\triangle h,mix}) \qquad (7.3.24)$$

where

$F^E_{\triangle h,flow}$ = hot channel flow subfactor

$F^E_{\triangle h,mix}$ = hot channel mixing subfactor

Typical values for these factors in PWRs are $F^E_{\triangle h,fab} = 1.08$, $F^E_{\triangle h,flow} = 1.08$, $F^E_{\triangle h,mix} = 0.92$, and $F^E_{\triangle h} = 1.08$ [7].

7.4 Core Thermal Design

The core thermal-hydraulic design is performed with an objective to insure that adequate cooling is provided for the reactor core under normal steady state operation, anticipated transient operation, and abnormal occurrences such as a loss-of-coolant accident. Operating limits are established based upon heat transfer limitations such as those discussed in Sec. 7.1. In order to satisfy this criterion, design bases are selected which in a general form for power reactors are as follows:

(i) No fuel damage during steady state operation and operational transients (Condition I).

(ii) No fuel damage from faults of moderate frequency (Condition II).

(iii) Only a small fraction of the fuel rods are damaged from Condition III events.

(iv) Reactor can be brought to a safe state from Condition IV events.

Condition I events are normal occurrences during reactor operation. Condition II events are faults which are expected to occur at moderate frequency during the lifetime of the reactor and will produce no core damage. Condition III events are highly infrequent faults which may occur during the reactor lifetime and produce a small amount of core damage. Condition IV events are limiting faults, such as a loss-of-coolant accident, which are unlikely to occur during the lifetime of many operating reactors. Reference [13] gives a discussion of the various types of events included under Conditions I - IV.

Examples of more specific design bases applied to power reactors under Conditions I or II are as follows:

(i) Maximum fuel temperatures do not exceed the melting limit at 112% design over power at any time. Unirradiated UO_2 has a melting temperature of 5080°F and high burnup data show a linear reduction to 4800°F at 43,000 MWd/mtU [14]. Consequently, the maximum fuel temperture at end of life might be set at 4700°F for burnups up to 43,000 MWd/mtU.

(ii) The minimum departure from nucleate boiling ratio (DNBR) will not fall below some prescribed value such as 1.3 for normal steady state operation, design overpower, and anticipated transient operation. The DNBR is defined as the ratio of the predicted DNB heat flux (i.e., Eqs. (6.2.40-6.2.42) to the anticipated operating local heat flux. The minimum DNBR is set at a value high enough that cladding damage due to DNB is highly unlikely.

(iii) During steady state or anticipated transient operation, no hydraulic instabilities (discussed in 4.6) occur.

(iv) A prescribed fraction of the reactor vessel flow rate (such as 0.95) will be available for fuel rod cooling. Some of the flow may pass through control rod guide tubes, between fuel assemblies, between core barrel and baffle, etc. and not be effective in removing heat from fuel rods.

Once the design bases are selected, the thermal-hydraulic analysis is performed on the hot channels to insure that design limitations are not exceeded.

Simplified Axial Temperature Distributions

For a given reactor power level, core flow rate, and core geometry, it is easy to determine average thermal parameters such as fuel volumetric heat generation q'''_{avg}, cladding heat flux, q''_{avg}, fuel rod linear heat rate, q'_{avg}, and coolant enthalpy rise, Δh_{avg}. These quantities are given by

$$q'''_{avg} = \frac{\text{Core Power}}{\text{Fuel Volume}} \tag{7.4.1a}$$

$$q'''_{avg} = \frac{\text{Core Power}}{\text{Active Clad Heat Transfer Area}} \tag{7.4.1b}$$

$$q'_{avg} = \frac{\text{Core Power}}{\text{Total Active Fuel Rod Length}} \tag{7.4.1c}$$

$$\Delta h_{avg} = \frac{\text{Core Power}}{\text{Effective Core Mass Flow Rate}} \tag{7.4.1d}$$

The average thermal parameters are scaled up for use at the hot channel and hot spot by employing hot-channel factors discussed in Sec. 7.3.

For a reactor having a cosine axial power shape, the volumetric heat generation rate and heat flux distribution in the hot channel is given by

$$q'''(z) = q'''_{max} \cos\left(\frac{\pi z}{L_o}\right) \tag{7.4.2}$$

and

$$q''_w(z) = q''_{max} \cos\left(\frac{\pi z}{L_o}\right) \tag{7.4.3}$$

where L_o is the fuel rod length plus twice the reflector savings. For purposes of calculating fuel and cladding temperatures q'''_{max} is given by

$$\begin{aligned} q'''_{max} &= q'''_{avg} \; F_Z^N \, F_R^N \, F_Q^E \\ &= q'''_{avg} \; F_Q^N F_Q^E \end{aligned} \tag{7.4.4}$$

and for calculating coolant enthalpy rises

$$q''_{max} = q''_{avg} \; F_R^N \, F_{\Delta h}^E \, F_Z^N = q''_{avg} \; F_{\Delta h}^T \; F_Z^N \tag{7.4.5}$$

The enthalpy rise in the hot channel is obtained from an energy balance on the channel. By assuming no mixing between subchannels

$$\dot{m}dh = q''_w(z) \, P_h dz \tag{7.4.6}$$

where P_h is the channel heated perimeter. Substituting Eqs. (7.4.3) and (7.4.5) into Eq. (7.4.6) and performing the integration from the channel inlet gives

$$h(z) - h_{in} = \frac{P_h}{\dot{m}}\left(q''_{avg} \; F_{\Delta h}^T \; F_Z^N\right)\left(\frac{L_o}{\pi}\right) \times \left[\sin\left(\frac{\pi z}{L_o}\right) + \sin\left(\frac{\pi L}{2L_o}\right)\right] \tag{7.4.7}$$

where the channel actual length ranges from -L/2 to L/2. Since $\dot{m} = \rho u A = GA$ where A is the channel flow cross section area, Eq. (7.4.7) is rewritten for a channel whose heated perimeter is the same as its wetted perimeter in the form

$$h(z) - h_{in} = \left(\frac{4}{D_e G}\right)\left(q''_{avg} \; F_{\Delta h}^T \; F_Z^N\right)\left(\frac{L_o}{\pi}\right)\left[\sin\left(\frac{\pi z}{L_o}\right) + \sin\left(\frac{\pi L}{2L_o}\right)\right] \tag{7.4.8}$$

If no phase changes take place in the fluid, $h(z) - h_{in}$ may be replaced by $c_p[T_b(z) - T_{in}]$.

The cladding outside wall temperature is obtained from the use of heat-transfer coefficients (discussed in Sec. 6.2)

$$T_w(z) = \frac{q''_w(z)}{h} + T_b(z) \tag{7.4.9}$$

Frequently a hot-channel factor for evaluating temperature differences between the wall and bulk coolant is defined by

$$F^T_{\Delta t} = \frac{\text{Maximum Film Temperature Difference at Core Hot Spot}}{\text{Average Film Temperature Difference}} \tag{7.4.10}$$

$F^T_{\Delta t}$ is evaluated from

$$F^T_{\Delta t} = F^N_R \, F^N_Z \, F^E_{\Delta t} \tag{7.4.11}$$

where $F^E_{\Delta t}$ is an engineering hot-channel factor accounting for variations in fuel density, fuel enrichment, fuel and cladding dimensions, and effects of sub-channel geometry and flow on heat-transfer coefficients. For purposes of calculating film temperature drops, q''_{max} in Eq. (7.4.3) is given by

$$q''_{max} = q''_{avg} \; F^N_R \, F^N_Z \, F^E_{\Delta t} \tag{7.4.12}$$

If no boiling occurs in the coolant, the wall temperature distribution is obtained from Eqs. (7.4.3), (7.4.8), (7.4.9) and (7.4.12) to yield

$$T_w(z) = T_{in} + \frac{q''_{avg} \; F^N_R \, F^N_Z \, F^E_{\Delta t} \cos\left(\frac{\pi Z}{L_o}\right)}{h_{nominal}} + \left(\frac{4}{D_e G c_p}\right)\left(q''_{avg} \; F^T_{\Delta h} \; F^N_Z\right)\left(\frac{L_o}{\pi}\right)\left[\sin\left(\frac{\pi Z}{L_o}\right) + \sin\left(\frac{\pi L}{2L_o}\right)\right] \tag{7.4.13}$$

The cladding inside wall temperature is of interest because this yields the maximum cladding temperature which may have a limitation. The axial distribution of the hot channel cladding inside wall temperature for fuel rods is obtained by combining Eqs. (6.3.30), (7.4.3) (7.4.4), and (7.4.13)

$$T_{sc}(z) = T_{in} + q''_{avg} \cos\left(\frac{\pi Z}{L_o}\right)\left[\frac{F^N_R \, F^N_Z \, F^E_{\Delta t}}{h_{nominal}} + \frac{F^N_Q \, F^E_Q (r_o + \delta_c)}{k_c} \ln\left(\frac{r_o + \delta_c}{r_o}\right)\right] + \left(\frac{4}{D_e G c_p}\right)\left(q''_{avg} \; F^T_{\Delta h} \; F^N_Z\right)\left(\frac{L_o}{\pi}\right)\left[\sin\left(\frac{\pi Z}{L_o}\right) + \sin\left(\frac{\pi L}{2L_o}\right)\right] \tag{7.4.14}$$

In a similar manner the fuel pellet surface temperature distribution is found by adding Eq. (6.3.27) to Eq. (7.4.14) after applying appropriate hot-channel factors to the fuel surface heat flux distribution. The fuel centerline temperature is obtained by adding Eq. (6.3.18), or Eq. (6.3.20) if the conductivity integral is used, to the fuel surface temperature distribution. The appropriate hot-channel factors must be applied to the average linear heat rate in order to obtain the hot channel linear heat rate distribution.

Figure 7.4 illustrates typical fuel, cladding, and coolant axial temperature distributions for a chopped-cosine axial power distribution. The fuel and cladding have maximum temperatures which are located downstream from the core midplane. The location of these maximum temperatures is found by differentiating the temperature distribution with respect to z and setting this quantity equal to zero. For the cladding wall temperature, differentiating Eq. (7.4.13) gives for the location of the maximum temperature

$$z_w = \frac{L_o}{\pi}\tan^{-1}\left[\left(\frac{4h}{D_e G c_p}\right)\left(\frac{L_o}{\pi}\right)\left(\frac{F^T_{\Delta h}}{F^N_R\, F^E_{\Delta t}}\right)\right] \qquad (7.4.15)$$

If phase changes occur in the fluid, one can only use $c_p[T_b(z) - T_{in}]$ to represent the fluid enthalpy change in Eq. (7.4.8) up to the point where $T_b = T_{sat}$. From that point onward $T_b = T_{sat}$. Once T_w exceeds T_{sat} by a few degrees it is necessary to consider the possibility of nucleate boiling when evaluating heat-transfer coefficients in Eq. (7.4.9). After the onset of nucleate boiling, wall temperatures can be evaluated using the Jens and Lottes correlation given by Eq. (4.5.21) or wall temperatures can be found from Eq. (6.2.36) with a great deal of effort. Because of the high efficiency of the heat transfer mechanism with nucleate boiling, wall temperatures will range from 5-30°F above T_{sat}.

Example 7.2

The following information is provided in the PSAR for the Bellefonte Nuclear Power Plant [15].

Core Power Output	3600 MWt
Nominal Systems Pressure	2250 psia
Core Mass Flow Rate	$1.50 \times 10^8 lb_m/hr$
Coolant Inlet Temperature	572.3°F
Active Heat Transfer Surface Area	63,991 ft^2
Rod Diameter	0.379 in. OD
Active Rod Length	143 in.
Pin Pitch	0.501 in.

Average Subchannel Mass Velocity	$2.46 \times 10^6 lb_m/ft^2$-hr
Percent Power Generated in Fuel and cladding	97.3%

Hot-Channel Factors:

$F_Z^N = 1.70 \qquad F_R^N = 1.55$

$F_Q^E = 1.025 \qquad F_{\Delta h}^E = 1.014$

Subchannel flow in hot channel is 85.9% of nominal flow. The engineering hot channel factors do not contain contributions due to the core flow subfactor or core mixing subfactor because these hot-channel factors were used with a computer model which calculated these effects directly.

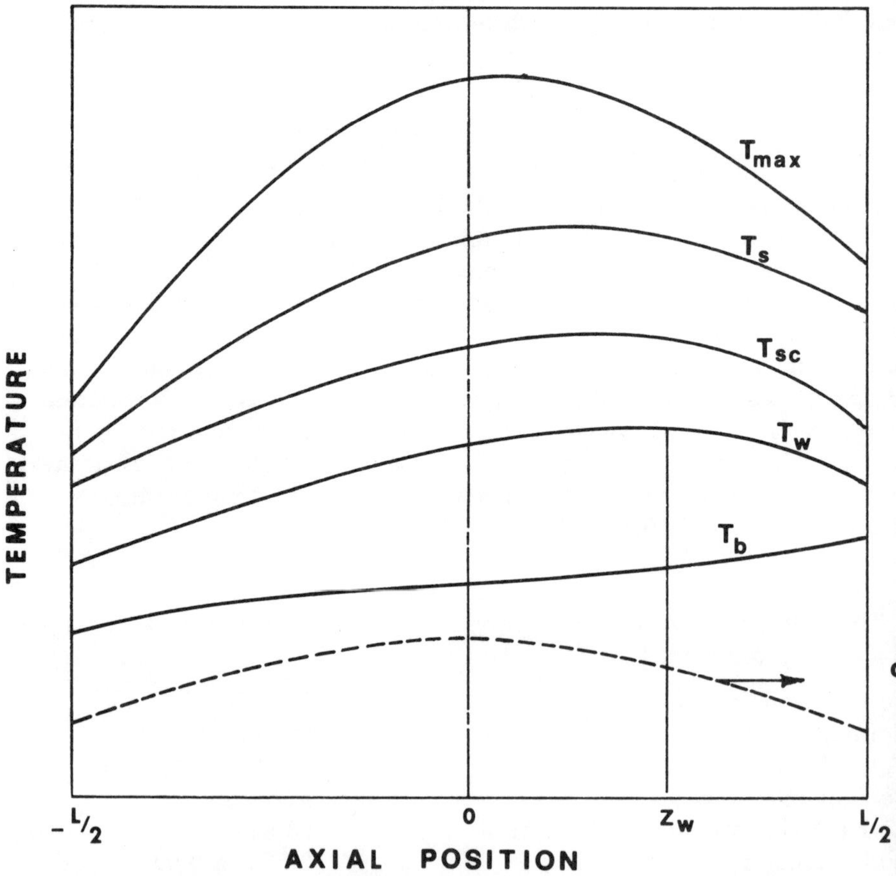

Fig. 7.4. Axial temperature distribution for fuel, cladding, and coolant (no boiling occuring in coolant).

Assuming an equal power split between the upper and lower half of the core and the maximum heat flux at the core midplane, determine the hot channel cladding wall temperature at the core midplane for 112% design over-power.

the average thermal parameters of interest are

$$\Delta h_{avg} = \frac{3600 \times 3.412 \times 10^6}{1.50 \times 10^8} = 81.9 \text{ Btu/lb}_m$$

$$q''_{avg} = (0.973)\,\frac{3600 \times 3.412 \times 10^6}{63{,}991} = 186{,}800 \text{ Btu/hr-ft}^2$$

At 112% overpower the hot channel enthalpy rise is

$$\begin{aligned}\Delta h_{hc} &= 1.12\,\Delta h_{avg}\, F_R^N\, F_{\Delta h}^E/\text{Flow Reduction}\\ &= (1.12)(81.9)(1.55)(1.014)/(0.859)\\ &= 168 \text{ Btu/lb}_m\end{aligned}$$

From the 1967 ASME Steam Tables, the enthalpy of water at 572.3°F and 2250 psia is 575.6 Btu/lb_m . Consequently, the hot channel exit enthalpy is 575.6 + 168 = 743.6 Btu/lb_m which corresponds to a two-phase mixture. The fluid enthalpy at the core midplane is 575.6 + 84 = 659.6 Btu/lb_m . For a pressure of 2250 psia, the fluid temperature is 630.3°F. The saturation temperature at 2250 psia is 652.7°F. Consequently, the fluid is subcooled 22.4°F at the core midplane.

At 112% overpower the maximum heat flux is

$$\begin{aligned}q''_{max} &= 1.12\, q''_{avg}\ F_R^N\, F_Z^N\, F_Q^E\\ &= (1.12)(186{,}800)(1.55)(1.70)(1.025)\\ &= 565{,}900 \text{ Btu/hr-ft}^2\end{aligned}$$

With this high a heat flux and the water being only 22.4°F subcooled, it would require a single-phase forced convection heat-transfer coefficient in excess of 20,000 Btu/hr-ft²-°F in order to prevent subcooled nucleate boiling. This would not occur in a pressurized water reactor so Eq. (4.5.21) will be used to evaluate the wall temperature.

$$\begin{aligned}T_w &= T_{sat} + \frac{60(q''_w/10^6)^{1/4}}{e^{p/900}}\\ &= 652.7 + \frac{60(0.5659)^{1/4}}{e^{2250/900}}\\ &= 657^\circ\text{F}\end{aligned}$$

The maximum surface cladding temperature quoted in Ref. [15] is 657°F.

Burnout Analysis

In water-cooled reactors considerable attention is given to analyzing burnout because the possibility of this occurrence may limit reactor power levels. The design basis is usually a minimum accepted value of the DNB ratio which is defined by

$$\text{DNBR} = \frac{\text{DNB heat flux predicted by applicable correlation}}{\text{Local reactor operating heat flux}} \qquad (7.4.16)$$

Since DNB heat fluxes strongly depend on local coolant enthalpy and decrease with increasing coolant enthalpy, the minimum DNBR usually occurs in the hot channel. This would be definite if maximum heat fluxes exist in the channel with the highest coolant enthalpy rise. For the hot channel, the DNBR is given by

$$\text{DNBR} = \frac{q''_{DNB}}{q''_w(\text{operating})} \qquad (7.4.17)$$

where the local operating heat flux is given by

$$q''_w(\text{operating}) = q''_{avg} \; F^N_R \; F^E_Q \; F_{loc} \; F_{axial} \; F_{op} \qquad (7.4.18)$$

where

F_{loc} = local pin peaking factor
F_{axial} = axial peaking factor as a function of channel position
F_{op} = ratio of reactor power to 100% power

The local pin peaking factor is due to circumferential variations in heat flux around the fuel pin due to cladding ovality or eccentrically located fuel pellets. The DNB heat flux is predicted by an appropriate correlation which for PWR's might be Eq. (6.2.41) along with Eqs. (6.2.40) and (6.2.42) or Fig. 6.11 for BWR's.

The DNBR varies along the hot channel length because the DNB heat flux depends upon local coolant velocities and enthalpies and the operating heat flux varies axially. Figure 7.5 illustrates typical distributions for q''_w , q''_{DNB} , and DNBR that exist in the hot channel. The minimum DNBR occurs downstream from the position of maximum heat flux because of the effect of increasing fluid enthalpy on decreasing q''_{DNB} . The design basis for water-cooled reactors is that the DNBR always exceed 1.3 for the design overpower condition and anticipated transients.

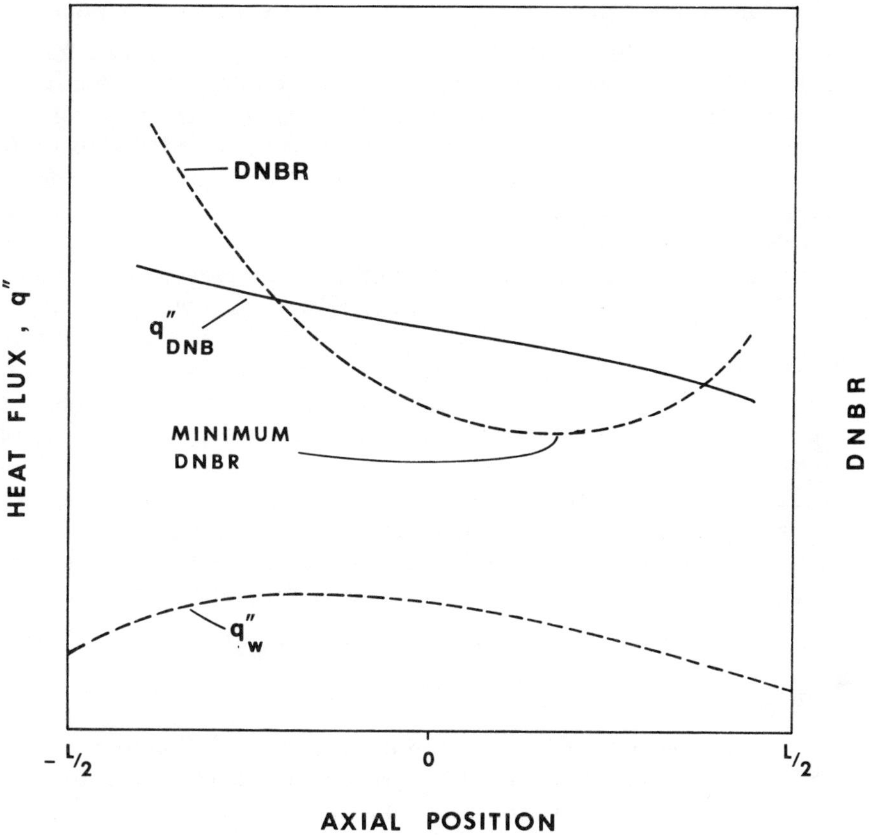

Fig. 7. 5. Channel axial distribution for q''_w, q''_{DNB}, and DNBR .

7.5 Computer Analysis

Because of the complexity of the heat transfer processes in reactor cores, it is necessary to employ numerical methods in order to obtain fuel and cladding temperature distributions; coolant density, mass velocity, enthalpy, and void fraction distributions; and DNBR distributions for flow channels under all anticipated operating conditions. The analysis is usually done with a series of computer codes because a single code to handle all processes in detail would require too much computer core space. The first analysis is a core physics study to determine core power distributions for the anticipated core operating modes. This study yields nuclear hot-channel factors if they are needed for subsequent hot channel analysis. Techniques employed in the core physics calculations are discussed in Chapter 5.

After core power distributions are determined, a thermal-hydraulic analysis is performed to determine core flow, pressure, coolant density, and DNBR distributions. Codes such as COBRA [16] and THINC [17] were developed to perform these calculations. These codes simultaneously solve the continuity, momentum, and energy equations in three dimensions for parallel flow channels. The mechanisms for interchannel mixing discussed in Sec. 7.2 are incorporated in these codes. Typical input data are channel inlet flow and enthalpy distributions, fuel rod axial heat flux distributions, outlet pressure, interchannel mixing correlations, etc.

In reactors where the fuel assemblies are separated by cans (e.g. BWR's) attention is generally focused on the high power assembly. The assembly is usually divided into subchannels having the configuration shown in Fig. 7.3. Each subchannel is then divided into a series of axial segments. The equations of continuity, momentum conservation, and energy conservation are then written in finite difference form for each segment of each subchannel.

The numerical version of the continuity equation is determined by considering the two parallel subchannels shown in Fig. 7.6. For steady state the mass flow rate into the volume element of the **kth** subchannel is equal to the mass flow rate out. Applying this principle gives

$$\dot{m}_k + w'_{mk}\Delta z = \dot{m}_k + \Delta\dot{m}_k + w'_{km}\Delta z + w_{km}\Delta z \tag{7.5.1}$$

where

$\dot{m}_k$ = mass flow rate into **kth** subchannel, lb_m/hr
w'_{mk}, w'_{km} = turbulent interchange between subchannels, $lb_m/hr\text{-}ft$
w_{km} = net flow from subchannel k to subchannel m, $lb_m/hr\text{-}ft$

The subscript km means the flow is from subchannel k to subchannel m. Since $w'_{km} = w'_{mk}$, Eq. (7.5.1) becomes

$$\frac{\Delta\dot{m}_k}{\Delta z} = -w_{km} \tag{7.5.2}$$

For N subchannels adjacent to subchannel k, Eq. (7.5.2) is written

$$\frac{\Delta\dot{m}_k}{\Delta z} = -\sum_{m=1}^{N} w_{km} \tag{7.5.3}$$

naturally, when there is a net flow from subchannel m to subchannel k, w_{km} is negative or $w_{mk} = -w_{km}$.

The momentum equation is derived by examining Fig. 7.7 which illustrates forces and flows acting upon the surface of subchannel k. By applying Newton's second law, $\Sigma\vec{F} = \dot{m}\nabla\vec{u}/g_c$, gives for the axial direction

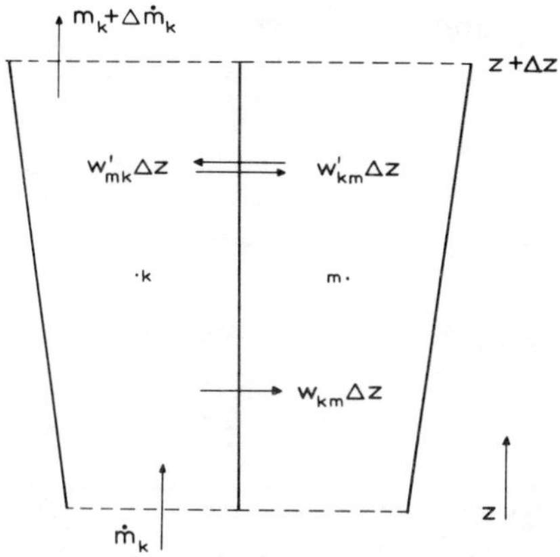

Fig. 7.6. Subchannel control volume for deriving continuity equation

$$p_k A_k + p_k \Delta(A_k) - F_k \Delta z - \rho_k A_k \frac{g}{g_c} K\Delta z - p_k A_k - \Delta(p_k A_k) = \frac{1}{g_c}$$

$$[\,\dot{m}_k u_k + \Delta(\dot{m}_k u_k) - \dot{m}_k u_k + w'_{km} u_k \Delta z - w'_{mk} u_m \Delta z + w_{km} u_k \Delta z\,] \tag{7.5.4}$$

where

F_k = friction force per unit length due to channel wall and spacers, lb_f/ft

p_k = pressure, lb_f/ft^2

A_k = flow cross section area, ft^2

ρ_k = fluid density in control volume, lb_m/ft^3

K = sine of subchannel angle of inclination

u_k, u_m = mean velocity in subchannels k and m, ft/hr.

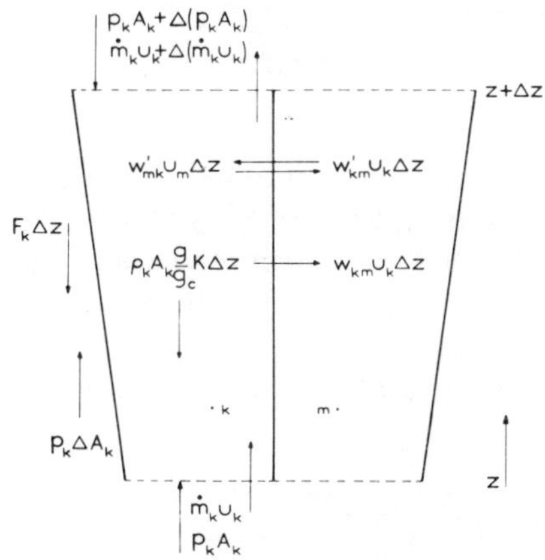

Fig. 7.7. Subchannel control volume for deriving momentum equation

Since $w'_{km} = w'_{mk}$ and $w_{km} = -\Delta \dot{m}_k/\Delta z$, Eq. (7.5.4) is rewritten

$$-A_k \Delta p_k = F_k \Delta z + \rho_k A_k \frac{g}{g_c} K \Delta z + \frac{w'_{km}}{g_c}(u_k - u_m)\Delta z + \frac{w_{km} u_k \Delta z}{g_c} + \frac{\Delta(\dot{m}_k u_k)}{g_c} \tag{7.5.5}$$

The far right term in Eq. (7.5.5) is expanded to give

$$\Delta(\dot{m}_k u_k) = u_k \Delta \dot{m}_k + \dot{m}_k \Delta u_k \tag{7.5.6}$$

Since $u_k = \dot{m}_k/(\rho_k A_k)$

$$\Delta u_k = \frac{1}{\rho_k A_k}\Delta \dot{m}_k + \dot{m}_k \Delta(\rho_k A_k)^{-1} = \frac{u_k}{\dot{m}_k}\Delta \dot{m}_k + \dot{m}_k \Delta(\rho_k A_k)^{-1} \tag{7.5.7}$$

and $\Delta \dot{m}_k = -w_{km}\Delta z$, Eq. (7.5.6) is rewritten

$$\Delta(\dot{m}_k u_k) = -u_k w_{km} \Delta z + \dot{m}_k^2 \Delta(\rho_k A_k)^{-1} - u_k w_{km} \Delta z \tag{7.5.8}$$

Substituting Eq. (7.5.8) into Eq. (7.5.5) and dividing by Δz and A_k gives

$$\frac{\Delta p_k}{\Delta z} = -\frac{F_k}{A_k} - \rho_k K \frac{g}{g_c} - \frac{1}{g_c A_k} w'_{km}(u_k - u_m) - \frac{u_k w_{km}}{g_c A_k} + \frac{2 u_k w_{km}}{g_c A_k} - \frac{G_k^2 A_k}{g_c}\frac{\Delta(\rho_k A_k)^{-1}}{\Delta z} \tag{7.5.9}$$

The term F_k/A_k due to friction is expressed as

$$\frac{F_k}{A_k} = \frac{G_k^2}{g_c}\left[\frac{f_w \Phi_k^2}{\rho_k 2 D_{e,k}} + \frac{K_k}{2\rho_k \Delta z}\right] \tag{7.5.10}$$

where

Φ_k^2 = two-phase friction multiplier or 1 if flow is single phase

K_k = loss coefficient due to spacer devices

In deriving Eq. (7.5.9) it was assumed that w_{km} was positive and the flow from subchannel k to m. If the flow was from subchannel m to k, then the term $-u_k w_{km}/(g_c A_k)$ should be $-u_m w_{km}/(g_c A_k)$. Therefore, the general form of Eq. (7.5.9) incorporating Eq. (7.5.10) is

$$\frac{\Delta p_k}{\Delta z} = -\frac{G_k^2}{g_c}\left[\frac{f_w \Phi_k^2}{\rho_k 2 D_{e,k}} + \frac{K_k}{2\rho_k \Delta z} + A_k \frac{\Delta(\rho_k A_k)^{-1}}{\Delta z}\right] - K \rho_k \frac{g}{g_c} - \frac{1}{g_c A_k}(u_k - u_m) w'_{km} + \frac{1}{g_c A_k}(2u_k - u^*) w_{km} \tag{7.5.11}$$

where u* is the effective axial velocity carried by diversion cross flow, ft/sec. The effective axial velocity is given by u_k if $w_{km} \geqslant 0$ and u_m if $w_{km} < 0$.

The control volume used to derive the energy equation is shown by Fig. 7.8. Under steady state the energy entering the volume element for subchannel k equals the energy leaving. Neglecting heat conduction between adjacent subchannels, the energy conservation equation is given by

$$\dot{m}_k h_k + w'_{mk} h_m \Delta z + q'_k \Delta z = w_{km} h_k \Delta z + w'_{km} h_k \Delta z + \dot{m}_k h_k + \Delta(\dot{m}_k h_k) \qquad (7.5.12)$$

where

q'_k = subchannel k linear heat rate, Btu/hr-ft

h_k = channel k fluid enthalpy, Btu/lbm

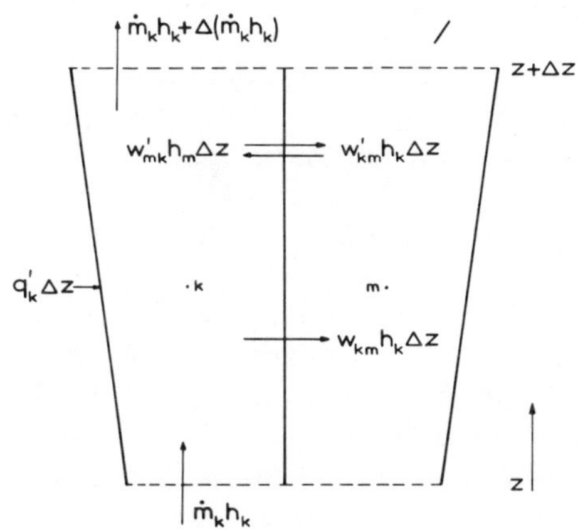

Fig. 7.8. Subchannel control volume for deriving energy equation.

Dividing by Δz , Eq. (7.5.12) is rearranged

$$\dot{m}_k \frac{\Delta h_k}{\Delta z} + h_k \frac{\Delta \dot{m}_k}{\Delta z} = q'_k + w'_{km}(h_m - h_k) - w_{km} h_k \qquad (7.5.13)$$

In deriving Eq. (7.5.13) it is assumed that $w_{km} > 0$ so the enthalpy of the far right term is h_k . If $w_{km} < 0$, the enthalpy of this term would be h_m because energy is then carried into the control volume.

Since $\Delta \dot{m}_k/\Delta z = -w_{km}$, the general form of Eq. (7.5.13) is given by

$$\dot{m}_k \frac{\Delta h_k}{\Delta z} = q'_k - (h_k - h_m)w'_{km} - (h^* - h_k)w_{km} \qquad (7.5.14)$$

where h* is the effective enthalpy carried through the gap by diversion cross flow or flow sweeping, Btu/lb_m. For uniform subchannel enthalpy, h* is h_k when $w_{km} > 0$ and h_m when $w_{km} < 0$. The additional terms required for including turbulent and cross flow exchange with other subchannels are analogous to those shown.

The solution of the conservation equations begins at the core inlet. The inlet conditions are assumed known and the conservation equations, together with appropriate heat transfer equations, boiling models, void models, and thermodynamic properties, are solved simultaneously to predict fluid conditions at the exit of the axial segment. These fluid conditions are then used as input to the next length step where the process is repeated. The calculation continues in this manner until the core exit is reached. The COBRA code [16] is a well documented example of this approach.

In cores without fuel assembly cans (i.e. PWR's), appreciable flow interchange can occur along the length of the core. Because there may be in excess of 50,000 coolant subchannels in a power reactor core, the computer analysis cannot be performed on all subchannels. The approach used in applying THINC is to perform a core-wide analysis letting each fuel assembly represent a flow channel [18]. The procedure is complicated by the fact that the inlet flows to the assemblies are not known. When the appropriate inlet flow distribution is assigned, the exit pressures from all assemblies are identical. For each fuel assembly the axial distributions of flow, enthalpy, temperature, pressure, and void fraction are computed. Interassembly flow and energy exchange with the hot fuel assembly are stored for later use as input data for the hot assembly analysis. The correct exit flow distribution may be determined by an iterative procedure [16] or by simultaneous solution of the conservation equation for all elements [17].

In the most sophisticated analysis the core-wide analysis is followed by a detailed analysis of the hot assembly [19]. The inlet flow and flows to and from the hot assembly determined in the core-wide analysis are used as input data for the hot assembly analysis. The hot assembly is divided into four elements (flow channels) and the analysis repeated. The interchannel flow and energy exchange are used for the final analysis on the hot quarter assembly which is divided into individual fuel rods and flow subchannels. Engineering hot-channel factors are applied to the hot assembly average conditions in order to obtain the thermal parameters for the hot channel.

Transient thermal-hydraulic analysis is done to assess the effects of sudden changes in primary pressure, reactor power level, and core flow on DNBR and flow stability. COBRA and THINC are also used to determine channel flows, pressures, enthalpies, void fractions, and temperature distributions during transients. Flow instabilities (which are discussed in Sec. 4.6) can be analyzed with the HYDNA or STABLE-3 programs.

REFERENCES

1. Editorial Staff, New acceptance criteria for emergency core-cooling systems of light-water-cooled nuclear power reactors, **Nuclear Safety** 15, 2, 173-183 (March-April 1974).
2. Hetsroni, G., The use of hydraulic models in nuclear reactor design, **Nucl. Sci. Engr.** 28, 1-11 (1967).
3. Rodgers, J. T. and N. E. Todreas, Coolant interchannel mixing in reactor fuel rod bundles single-phase coolants, **Heat Transfer in Rod Bundles**, ASME, New York, 1-56, 1968.
4. Rodgers, J. T. and R. G. Rosehart, Turbulent interchange mixing in fuel bundles, **Trans Conf. Applied Mechanics**, University of Waterloo, Canada (1969).
5. Beus, S. G., Two-phase turbulent mixing model for flow in rod bundles, WAPD-T-2438, Bettis Atomic Power Laboratory, Pittsburgh (1970).
6. LeTourneau, B. W. and R. E. Grimble, Engineering hot-channel factors for nuclear reactor design, **Nucl. Sci. Engr.** 1, 5, 359 (October 1956).
7. Weisman, J., **Elements of Nuclear Reactor Design**, Elsevier, New York, 1977.
8. Tong, L. S. and J. Weisman, **Thermal Analysis of Pressurized Water Reactors**, American Nuclear Society, LaGrange Park, Illinois, 1970.
9. El-Wakil, M. M., **Nuclear Heat Transport**, International Textbook Co., Scranton, Penn., 1971.
10. Fenech, H., and H. M. Gueron, The synthesis method of uncertainty analysis in nuclear reactor thermal design, **Nucl. Sci. Engr.** 31, 505-512 (1968).
11. Atherton, R., The evolution of hot-channel factors of Shippingport reactor cores, **Reactor and Fuel Processing Tech.** 10, 4, 261-270 (1967).
12. Antognetti, G., et al., Statistical methods for hot-channel and hot spot calculations, EUR 1703.e., EURATOM (1964).
13. The safety of nuclear power reactors (light water-cooled) and related facilities, WASH-1250, U.S.A.E.C. (July, 1973).

14. Christensen, J. A., Allio, R. J., and A. Biancheria, Melting point of irradiated uranium dioxide, WCAP-6065, Westinghouse (February 1965).
15. Bellefonte Nuclear Plant, Preliminary safety analysis report, **Vol.** 4, Tennessee Valley Authority (1973).
16. Rowe, D. S., COBRA-IIIC: A digital computer program for steady state and transient thermal-hydraulic analysis of rod bundle nuclear fuel elements, BNWL-1695, Pacific Northwest Laboratory (1973).
17. Hochreiter, L. E., Chelmer, H., and P. T. Chu, THINC-IV: An improved program for thermal-hydraulic analysis of rod bundle cores, WCAP-7956, Westinghouse (June 1973).
18. Sequoyah Nuclear Plant, Final safety analysis report, **Vol** 4, Tennessee Valley Authority (1974).
19. Chelmer, H., Weisman, J., and L. S. Tong, Subchannel thermal analysis of rod bundle cores, WCAP-7015, Revision 1 (January 1969).

PROBLEMS

7.1 A boiling water reactor operates at a pressure of 1,000 psia. The fuel rods are 0.45 in. in diameter and are arranged with a pitch of 0.65 in. The coolant inlet temperature is 520°F and the inlet velocity is 10 ft/sec. The fuel rod length is 12 feet and you may assume a uniform heat flux of 200,000 Btu/hr–ft^2 on the cladding surface. Plot the turbulent interchange (w'_{km}) as a function of rod position. Use the Bankoff void-quality model to determine slip ratios.

7.2 A cubical shaped graphite-moderated reactor may be assumed to have the neutron flux distribution of a bare, homogeneous reactor. If the engineering hot channel factors are $F_Q^E = 1.05$, $F_{\Delta h}^E = 1.08$, and $F_{\Delta t}^E = 1.03$, , what are the values of F_Q^T ,$F_{\Delta h}^T$, and $F_{\Delta t}^T$.

7.3 A fuel pellet manufacturer maintains that, to a confidence limit of 99.865%, variations in fuel enrichment, fuel density, and fuel pellet diameter don't exceed 0.5%, 2%, and 1%, respectively. Using the statistical method, calculate F_Q^E to a confidence level of 97.725%.

7.4 The reactor described in problem 7.1 contains 95% theoretical density UO_2 fuel pellets with a diameter of 0.370 in., clad with zircaloy-4 with a thickness of 0.0265 in. The coolant flow rate entering the reactor core is 352,000 GPM and the coolant inlet temperature is 554°F. The core pressure is 2300 psia. If $F_Q^T = 2.3$, $F_{\Delta h}^T = 1.4$, and $F_{\Delta t}^T = 2.35$, calculate the maximum UO_2 temperature assuming the hot spot is at the core midplane.

7.5 Using the parameters for the Bellefonte Nuclear Power Plant given in Example 7.2, calculate the DNBR at 112% design overpower at the hot channel core midplane using Eqs. (6.2.40-6.2.42). Assume $q''(z) = q''_{max} \sin(\pi z/143)$ where z is axial position in inches.

7.6 A reactor cooled with sodium is to produce 1000 MW of thermal power. The cylindrical fuel elements are 10 ft. long and 0.50 in. in diameter. The coolant flows through an annular channel whose outer diameter is 1.0 in. The coolant inlet temperature is 750°F and inlet velocity is 25 ft/sec. The maximum permissible cladding surface temperature is 1200°F. Assume the reactor is a bare right cylinder and that the axial and radial neutron flux distributions are approximated by cosine and J_o distributions, respectively. In addition, assume all engineering hot channel factors are unity. Calculate the following:

a. Hot channel outlet temperature.
b. Average channel outlet temperature.
c. Approximate number of coolant channels required.

7.7 A reactor fuel element consisting of uranium metal clad with stainless steel and cooled by Na is given below

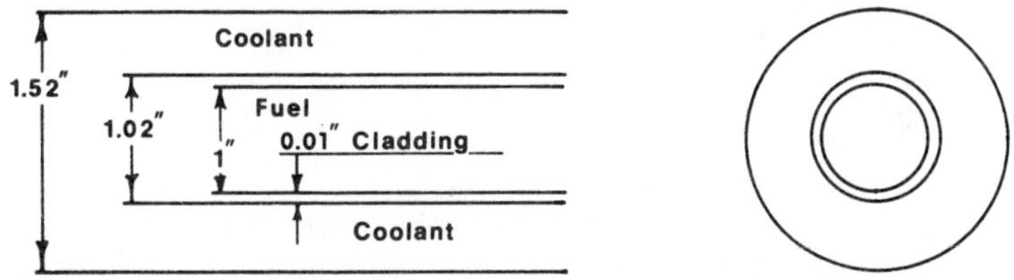

Each rod produces 0.2 MW of thermal power and is 10 ft long. The coolant inlet temperature is 600°F and the channel mass flow rate is $8\ lb_m/sec$. The fuel enrichment is varied axially so the rod wall temperature is kept constant along the entire channel length. Neglect bond resistance between fuel and cladding and axial heat conduction in the fuel. Determine the coolant outlet temperature, wall temperature, and maximum fuel center-line temperature. Where does the maximum fuel center-line temperature occur?

7.8 The flow channel for a high flux reactor is an annulus with a 2 in. inside diameter and 2.4 in. outside diameter. The active fuel element length is 3 ft. The reactor is cooled and moderated with heavy water with an inlet temperature of 130°F and inlet velocity of 75 ft/sec. Heat is applied to both sides of the channel and the axial heat flux distribution is given by

$$q''_w = 8 \times 10^5 \cos(\pi z/L_o), \text{ Btu/hr-ft}^2$$

where L_o is 4 ft. Assuming no boiling takes place in the channel, calculate the maximum cladding surface temperature.

7.9 For the subchannel geometry and coolant inlet conditions of problem 7.1, a BWR has an axial heat flux distribution given by

$$q''_w(z) = q''_o e^{-(\pi z/12)} \sin(\pi z/12)$$

where q''_o is a constant and z = 0 is the channel entrance. The exit quality is 12%. Find the following

a. Height at which the heat flux is a maximum, q''_{max}

b. The value of q''_o in terms of q''_{max}

c. The position of bulk boiling

7.10 A PWR at 2300 psia has 12 ft. long, 0.43 in. diameter rods arranged on a pitch of 0.55 in. Two adjoining subchannels have average linear heat rates of 6 and 8 kW/ft. The water mass velocity in both subchannels is 2×10^6 lb_m/hr-ft^2 and the average bulk temperature is 580°F. Calculate the enthalpy rise in both subchannels (a) assuming no interchannel mixing and (b) accounting for turbulent interchange between the two subchannels.

CHAPTER 8

STRESS ANALYSIS IN NUCLEAR REACTOR SYSTEMS

8.1 Introduction

The design of nuclear reactor systems demands that all components have an extremely low probability of failure over many years of operation at variable reactor power levels. Nuclear reactors in operation today have intense volumetric heat generation rates, heat transfer surfaces are subject to high-transfer rates, and for many systems large internal pressures are required in order to prevent coolant boiling or raise coolant outlet temperatures. All of these factors contribute to considerable stresses, both thermal and mechanical, throughout all of the major reactor system components. As a consequence, it is very important to be able to accurately predict the state of stress in reactor system components so as to insure that component failure does not occur.

High volumetric heat generation rates and heat fluxes in reactor systems cause large temperature gradients in many system components. These temperature differences produce expansion of materials and if no restraints, both internal and external, are placed on materials, they expand in a stress-free manner. However, interconnecting components produce external restraints on expansion and, therefore, thermal stresses result in reactor components. In addition, internal heat generation, transient behavior, and heat flow through non-simply connected bodies cause internal restraints on materials and resultant thermal stresses due to heating may place upper limits on reactor steady-state power levels which are lower than thermal or hydraulic constraints discussed in previous chapters. In addition, potential failure due to thermal stresses may place constraints on the rate of raising or lowering reactor power levels.

Stress analysis in nuclear reactor systems is usually very complex because of the unusual environment involved. Many reactor systems employ unusual materials in which their behavior to heating and neutron and gamma irradiation is not well known. Neutron and gamma irradiation over extended periods cause swelling and imbrittleness of many materials. Irradiation and temperature changes produce changes in many important physical properties of reactor system materials.

An additional complicating factor in reactor system stress analysis is the unsteadiness of the reactor core. Internal pressures on fuel element cladding are continuously increasing due to the buildup of fission product gases. Neutron flux distribution and, consequently, heat sources are changing in magnitude due to control rod movement and fuel depletion. Consequently, positions of maximum stress in fuel elements may be changing location over extended periods of reactor operation. Material damage due to gamma and neutron fluxes usually increases with increasing integrated flux times.

A considerable amount of analysis is performed to determine the state of stress and its consequences in nuclear reactor systems. A brief summary of the steps involved is as follows:

(i) Physics analysis to determine material irradiations and heat source distributions throughout the reactor system.

(ii) Thermal-hydraulic analysis to determine temperature distributions throughout the reactor system. Step i is incorporated in this analysis.

(iii) Stress analysis which requires temperature distributions in the system and may require irradiation history of irradiated materials.

These steps are interrelated because displacements produced by thermal expansion may alter heat source and temperature distributions. Situations requiring simultaneous solution to the three steps outlined are most complicated. Transient analysis also is very difficult to treat. In addition, determining stresses in three-dimensional bodies or configurations of low symmetry can be very complex.

The treatment of stresses in reactor systems requires a thorough knowledge of the behavior of the materials under stresses. For materials which are elastic, the governing stress equations are linear and thermoelastic stresses may be investigated independently of the mechanical stresses produced by external forces. The final values of stress are obtained by summing the mechanical and thermoelastic stresses. For plastic materials some of the stress equations are nonlinear and, consequently, thermoplastic and mechanical stresses must be treated simultaneously.

The methods employed for analyzing stresses in elastic or plastic materials are suitable for determining averaged macroscopic stresses, or so called stresses of the first kind. Under these conditions the properties of materials under consideration must be isotropic. Consequently, the theory we shall use is most likely unsuitable for analyzing a single crystal of some material. Stresses of the second kind, which are found in anisotropic materials, are most difficult to treat and will not be discussed in this book. Occasions may arise in which it is necessary to analyze single fuel crystals.

Thermal stresses in reactor components have their smallest possible values if material properties have the following trends [1]:

(i) small modulus of elasticity E, psi
(ii) small coefficient of thermal expansion a, $(^{\circ}F)^{-1}$
(iii) large thermal conductivity k, Btu/hr-ft-$^{\circ}$F
(iv) small Poisson ratio ν

The complexity of nuclear reactor systems creates considerable and, in some cases, practically insurmountable difficulties in determining the state of stress in its components. Frequently stress analysis is carried out by analyzing simplified models of the structure and thus obtaining approximate solutions to the state of stress. These computations are of great value to the reactor designer because they give indications of the maximum allowable dimensions of structures, upper limits on volumetric heat generation rates, and locate regions of potential weakness which may require further reinforcement.

The approach used to develop insight into the treatment of reactor system stresses is to derive the fundamental stress equations and then analyze a few problems of interest to reactor system design. It is hopeless in this book to present more than a cursory treatment and interested readers are urged to consult one or more of the excellent references on the subject of stress analysis [1, 2, 3, 4].

8.2 General Considerations of Stresses

In order to develop a theory for determining the state of stress in materials, it is necessary to precisely define what is meant by a stress. We can examine a body in equilibrium which has a variety of forces acting upon it. By equilibrium we imply that the body is at rest and thus the vector sum of the forces and moments acting upon the body are zero. Figure 8.1(a) illustrates the body under consideration and we can examine the resultant forces on a small cross sectional area δA, which is represented by the line A-A on the body. The forces on δA, shown in Fig. 8.1, are classified as normal forces given by δF_σ and tangential forces given by δF_τ. A normal stress is defined by

$$\sigma \equiv \lim_{\delta A \to 0} \frac{\delta F_\sigma}{\delta A} \tag{8.2.1}$$

and a shear stress is defined by

$$\tau \equiv \lim_{\delta A \to 0} \frac{\delta F_\tau}{\delta A} \tag{8.2.2}$$

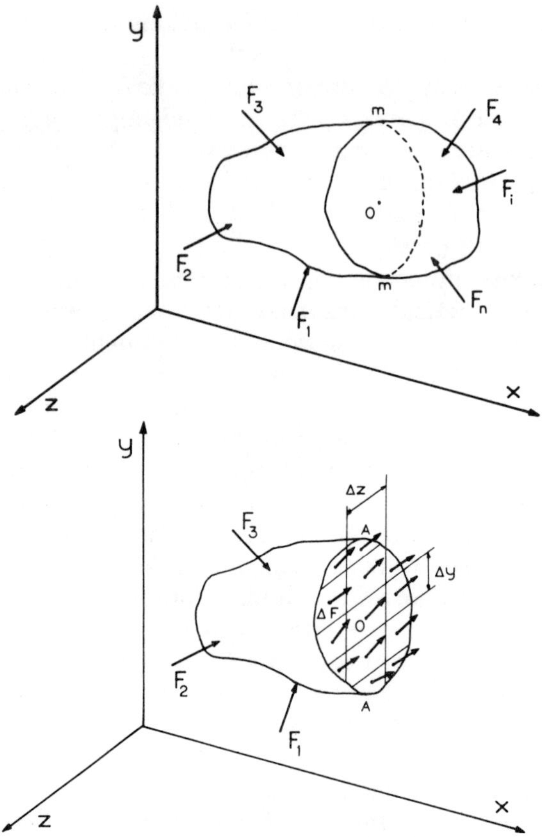

Fig. 8.1(a). Internal forces acting on A-A

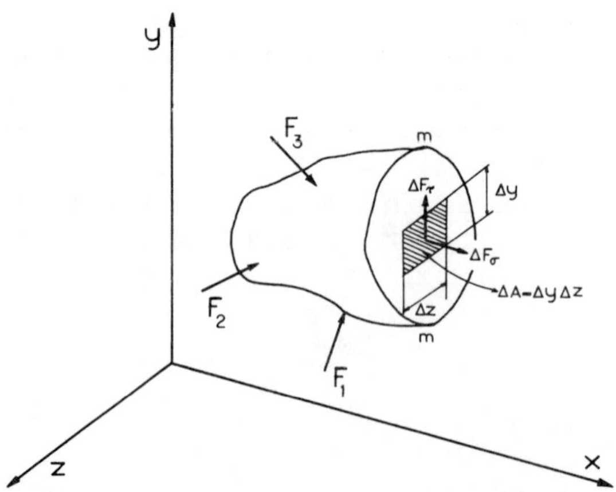

Fig. 8.1(b). Rectangular components of the force vector $\mathbf{\Delta F}$ acting on the small area centered on point 0

In the discussion of stresses it is convenient to have a coordinate system in which to refer both direction and surface on which both normal and shear stresses act. With orthogonal coordinates, two coordinates would be parallel to a surface and the third coordinate normal to a surface. We shall assign positive and negative signs on both surfaces and stresses. Surfaces are designated positive when an outer directed normal is in a positive coordinate direction. A surface is negative when an outer directed normal is in a negative coordinate direction. Surfaces are labeled by the coordinate which is normal to them; i.e., an x-surface is normal to the x-coordinate.

The following convention is used to describe stresses [3]:

(i) Normal Stresses. A normal stress is positive if it acts in a positive direction on a positive surface or a negative direction on a negative surface. The converse defines negative stresses. From this definition, positive stresses produce tension and negative stresses produce compression. Normal stresses are subscripted first by the surface upon which they act and second by the coordinate direction. Consequently, a normal stress acting upon an x-surface is designated σ_{xx} . For rectangular coordinates, the normal stresses are σ_{xx} , σ_{yy} , and σ_{zz} .

(ii). Shear Stresses. A shear stress has a positive sign when it acts in a positive coordinate direction on a positive surface or a negative coordinate direction on a negative surface. Negative shear stresses are defined by the converse. Shear stresses are subscripted first by the surface upon which they act and second by the direction in which they act. Consequently a shear stress acting on an x-surface acting in the y-coordinate direction is designated τ_{xy} . For rectangular coordinates the six shear stresses are τ_{xy} , τ_{xz} , τ_{yx} , τ_{yz} , τ_{zx} , and τ_{zy} . Figure 8.2 illustrates positive and negative normal and shear stresses.

The nine stresses components determine the state of stress at a point in a material. The components represent the stress tensor which is discussed in the treatment of fluid mechanics. The stress tensor

$$T_{ij} = \begin{bmatrix} \sigma_{xx} & \tau_{xy} & \tau_{xz} \\ \tau_{yx} & \sigma_{yy} & \tau_{yx} \\ \tau_{zx} & \tau_{zy} & \sigma_{zz} \end{bmatrix} \qquad (8.2.3)$$

$$i = 1,2,3$$
$$j = 1,2,3$$

is symmetric because equilibrium requirements demand that

$$\tau_{xy} = \tau_{yx}, \quad \tau_{xz} = \tau_{zx}, \quad \tau_{yz} = \tau_{zy}.$$

When analyzing the state of stress at a point in a material, the magnitude and sign of the stress tensor components depend upon

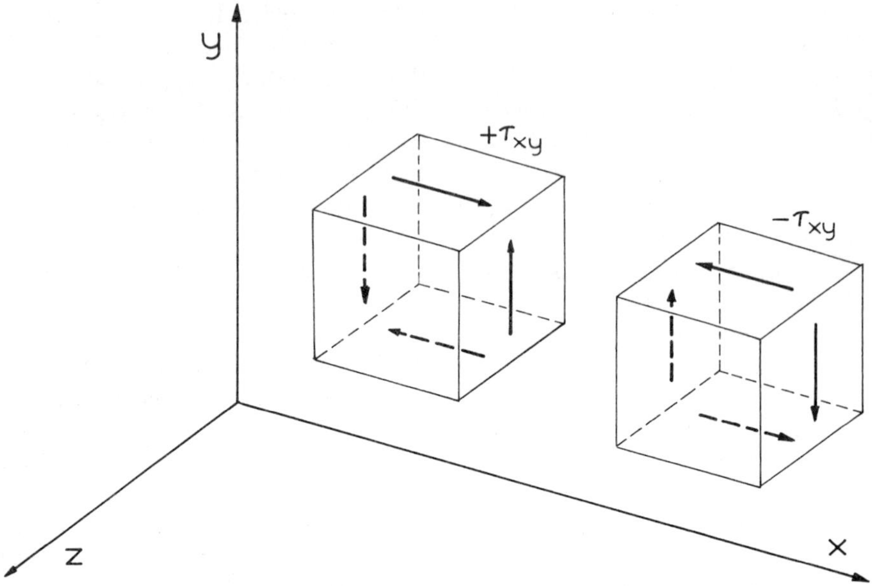

Fig. 8.2. Definition of positive and negative τ_{xy}

orientation of the coordinate system. This is apparent by examining Fig. 8.1, since the magnitude and sign of stresses on δA depend upon orientation of δA.

The directional properties of shear and normal stresses is best understood by examining the equilibrium requirements for plane stress. Plane stress means that all stresses lie in the same plane. Figure 8.3 illustrates the examples under consideration.

One of the requirements for equilibrium of a body is the sum of the moments about any point must be zero $(\Sigma\vec{M} = 0)$. From our plane stress element shown in Fig. 8.3(a), the equilibrium requirements are

$$(\tau_{xy}\,dx)dy = (\tau_{yx}\,dy)dx \tag{8.2.4}$$

or

$$\tau_{xy} = \tau_{yx} \tag{8.2.5}$$

The result given by Eq. (8.2.5) is that for plane stress the shear stress components on perpendicular faces are equal in magnitude and directed relative to each other in the manner shown in Fig. 8.3(a). This result also holds for the more general three-dimensional case. Consequently, $\tau_{xy} = \tau_{yx}$, $\tau_{xz} = \tau_{zx}$, and $\tau_{yz} = \tau_{zy}$.

The next consideration is to determine relationships between known stresses on x- and y-faces and stresses on a face whose normal x′ lies in the xy plane and makes an arbitrary angle θ with the x-axis. Equilibrium requirements on the small wedge shown in Fig. 8.3(b) are that the sum of the forces must be zero $(\Sigma \vec{F} = 0)$. Consequently,

$$\Sigma F_{x'} = \sigma_{x'} \overline{MN} - (\sigma_x \overline{MP}) \cos\theta - (\tau_{xy} \overline{MP}) \sin\theta - (\sigma_y \overline{NP}) \sin\theta - (\tau_{xy} \overline{NP}) \cos\theta = 0 \tag{8.2.6}$$

$$\Sigma F_{y'} = \tau_{x'y'} \overline{MN} + (\sigma_x \overline{MP}) \sin\theta - (\tau_{xy} \overline{MP}) \cos\theta - (\sigma_y \overline{NP}) \cos\theta + (\tau_{xy} \overline{NP}) \sin\theta = 0 \tag{8.2.7}$$

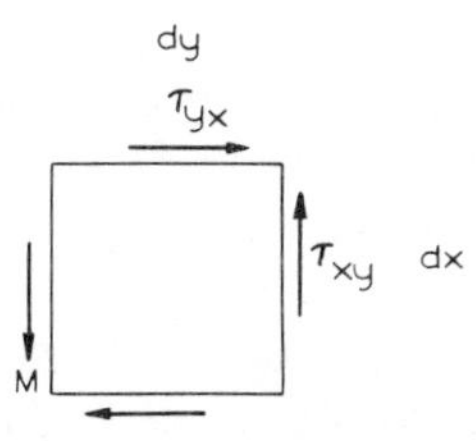

Fig. 8.3(a). Equilibrium requirements of shear stresses

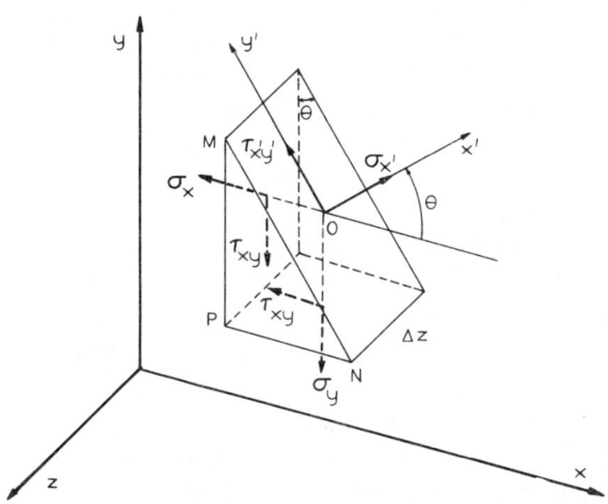

Fig. 8.3(b). Stress components acting on faces of a small wedge.

By using trigonometric relations between sides of a triangle, Eqs. (8.2.6) and (8.2.7) are simplified to give

$$\sigma_{x'} = \sigma_x \cos^2\theta + \sigma_y \sin^2\theta + 2\tau_{xy} \sin\theta \cos\theta \tag{8.2.8}$$

$$\tau_{x'y'} = (\sigma_y - \sigma_x) \sin\theta \cos\theta + \tau_{xy} (\cos^2\theta - \sin^2\theta) \tag{8.2.9}$$

From the results given by Eqs. (8.2.8) and (8.2.9), it is apparent that, if the stress components are known on two perpendicular faces, then stress components are known on all faces whose normals lie in the plane. By substituting $\theta + 90^\circ$ for θ in Eq. (8.2.8) we have the normal stress $\sigma_{y'}$, acting on a face perpendicular to the y'-axis.

$$\sigma_{y'} = \sigma_x \sin^2\theta + \sigma_y \cos^2\theta - 2\tau_{xy} \sin\theta \cos\theta \tag{8.2.10}$$

If stress components for all possible orientations of faces are known at a point, then the **state of stress** is known at the point in question. Consequently, for plane stress one must know two normal stresses and one shear stress in order to know the state of stress.

Equations (8.2.8), (8.2.9), and (8.2.10) are rearranged by use of trigonometric double angle identities to give

$$\sigma_{x'} = \frac{\sigma_x + \sigma_y}{2} + \frac{\sigma_x - \sigma_y}{2} \cos 2\theta + \tau_{xy} \sin 2\theta \tag{8.2.11}$$

$$\sigma_{y'} = \frac{\sigma_x + \sigma_y}{2} - \frac{\sigma_x - \sigma_y}{2} \cos 2\theta - \tau_{xy} \sin 2\theta \tag{8.2.12}$$

$$\tau_{x'y'} = -\frac{\sigma_x - \sigma_y}{2} \sin 2\theta + \tau_{xy} \cos 2\theta \tag{8.2.13}$$

These equations describe a circle called **Mohr's circle for stress** (see Fig. 8.4). Normal stresses are represented horizontally and shear stresses vertically. The Mohr's circle describes the state of stress at a point in a material in plane stress. Any line which passes through the center of the Mohr's circle has intersections on the circle which describe stresses on two mutually perpendicular planes. By knowing three independent stresses, one can construct a Mohr's circle and by inspection determine stresses at a point with respect to any coordinate system.

The Mohr's circle is only applicable for plane stress. However, the approach used can be extended to three dimensions and the state of stress at a point is represented by the surface of an ellipsoid. One must now have given six independent stress components, preferably normal and shear stresses on three mutually perpendicular planes.

Construction of a Mohr's circle is rather elementary using a graphical technique. The following steps are used for this process:

(i) σ_x, σ_y, and τ_{xy} have positive or negative signs according to the previously adopted convention.

(ii) Lay off a coordinate system with shear stresses (τ) as the

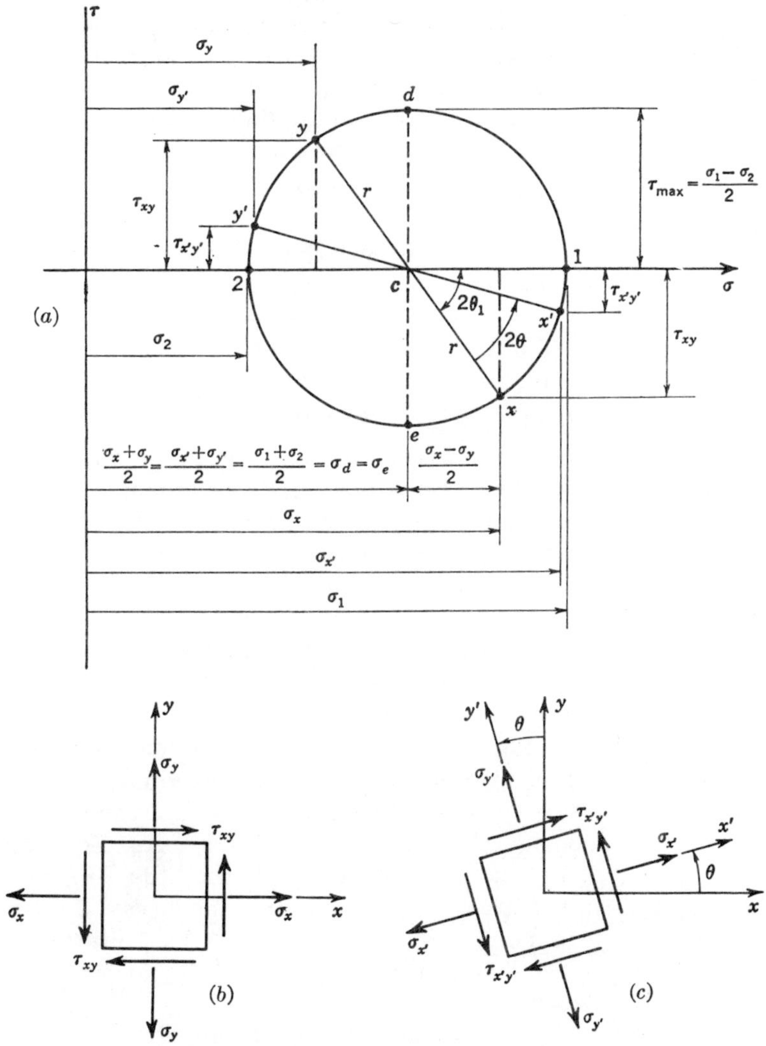

Fig. 8.4. Mohr's circle for plane stress [3].

ordinate and normal stresses (σ) as the abcissa (see Fig. 8.4). Normal stresses to the left of the τ-axis are negative.

(iii) Draw two vertical lines on the coordinate system which are parallel to the τ-axis and intersect the σ-axis at σ_x and σ_y.

(iv) A positive shear stress (τ_{xy}) on the x-face is located downward on the line through σ_x and a negative shear stress is located upward. This identifies a point of the $\sigma-\tau$ plane which gives the stresses on the x-face (see Fig. 8.4).

(v) A positive shear stress (τ_{yx}) on the y-face is located upward on the line through σ_y and a negative shear stress is located downward. This identifies a point on the $\sigma-\tau$ plane which gives the stresses on the y-face (see Fig. 8.4).

(vi) Draw a line between points x and y determined by steps iii, iv, and v (see Fig. 8.4). This line describes the diameter of the Mohr's circle. The center of the Mohr's circle is located on the σ-axis at point $\bar{0}$ which has coordinates $[(\sigma_x + \sigma_y)/2,0]$.

(vii) The radius of the Mohr's circle is given by

$$r = \sqrt{(\sigma_y - \bar{0})^2 + (\tau_{xy})^2} \tag{8.2.14}$$

The Mohr's circle is laid off by using a compass.

(viii) In order to find the stresses ($\sigma_{x'}, \sigma_{y'}$, and $\tau_{x'y'}$) on mutually perpendicular planes obtained by rotating the x'-axis an angle θ from the x-axis, contruct a line passing through point $\bar{0}$ which is rotated an angle 2θ from the line x-y. The direction of the angle of rotation is the same as the angle of rotation from the x-axis.

(ix) The magnitudes of the stress components are found by considering the geometry of circles and triangles. Frequently it is necessary to make sketches of a Mohr's circle in order to determine desired stress components.

Great utility is found in the use of Mohr's circle for analyzing plane stress. Almost by inspection one can determine maximum or minimum normal and shear stresses once three independent stresses are given. This would be a considerably more difficult task using Eqs. (8.2.11), (8.2.12), and (8.2.13). It is repeated again that the Mohr's circle represents the state of stress at a point $\bar{0}$ for plane stress. If for a fixed coordinate system a point adjacent to point $\bar{0}$ has different stress components σ_x, σ_y, and τ_{xy} , then its state of stress is given by a different Mohr's circle.

It is important to understand the multidirectional character of stresses, since yielding of materials is usually dependent upon maximum values of shear or normal stresses. Calculations performed with respect to a given coordinate system may not yield maximum shear or normal stresses and it is then necessary to analyze the Mohr's circle in order to determine maximum stresses.

Of particular interest is a coordinate system whose orientation yields zero shear stresses. This set of axes is called the **principal axes of stress**. The normal stresses associated with this set of axes are called **principal stresses**. By inspection of Fig. 8.4 it is obvious that principal stresses are the maximum and minimum normal stresses at a point and are designated σ_1, σ_2, respectively. In addition, the maximum normal stress is always greater than or equal to the maximum shear stress. In the analysis of simple bodies with high degrees of symmetry, like a rod, it is convenient to orient the coordinate system along the principal axes.

For plane stress, the maximum shear stress is given by

$$\tau_{Max} = \frac{\sigma_1 - \sigma_2}{2} \tag{8.2.15}$$

Since τ_{Max} is a common basis for establishing maximum stresses in materials, the difference in the maximum and minimum principal stresses is given a special designation—the stress intensity S.

$$S \equiv \sigma_1 - \sigma_2 \tag{8.2.16}$$

The concept of using Mohr's circle is extended to three dimensions as shown in Ref. 3. This technique is rather complicated and we will restrict our considerations to situations which are usually approximated by plane strain or in which the obvious location of a three-dimensional coordinate system is the principal axes.

8.3 General Considerations of Deformations [Strains]

Bodies in which temperature changes occur or forces act upon undergo displacements which are classified as (1) rigid body motions or (2) deformations. Rigid body motions involve either translations or rotations of the body as a whole while deformations involve a motion of points within a body relative to each other. Temperature changes produce displacements of the latter kind.

Deformations in a body are called strains and are usually described with respect to a coordinate system. Strains at a point in a body are represented in terms of displacements of straight lines. It is assumed that the point under consideration is small enough that straight lines remain straight after a strain occurs.

The discussion of deformations or strains is simplified by considering a body whose particles all lie in the same plane and deform only in this plane. This situation is called **plane strain**. The deformation at a point is then described by considering the motion of two lines which are initially perpendicular to each other. Figure 8.5 illustrates the point under consideration. Strains are described as **normal strains** which give a measure of the relative contraction or elongation of a line and **shear strains** which give a measure of the relative rotation of two lines.

Normal strains are defined by the fractional change in length of line segments. Referring to Fig. 8.5 the normal strains in the x- and y-directions are given by

$$\epsilon_x \equiv \lim_{\Delta x \to 0} \frac{O'C' - OC}{OC}$$

$$\epsilon_y \equiv \lim_{\Delta y \to 0} \frac{O'E' - OE}{OE} \qquad (8.3.1)$$

From the definition of normal strains elongations are positive and contractions negative.

Shear strains are specified by the change in angle between two axes which are perpendicular to each other in the undeformed body. The symbol for a shear strain is γ and it is usually double subscripted to indicate the original axes. From the example shown in Fig. 8.5, the shear strain is defined by

$$\gamma_{xy} \equiv \lim_{\substack{\Delta x \to 0 \\ \Delta y \to 0}} \tan(\angle COE - \angle C'O'E') \qquad (8.3.2)$$

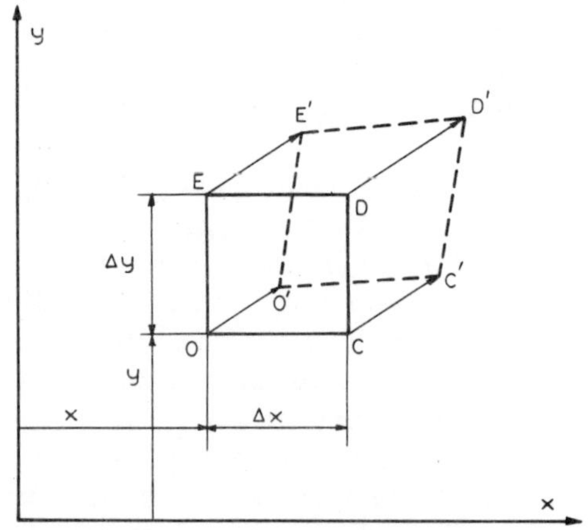

Fig. 8.5. Deformation in the xy plane of a small element of a continuous body.

Since shear strains of interest in engineering are usually quite small (less than 0.01), it is adequate to define shear strains by the change in angle in radians, or

$$\gamma_{xy} \equiv \lim_{\substack{\Delta x \to 0 \\ \Delta y \to 0}} (\pi/2 - C'O'E') \tag{8.3.3}$$

Shear strains are positive when the rotation is such that the first and third quadrants of the coordinate system become smaller and negative when the first and third quadrants become larger. The shear strain in Fig. 8.5 is positive. Positive shear stresses cause positive shear strains and negative shear stresses cause negative shear strains.

The magnitudes and signs of normal and shear strains depend upon orientation of the coordinate system. By using similar considerations employed in analyzing stresses, a Mohr's circle of strain is used to determine the state of strain at a point. One simply substitutes ϵ_x, ϵ_y, and γ_{xy} for σ_x, σ_y, and τ_{xy} in Eqs. (8.2.11), (8.2.12), and (8.2.13). Mohr's circle of strain is not used as frequently as the Mohr's circle for stress.

Extension of the concept of strains to three dimensions is readily apparent. One simply introduces one more normal strain componnt ϵ_z and two more shear strain components γ_{xz} and γ_{yz}. In a mathematical sense strains are a second-order tensor like stresses. The tensor is symmetric because $\gamma_{xy} = \gamma_{yx}$, $\gamma_{xz} = \gamma_{zx}$, and $\gamma_{yz} = \gamma_{zy}$.

8.4 General Considerations of Force-Deformation Relations

Determining relationships between forces and deformation is approached in several ways. One can rely upon experimental observations made on an atomic scale and use a theoretical extension to a macroscopic level. Another approach is to use experimental observations on a macroscopic level. Up to this time the latter approach is the most successful in yielding information of value for engineering stress analysis.

Materials behave in different fashions when deformed by the application of stresses. Some materials return to their original shape after the stress is removed. This type of deformation is called an **elastic deformation**. Other materials remain permanently deformed after a stress is removed. If the deformation under stress is independent of time and depends upon the magnitude of stress, then materials which remain permanently deformed after a stress is removed have undergone **plastic deformation**. Materials which are plastically deformed may require a greater load to produce a further plastic deformation and these materials have undergone **strain-hardening**.

With an increasing application of a load upon a material, failure due to fracture ultimately takes place. A **ductile** material is defined as one which undergoes a considerable amount of plastic strain compared to elastic strain before failure occurs. A **brittle** material is defined as one which undergoes very little strain due to stress before failure.

Another phenomenon which occurs in materials under continuous stress is an increase in strain with time. The time-dependent portion of the strain is called **creep**. The effects of creep may limit allowable stresses and lifetime of fuel elements and other reactor materials. Another time dependent behavior of materials is a slow recovery to the original shape after stresses are removed. This phenomenon is called the **elastic aftereffect** or **recovery**.

The most common force-deformation experiment performed on materials is the tensile test. This test entails applying a load along the axis of a test specimen and measuring the amount of axial strain for different loads. The experimental results are then represented graphically as a stress-strain curve. Figure 8.6 illustrates a typical stress-strain curve which identifies various stresses of interest.

Initially upon loading a body, stress is very closely proportional to strain. The point at which stress starts to depart from this behavior is called the **proportional limit** and this position is designated σ_p on Fig. 8.6. Up to the proportional limit, we can mathematically represent the stress-strain behavior for a tensile test by

$$\epsilon_z = \frac{\sigma_z}{E} \tag{8.4.1}$$

where E has units of psi and is called the **modulus of elasticity** or Young's modulus. Another limiting stress of interest is the **elastic limit** which represents the greatest stress in a material which doesn't produce a permanent strain upon removing the stress. Figure 8.7(a) illustrates a material which is stressed elastically beyond the proportional limit. In Fig. 8.7(b) a material is stressed beyond the elastic limit and the stress then removed. Further loading produces the stress-strain curve illustrated with the loading and unloading curves having slopes similar to the initial elastic curve. It should be noticed that the proportional and elastic limits are extended after permanent plastic deformation due to strain-hardening. Consequently, these limits depend upon the stress-strain history of the material. The total strain OB in this example is represented by a plastic strain OA and an elastic strain AB.

Since it is rather difficult to precisely determine proportional and elastic limits in materials, it is customary to measure stresses which

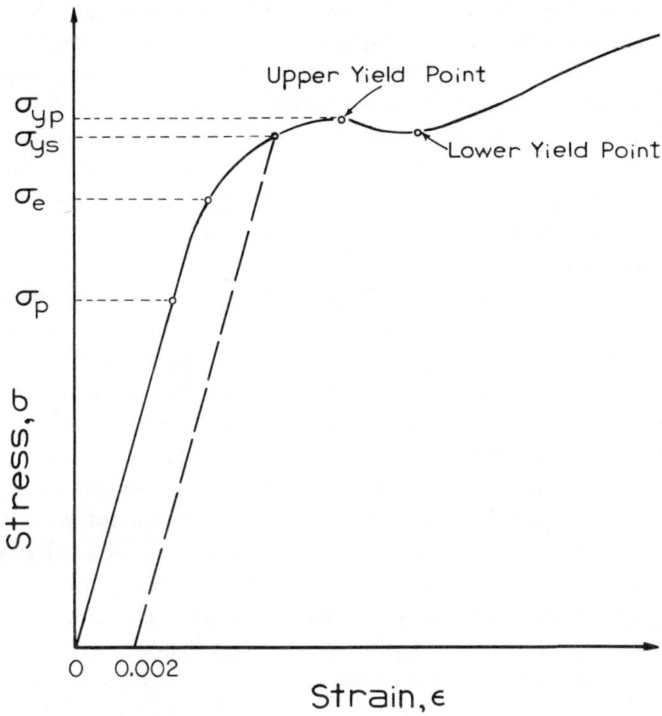

Fig. 8.6. Stress-strain curve definitions.

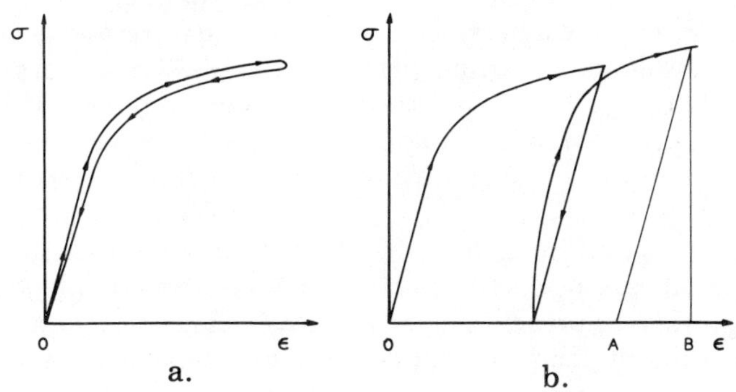

Fig. 8.7. Effect of unloading and reloading (a) Elastic. (b) Plastic.

produce a specified amount of permanent strain once the load is removed. This limiting stress is called the **offset yield strength** or **yield strength** σ_{ys} and the permanent or plastic strain is usually 0.2 percent. The yield strength is determined by drawing a line parallel to the initial tangent of the stress-strain curve which passes through the abscissa at the specified plastic strain. The intersection of this line with the stress-strain curve is the yield strength. In brittle materials it may not be possible to specify yield strength since failure occurs before plastic strain occurs.

In some materials plastic deformations occur rather abruptly after the elastic limit and there is an increase in strain with no apparent increase or even a decrease in stress. The **yield point** is defined as the stress at which strain increases without a further increase in stress. Two types of yield points are described for a material. The **upper yield point** marks a point of subsequent increase in strain with little increase in stress. Many common types of steels display yield points.

The tensile test describes the stress-strain behavior for a body under tension. Many materials, notably glass, have different stress-strain behavior under compression. For most ductile materials, the stress-strain behavior for small strains is nearly the same [3].

In a simplified approach to stress analysis, it is customary to idealize the stress-strain curves for materials so as to obtain simple mathematical equations. Figure 8.8 illustrates some idealized models for stress-strain behavior which are quite useful for analysis. A **rigid** material (shown in Fig. 8.8(a)) is one in which there is no strain from an applied stress. A **linearly elastic** material is one in which strain is proportional to stress to the point of failure. A **rigid-plastic** material is one in which elastic or time-dependent strains are neglected. If stresses occur up to the plastic region, permanent deformations exist upon removing the load. For rigid-plastic materials, strain-hardening may be neglected (Fig. 8.8(c)) or strain-hardening may be assumed (Fig. 8.8(d)). Many materials are approximated by an elastic-plastic behavior. Strain-hardening may be neglected (Fig 8.8(e)) or assumed not negligible (Fig 8.8(f)).

In the stress problems treated in the following sections, it is assumed that materials are perfectly elastic or elastic-perfectly plastic. In many instances this approach is conservative. As an example, thermal stresses without load restraints tend to relieve themselves due to plastic flow. Assuming materials to be elastic overestimates thermal stresses.

8.5 The general Equations of Elasticity

In stress analysis three fundamental principles are employed which yield sets of equations whose solutions determine the state of stress and strain in a material. In the general sense these equations are quite formidable; however, in the treatment of many important problems considerable simplification is possible and relatively simple solutions are found. The fundamental principles used to derive the stress equation are

(i) Study of forces and equilibrium requirements. For equilibrium of a body we have $\Sigma\vec{F} = 0$ and $\Sigma\vec{M} = 0$

(ii) Study of deformations in a material and conditions of geometric fit. This entails using principles of geometry to develop relations between deformations or strains in different coordinate directions which are satisfied if failure doesn't occur in the material.

(iii) Application of force-deformation relations based upon experimental observations.

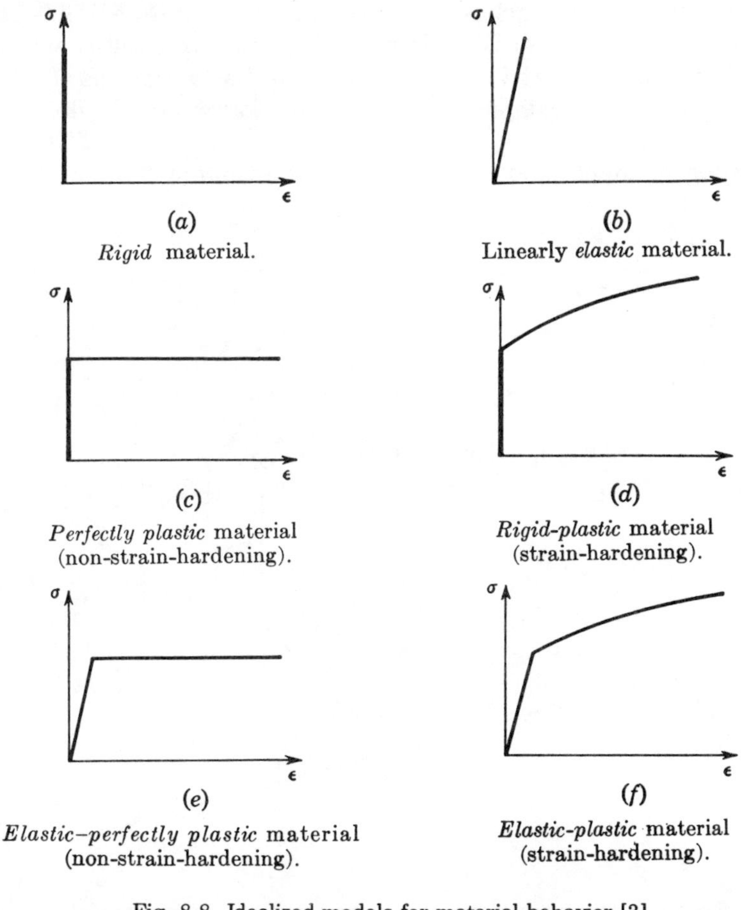

(*a*) *Rigid* material.

(*b*) Linearly *elastic* material.

(*c*) *Perfectly plastic* material (non-strain-hardening).

(*d*) *Rigid-plastic* material (strain-hardening).

(*e*) *Elastic–perfectly plastic* material (non-strain-hardening).

(*f*) *Elastic-plastic* material (strain-hardening).

Fig. 8.8. Idealized models for material behavior [3].

Application of these principles is of the greatest importance to understanding the approach used in stress analysis. As a consequence, the rest of this section is devoted to deriving the equations of elasticity by employing basic principles. A more detailed treatment is given in Ref. 3 and individuals are encouraged to consult this book for a more lucid discussion.

Equilibrium Requirements

We have previously stated that, for a body to be in equilibrium, the vector sum of the forces and moments at any point in a body are zero. As a consequence of this requirement certain relationships are satisfied with regard to the spatial rate of change of stress components. The equilibrium equations are derived by considering forces and moments acting upon a small volume element. We consider the case of plane stress because the approach is somewhat simpler. Figure 8.9 illustrates the volume element under consideration for plane stress. It is assumed that the volume element is small enough so that variations in stress components are approximated by first-order terms of a Taylor series expansion.

Summing moments about the center of the volume element gives

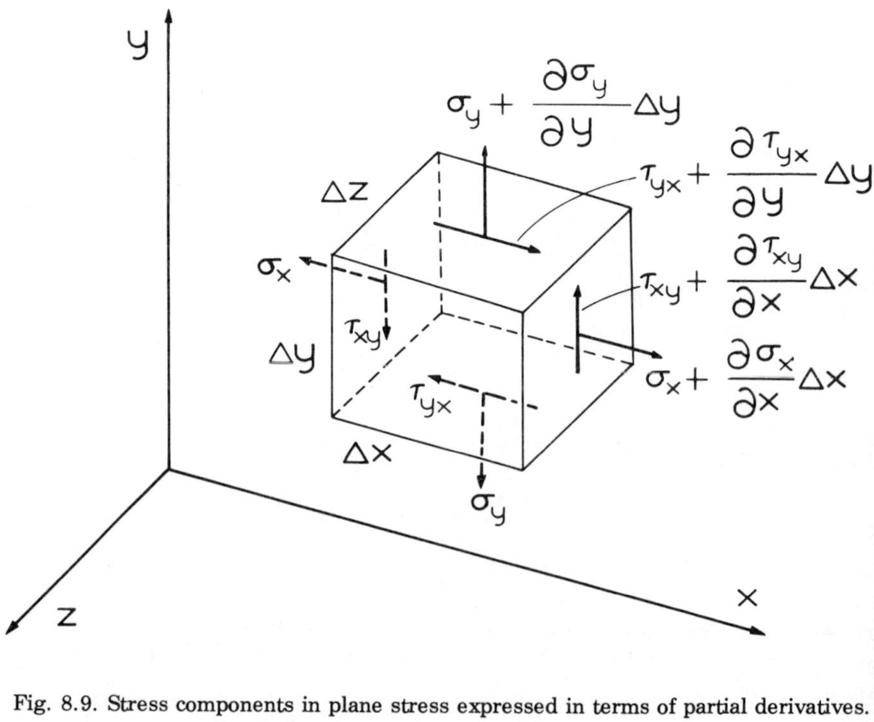

Fig. 8.9. Stress components in plane stress expressed in terms of partial derivatives.

$$\Sigma M = (\tau_{xy}\, \Delta y \Delta z)\frac{\Delta x}{2} + \left[\left(\tau_{xy} + \frac{\partial \tau_{xy}}{\partial x}\Delta x\right)\Delta y \Delta z\right]\frac{\Delta x}{2} - (\tau_{yx}\, \Delta x \Delta z)\frac{\Delta y}{2}$$

$$- \left[\left(\tau_{yx} + \frac{\partial \tau_{yx}}{\partial y}\Delta y\right)\Delta x \Delta z\right]\frac{\Delta y}{2} = 0 \qquad (8.5.1)$$

Simplifying Eq. (8.5.1) gives

$$\tau_{xy} + \frac{\partial \tau_{xy}}{\partial x}\Delta x - \tau_{yx} - \frac{\partial \tau_{yx}}{\partial y}\Delta y = 0 \qquad (8.5.2)$$

In the limit as Δx and Δy go to zero, the only terms left in Eq. (8.5.2) are

$$\tau_{yx} = \tau_{xy} \qquad (8.5.3)$$

which is the same result previously stated for plane stress.

Summing forces in the x- and y-directions gives

$$\Sigma F_x = \left(\sigma_x + \frac{\partial \sigma_x}{\partial x}\Delta x\right)\Delta y \Delta z + \left(\tau_{yx} + \frac{\partial \tau_{yx}}{\partial y}\Delta y\right)\Delta x \Delta z - \sigma_x\, \Delta y \Delta z$$

$$- \tau_{yx}\, \Delta x \Delta z = 0 \qquad (8.5.4)$$

$$\Sigma F_y = \left(\sigma_y + \frac{\partial \sigma_y}{\partial y}\Delta y\right)\Delta x \Delta z + \left(\tau_{xy} + \frac{\partial \tau_{xy}}{\partial x}\Delta x\right)\Delta y \Delta z - \sigma_y\, \Delta x \Delta z$$

$$- \tau_{xy}\, \Delta y \Delta z = 0 \qquad (8.5.5)$$

Equations (8.5.4) and (8.5.5) simplify to

$$\frac{\partial \sigma_x}{\partial x} + \frac{\partial \tau_{yx}}{\partial y} = 0 \qquad (8.5.6)$$

$$\frac{\partial \sigma_y}{\partial y} + \frac{\partial \tau_{xy}}{\partial x} = 0 \qquad (8.5.7)$$

Equations (8.5.3), (8.5.6), and (8.5.7) are the equilibrium equations for plane stress. These equations are extended to three dimensions by adding the missing normal and shear stress components and setting the sum of forces in the three coordinate directions equal to zero. The more general equilibrium equations also contain terms for body forces due to gravity, magnetic fields, etc., which are distributed over the volume element. The general equilibrium equations in rectangular coordinates are

$$\frac{\partial \sigma_x}{\partial x} + \frac{\partial \tau_{xy}}{\partial y} + \frac{\partial \tau_{xz}}{\partial z} + X = 0 \qquad (8.5.8)$$

$$\frac{\partial \tau_{yx}}{\partial x} + \frac{\partial \sigma_y}{\partial y} + \frac{\partial \tau_{yz}}{\partial z} + Y = 0 \qquad (8.5.9)$$

$$\frac{\partial \tau_{zx}}{\partial x} + \frac{\partial \tau_{zy}}{\partial y} + \frac{\partial \sigma_z}{\partial z} + Z = 0 \qquad (8.5.10)$$

$$\tau_{xy} = \tau_{yx} \ ; \quad \tau_{xz} = \tau_{zx} \ ; \quad \tau_{yz} = \tau_{zy} \qquad (8.5.11)$$

where X, Y, and Z are body forces per unit volume. Equation (8.5.11) is eliminated from the equilibrium relations by making appropriate substitutions into Eqs. (8.5.8), (8.5.9), and (8.5.10).

In spherical coordinates (r, θ) the equilibrium equations between the stress components σ_r , σ_θ , and $\tau_{r\theta} = \tau_{\theta r}$ are given by

r-direction
$$\frac{\partial \sigma_r}{\partial r} + \frac{1}{r}\frac{\partial \tau_{r\theta}}{\partial \theta} + \frac{1}{r}(\sigma_r - \sigma_\theta) + R = 0 \qquad (8.5.12)$$

θ -direction
$$\frac{\partial \tau_{r\theta}}{\partial r} + \frac{1}{r}\frac{\partial \sigma_\theta}{\partial \theta} + \frac{2}{r}\tau_{r\theta} + \theta = 0 \qquad (8.5.13)$$

where R and θ are body forces per unit volume acting in the r- and θ -directions, respectively. For cylindrical coordinates with angular symmetry (r,z), the equilibrium equations between the stress components, σ_z , σ_r , σ_θ , and $\tau_{rz} = \tau_{zr}$ are given by

r-direction
$$\frac{\partial \sigma_r}{\partial r} + \frac{\partial \tau_{rz}}{\partial z} + \frac{1}{r}(\sigma_r - \sigma_\theta) + R = 0 \qquad (8.5.14)$$

z-direction
$$\frac{\partial \tau_{rz}}{\partial r} + \frac{\partial \sigma_z}{\partial z} + \frac{1}{r}\tau_{rz} + Z = 0 \qquad (8.5.15)$$

θ -direction
$$\frac{\partial \sigma_\theta}{\partial \theta} = 0 \qquad (8.5.16)$$

The normal stress components σ_r , σ_θ , and σ_z represent components in the radial, angular, and axial directions, respectively. τ_{rz} represents shear stresses acting on a radial face in the z-direction or vice versa.

Geometric Compatibility

The conditions for geometric compatibility imply that certain relations must exist between **strains** and **displacements** of a point in order for there to be no failures or holes in the material. The treatment of plane strain (all strains lie in the same plane) is somewhat simpler to analyze, so this approach is employed and we then extend the treatment to three dimensions.

The displacements of any point on a deformed plane are expressed mathematically by displacement functions for each coordinate direction. We define the following functions to represent displacements of a point.

u = u(x,y) = displacement of a point in the x-direction

v = v(x,y) = displacement of a point in the y-direction

Naturally in a three-dimensional sense these functions are dependent upon three coordinate directions and there is a function w(x,y,z) to represent displacements in the z-direction.

The geometric compatibility requirement that no holes or void be created in the material implies that u and v (also w) be continuous functions. As an example, if for a fixed value of y, u is plotted as a function of x. If there is a discontinuity in u, there must be a break in the material at the discontinuity.

Figure 8.10 illustrates a small element which has undergone

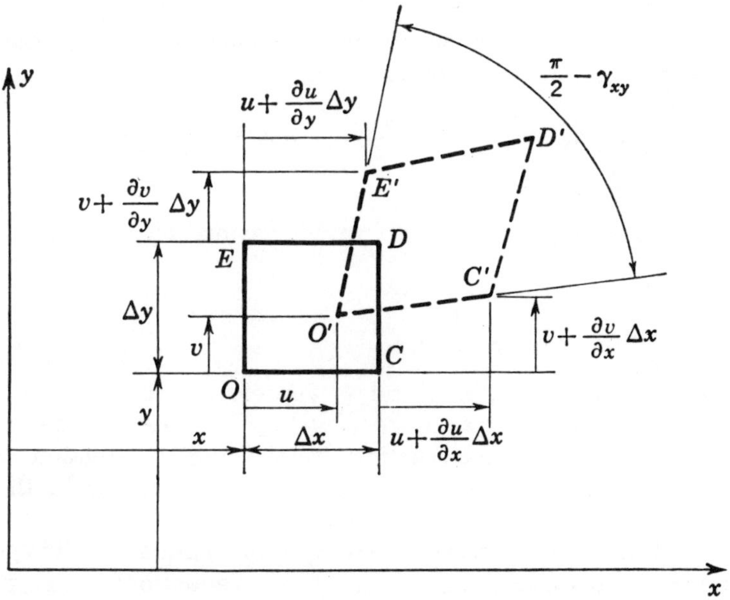

Fig. 8.10. Plane strain deformation expressed in terms of the components u and v and their partial derivatives.

deformation due to stresses or temperature changes. By keeping strains small and distances OC and OE small, displacements of points E and C are represented in terms of first-order Taylor series expansions of displacements of 0. From previous definitions of strain components, ϵ_x , ϵ_y , and γ_{xy} , the relations between strains and displacements follows.

$$\epsilon_x = \lim_{\Delta x \to 0} \frac{O'C' - OC}{OC} = \lim_{\Delta x \to 0} \frac{\left(\Delta x + \frac{\partial u}{\partial x}\Delta x\right) - \Delta x}{\Delta x} = \frac{\partial u}{\partial x} \quad (8.5.17)$$

$$\epsilon_y = \lim_{\Delta y \to 0} \frac{O'E' - OE}{OE} = \lim_{\Delta y \to 0} \frac{\left(\Delta y + \frac{\partial v}{\partial y}\Delta y\right) - \Delta y}{\Delta} = \frac{\partial v}{\partial y} \quad (8.5.18)$$

$$\gamma_{xy} = \lim_{\substack{\Delta x \to 0 \\ \Delta y \to 0}} (\pi/2 - < C'O'E') = \lim_{\substack{\Delta x \to 0 \\ \Delta y \to 0}} \left[\frac{\pi}{2} - \left(\frac{\pi}{2} - \frac{\frac{\partial v}{\partial x}\Delta x}{\Delta x} - \frac{\frac{\partial u}{\partial y}\Delta y}{\Delta y}\right)\right]$$

$$= \frac{\partial u}{\partial y} + \frac{\partial v}{\partial x} \quad (8.5.19)$$

The approach employed is extended to three dimensions and the rest of the compatibility relations are

$$\epsilon_z = \frac{\partial w}{\partial z} \; ; \; \gamma_{xz} = \frac{\partial w}{\partial x} + \frac{\partial u}{\partial z} \; ; \; \gamma_{yz} = \frac{\partial v}{\partial z} + \frac{\partial w}{\partial y} \quad (8.5.20)$$

The relations between strain and displacement components for spherical coordinates with central symmetry are

$$\epsilon_r = \frac{\partial u_r}{\partial r} \; ; \; \epsilon_\theta = \frac{1}{r}\frac{\partial u_\theta}{\partial \theta} + \frac{u_r}{r} \; ; \; \gamma_{r\theta} = \frac{1}{2r}\left(\frac{\partial u_r}{\partial \theta} - u_\theta + r\frac{\partial u_\theta}{\partial r}\right) \quad (8.5.21)$$

and for cylindrical geometry with angular symmetry.

$$\epsilon_r = \frac{\partial u_r}{\partial r} \; ; \; \epsilon_\theta = \frac{u_r}{r} \; ; \; \epsilon_z = \frac{\partial u_z}{\partial z} \; ; \; \gamma_{rz} = \frac{\partial u_r}{\partial z} + \frac{\partial u_z}{\partial r} \quad (8.5.22)$$

The normal strains ϵ_r , ϵ_θ , and ϵ_z represent strains in the radial, angular, and axial direction, respectively. $\gamma_{r\theta}$ represents a shear strain on an angular face acting in the radial direction or vice versa. γ_{rz} is the shear strain on a radial face acting in the axial direction or vice versa.

By double differentiations of the compatibility equations it is possible to eliminate displacements from these equations and develop second-order differential equations yielding required relationships between normal and shear strains. This approach adds to the mathematical complexity of stress analysis and is not used in the approach developed in this text.

In the solution of stress problems it is necessary to know certain boundary conditions on the surface of the body under consideration. Frequently, the displacement of surfaces is a boundary condition readily determined. Consequently, the compatibility equations developed, Eqs. (8.5.17) through (8.5.22) are used for determining displacements of points in a body.

Stress-Strain Relations

In the discussion of the tensile test, it is pointed out that, under initial loading, strain is closely proportional to stress. As a consequence, for axial loading we wrote

$$\epsilon_z = \frac{\sigma_z}{E}$$

The modulus of elasticity E is a function of the type of material, amount of strain in the material, temperature, and possibly whether the stress is positive or negative. Typical values of E are given in Table 8.1.

Experience has shown that axial elongations in a material produce lateral contractions and axial contractions produce lateral elongations. Lateral deformations are a fixed fraction of the axial strain and this fraction is given by **Poisson's ratio** which has the symbol υ . For axial loading lateral strains are independent of direction in an isotropic material and therefore

$$\epsilon_x = \epsilon_y = -\upsilon\, \epsilon_z = -\upsilon \frac{\sigma_z}{E} \tag{8.5.23}$$

Poisson's ratio usually ranges from about 0.3 for metals to 0.5 for rubber. For υ equal to 0.5 there is no change in volume due to stresses.

Because of the linearity of the stress-strain relations, an increment of stress produces the same increment of strain independent of prior stresses. A normal stress in another coordinate direction produces normal strains which are directly added to strains produced by σ_z . Normal stresses produce only normal strains.

Shear stresses produce their corresponding shear strain components and for elastic materials they are related by

$$\gamma_{xy} = \frac{\tau_{xy}}{G} \quad ; \quad \gamma_{xz} = \frac{\tau_{xz}}{G} \quad ; \quad \gamma_{yz} = \frac{\tau_{yz}}{G} \tag{8.5.24}$$

where G is called the **shear modulus**. It can be proven that $G = E/(2+2\upsilon)$. Shear stresses produce only shear strains.

For a linear elastic, isotropic material with all components of stress

Table 8.1

Elastic constants for isotropic materials at room temperature [3].

Material	Composition	Modulus of elasticity E, 10^6 psi	Poisson's ratio, ν	Shear modulus G, 10^6 psi	Coeff. of linear expansion α, $10^{-6}/°F$	Density, lb/in.^{3}
Aluminum*	Pure and alloy	9.9–11.4	0.32–0.34	3.7–3.85	11.1–13.4	0.096–0.104
Brass*, †	60–70% Cu, 40–30% Zn	14.5–15.9	0.33–0.36	5.3–6.0	11.0–11.6	0.302–0.307
Copper*, †, ‡		17–18	0.33–0.36	5.8–6.7	9.2–9.4	0.323–0.324
Iron, cast†, ‡	2.7–3.6% C	13–21	0.21–0.30	5.2–8.2	5.8	0.251–0.265
Steel*, †	Carbon and low alloy	28–32	0.26–0.29	11.0–11.9	5.5–7.1	0.279–0.284
Stainless steel**, ‡	18% Cr, 8% Ni	28–30	0.30	10.6	8.3–9.4	0.276–0.286
Titanium*, ‡	Pure and alloy	15.4–16.6	0.34	6.0	4.9	0.163
Glass§	Various	7.2–11.5	0.21–0.27	3.8–4.7	3.3–5.3	0.086–0.14
Methyl methacrylate‖		0.35–0.5			50	0.042
Polyethylene‖		0.02–0.055			100	0.033
Rubber¶		0.00011–0.00060	0.50	0.00005–0.00020	70–110	0.036–0.045

* C. J. Smithells, "Metals Reference Book," Interscience Publishers, Inc., New York, 1955.
† L. S. Marks, "Mechanical Engineers Handbook," McGraw-Hill Book Company, Inc., New York, 1958.
‡ "Metals Handbook," American Society for Metals, Cleveland, 1948.
§ G. W. Morey, "Properties of Glass," Reinhold Publishing Corporation, New York, 1954, p. 16.
‖ *Modern Plastics*, Encyclopedia Issue, vol. 34, 1956.
¶ U.S. Rubber Co., "Engineering Properties of Rubber," Fort Wayne, Ind., 1950.
** C. L. Mantell, "Engineering Materials Handbook," McGraw-Hill Book Company, Inc., New York, 1958.

present, the stress-strain relations are given by

$$\epsilon_x = \frac{1}{E}[\sigma_x - \upsilon(\sigma_y + \sigma_z)]$$

$$\epsilon_y = \frac{1}{E}[\sigma_y - \upsilon(\sigma_x + \sigma_z)]$$

$$\epsilon_z = \frac{1}{E}[\sigma_z - \upsilon(\sigma_x + \sigma_y)]$$

$$\gamma_{xy} = \frac{\tau_{xy}}{G} \quad ; \quad \gamma_{xz} = \frac{\tau_{xz}}{G} \quad ; \quad \gamma_{yz} = \frac{\tau_{yz}}{G} \tag{8.5.25}$$

The six equations given by Eq. (8.5.25) are called the generalized Hooke's law. These equations are independent of the orthogonal coordinate system used. For spherical or cylindrical coordinates one simply substitutes the appropriate strains and stresses.

Equation (8.5.25) is not general enough for analysis of reactor systems because additional strains may be produced by heating and irradiation. These strains are always normal strains. For temperature changes, the normal strain at a point in any direction is given by

$$\epsilon = a(T - T_o) \tag{8.5.26}$$

where a is the linear coefficient of thermal expansion which is defined by $a \equiv (1/L)dL/dT$. T_o is a reference temperature at which the normal strain is said to be zero. The normal strains due to irradiation are designated ψ and depend upon type of radiation, type of material, and generally increase with irradiation time. Due to the linearity of the stress-strain relations, the effect of temperature and irradiation are added linearly to Eq. (8.5.25).

$$\epsilon_x = \frac{1}{E}[\sigma_x - \upsilon(\sigma_y + \sigma_z)] + a(T - T_o) + \psi$$

$$\epsilon_y = \frac{1}{E}[\sigma_y - \upsilon(\sigma_x + \sigma_z)] + a(T - T_o) + \psi$$

$$\epsilon_z = \frac{1}{E}[\sigma_z - \upsilon(\sigma_x + \sigma_y)] + a(T - T_o) + \psi$$

$$\gamma_{xy} = \frac{\tau_{xy}}{G} \quad ; \quad \gamma_{xz} = \frac{\tau_{xz}}{G} \quad ; \quad \gamma_{yz} = \frac{\tau_{yz}}{G} \tag{8.5.27}$$

In summary we have derived a set of 18 equations containing 18 unknowns. There are six equilibrium equations in terms of nine stress components, there are six compatibility equations containing three displacement components and six strains, and finally we have six stress-strain equations relating stresses and strains.

Simultaneous solution of the equations of elasticity is a rather tedious task. However, by a judicious choice of coordinate system many of the unknowns are zero. As an example, if the coordinate system is the principal axes, all shear stresses and shear strains are zero and the set of equations is reduced to nine equations containing nine unknowns. In many practical cases the number of equations and unknowns are greatly reduced by taking advantage of symmetry and properties known from conditions of the problem.

In order to determine the state of stress in a body it is necessary to know certain boundary conditions on the body surface. Typical boundary conditions are specification of forces acting on certain surfaces of the body, the condition that no forces act on certain surfaces, or that certain surfaces are restrained in given positions. As an example, for a specified temperature change a rectangular slab is allowed to expand free in the x-direction without any forces or restraints applied on the x-faces, while the y- and z-faces are held in their original position. Boundary conditions for this problem are σ_x is zero on the x-surface and v and w are zero on the y- and z-surfaces. Selection of proper boundary conditions requires a deep insight into the physical processes involved.

In many stress problems involved in reactor analysis, heating is a major contribution to the presence of stresses. Stresses due to temperature changes are called **thermal stresses**. The analysis of thermal stresses is somewhat simplified due to the linearity of the equations of elasticity. One can solve for the state of stress in a material assuming there are no external forces acting upon the body. Stresses found from this analysis are due solely to temperature changes. Next one applies the known external forces, and restraints on boundaries if necessary, and determines the stresses due to these factors. The state of stress in the material is then found by linear addition of the two types of stress contributions. In summary, in the analysis of thermal stresses in elastic material

$$\sigma = \sigma(\text{loads}) + \sigma(\text{thermal effects}) \tag{8.5.28}$$

No general approach exists for solving all stress problems that occur in nuclear reactor systems. There is too much reliance on good engineering judgment on the orientation of coordinates, specifying zero values for some normal and shear stresses, and specifying appropriate boundary conditions. The following procedure may be helpful in solving some of the simplified problems in reactor stress analysis:

(i) Find a suitable coordinate system for analyzing the problem. The geometry of the material will probably dictate the origin and type of coordinate system, i.e., using cylindrical coordinates for analyzing fuel rods.

(ii) Obtain the temperature distribution in the material from a heat transfer analysis.

(iii) Write down the appropriate equilibrium equations for the coordinate system chosen. From physical insight ascertain whether some of the stresses are zero or constants.

(iv) Write down the appropriate equations for geometric fit. These equations will relate strains to derivatives of displacements.

(v) Write down the stress-strain relations and substitute the appropriate temperature distributions found from step (ii) for the terms $\alpha(T - T_o)$.

(vi) Substitute the expressions for step (iv) into step (v) so as to eliminate the strains from the stress-strain relations.

(vii) Determine appropriate boundary conditions on the surface of the material under consideration. These boundary conditions are usually stresses at the surface or displacements of the surface. Surfaces that are completely unrestrained may have zero stresses on the surface while surfaces that are completely restrained have zero displacements of the surface.

(viii) Solve the set of equations and apply appropriate boundary conditions.

The results of a complete stress analysis yields the state of stress throughout the material considered. One still has to consider whether these stresses are sufficient to cause yielding of the material. Unfortunately, most of the tests used to determine yielding of materials are the simple tensile test in which a load is applied in the axial direction and no loads exist in lateral directions. Consequently, a criterion is needed for predicting possible failure when stresses exist in several coordinate directions. In a tensile test, if one of the coordinate axes is located in the direction of loading, the coordinates are principal axes of stress (see Sec. 8.2). Consequently, if Y denotes the stress at yielding, the principal stresses are $\sigma_1 = Y$, $\sigma_2 = 0$,and $\sigma_3 = 0$. For cases of two- or three-dimensional state of stress, the **Mises yield criterion** is that yielding can occur when the root mean square of the differences between principal stresses reaches the same value it has when yielding occurs in a tensile test [3]. This criterion is stated mathematically as

$$\sqrt{\frac{1}{2}\left[(\sigma_1 - \sigma_2)^2 + (\sigma_2 - \sigma_3)^2 + (\sigma_3 - \sigma_1)^2\right]} = Y \qquad (8.5.29)$$

There is reasonable experimental verification of this criterion. Figure 8.11 illustrates yielding under combined stresses in a thin-walled cylinder. In a thin-walled cylinder the radial stress ($\sigma_r = \sigma_3$) is small compared to the tangential stress ($\sigma_\theta = \sigma_2$) and the axial stress ($\sigma_z = \sigma_1$) .

In the United States the majority of the stress analysis of nuclear components is governed by the ASME Boiler and Pressure Vessel Code [5] which uses the maximum shear stress criteria for establishing maximum stresses in materials. The maximum shear stress criteria is that failure occurs when

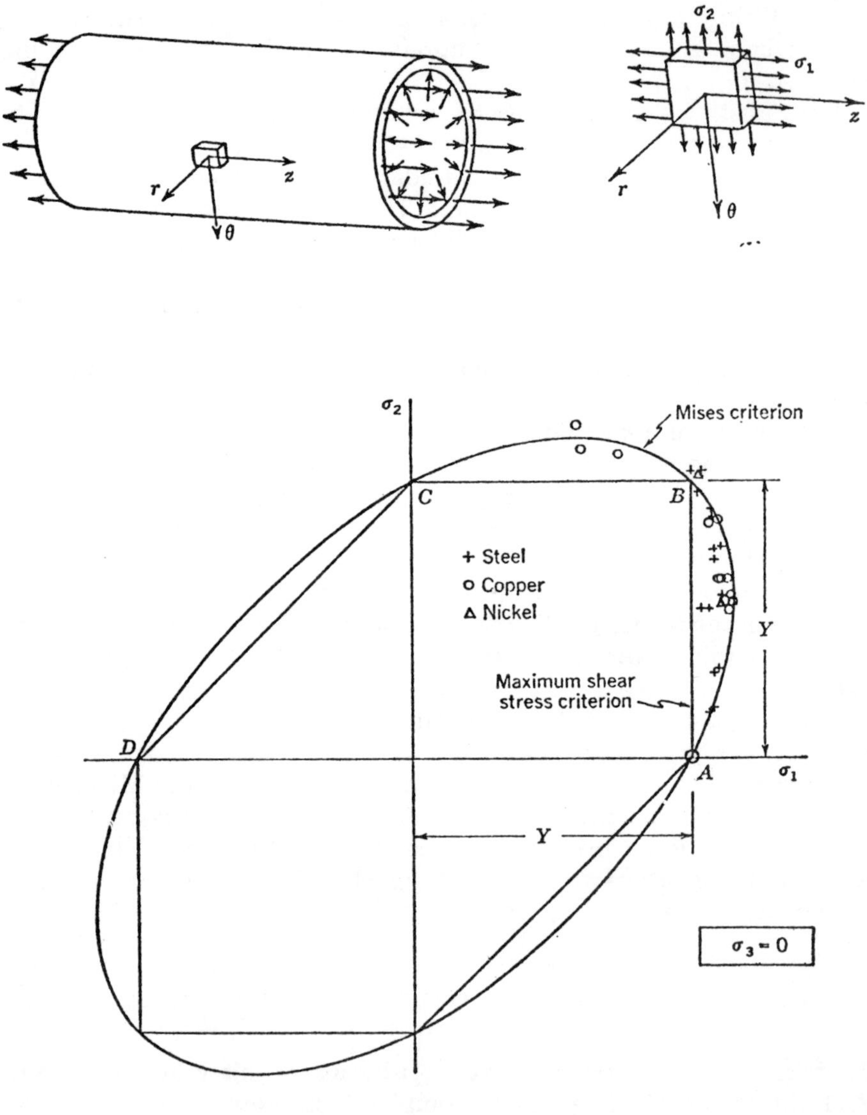

Fig. 8.11. Yielding of thin-walled tubes under combined stress [3].

$$\tau_{max} = Y/2 \tag{8.5.30}$$

Since

$$\tau_{max} = \frac{\sigma_{max} - \sigma_{min}}{2} = \frac{S}{2} \tag{8.5.31}$$

where σ_{max} and σ_{min} are maximum and minimum normal (principal) stresses. Because the stress intensity is the difference between principal stresses, the failure criteria is

$$S = Y \tag{8.5.32}$$

Consequently the ASME code allows the stress intensity to only approach some fraction (usually 1/3) of Y in order to insure failure not occur.

Other factors also need to be considered when analyzing material failure. Cycling of loads may cause failure at stresses below Y because of propagation of cracks during cycling. At high temperatures plastic deformations (strains) increase with time due to creep. This is important for stainless steel above 1000°F and zircaloy above 500°F [6]. Figure 8.12 shows anticipated time dependence of strain for a material under tension. In the linear portion of the curve, the strain may be correlated by [6]

$$\epsilon = A \sigma^m t \tag{8.5.33}$$

where A and m are constants dependent upon material and temperature. A is low at room temperatures, but increases rapidly above a critical temperature. The failure criteria is to stay below a **creep limit** which is the stress to which a material is subjected without creep exceeding a specific amount after a given time at operating temperature (i.e., creep strain of 0.01 at 100,000 hours). At temperatures at which steel vessels and pipes operate in water-cooled reactors, the creep limit is not a problem. However, for zircaloy the creep limit is a consideration in fuel-element design. Creep is also a problem for stainless steel in HTGRs.

9.6 Stresses in Fuels and Cladding

Because of the complexity of the equations of elasticity, simple

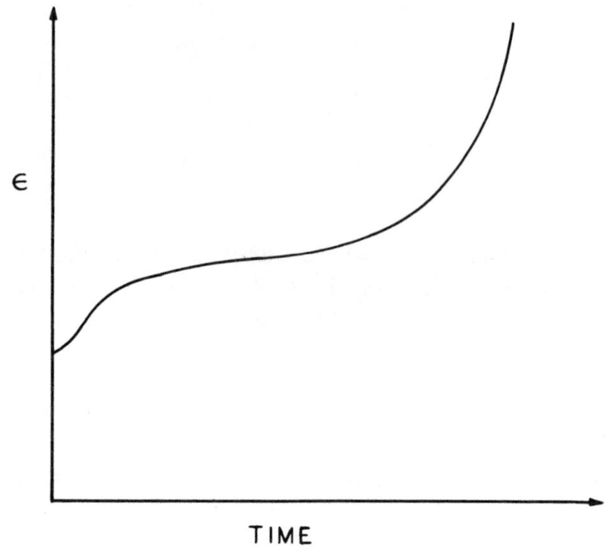

Fig. 8.12. Creep of metals under stress.

solutions are usually found for bodies with high symmetry and simple boundary conditions. In order to develop some understanding for approaches used in solving stress problems and relative importance of different types of stresses, a few problems are examined.

Conditions for Zero Thermal Stresses

It is of interest to know what conditions are necessary to insure that no thermal stresses exist in a body. From our compatibility and stress-strain relations for plane stress, we can write

$$\epsilon_x = \frac{\partial u}{\partial x} = \frac{1}{E}[\sigma_x - \upsilon\sigma_y] + a(T - T_o) \quad (8.6.1a)$$

$$\epsilon_y = \frac{\partial v}{\partial y} = \frac{1}{E}[\sigma_y - \upsilon\sigma_x] + a(T - T_o) \quad (8.6.1b)$$

$$\gamma_{xy} = \frac{\partial v}{\partial x} + \frac{\partial u}{\partial y} = \frac{\tau_{xy}}{G} \quad (8.6.1c)$$

If the body is free to expand in an unrestrained manner and we assume no stresses, then

$$\frac{\partial u}{\partial x} = a(T - T_o) \qquad (8.6.2a)$$

$$\frac{\partial v}{\partial y} = a(T - T_o) \qquad (8.6.2b)$$

$$\frac{\partial v}{\partial x} + \frac{\partial u}{\partial y} = 0 \qquad (8.6.2c)$$

Differentiating Eq. (8.6.2a) twice with respect to y and Eq. (8.6.2b) twice with respect to x gives

$$\frac{\partial^2}{\partial x^2}\left(\frac{\partial v}{\partial y}\right) = a\frac{\partial^2 T}{\partial x^2} \qquad (8.6.3a)$$

$$\frac{\partial^2}{\partial y^2}\left(\frac{\partial u}{\partial x}\right) = a\frac{\partial^2 T}{\partial y^2} \qquad (8.6.3b)$$

Adding and rearranging Eq. (8.6.3) gives

$$\frac{\partial^2}{\partial x\,\partial y}\left(\frac{\partial u}{\partial y} + \frac{\partial v}{\partial x}\right) = a\left(\frac{\partial^2 T}{\partial x^2} + \frac{\partial^2 T}{\partial y^2}\right) \qquad (8.6.4)$$

However, inspection of Eq. (8.6.2c) shows the left side of Eq. (8.6.4) is zero. Therefore, a condition for no thermal stresses for plane stress is

$$\nabla^2 T = 0 \qquad (8.6.5)$$

Equation (8.6.5) tells us that in order to have no thermal stresses in a plane, the body must be at steady state and contain no heat sources or sinks. Any stresses that result in the body are due to imposed boundary conditions and are not due to temperature gradients within the body. Further requirements to assure no stresses for the two-dimensional body are that there be no stresses on the surface and that the body be simply connected. For two-dimensions, simply connected means the boundaries of the body can be drawn without lifting the pencil from the paper.

The arguments for no stresses in two-dimensions are extended to three-dimensions. A requirement is that the temperature distribution be linear such as $T - T_o = ax + by + cz$ or $axyz$

Thermal Stresses in Fuel or Control Plates

The thin plate illustrated in Fig. 8.13 represents a control or fuel element of a nuclear reactor shaped like a plate whose thickness "2a" is small compared with its width or length. In the absence of a neutron flux, the entire plate is supposed to have no stresses or strains and be at uniform temperature T_o . In the presence of a neutron flux, heat is generated inside the plate; its temperature is raised to T, a function of x, y, and z; and a system of stresses and strains is set up. The object of thermal stress analysis in this case is to determine how these stresses and strains vary in position.

In a uniform neutron flux, the plate temperature at distances greater than "2a" away from ends and edges will be a function of x only and is obtained by solving the one-dimensional heat-conduction equation

$$-k \frac{d^2T}{dx^2} = q''' \tag{8.6.6}$$

where q''' is the volumetric heat generation rate. For uniform q''' the temperature distribution in the fuel material is given by

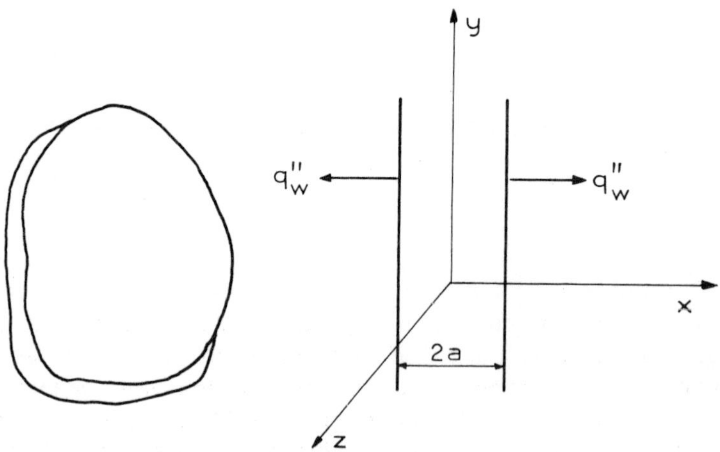

Fig. 8.13. Infinite flat plate with internal heat generation

$$T(x) = T_{max} - \frac{q''' x^2}{2k} \tag{8.6.7}$$

where T_{max} is the maximum temperature in the fuel material which is at the center ($x = 0$).

We wish to find stresses in the plate at distances away from the end or edge when there are no restraints on the plate faces. Exact solution of the equations of elasticity for this case is difficult because the plate faces, originally plane, become curved owing to the greater expansion at the plate center where temperatures are higher.

To make the problem tractable we solve instead the problem in which the y- and z-faces of the plate are restrained in their original positions, so that

$$\epsilon_y = \epsilon_z = 0 \tag{8.6.8}$$

In the position of the plate in which T is a function of x, ϵ_x depends only on x, and there is no shearing strain because the only displacement possible is in the x-direction.

For the conditions formulated, shear stresses are zero because there are no shear strains. Consequently, the equilibrium equations are reduced to

$$\frac{\partial \sigma_x}{\partial x} = 0 \quad ; \quad \frac{\partial \sigma_y}{\partial y} = 0 \quad ; \quad \frac{\partial \sigma_z}{\partial z} = 0 \tag{8.6.9}$$

Since σ_x is zero on the plate faces, σ_x is zero everywhere in the material. In addition, for conditions of no strains in the y- and z-directions, σ_y and σ_z are functions only of x and are equal to each other. This is because we should be able to translate the coordinate system in the y-z plane with no changes or rotate the coordinate system about the x-axis with no apparent change. Therefore

$$\sigma_y = \sigma_z = f(x) \tag{8.6.10}$$

The Hooke's law equations reduce to

$$\epsilon_x = -\frac{\upsilon}{E} [\sigma_y + \sigma_z] + a(T - T_o) \tag{8.6.11}$$

$$\epsilon_y = 0 = \frac{1}{E}[\sigma_y - \upsilon\sigma_z] + a(T - T_o) \tag{8.6.12}$$

$$\epsilon_z = 0 = \frac{1}{E}[\sigma_z - \upsilon\sigma_y] + a(T - T_o) \tag{8.6.13}$$

Equations (8.6.12) and (8.6.13) are easily solved for σ_y and σ_z

$$\sigma_y = \sigma_z = -\frac{Ea(T - T_o)}{1 - \upsilon} \tag{8.6.14}$$

From the temperature distribution (Eq. 8.6.7) in the material we have

$$\sigma_y = \sigma_z = -\frac{Ea}{1 - \upsilon}\left[T_{max} - T_o - \frac{q''' x^2}{2k}\right] \tag{8.6.15}$$

The strain in the x-direction is then given by

$$\epsilon_x = \frac{1 + \upsilon}{1 - \upsilon}a(T - T_o) = \frac{1 + \upsilon}{1 - \upsilon}a\left[T_{max} - T_o - \frac{q''' x^2}{2k}\right] \tag{8.6.16}$$

The minus sign on Eq. (8.6.15) means the plate is under compression in the y- and z-directions.

The average compressive stress normal to the plate in the y- or z-directions is

$$\overline{\sigma_y} = \frac{1}{2a}\int_{-a}^{a}\sigma_y dx = \frac{Ea}{2a(1 - \upsilon)}\int_{-a}^{a}\left[T_{max} - T_o - \frac{q''' x^2}{2k}\right] dx$$

$$= -\frac{Ea}{1 - \upsilon}\left[T_{max} - T_0 - \frac{q''' a^2}{6k}\right] \tag{8.6.17}$$

Next we imagine that the y- and z-faces of the plate are allowed to move apart, while remaining plane and parallel, until the average compressive stresses in the y- and z-directions drop to zero. Under these conditions, from the principle of superposition

$$\sigma'_y = \sigma_y - \overline{\sigma_y} = \frac{Ea\, q''' a^2}{2(1 - \upsilon)k}\left[\left(\frac{x}{a}\right)^2 - \frac{1}{3}\right] \tag{8.6.18}$$

with an identical equation for σ_z'. It should be noted that Eq. (8.6.18) does not contain either T_o or T_{max}. This is to be expected, since these temperature distributions satisfy $\nabla^2 T = 0$ and for no restraints there is no thermal stresses. However, if surfaces are restrained, T_o and T_{max} appear as a contributor to the stresses as shown in Eq. (8.6.15).

Finally, **all** restraints are removed from the y- and z-faces of the plate. Near these faces, a complex system of stresses and strains are set up in order to satisfy the boundary conditions that $\sigma_y' = 0$ at the y-face and $\sigma_z' = 0$ at the z-face. But when we consider the plate at distances substantially greater than "2a" from these faces, **Saint Venant's principle** permits us to state that the stress system is not changed from Eq. (8.6.18) because the total force at each face of the plate is not changed. Saint Venant's principle states that, in thin plates or long narrow rods, at distances away from the edge or end substantially greater than the thickness, the pattern of stresses and displacements depends only on the total applied force to the edges or ends and not at all on the detailed way in which stresses are applied to the edges or ends. Consequently, Eq. (8.6.18) is the desired solution for thermal stresses in a plate, heated internally by neutron absorption or an electric current and unrestrained on all faces. The Mohr's circle of stress for this case of plane stress is a point. Thus there are no shear stresses and all stresses in any direction lying on the y-z plane are the same.

Figure 8.14 illustrates the stress distribution across the flat plate. The edges of the plate are in tension and the center is in compression. The maximum tensile stress (x = a) is given by

$$(\sigma_t)_{max} = \frac{E\alpha a^2 q'''}{3(1 - \upsilon)k} \tag{8.6.19}$$

and the maximum compressive stress (x = 0) is given by

$$(\sigma_c)_{max} = -\frac{E\alpha a^2 q'''}{6(1 - \upsilon)k} \tag{8.6.20}$$

Thus the tensile stress on the surface is the largest stress in the body.

By using Eq. (8.6.19), we can examine from a thermal stress point of view the relative merits of some reactor materials. Table 8.2 illustrates the ratio $\sigma_{yield}/[E\alpha/k(1 - \upsilon)]$ for several materials. The larger the value of this ratio, the larger the permissible volumetric heat generation rate before yielding. As shown, graphite has very good combined properties for resisting thermal stresses.

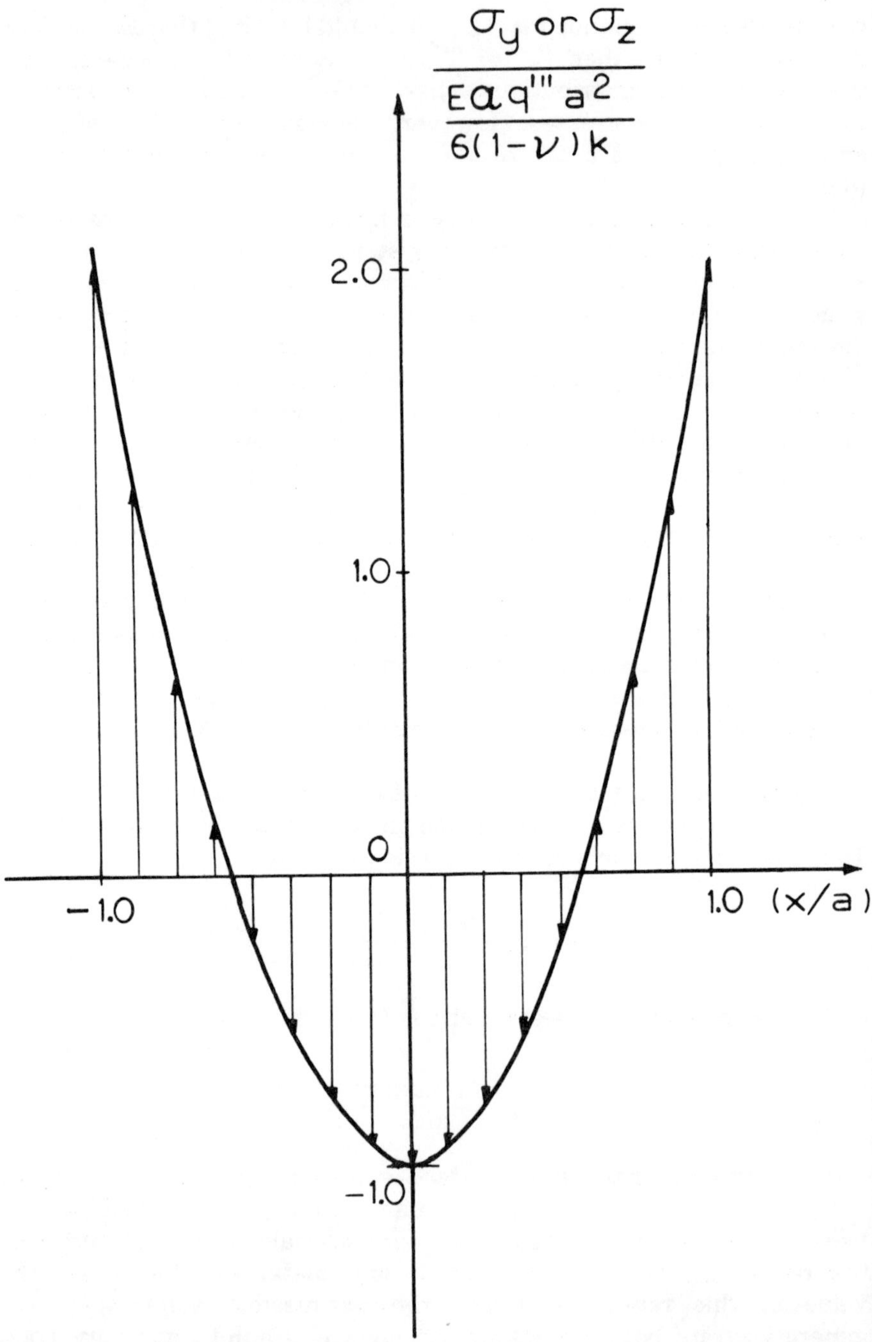

Fig. 8.14. Thermal stress distributions in an unrestrained infinite flat plate with internal heat generation.

Table 8.2

Thermal stress performance of reactor materials [7].

Material	$\frac{E\alpha}{k(1-\nu)}$	σ_{yield} (psi)	$\frac{\sigma_{yield}}{\frac{E\alpha}{k(1-\nu)}}$
Aluminum (2S)	1.4	3,500	2,500
Beryllium (extruded)	6.0	41,000	6,800
BeO	14.6	15,000	1,025
Graphite	0.1	3,200	32,000
Stainless Steel (Type 347)	33.8	69,000	2.040
Uranium	24.8	29,000	11,170
22% U-Al	2.0	8,100	4,050
Zirconium	4.6	15,600	3,400

Thermal Stresses in an Axially Symmetric Cylinder

Next we consider a cylinder of radius, R, initially without stress or strain at T_o, and then heated to some axially symmetric temperature T(r), which is independent of θ and z. An example would be a cylindrical fuel or control rod at a substantial distance away from the end. We wish to find the stresses due to temperature changes when the rod is unrestrained at the ends. To do this we first consider the rod with ends fixed and then move the ends apart while remaining parallel until the mean compressive stress is zero. Finally the ends are completely freed and stresses away from the ends are found using Saint Venant's principle.

For an axially symmetric cylinder the heat conduction equation is given by

$$-\frac{k}{r}\frac{d}{dr}\, r\left(\frac{dT}{dr}\right) = q''' \tag{8.6.21}$$

For uniform heat generation the solution to Eq. (8.6.21) is

$$T(r) = T_{max} - \frac{q''' r^2}{4k} \tag{8.6.22}$$

where T_{max} is the maximum fuel temperature at the centerline.

Under the imposed condition of end restraints, $\epsilon_z = 0$ and all shearing stresses or strains are zero. The variables left to be determined are σ_r, σ_θ, σ_z, ϵ_θ, and ϵ_r.

For conditions of no body forces, the equilibrium equations are given by

$$\frac{\partial \sigma_r}{\partial r} + \frac{1}{r}(\sigma_r - \sigma_\theta) = 0 \qquad (8.6.22a)$$

$$\frac{\partial \sigma_z}{\partial z} = 0 \qquad (8.6.22b)$$

Equation (8.6.22b) is used to show that σ_z is not a function of z. In addition from symmetry of the problem, σ_z is not a function of θ and, therefore, depends only upon r. From symmetry arguments, one can also conclude that σ_r and σ_θ are functions only of r. Therefore, since all stresses are functions of only the radial coordinate, the strains are also functions of only the radial coordinate because of their relations to stresses through the stress-strain equations.

For the axially symmetric cylinder, the geometric compatibility equations are given by

$$\epsilon_r = \frac{du_r}{dr} \quad ; \quad \epsilon_\theta = \frac{u_r}{r} \qquad (8.6.23)$$

We can eliminate u_r from these equations and obtain a relation between the two strains.

$$\epsilon_r = \frac{d}{dr}(r\epsilon_\theta) \qquad (8.6.24)$$

The last set of equations required for the analysis of the cylinder are the Hooke's law relations

$$\epsilon_r = \frac{1}{E}[\sigma_r - \upsilon(\sigma_\theta - \sigma_z)] + a(T -- T_o) \qquad (8.6.25)$$

$$\epsilon_\theta = \frac{1}{E}[\sigma_\theta - \upsilon(\sigma_r + \sigma_z)] + a(T - T_o) \qquad (8.6.26)$$

$$\epsilon_z = 0 = \frac{1}{E}[\sigma_z - \upsilon(\sigma_r + \sigma_\theta)] + a(T - T_o) \qquad (8.6.27)$$

We now have a sufficient number of equations for solving the unknown stresses (σ_r , σ_θ , and σ_z) and strains (ϵ_r and ϵ_θ). These equations are Eqs. (8.6.22a), (8.6.24), (8.6.25), (8.6.26), and (8.6.27). In addition, we need Eq. (8.6.22) for the temperature distribution. It should be noted that the equations of elasticity developed are independent of temperature distribution as long as it depends only on the radial coordinate. Since the rod is unrestrained in the radial direction, σ_r is zero on the surface of the rod ($r = R$). A further boundary condition required is that σ_r is finite at $r = 0$.

Equation (8.6.27) is solved for σ_z and Eq. (8.6.22a) is solved for σ_θ. With these relations σ_θ and σ_z are eliminated from Eqs. (8.6.25) and (8.6.26) and the two strains are then given in terms of σ_r. The rearranged equations are then substituted into Eq. (8.6.24) to give the following differential equation for σ_r :

$$r\frac{d^2\sigma_r}{dr^2} + 3\frac{d\sigma_r}{dr} + \frac{Ea}{1-\upsilon}\frac{dT}{dr} = 0 \tag{8.6.28}$$

By multiplying Eq. (8.6.28) by r^2, substituting the expression for dT/dr from Eq. (8.6.22), and rearranging, Eq. (8.6.28) becomes

$$\frac{d}{dr}\left(r^3\frac{d\sigma_r}{dr}\right) = \frac{Eaq'''}{(1-\upsilon)2k}r^3 \tag{8.6.29}$$

Integration of Eq. (8.6.29) twice yields

$$\sigma_r = -\frac{Eaq'''}{16(1-\upsilon)k}[R^2 - r^2] \tag{8.6.30}$$

By use of Eq. (8.6.22a) the tangential stress is give by

$$\sigma_\theta = -\frac{Eaq'''}{16(1-\upsilon)k}[R^2 - 3r^2] \tag{8.6.31}$$

and from Eqs. (8.6.22) and (8.6.27) the axial stress is given by

$$\sigma_z = -\frac{\upsilon\ Eaq'''}{8(1-\upsilon)k}[R^2 - 2r^2] - Ea\left[T_{max} - \frac{q'''r^2}{4k} - T_o\right] \tag{8.6.32}$$

It should be noted that only σ_z is dependent upon temperature T_o. This is to be expected because restraints occur only in the axial direction. The average axial compression stress in the fully restrained condition is given by

$$\bar{\sigma}_z = \frac{1}{\pi R^2}\int_o^R 2\pi r\sigma_z\,dr = -Ea\left[T_{max} - \frac{q'''R^2}{8k} - T_o\right] \tag{8.6.33}$$

The restraints on the end of the rod are now removed so the average axial compressive stress is zero. By superposition

$$\sigma_z' = \sigma_z - \bar{\sigma}_z = - \frac{E\alpha q'''}{8(1 - \upsilon)k} [R^2 - 2r^2] \tag{8.6.34}$$

Finally all restraints are removed from the end of the rod. By applying Saint Venant's principle, Eq. (8.6.34) represents the axial stress for an unrestrained rod at distances substantially greater than R from the rod end.

Figure 8.15 illustrates the distribution of stresses in an unrestrained cylinder. The greatest tensile stress occurs on the rod surface in the

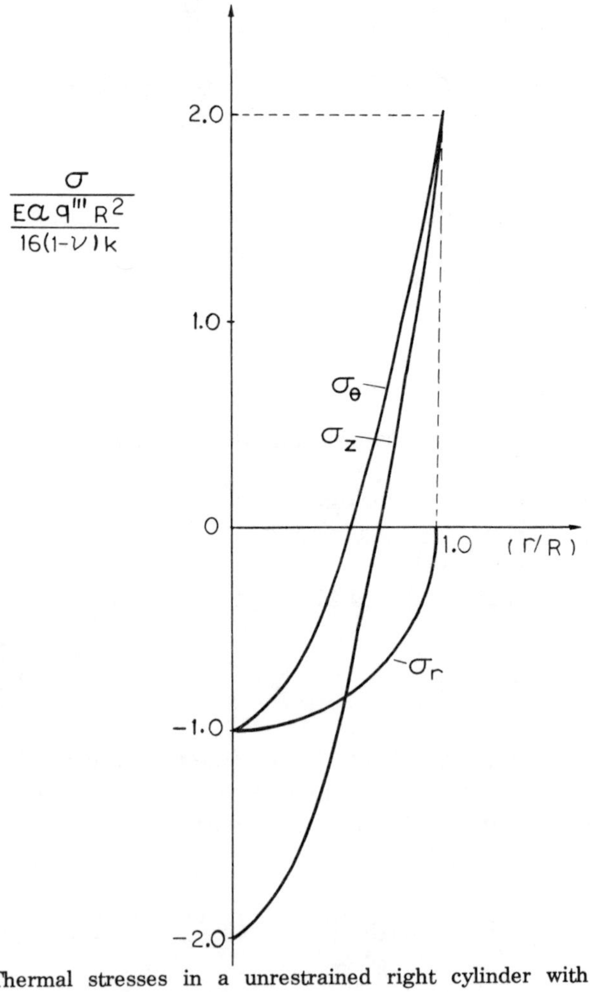

Fig. 8.15. Thermal stresses in a unrestrained right cylinder with internal heat generation.

axial and tangential directions. The greatest compressive stress is at the center of the rod in the axial direction. Since the coordinate system chosen is the principal axes, the Mises yield criterion can be used to determine the critical region where failure is likely to occur. By this criterion yielding occurs on the rod surface whenever σ_z or σ_θ equal the yield stress.

In ductile materials, when stresses exceed the elastic limit, plastic strains take place and actual stresses are much smaller than predicted by using the equations of elasticity. The plastic flow in a rod occurs first on the surface and proceed inward thus causing smaller stresses than our analysis indicates. Once plastic flow has occured in a material, thermal stresses are not relieved in the same manner as elastic stresses as the heating is reduced. If heating is eliminated in a material which has yielded plastically, **residual stresses** are left in the material which initially is stress free. With successive cycles of heating and cooling, residual stresses may build up to a value where failure may take place.

Thermal Stresses in an Axially Symmetric Hollow Cylinder

In hollow cylinder cladding materials, thermal stresses occur because of temperature gradients in the material. In a hollow cylinder without internal heat generation, the temperature distribution is given by

$$T(r) - T_w = \Delta T \frac{\ln R/r}{\ln R/r_o} \tag{8.6.35}$$

where T_w is the outer cladding wall temperature, ΔT is the temperature difference between inner and outer surfaces, and R and r_o are the outer and inner radii, respectively.

The same approach is used to solve for thermal stresses in the unrestrained hollow cylinder as used for the solid cylinder. The mathematics is slightly more complicated because of the logarithmic temperature distribution. For no restraints in either the radial or axial directions, the thermal stresses are given by

$$\sigma_r = -\frac{E\alpha\Delta T}{2(1-\upsilon)\ln R/r_o}\left[\ln R/r + \frac{r_o^2}{R^2 - r_o^2}\left(1 - \frac{R^2}{r^2}\right)\ln\frac{R}{r_o}\right] \tag{8.6.36}$$

$$\sigma_\theta = \frac{E\alpha\Delta T}{2(1-\upsilon)\ln R/r_o}\left[1 - \ln\frac{R}{r} - \frac{r_o^2}{R^2 - r_o^2}\left(1 + \frac{R^2}{r^2}\right)\ln\frac{R}{r_o}\right] \tag{8.6.37}$$

$$\sigma_z = \frac{E\alpha\Delta T}{2(1-\upsilon)\ln R/r_o}\left[1 - 2\ln\frac{R}{r} - \frac{2r_o^2}{R^2 - r_o^2}\ln\frac{R}{r_o}\right] \tag{8.6.38}$$

In reactor cladding materials ΔT is positive and σ_r is compressive throughout the material and zero on the surfaces. The other stresses σ_θ and σ_z are compressive near the inner surface and tensile near the outer surface. The largest values of the compressive and tensile stresses occur at the inner and outer surfaces. At the inner surface $(r = r_o)$

$$(\sigma_\theta)_{r,o} = (\sigma_z)_{r,o} = \frac{E\alpha\Delta T}{2(1 - \nu) \ln R/r_o} \left[1 - \frac{2R^2}{R^2 - r_o^2} \ln \frac{R}{r_o}\right] \tag{8.6.39}$$

and at the outer surface $(r = R)$

$$(\sigma_\theta)_R = (\sigma_z)_R = \frac{E\alpha\Delta T}{2(1 - \nu) \ln R/r_o} \left[1 - \frac{2r_o^2}{R^2 - r_o^2} \ln \frac{R}{r_o}\right] \tag{8.6.40}$$

For thin-walled cylinders in which $R/r_o = 1 + m$, Eqs. (8.6.39) and (8.6.40) are approximated by [8]

$$(\sigma_\theta)_{r,o} = (\sigma_z)_{r,o} = -\frac{E\alpha\Delta T}{2(1 - \nu) \ln R/r_o} (1 + m/3) \tag{8.6.41}$$

$$(\sigma_\theta)_R = (\sigma_z)_R = \frac{E\alpha\Delta T}{2(1 - \nu) \ln R/r_o} (1 - m/3) \tag{8.6.42}$$

Further stresses may exist in hollow cylinder cladding materials due to both internal and external forces acting upon the surfaces of the cylinder. The internal forces may be due to expansion of fuel pellets and the build up of fission product gas pressure. External forces are usually due to hydrostatic pressure of the reactor coolant. Stresses due to these pressures are determined by use of the equilibrium, geometric compatibility, and stress-strain equations. For a hollow cylinder of inner radius r_o and outer radius R subjected to an internal pressure p_o and external pressure p, the boundary conditions on the radial stress are

$$\sigma_r = -p_o \quad \text{at} \quad r = r_o \quad \text{and} \quad \sigma_r = -p \quad \text{at} \quad r = R \tag{8.6.43}$$

For a hollow cylinder with no heat transfer through the cylinder walls, Eq. (8.6.28) with $dT/dr = 0$ is used for determining radial stresses. The solution of the homogeneous portion of Eq. (8.6.28) is

$$\sigma_r = c_1 + c_2/r^2 \tag{8.6.44}$$

The integration constants are evaluated by applying the boundary conditions Eq. (8.6.43). This gives

$$\sigma_r = -\frac{p_o}{1-(r_o/R)^2}\left[\left(\frac{r_o}{r}\right)^2 - \left(\frac{r_o}{R}\right)^2 + \frac{p}{p_o}\left(1-(r_o/r)^2\right)\right] \tag{8.6.45}$$

Substituting Eq. (8.6.45) into Eq. (8.6.22a) gives for the tangential stress

$$\sigma_\theta = \frac{p_o}{1-(r_o/R)^2}\left[(r_o/r)^2 + (r_o/R)^2 - \frac{p}{p_o}\left(1+(r_o/r)^2\right)\right] \tag{8.6.46}$$

For conditions in which the cylinder is free to move in the axial direction, a force balance on the cylinder end gives for the axial stress

$$\sigma_z = \frac{p_o}{1-(r_o/R)^2}\left[(r_o/R)^2 - p/p_o\right] = \text{constant} \tag{8.6.47}$$

For a thick-walled cylinder with no external pressure ($p = 0$), the maximum stresses are at $r = r_o$

$$(\sigma_r)_{max} = -p_o \text{ and } (\sigma_\theta)_{max} = \frac{p_o[1+(r_o/R)^2]}{1-(r_o/R)^2} \tag{8.6.48}$$

For a thin-walled cylinder with $R = r_o(1 + m)$ where m is small compared to one, the tangential (hoop) stress at a large distance from the end is

$$\sigma_\theta = \frac{p_o(1-m)}{m} \approx \frac{p_o r_o}{t} \tag{8.6.49}$$

where t is the cylinder thickness.

Tangential stresses in cylinders under pressure are much larger than radial stresses. For conditions of combined heat transfer and internal or external pressure, the stresses are obtained by linear addition of thermal stresses and stresses due to hydrostatic pressure.

Thermal Stresses in a Sphere

Next consider a sphere of radius R, initially without stress or strain at T_o, and then heated to some spherically symmetric temperature distribution T(r) which is independent of the two angular coordinates. An example would be a spherical fuel particle which is proposed for use in pebble bed nuclear reactors. We find the stress distributions for conditions of no restraints on the sphere surface or $\sigma_r = 0$ at r = R.

For steady state, the heat conduction equations with spherical symmetry is given by

$$\frac{d^2T}{dr^2} + \frac{2}{r}\frac{dT}{dr} + \frac{q'''}{k} = 0 \tag{8.6.50}$$

For uniform heat generation, the solution to Eq. (8.6.50) is

$$T = T_{max} - \frac{q''' r^2}{6k} \tag{8.6.51}$$

where T_{max} is the maximum temperature at the sphere center.

From symmetry there are no shear stresses, only normal stresses which are divided into radial and tangential components. The tangential stresses are the same from symmetry. The variables to be determined are σ_r, σ_θ, ϵ_r, and ϵ_θ.

For no body forces or shear stresses, the equilibrium equations for the symmetric sphere are given by

$$\frac{d\sigma_r}{dr} + \frac{2}{r}(\sigma_r - \sigma_\theta) = 0 \tag{8.6.52}$$

$$\frac{d\sigma_\theta}{d\theta} = 0 \tag{8.6.53}$$

Equation (8.6.53) is used to infer that σ_θ is independent of θ and we can then show that all the other variables are independent of θ. This is an expected result because physically one would expect all stresses and strains to be only a function of the radial coordinate.

For spherical symmetry, the geometric compatibility equations are given by

$$\epsilon_r = \frac{du_r}{dr} \; ; \quad \epsilon_\theta = \frac{u_r}{r} \tag{8.6.54}$$

Eliminating u_r from these equations gives

$$\epsilon_r = \frac{d}{dr}(r\epsilon_\theta) \tag{8.6.55}$$

The Hooke's law relations for the sphere are given by

$$\epsilon_r = \frac{1}{E}[\sigma_r - 2\upsilon\sigma_\theta] + a(T - T_o) \tag{8.6.56}$$

$$\epsilon_\theta = \frac{1}{E}[\sigma_\theta - \upsilon(\sigma_r + \sigma_\theta)] + a(T - T_o) \tag{8.6.57}$$

We now have a set of four equations (Eqs. 8.6.52, 8.6.55, 8.6.56, and 8.6.57) in terms of four unknowns (σ_r , σ_θ , ϵ_r , and ϵ_θ). In addition, we need Eq. (8.6.51) for the temperature distribution. It is desirable to eliminate σ_θ from these equations because the only known boundary conditions for the unrestrained sphere is $\sigma_r = 0$ at r = R. In addition, we require that σ_r is finite at the sphere center.

Equation (8.6.52) is solved for σ_θ and used to eliminate σ_θ from Eqs. (8.6.56) and (8.6.57). Substituting the new forms of Eqs. (8.6.56) and (8.6.57) into Eq. (8.6.55) and rearranging gives

$$r\frac{d^2\sigma_r}{dr^2} + 4\frac{d\sigma_r}{dr} + \frac{2Ea}{(1-\upsilon)}\frac{dT}{dr} = 0 \tag{8.6.58}$$

By multiplying Eq. (8.6.58) by r^3 , substituting the expression for dT/dr from Eq. (8.6.51), and rearranging, Eq. (8.6.58) becomes

$$\frac{d}{dr}\left(r^4\frac{d\sigma_r}{dr}\right) = \frac{2Eaq'''}{3(1-\upsilon)k}r^4 \tag{8.6.59}$$

Integration of Eq. (8.6.59) twice and substituting the boundary conditions σ_r is finite at r = 0 and $\sigma_r = 0$ at r = R gives

$$\sigma_r = -\frac{Eaq'''}{15(1-\upsilon)k}[R^2 - r^2] \tag{8.6.60}$$

By use of Eq. (8.6.52) the tangential stress is given by

$$\sigma_\theta = \frac{Eaq'''}{15(1-\upsilon)k}[2r^2 - R^2] \tag{8.6.61}$$

It should be noted that neither stress component is dependent upon T_{max} or T_o . This is because no restraints were put on the surface of the sphere.

Figure 8.16 illustrates the distribution of stresses in the unrestrained sphere. The greatest tensile stress occurs on the sphere surface in the angular direction. The greatest compressive stress occurs at the sphere center with both radial and angular stresses being the same. Since the coordinate system is the principal axes, the Mises yield criterion can be used to determine the region where failure is most likely to occur. By this criterion yielding will occur on the surface of the sphere whenever the tangential stress equals the yield stress.

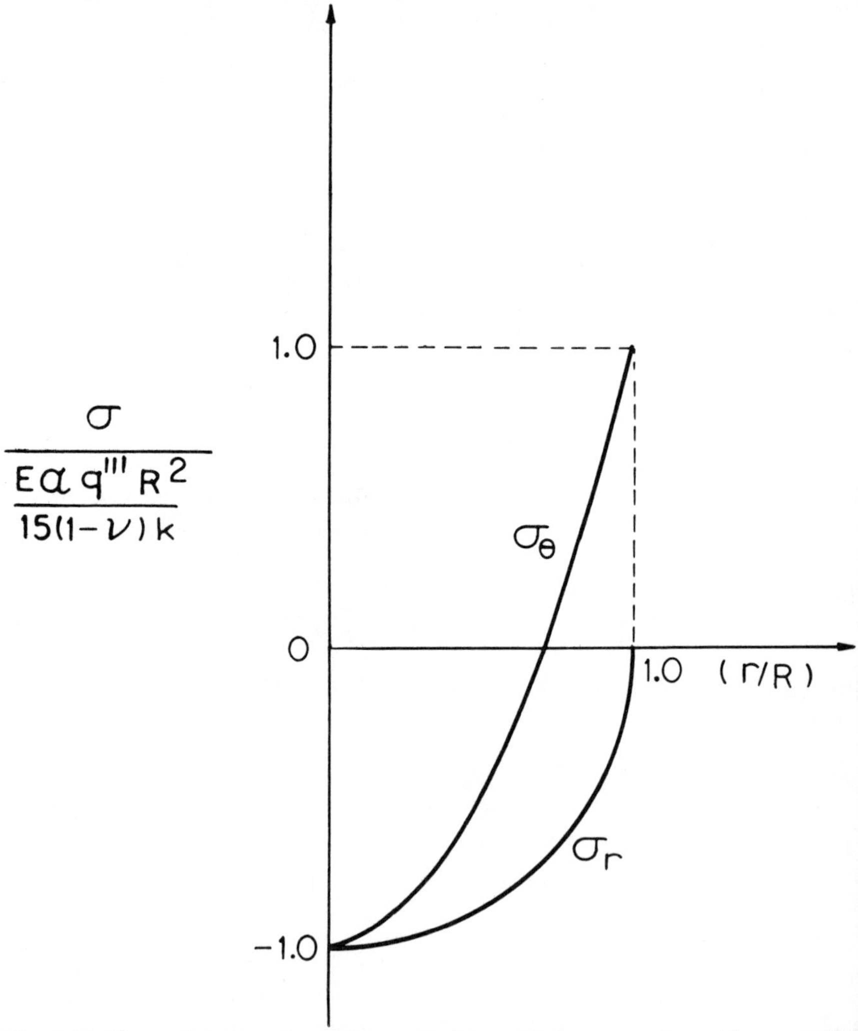

Fig. 9.16. Thermal stresses in a sphere with internal heat generation and no restraints

Example 8.1

We can examine the power limits placed upon a sphere of uranium metal in order to prevent yielding. For uranium metal the yield stress is 29,000 psi. The maximum stress at the surface is given by

$$(\sigma_\theta)_{max} = \frac{Ea q''' R^2}{15(1 - \upsilon)k} \qquad (8.6.62)$$

For uranium metal $Ea/k(1 - \upsilon) = 24.8 \frac{psi}{Btu/ft\text{-}hr}$. Therefore

$$q'''_{max} = \sigma_{yield} \frac{15(1 - \upsilon)k}{Ea R^2} \qquad (8.6.63)$$

For a sphere of radius 1/2 in., the maximum volumetric heat generation rate is

$$q'''_{max} = \frac{29{,}000 \times 15}{24.8 \times (1/24)^2}$$

$$= 1.01 \times 10^7 \text{ Btu/hr-ft}^3$$

The power produced in the sphere is given by

$$\text{Power} = q'''_{max}(4/3)\pi R^3 \qquad (8.6.64)$$

$$= 3060 \text{ Btu/hr} = 0.9 \text{ kW}$$

For the example given, the sphere is free to expand without restraints on the surface. However, radial forces may be exerted on the surface of the sphere due to pressure from a coolant. The elastic stresses due to these forces are obtained by solving Eq. (8.6.58) with $dT/dr = 0$. The boundary conditions are now σ_r is finite at $r = 0$ and $\sigma_r = -p$ at $r = R$. The solution to Eq. (8.6.58) is simply

$$\sigma_r = -p \qquad (8.6.65)$$

From Eq. (8.6.52), the solution for the tangential stress is

$$\sigma_\theta = -p \qquad (8.6.66)$$

With both internal heat generation and hydrostatic pressure, the radial stress is found by adding Eqs. (8.6.60 and 8.6.65) and the tangential stress is found from adding Eqs. (8.6.61 and 8.6.66).

9.7 Stresses in Pressure Vessels

Pressure vessels are vital components in nuclear reactor systems and, consequently, a great deal of attention is given to their design. In the United States the techniques for stress analysis and the application of design limits is governed by the ASME Boiler and Pressure Vessel Code, Section III—Nuclear Vessels [5]. The boiler code is a formidable document and only a few examples of analysis are presented in this section.

Thin-wall Vessel

Consider a thin-walled cylinder of radius R and wall thickness t with an internal pressure p_o placed in an environment with negligible external (atmospheric) pressure. The stresses are found by considering the forces acting upon the cylindrical section shown in Fig. 8.17.

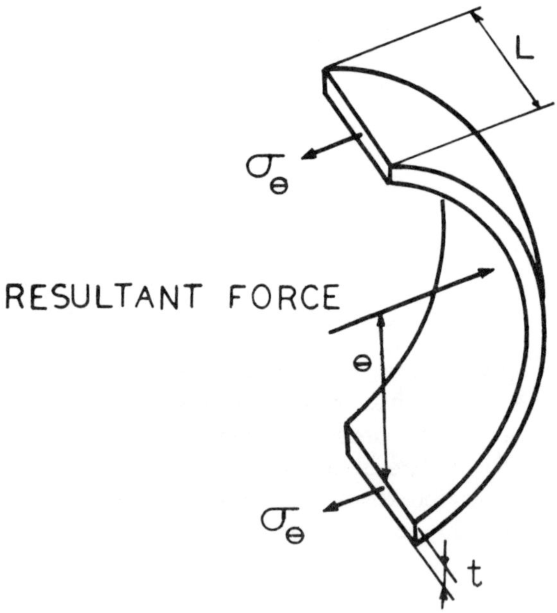

Fig. 8.17. Stresses in thin-walled cylinder

The horizontal force due to pressure acting on a differential area da is $p_o \sin\theta$ da. Since the differential area is given by $LRd\theta$, the total horizontal force is

$$F = p_o LR \int_0^{\pi} \sin\theta \, d\theta = p_o 2LR \tag{8.7.1}$$

This force is resisted by the tangential stress σ_θ acting over the cylinder wall cross-section or

$$2\sigma_\theta tL = p_o 2LR \tag{8.7.2}$$

which gives for the tangential stress

$$\sigma_\theta = \frac{p_o R}{t} \tag{8.7.3}$$

The tangential stress is uniform over the cross-section and, consequently, is not reduced by yielding of the material. A stress of this nature is called a **general** or **primary membrane stress** and is given a symbol p_m .

There is also a radial and longitudinal (axial) stress acting on the cylinder wall. At the outer surface $\sigma_r = 0$ and at the inner surface there is a compressive stress $\sigma_r = -p_o$. The average radial stress is given by $-p_o/2$. A force balance on the vessel head gives the axial stress or

$$2\pi Rt\sigma_z = p_o \pi R^2 \tag{8.7.4}$$

which gives

$$\sigma_z = \frac{p_o R}{2t} \tag{8.7.5}$$

The coordinate system chosen is the principal axes and the maximum normal stress is the tangential stress and the minimum normal stress is the radial stress. Therefore, the stress intensity is given by

$$S = \frac{p_o R}{t} - (-p_o/2) = \frac{p_o(D + t)}{2t} \tag{8.7.6}$$

where D is the vessel diameter.

Example 9.2

Consider a pressure vessel with the following characteristics:

$R = 90$ in. ; $t = 4.5$ in. ; $p_o = 1{,}000$ psig

The tangential stress is given by

$$\sigma_\theta = \frac{1000 \times 90}{4.5} = 20{,}000 \text{ psi}$$

The average radial stress is

$$\sigma_r = \frac{1000}{2} = -500 \text{ psi}$$

The axial stress is

$$\sigma_z = \frac{1000 \times 90}{2 \times 4.5} = 10{,}000 \text{ psi}$$

Consequently, the stress intensity is

$$S = 20{,}000 - (-500) = 20{,}500 \text{ psi}$$

If we apply the Mises criteria, we obtain

$$\sqrt{\tfrac{1}{2}[(20{,}500)^2 + (10{,}500)^2 + (10{,}000)^2]} = 17{,}759 \text{ psi}$$

Therefore, if Y = 20,000 psi, the maximum shear stress theory predicts failure while the Mises criteria does not. Consequently, from this example the maximum shear stress criteria is more conservative.

Thick-wall Vessel

Most nuclear pressure vessels are thick-walled and the preceding thin-walled vessel equations do not apply. For thick-walled vessels the inner and outer portion of the vessels carry different loads. The radial and tangential stresses are obtained from Eq. (8.6.45) and (8.6.46) by letting the external pressure p = 0.

$$\sigma_r = -p_o \left(\frac{r_o^2}{R^2 - r_o^2}\right)\left(\frac{R^2}{r^2} - 1\right) \tag{8.7.7}$$

$$\sigma_\theta = p_o \left(\frac{r_o^2}{R^2 - r_o^2}\right)\left(\frac{R^2}{r^2} + 1\right) \tag{8.7.8}$$

where r_o is the cylinder inside radius and R is the outside radius. The axial stress is obtained from a force balance on the vessel head which yields

$$\sigma_z = p_o \left(\frac{r_o^2}{R^2 - r_o^2}\right) \tag{8.7.9}$$

The tangential and radial stresses are at their maximum when $r = r_o$. Consequently, the maximum stress intensity $S_{max} = \sigma_\theta - \sigma_r$ occurs at $r = r_o$.

$$S_{max} = p_o \left[\frac{R^2 + r_o^2}{R^2 - r_o^2} + 1\right]$$

$$= p_o \frac{RD}{(R + r_o)t} \tag{8.7.10}$$

where t is the vessel wall thickness and D is the vessel outside diameter.

For example 8.2 conditions

$$S_{max} = 21{,}512$$

Therefore, the maximum stress intensity for a thick-walled cylinder is larger than a thin-walled cylinder because a greater load is carried at the inside surface.

The difference between the tangential stress for a thick-walled cylinder computed from Eq. (8.7.8) and that of a thin-walled cylinder computed from Eq. (8.7.3) is called the **secondary tangential stress**. Similarly, the difference between the radial stress computed from Eq. (8.7.7) and the average thin-walled radial stress $\sigma_r = -p_o/2$ is called the **secondary radial stress**. Any stress which is necessary to satisfy continuity of structure is called a secondary stress. Secondary stresses normally do not cause vessel failure providing primary stresses are below the yield stress. Local yielding tends to relieve secondary stresses and, consequently, these stresses are self equilibrating.

Spherical Heads

The spherical head is the simplest top which can be attached to a cylindrical pressure vessel. Stresses in the head are found by examining forces on the element shown in Fig. 8.18. We first examine a

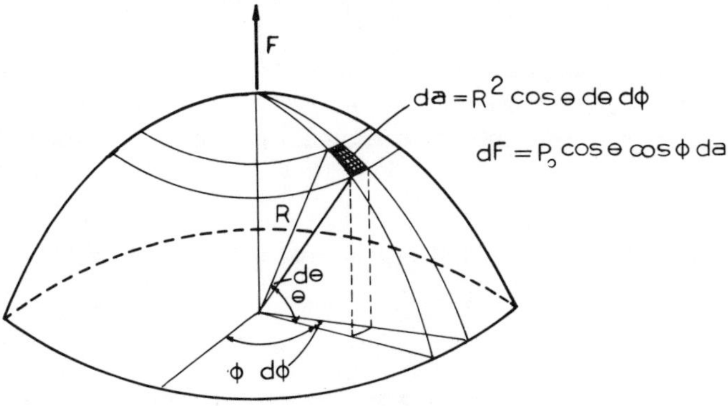

Fig. 8.18. Stresses in cylindrical head

thin-walled head of radius R with internal pressure p_o so as to determine primary membrane stresses. For a radial angle θ and azimuthal angle ϕ, the vertical force dF acting on da is $dF = p_o \cos\theta \cos\phi \; da$. Since $da = R^2 \cos\theta \; d\theta \; d\phi$ we have for the vertical force F

$$F = 4R^2 p_o \int_0^{\pi/2} \int_0^{\pi/2} \cos^2\theta \cos\phi \; d\theta \; d\phi = \pi R^2 p_o \tag{8.7.11}$$

The force on the head is resisted by a tangential stress

$$2\pi R t \sigma_\theta = \pi R^2 p_o \tag{8.7.12}$$

where t is the head wall thickness. Consequently, the tangential stress is given by

$$\sigma_\theta = \frac{p_o R}{2t} \tag{8.7.13}$$

which is the same result for the axial stress and one-half the tangential stress in the cylinder wall.

Two of the principal stresses in the spherical head are tangential, both having the same value. (If you have doubts about tangential stresses, imagine a basketball.) The tangential stresses have to be independent of orientation. The other stress is a radial stress which is $-p_o$ on the inside surface and 0 on the outside. The average radial stress $-p_o/2$. The stress intensity in a thin-walled hemisphere is

$$S = \frac{p_o R}{2t} - \left(-\frac{p_o}{2}\right) = \frac{p_o}{2}\left(\frac{R}{t} + 1\right) \tag{8.7.14}$$

Stresses in a thick-walled hemisphere are obtained in the same manner as for a thick-walled cylinder. For a thick-walled hemisphere of inside radius r_o and outside radius R, the tangential stress is given by [6]

$$\sigma_\theta = \frac{p_o R^3}{2} \frac{[(r_o/r)^3 + 2(r_o/R)^3]}{(R^3 - r_o^3)} \tag{8.7.15}$$

and the radial stress is

$$\sigma_r = -p_o R^3 \frac{[(r_o/r)^3 - (r_o/R)^3]}{R^3 - r_o^3} \tag{8.7.16}$$

The maximum stresses occur at the inside surface where $r = r_o$. Consequently, the maximum stress intensity is given by

$$S_{max} = \frac{1.5 p_o R^3}{R^3 - r_o^3} \tag{8.7.17}$$

Hemispherical heads have the smallest stress levels of any head that is attached to a cylindrical pressure vessel. However, this type of head adds considerably to overall heights of pressure vessels which is a disadvantage. Consequently, dished (ellipsoidal) or torispherical heads are frequently used because they are shorter and of less weight. For small pressure vessels flat heads are frequently used because of their ease of fabrication; however, they must have considerably thicker walls because of their greater stresses. Techniques for analyzing stresses in these vessel heads are given in Refs. [9] and [10].

Stresses at Discontinuites

A discontinuity is created at the junction of a pressure vessel head and cylinder which causes additional stresses because of different behavior of the two pieces. As shown in the preceding analysis, for the same wall thickness the tangential stresses in the hemispherical head is one-half that in the cylinder. Consequently, when joining these pieces additional stresses are created in order to maintain continuity of structure. In the design of large pressure vessels the head wall thickness is thinner than the cylinder so as to attempt to equalize deflections of structures; however, the vessel head thickness is made the same as the cylinder at the point of attachment. When both head and cylinder are designed for minimum thickness the discontinuity stresses can increase the maximum stress up to 25% [6]. This increase is a maximum at the joint and is attenuated with increasing distance from the joint.

Discontinuity stresses are analyzed by considering stresses and moments at the junction shown in Fig. 8.19. To maintain continuity of structure, self-equilibrating shear forces Q_o and moments M_o are created in each piece. These quantities are analyzed in the following manner [5, 6]:

(1) Imagine vessel is separated at the discontinuity.

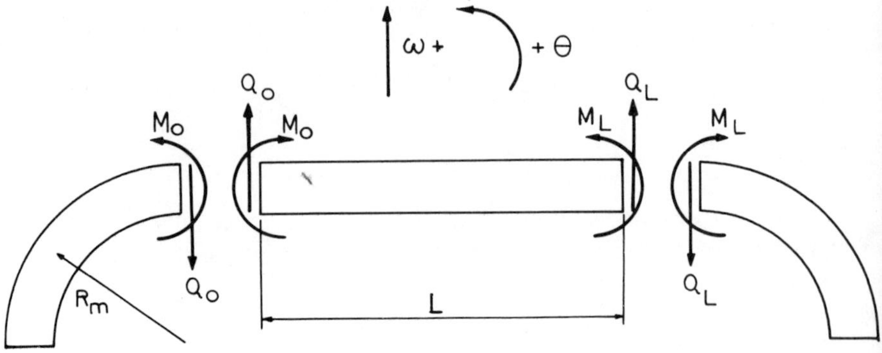

Fig. 8.19. Forces and moments at head-cylinder junction

(2) Calculate the lateral displacements ω and rotation θ at the junction due to edge loads Q_o and M_o.

For a hemisphere, these displacements are given by

$$\omega_h = -\frac{2R_m\lambda}{Et}Q_o + \frac{2\lambda^2}{Et}M_o \qquad (8.7.18)$$

$$\theta_h = -\frac{2\lambda^2}{Et}Q_o + \frac{4\lambda^3}{R_m Et}M_o \qquad (8.7.19)$$

where

$R_m = R_i + t/2$

t = thickness

E = Young's Modulus

$\lambda = \beta R_m$

$\beta = \left[3(1 - \upsilon^2)/R_m^2 t^2\right]^{1/4}$

Q_o = radial shearing force per unit of circumference, pounds/inch

M_o = bending moment per unit of circumference, inch-pounds/inch

For the cylinder the displacements are given by

$$\omega_c = [B_{11}/(2\beta^3 D)]Q_o + [B_{12}/(2\beta^2 D)]M_o + [G_{11}/(2\beta^3 D)]Q_L$$

$$+ [G_{12}/2\beta^2 D)]M_L \qquad (8.7.20)$$

$$-\theta_c = [B_{12}/(2\beta^2 D)]Q_o + [B_{22}/(2\beta D)]M_o + [G_{12}/(2\beta^2 D)]Q_L + [G_{22}/(2\beta D)]M_L \tag{8.7.21}$$

where

$$B_{11} = (\sinh 2\beta L - \sin 2\beta L)/2(\sinh^2 \beta L - \sin^2 \beta L)$$

$$B_{12} = (\cosh 2\beta L - \cos 2\beta L)/2(\sinh^2 \beta L - \sin^2 \beta L)$$

$$B_{22} = (\sinh 2\beta L + \sin 2\beta L)/(\sinh^2 \beta L - \sin^2 \beta L)$$

$$G_{11} = \frac{-(\cosh \beta L \sin \beta L - \sinh \beta L \cos \beta L)}{\sinh^2 \beta L - \sin^2 \beta L}$$

$$G_{12} = -2 \sinh \beta L \sin \beta L/(\sinh^2 \beta L - \sin^2 \beta L)$$

$$G_{22} = \frac{-2(\cosh \beta L \sin \beta L + \sinh \beta L \cos \beta L)}{\sinh^2 \beta L - \sin^2 \beta L}$$

$$\beta = \left[3(1 - \upsilon^2)/R_m^2 t^2\right]^{1/4}$$

$$L = \text{cylinder length}$$

$$D = Et^2/12(1 - \upsilon^2)$$

Similar equations apply at the other end x = L. When both ends have the same hemisphere attached, $Q_L = Q_o$, $M_L = M_o$, and $\omega_L = \omega_o$, $\theta_L = \theta_o$

(3) Calculate displacements of the head, ω_{hp}, and cylinder, ω_{cp}, due to internal pressure. For a thin-walled hemisphere

$$\omega_{hp} = (1 - \upsilon)\frac{p_o R_i^2}{2Et} \tag{8.7.22}$$

$$\theta_{hp} = 0 \tag{8.7.23}$$

For a thin-walled cylinder

$$\omega_{cp} = \left(1 - \frac{\upsilon}{2}\right)\frac{p_o R_i R_m}{Et} \tag{8.7.24}$$

$$\theta_{cp} = 0 \tag{8.7.25}$$

(4) Determine the displacements of each element caused by temperature distributions. (If all parts are at same temperature and of same material, temperature deformations need not be considered.)

(5) Equate total lateral displacements and rotations of adjacent elements at each junction. For no temperature effect at point 0,

$$\omega_h + \omega_{hp} = \omega_c + \omega_{cp} \tag{8.7.26}$$

$$\theta_h = \theta_c \tag{8.7.27}$$

(6) Solve these equations for Q_o and M_o (and Q_L and M_L if heads are different).

(7) Compute the discontinuity stresses at each junction due to Q_o and M_o. For a cylinder, these stresses are given by

$$\sigma_z(z) = \pm\ 6M(z)/t^2 \tag{8.7.28}$$

$$\sigma_\theta(z) = E\omega(z)/R_m \pm 6M(z)/t^2 \tag{8.7.29}$$

$$\sigma_r = 0 \tag{8.7.30}$$

where M(z) and $\omega(z)$ are at location z which would be 0 or L. The $\pm$ sign means the upper sign refers to inner surface and the lower sign to the outer surface. The value of $\omega(z)$ that goes in the formula for σ_θ is the radial displacement due to M_o, Q_o, etc.; the other displacements should not be used. More complicated expressions are available for the stresses in the hemisphere [5].

(8) Compute the total stresses.

Total stresses are obtained by combining discontinuity stresses from step 7 with stresses from other loadings, i.e., pressure. Calculate total principal stresses on the inside and outside surfaces.

(9) Compute stress intensities at each location from principal stresses from 8 and match with code requirements.

Example 8.3 [6]

Calculate stresses in a cylinder with hemispherical heads and an inside pressure of 950 psig. The pressure vessel has the following characteristics:

Head R_i = 30 in. inside radius
t = 1.375 in.

Shell R_i = 30 in.
t = 1.375 in.
L = 10 in.
$E = 29 \times 10^6$ psi
$\upsilon = 0.3$
$\beta = 0.198, \quad \lambda = 6.07, \quad R_m = 30.688$

(1-2). From steps 1 and 2 the displacements are as follows:
For the head

$$\omega_h = [-9.35Q_o + 1.85M_o] \times 10^{-6}$$

$$\theta_h = [-1.85Q_o + 0.732M_o] \times 10^{-6}$$

For the cylinder

$$B_{11} = 1.14, \quad B_{12} = 1.14, \quad B_{22} = 2.3$$

$$G_{11} = -0.318, \quad G_{12} = -0.555, \quad G_{22} = -0.332$$

Consequently, the displacements are

$$\omega_c = [10.7Q_o + 2.115M_o - 3.81Q_o - 1.025M_o] \times 10^{-6}$$

$$= [6.89Q_o - 1.09M_o] \times 10^{-6}$$

$$\theta_c = -[2.115Q_o + 0.782M_o - 1.026Q_o - 0.123M_o] \times 10^{-6}$$

$$= -[1.09Q_o + 0.569M_o] \times 10^{-6}$$

(3) Applying step 3 using the thin-shell equations, the displacements due to pressure are

$$\omega_{hp} = \frac{(0.7)(950)(30)^2}{(2)(29 \times 10^6)(1.375)} = 7504 \times 10^{-6} \text{ in.}$$

$$\omega_{cp} = \frac{(0.85)(950)(30)(30.688)}{(29 \times 10^{-6})(1.375)} = 18{,}664 \times 10^{-6} \text{ in.}$$

(4) Assuming no temperature gradients gives

(5) $\omega_h + \omega_{hp} = \omega_c + \omega_{cp}$

$$\theta_h = \theta_c$$

(6) or

$$-9.35Q_o + 1.85M_o + 7504 = 6.89Q_o + 1.09M_o + 18{,}644$$

$$-1.85Q_o + 0.732M_o = -1.09Q_o - 0.569M_o$$

Simplifying

$$-16.24Q_o + 0.76M_o = 11{,}140$$

$$-0.76Q_o + 1.30M_o = 0$$

Which yields

Q_o = –705 lb/in. (– sign means original direction of Q_o is wrong)

M_o = –412 in–lb/in. (–sign means original M_o direction is wrong)

For the cylinder, the displacement is

$$\omega_o = -5{,}306 \times 10^{-6} \text{ in.}$$

(7) For the cylinder, the discontinuity stresses are

$$\sigma_z = \pm \frac{6M_o}{t^2} = \begin{matrix} -1307 (\text{psi inner surface}) \\ +1307 (\text{psi outer surface}) \end{matrix}$$

$$\sigma_\theta = \frac{E\omega_o}{R_m} \pm \frac{6\upsilon M_o}{t^2}$$

$$= \frac{29 \times 10^6}{30.688}(-5306 \times 10^{-6}) \pm \frac{(6)(0.3)(-412)}{(1.375)^2}$$

$$= \begin{matrix} -5406 (\text{psi inner surface}) \\ -4622 (\text{psi outer surface}) \end{matrix}$$

$$\sigma_r = 0$$

(8) The pressure stresses are given by

$$\sigma_\theta = p_o \left(\frac{R_i^2}{R_o^2 - R_i^2}\right)\left(\frac{R_o^2}{r^2} + 1\right)$$

$$= 950 \left(\frac{30^2}{(31.325)^2 - 30^2}\right)\left(\frac{(31.375)^2}{r^2} + 1\right)$$

$$= \begin{matrix} 21{,}477 \text{ psi (inner surface)} \\ 20{,}262 \text{ psi (outer surface)} \end{matrix}$$

$$\sigma_z = \frac{p_o}{(R_o^2/R_i^2 - 1)} = 10{,}127 \text{ psi}$$

$$\sigma_r = \begin{matrix} -950 \text{ psi (inner surface)} \\ 0 \text{ psi (outer surface)} \end{matrix}$$

Consequently, the combined stresses on the cylinder are as follows:

Inner surface

$$\sigma_r = -950 + 0 = -950 \text{ psi}$$
$$\sigma_\theta = 21{,}477 - 5406 = 16{,}071 \text{ psi}$$
$$\sigma_z = 10{,}127 - 1307 = 8{,}820 \text{ psi}$$
$$S_i = 17{,}021 \text{ psi}$$

Outer surface

$$\sigma_r = 0$$
$$\sigma_\theta = 20{,}262 - 4622 = 15{,}640 \text{ psi}$$
$$\sigma_z = 10{,}127 + 1307 = 11{,}434 \text{ psi}$$
$$S_o = 15{,}640 \text{ psi}$$

Note, discontinuity stresses reduced stresses in cylinder portion of the pressure vessel. However, these stresses increase the level of stresses in the head portion.

8.8 Nuclear Vessel Design

As previously mentioned, the design of pressure vessels and piping is governed by the ASME Boiler and Pressure Vessel Code, Section III—Nuclear Vessels [5], which specifies method of failure by the maximum shear stress theory. Their design approach is based upon a number of definitions.

(1) Normal stress—component normal to the plane of reference.

(2) Membrane stress (P_m) —component of a normal stress which is uniformly distributed and equal to the average stress across the thickness of the section considered, P_m .

(3) Bending stress (P_B) —variable components of normal stress which is the variation from the average (membrane) stress across the section considered, P_B .

(4) Primary stress—any normal or shear stress developed by a load which is necessary to satisfy laws of equilibrium of external and internal forces and moments. Primary stresses which exceed yield strength will result in distortion or failure. Thermal stress is not a primary stress. Examples of primary stresses are,

a) general membrane stress in a cylinder or sphere due to pressure or distributed live loads,

b) bending stress in central portion of a flat head.

(5) Secondary stress (Q)—normal or shear stress developed by constraint of adjacent material or by self-constraint of the structure. It is self-limiting. Local yielding and minor distortions can satisfy conditions which cause stress to occur.

Examples are:

a) general thermal stress,

b) bending stress at a gross structural discontinuity

(6) Local primary membrane stress (P_L) —case in which a membrane stress produced by pressure or other mechanical loading and associated with a primary and/or a discontinuity effect produces excessive distortion in the transfer of load to other portions of the structure. A stressed region is local if the distance which stress intensity exceeds $1.1\ S_m$ does not extend in the meridional direction more than $0.5\sqrt{Rt}$. Example is the membrane stress in a shell produced by external load and moment at a permanent support or nozzle connection.

(7) Peak stress (F)—increment of stress additive to the primary plus secondary stresses by reason of local discontinuities or local thermal stress. This stress does not cause distortion but may be source of fatigue crack or a brittle fracture. Examples are:

a) thermal stress in austenitic steel cladding of a carbon steel component,

b) thermal stress which may cause fatigue but not distortion,

c) stress at a local discontinuity,

d) surface stress produced by thermal shock.

(8) Load stress—stress from internal pressure, applied load, or gravity.

(9) Thermal stress—self-balancing stress produced by non-uniform

temperature or different expansion coefficients. Two types are

a) General thermal stress—associated with distortion of structure in which it occurs. If stress is twice yield strength, elastic analysis may be invalid and successive thermal cycles may produce incremental distortion.

Examples are:
1) stress from axial temperature distribution in cylinder,
2) stress from temperature differences in shell and nozzle,
3) stress from radial temperature distribution in cylinder.

b) Local thermal stress—associated with almost complete suppression of the differential expansion, thus no distortion.

Examples are:
1) stress in cladding material with different expansion coefficient,
2) stress in small hot spot.

(10) Total stress—sum of primary, secondary, and peak stresses.
(11) Stress cycle—condition in which stress differences go from an initial value through an algebraic maximum and then an algebraic minimum and then return to initial value.

Loading Criteria

In the analysis of nuclear vessels the following types of loading must be taken into account:

(1) Internal and external pressure.

(2) Weight of component and normal contents under operating or test conditions including additional pressure due to static and dynamic head of liquids.

(3) Superimposed loads such as other components, operating equipment, insulation, corrosion-resistant or erosion-resistant linings, and piping.

(4) Wind loads, snow loads, vibrations, and earthquake loads where specified.

(5) Reactions of supporting lugs, rings, saddles, or other types of supports.

(6) Temperature effects.

Design Procedures

The following design procedures are recommended for nuclear vessels:

(1) Construct an orthogonal set of coordinates and designate 3 normal and 3 shear stress components.

(2) Calculate stress components for each type of loading

a) General primary-membrane stress, P_m

b) Local primary-membrane stress, P_L

c) Primary bending stress, P_b

d) Secondary stresses, Q

e) Peak stress, F

(3) Translate stress components to principal axes and determine principal stresses, σ_1, σ_2, σ_3.

(4) Calculate stress differences given by

$$S_{12} = \sigma_1 - \sigma_2$$

$$S_{23} = \sigma_2 - \sigma_3$$

$$S_{31} = \sigma_3 - \sigma_1$$

The stress intensity is the largest absolute value of these differences, S.

(5) For a cycle the principal axes may change direction with time. Therfore, it may be necessary to calculate stress intensities as a function of time. Find the largest absolute magnitude of the stress intensity. The alternating stress intensity S_a is 1/2 of this magnitude.

Design Limitations

The code establishes allowable values of the stress intensity, symbolized by S_m, for the structural material as a function of temperature. These values are about 1/3 the yield stress of the material at that temperature. Table 8.3 gives some typical values.

The primary membrane stress intensity in a cylinder is given by

$$S_{pm} = \frac{p_o(D_i + t)}{2t} \quad \text{(no outside pressure)} \tag{8.8.1}$$

and a sphere

$$S_{pm} = \frac{p_o}{2}\left(\frac{R_i + t}{t}\right) \quad \text{(no outside pressure)} \tag{8.8.2}$$

Table 8.3

Design stress-intensity values [5]

Material (Spec. No.)	Grade or type	Nominal Composition	S_m (psi) For metal temperatures not exceeding 100°F	300°F	500°F	700°F	800°F
Carbon steel							
SA 442	55	C—Si	18,300	17,700	16,200	14,400	—
SA 516	55	C—Si	18,300	17,700	16,200	14,400	—
Low-alloy steel							
SA 302	A	Mn—1/2% M_0	25,000	25,000	25,000	24,400	—
SA 387	A	1/2%Cr—1/2% M_0	21,700	21,700	21,700	21,700	—
SA 240	410S	13% Cr	20,000	18,400	17,800	16,400	16,100
High-alloy steel							
SA 240	304	18% Cr, 8% Ni	20,000	19,800	16,400	15,100	14,800
SA 240	316	18% Cr, 10% Ni, 2% M_0	20,000	20,000	15,100	13,500	15,600
SA 240	347	18% Cr, 10% Ni, Cb	20,000	20,000	19,400	17,600	17,100

(1) Since the primary membrane stress intensity cannot exceed S_m, we can calculate minimum wall thickness as follows. For a cylinder

$$t_{min} = p_o R_i / (S_m - 0.5 p_o) \tag{8.8.3}$$

where p_o is the design pressure. (Design pressures will be more than operating pressures.)

For a hemisphere

$$t_{min} = p_o R_i / (2S_m - p_o) \tag{8.8.4}$$

These thicknesses apply to design loads.

The shell thickness calculated may not be adequate since consideration was not given to secondary stresses, bending stresses at discontinuities, thermal stresses, peak stresses, etc. These stresses must be accounted for by the ASME code, but since they are self-limiting a larger margin of safety is required.

(2) The stress intensity from primary local membrane stresses must be less than 1.5 S_m for design loads.

(3) The stress intensity from the sum of the local primary membrane stresses and the primary bending stresses must be less than 1.5 S_m for the design load.

(4) The stress intensity from the highest value at any point across a

thickness of a section of the general or local primary-membrane stresses plus primary bending stresses plus secondary stresses produced by operating conditions cannot exceed 3 S_m .

(5) For cyclic operation, the alternating stress intensity cannot exceed S_a as given by a graph of S_a as a function of number of cycles for the material under consideration. Figure 8.20 illustrates a design curve for cyclic operation of several types of steels. The allowable range of stress intensity is 2 S_a . This stress intensity must consider all possible stresses at a point including primary, secondary, and peak stresses. If Young's modulus is different from 30×10^6 , calculate new S_a from

$$S_a = S_{alt} \times \frac{30 \times 10^6}{E_{mat}} \tag{8.8.5}$$

for use of the chart.

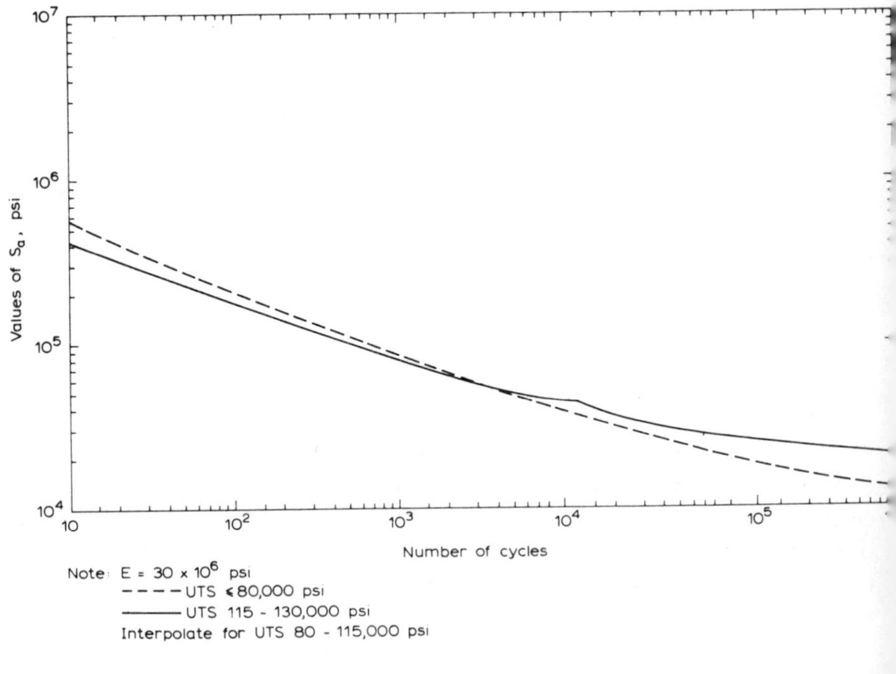

Fig. 8.20. Design fatigue curves for carbon, low-alloy and high-tensile steels, for metal temperatures not exceeding 700°F (ASME Boiler and Pressure Vessel Code, Section III - Nuclear Vessels) [6].

Example 8.4

If the pressure vessel in Example 8.3 is SA 387 low alloy steel at 500°F, will the cylinder meet ASME Code requirements?

$$S_m = 21{,}700 \text{ psi from Table 9.3}$$

Requirement 1 $t_{min} = \dfrac{pR}{S_m - 0.5p} = \dfrac{950 \times 30}{21{,}700 - 475} = 1.34$ in.

Since t = 1.375 in., the first criteria is met.

Requirement 2. Local primary membrane is equal to general primary membrane so criteria is met.

Requirement 3. There are no primary bending stresses.

Requirement 4. The maximum stress intensity is 17,021 psi at the inner surface which is less than $3\ S_m$, so the criteria is met.

Consequently, the ASME Code criteria is satisfied for the cylinder. It is also necessary to apply the code to the cylinder heads.

Piping Design

The ASME Code applies a different set of criteria to the analysis of piping in nuclear power plants which is too lengthy for presentation in this book. Protection against primary membrane stresses for straight pipes requires that

$$t_m = \frac{p_o D_o}{2(S_m + 0.4\, p_o)} + a \tag{8.8.6}$$

where

D_o = outside diameter

a = additional thickness to compensate for material removed in threading, provide for corrosion or erosion, and provide structural strength during erection.

There are other requirements due to loading, earthquakes, etc. which are too complex to treat and are discussed in the ASME Code. As an example

$$1.5\ S_m \geqslant B_1 \frac{p_o D_o}{2t} + B_2 \frac{D_o}{2I} M_i \tag{8.8.7}$$

M_i = resultant moments due to loads

B_1, B_2 = stress indices

Cladding Design

The theory and design approach recommended by the ASME code for pressure vessels is also usually followed in the stress analysis of reactor fuel cladding. Table 8.4 lists stress intensity limitations for a cladding with a yield stress (S_y) or ultimate tensile strength (S_u) . Methods for determining cladding stresses are discussed in Sec. 8.6.

Table 8.4

Reactor fuel cladding intensity limitation [6]

Category	Yield Strength S_y	Ultimate Tensile Strength S_u
General Primary Membrane	2/3 S_y	1/2 S_u
Local Primary Membrane	S_y	1/2–3/4 S_u
Primary Membrane plus Bending	S_y	1/2–3/4 S_u
Primary plus Secondary	2 S_y	1.0–1.5 S_u

8.9 Neutron Irradiation Effects on Design

Brittle fracture of materials is a failure due to crystal cleavage which results in essentially no yielding. Such a failure is similar to the shattering of glass which would be a catastrophic failure for a reactor vessel. Consequently, in the analysis of nuclear components attention must be given to insure that conditions for brittle fracture failure mechanisms can never occur.

In the absence of radiation, the steels used in pressure vessels follow an elastic or elastic-plastic behavior at room or elevated temperatures. However, at sufficiently low temperatures steels may have brittle stress-strain properties. The conditions for a brittle failure are [6]

(1) A flaw (crack or notch) must be present

(2) A tensile stress of sufficient magnitude must exist

(3) The metal must be at sufficiently low temperature.

The behavior under brittle fracture is understood by examining the stress-temperature diagram for crack initiation and arrest shown by Fig. 8.21. As temperature decreases, both the yield strength (curve B) and the tensile (ultimate) strength (curve A) increases. However, the increase in tensile strength is less than that of the yield strength. At some low temperature, about 10°F for carbon steel, yield and tensile strengths start to coincide. At this temperature and below, there is no yielding before failure and consequently, the failure is a brittle failure. The temperature at which the yield and tensile strengths coincide is called **nil ductility temperature** (NDT) [6].

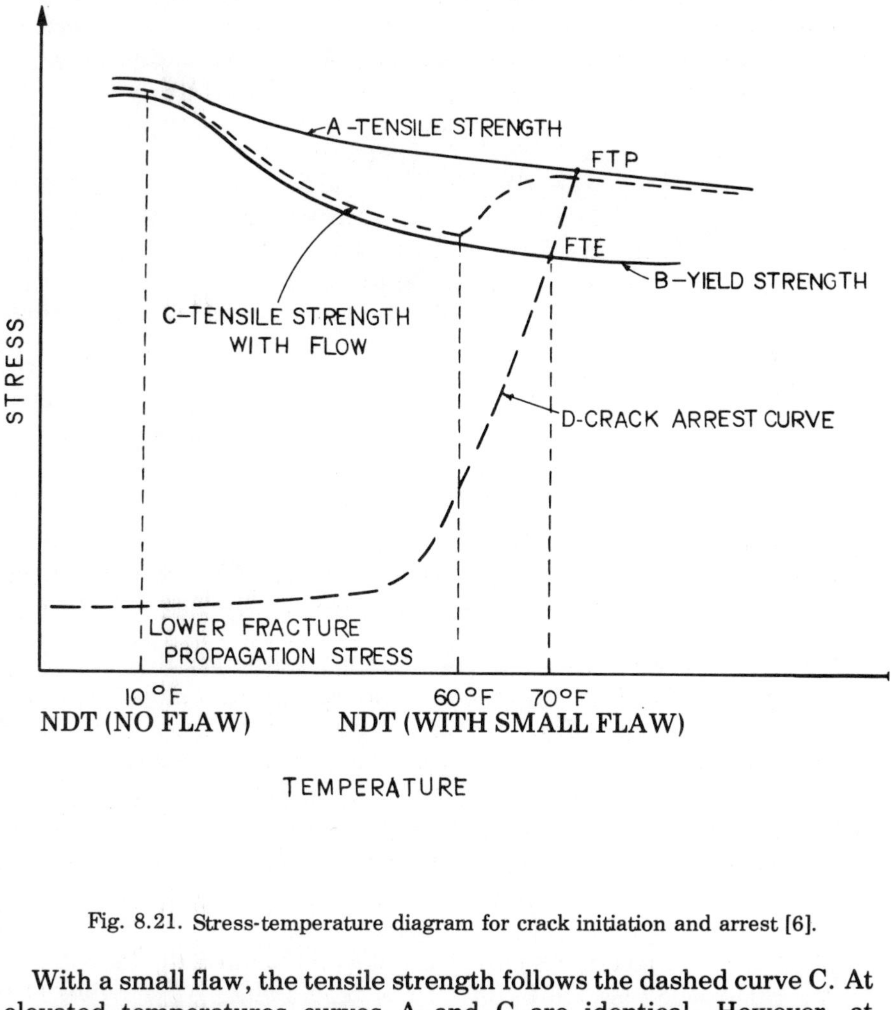

Fig. 8.21. Stress-temperature diagram for crack initiation and arrest [6].

With a small flaw, the tensile strength follows the dashed curve C. At elevated temperatures curves A and C are identical. However, at temperatures about 50°F above the NDT for unflawed materials, the tensile strength curve drops to the yield curve and then follows the yield curve. At the point where the tensile strength curve meets the yield strength curve for a flawed material, we have a new NDT temperature. If a flaw exists, any failure below the new NDT is a brittle failure.

In general, brittle failures occur because a flaw or crack is propagated throughout the material. For crack propagation to occur appropriate stress and temperature conditions must exist. There is a stress level (of the order of 6 500 psia in steels) below which a crack propagation will not occur. This is called the **lower fracture propagation stress**. As temperature is increased a higher stress is required before a crack can propagate. The relationship between temperature and stress required for a crack to propagate is called the **crack arrest curve** (curve D). At temperatures above those indicated by the crack arrest curve, crack propagation will not occur. The intersection of this curve with the yield curve is called the **fracture transition elastic** (FTE). The FTE temperature is the highest temperature at which crack propagation can occur under purely elastic loads. The intersection of the crack arrest curve with the tensile (ultimate) strength curve (curve A) is called the **fracture transition plastic** (FTP). Above the FTP temperature, failure is entirely by shear (no brittle failure can occur).

For no brittle failure to occur, we must operate above the NDT temperature. For greater safety, it is desirable to operate above the FTE temperature so that there can be no fracture propagation (no brittle fracture) under purely elastic loads. It is found that the FTE temperature is about 60°F above the NDT temperature, consequently, it has been standard practice to limit the minimum system temperature while under pressure to the temperature corresponding to the NDT + 60°F [6].

Pressure vessels undergoing fast neutron irradiation experience an increase in the NDT. It has been observed that irradiation causes a decrease in ductility and yield. The irradiated material is more brittle than not irradiated and one can expect an increase in NDT.

The increase in NDT is a function of the exposure (nvt) of fast neutrons above 1 Mev. At any time in the reactor lifetime, the NDT of an irradiated material, $\mathrm{NDT}(\theta)$, is given by [6]

$$\mathrm{NDT}(\theta) = \mathrm{NDT}(0) + \Delta\mathrm{TT} \tag{8.9.1}$$

where

$\mathrm{NDT}(0)$ = nil ductility temperature of unirradiated material
$\Delta\mathrm{TT}$ = change in nil ductility temperature from irradiation

The value of $\Delta\mathrm{TT}$ is a function of the fast neutron exposure and temperature of irradiaton. Table 8.5 lists several values of $\Delta\mathrm{TT}$ for typical pressure vessel steels.

Table 8.5

Change of nil ductility temperature of pressure vessel steels with neutron irradiation [6].

(nvt)	ΔTT	
above 1 MEV	Irradiation at 450°F	at 550°F
10^{18}	105	40
10^{19}	270	175
10^{20}	530	360

The increase in NDT with irradiation can lead to operating restrictions on reactor pressure vessels. As an example, until a temperature exceeding the new NDT is attained, a stress exceeding the lower fracture propagation stress should not be applied to a pressure vessel. Therefore, the maximum stress induced by a system pressure slightly greater than the coolant vapor pressure at $NDT(\theta)$ must not exceed the lower fracture propagation stress. This limits the ΔTT and neutron exposure of pressure vessels and, consequently, shielding against fast neutron irradiation is required.

To assure that the change in NDT is as expected, samples of pressure vessel steel are placed in the thermal shield region. These samples are periodically withdrawn and any change in NDT measured.

Example 8.5

A typical fast neutron flux in the pressure vessel wall of a thermal reactor is $6 \times 10^{10} n/cm^2\text{-sec}$. If the vessel has a 75% load factor for 40 years, what temperature should the coolant be before operation at full pressure at end of life. Irradiation temperature is 500°F.

$$\text{sec} = 3600 \times 24 \times 365 \times 40 \times 0.75 = 0.965 \times 10^9 \text{ sec}$$
$$\text{nvt} = 6 \times 10^{10} \times 0.965 \times 10^9 = 5.8 \times 10^{19} \text{ n/cm}^2$$
$$\Delta TT = 320°F \text{ (note dependence is non-linear)}$$

If we take NDT(0) as 10°F, then $NDT(\theta) = 330°F$. To remain above the FTE point, it is necessary to operate above 390°F. Therefore, at end of plant life the vessel (coolant) needs to be at 390°F before full system pressure is applied.

Experiments performed at the SKODA Works show that the difference between the FTE temperature and NDT depends upon the type of material [10]. Figure 8.22 shows this temperature difference as a function of yield stress. It should be noted that for steels of medium strength, which is the case for most steels, this temperature difference is higher than the 60°F shown on Fig. 8.21. This comparison suggests that increasing the yield stress of the material (and also the design stress) increases the difference FTE - NDT. Consequently, higher FTE temperatures should be used when specifying operating temperatures for vessels under pressure.

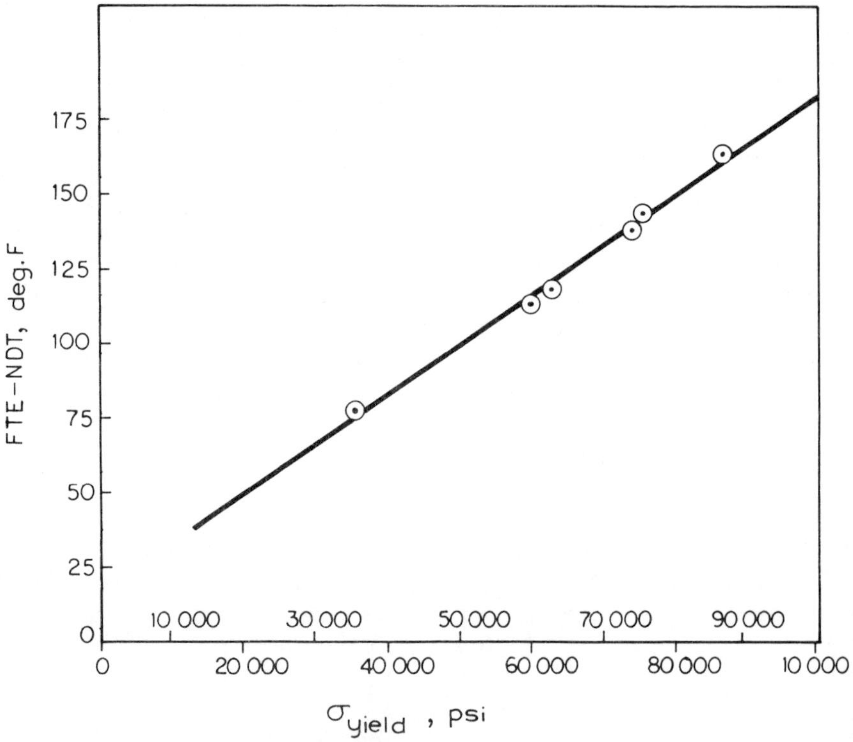

Fig. 8.22. Schematic representation of defect analysis diagram for steels [10]

REFERENCES

1. Gol'denblat, I. I. and N. A. Nikolaenko, **Calculation of Thermal Stresses in Nuclear Reactors**, Consultants Bureau, New York, 1964.
2. Timoshenko, S. and J. N. Goodier, **Theory of Elasticity**, McGraw-Hill, New York, 1951.

3. Crandall, S. H. and N. C. Dahl, **An Introduction to the Mechanics of Solids**, McGraw-Hill, New York, 1959.
4. Smith, C. O., Stress analysis notes, ORSORT Notes, ORNL (July 1964).
5. **ASME Boiler and Pressure Vessel Code, Section III—Nuclear Vessels**, Am. Soc. Mech. Eng., New York, 1972.
6. Weisman, J., **Elements of Nuclear Reactor Design**, Elsevier, New York, 1977.
7. Bonilla, C. F., **Nuclear Engineering**, McGraw-Hill, New York, 1957.
8. Brownell, L. E. and E. H. Young, **Process Equipment and Design**, Wiley, New York, 1968.
9. **ASME Boiler and Pressure Vessel Code, Section VIII—Pressure Vessels, Div. 1**, Am. Soc. Mech. Eng., New York, 1972.
10. Brumovsky, M. et al., Initiation and arrest-two approaches to pressure vessel safety, ZJE-194, SKODA Works, Plzen, Czechoslovakia (1976).

PROBLEMS

8.1 Sketch the Mohr's circle for stress for each of the following cases of plane stress.

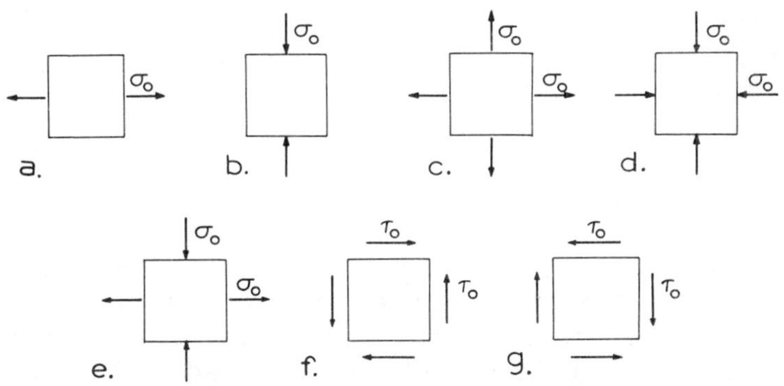

8.2 Find the principal stresses and the orientation of the principal axes of stress for the following cases of plane stress. What is the stress intensity?

a.	b.	c.
σ_x = 4,000 psi	σ_x = 14,000 psi	σ_x = -12,000 psi
σ_y = 0	σ_y = 2,000 psi	σ_y = 5,000 psi
τ_{xy} = 8,000 psi	τ_{xy} = -6,000 psi	τ_{xy} = 10,000 psi

8.3 A rod of length L is placed between two immovable walls which touch each end. At temperature T_o the rod is stress free. A heat source is now placed in the rod such that its temperature distribution is given by

$$T - T_o = 4\Delta T_{Max}\frac{X}{L}(1 - X/L)$$

What is the stress at the rod midpoint?

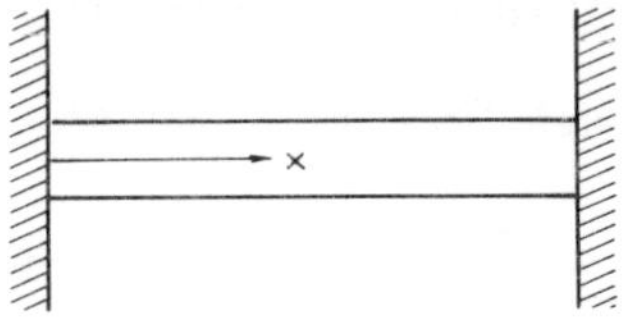

Hint: ϵ_x is not zero over most of the rod length

8.4 A pebble bed reactor contains uranium carbide fuel spheres of radius 1/4 inch. These spheres are cooled with helium at a pressure of 1000 psi. For a uniform volumetric heat generation rate of 2×10^7 Btu/hr–ft^3 in the spheres, determine if yielding will occur. If there is yielding in the spheres, **where does it occur** and **why**? The yield stress of uranium carbide is 10,000 psi. For uranium carbide $E\alpha/k(1 - \upsilon) = 20$ psi/Btu/hr–ft.

8.5 Determine maximum axial, radial and tangential stresses in the cladding of a PWR fuel rod of 0.37 in. inside diameter, 0.023 in. thickness, and 12 ft. length. The stresses will be due to fission product gas buildup and thermal expansion. The maximum linear heat rate is 15 kW/ft and the cladding is zircaloy with a thermal conductivity of 7.4 Btu/hr-ft-°F. The fuel enrichment is 3 percent U^{235} in the UO_2 fuel. Assume the reactor operating time is 10,000 hours and 0.5 gas atoms are produced per fission and one-half the gas atoms are released to a volume of 1.5 cu. in. Assume the gas temperature is 1000°F and $\alpha E/k(1 - \upsilon)$ is 4.6. The fuel rod is initially filled with helium at 50 psi and room temperature. Assume a cosine power shape and an external pressure of 2300 psia.

8.6 If the yield stress for the zircaloy cladding in problem 8.5 is 16,000 psi, determine if the ASME criteria for cladding is satisfied.

CHAPTER 9

FLUID TRANSIENTS

C. Samuel Martin

9.1 Introduction

In the context of the material reported in this chapter, fluid transients, or occasionally the term pressure transients, refers to wave or pressure pulse propagation in single-phase or two-phase fluid media. If the fluid is a single-phase liquid, and the elastic or compressible effects of the liquid are important, then the phenomenon is termed waterhammer, in contrast to surging if the liquid compressibility is neglected. Because of the dramatic increase in the elasticity of the resulting mixture, waterhammer is not as likely to occur in a dispersed two-phase mixture. However, in the event of dramatic gas compression or vapor pocket collapse, the occurrence of large isolated gas or vapor cavities in liquid pipelines can create severe waterhammer pressures.

The classical causes of waterhammer in engineering piping systems are (1) operation of valves, (2) starting and stopping of hydraulic machinery, (3) auto-oscillation of valves or hydraulic components, and (4) collapse of vapor pockets. In nuclear piping and circulating water systems waterhammer and pressure surges are known to be caused by a number of controlled and uncontrolled events. Some examples of waterhammer in nuclear piping systems are as follows:

1. Collapse of steam pocket in the feedwater line of a PWR
2. Injection of Emergency Core Coolant (ECC) liquid into the cold legs of a PWR
3. Filling of voided lines such as spray cooling with High Pressure Coolant Injection (HPCI) of a BWR
4. Pressure surges in blowdown lines of a BWR
5. Starting of Service Water Pumps with entrapped air or vapor
6. Sodium-water reaction in a LMFBR
7. Pump startup and power failure in circulating water systems

In many industrial examples of waterhammer in piping classical theories can be applied with confidence. Because of the potential of plastic deformation, steam pocket formation and collapse, existence of entrapped gas, and coupled fluid-structure interaction resulting from severe waterhammer, classical methods can not always be employed to nuclear piping systems.

Some of the factors limiting application of classical plane wave waterhammer theory to nuclear piping are:

1. Fluid Resistance
   a. Frequency-dependent friction
   b. Minor losses
2. Structural Interaction
   a. Line motion
   b. Plastic deformation
3. Two- and Three-Dimensional Effects
   a. Distortion of plane waves at fittings
   b. Vapor cavity collapse
4. Two-Phase Flow Effects
   a. Flashing (Cavitation)
   b. Condensation
   c. Non-equilibrim thermodynamics
   d. Effect of free gas
   e. Release of dissolved gas
5. Dynamic Effects of Turbomachinery
   a. Dynamic response
   b. Two-phase flow effects

In the following sections the basic theory of waterhammer is presented. The reader is encouraged to gain an understanding of the basic concepts of wave propagation in a single-phase liquid before undertaking the more complicated situations. Much more detailed accounts of analyses are given in books by Streeter and Wylie [1], Wylie and Streeter [2], Fox [3], and Chaudry [4]. The purpose of this chapter is simply to introduce the reader to simple concepts and analysis of waterhammer and examples of waterhammer in nuclear systems.

9.2 Basic Concepts

Waterhammer is a pressure wave, usually resulting from relatively rapid changes in the rate of flow in a pipe. Depending upon whether the flow is reduced or increased, and whether the change in flow occurs upstream or downstream in a pipe, the pressure may either rise or fall. From the standpoint of visualization of the phenomenon, as well as for the ease of analysis, waterhammer pressure variation in a pipe is frequently considered to be comprised of a series of abrupt waves, each of difference amplitude, and each travelling at the speed of sound in the pipe.

Wave Speed

Since waterhammer pressure variation throughout a pipe may be visualized as being comprised of a series of abrupt or discontinuous pressure changes, the relationship between pressure change and velocity change must be known. Referring to Fig. 9.1, in which a pressure pulse is travelling upstream past a section of pipe, the basic relationship between pressure change and velocity change is formulated. The undisturbed conditions correspond to p and V. From the standpoint of the derivation of the basic relationship, it is convenient to transform the unsteady flow situation shown on Fig. 9.1 to a so-called steady-flow counterpart, simply by adding to all velocities the wave propagation velocity, which is assumed to be a constant. The same effect could be produced by imagining that the observer is riding along with the wave at speed a relative to the fluid.

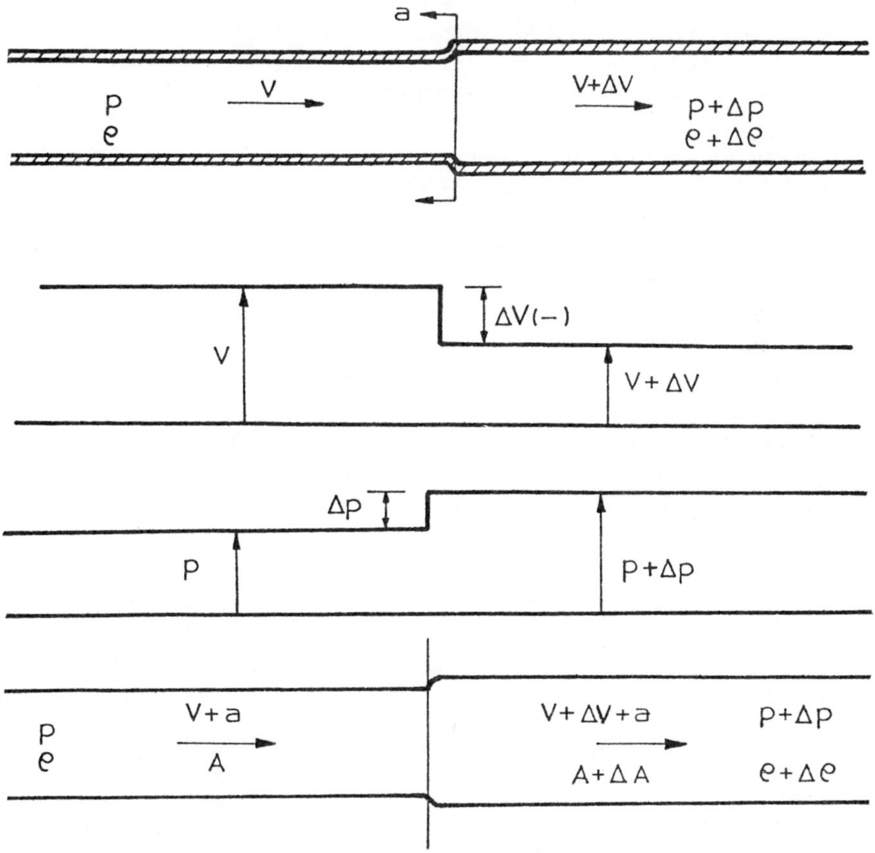

Fig. 9.1. Definition sketch of simple wave

The mass balance equation for the transformed steady flow is given by

$$d\,[\rho(V + a)\,A] = 0 \tag{9.2.1}$$

in which ρ is the mass density of the fluid and A is the cross-sectional area of the conduit. If the Mach number is defined as

$$M \equiv \frac{V}{a} \tag{9.2.2}$$

then Eq. (9.2.1) is expressed

$$\frac{dV}{a} = -(1 + M)\left(\frac{d\rho}{\rho} + \frac{dA}{A}\right) \tag{9.2.3}$$

The fluid compressibility is related to the bulk modulus

$$K = \frac{dp}{d\rho/\rho} \tag{9.2.4}$$

Referring to Fig. 9.2, the relative area change dA/A is expressed in terms of the pipe radius R

$$\frac{dA}{A} = 2\frac{dR}{R} \tag{9.2.5}$$

or

$$\frac{dA}{A} = -2\frac{de}{e} \tag{9.2.6}$$

by invoking Tresca's condition.

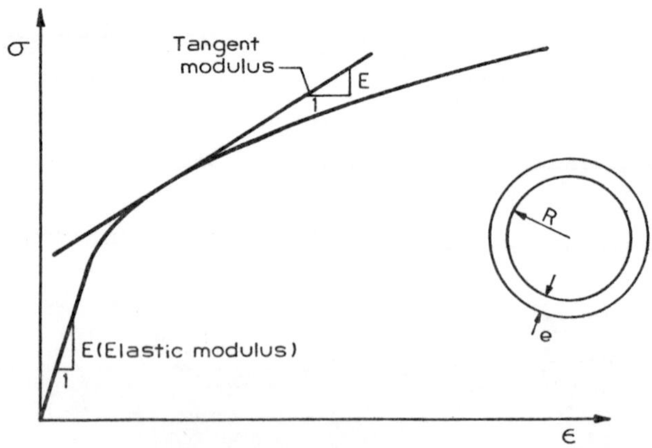

Fig. 9.2. Definition of tangent modulus and Young's modulus

By defining the tangent elastic modulus

$$E = \frac{d\sigma}{d\epsilon} \tag{9.2.7}$$

in which σ is the circumferential pipe stress

$$\sigma = \frac{pR}{e} \tag{9.2.8}$$

and ϵ is the radial strain, and assuming that $M << 1$, Eq. (9.2.3) is expressed in terms of the wave speed

$$a = -\frac{K}{1 + \frac{D}{e}\frac{K}{E - 2\sigma}}\frac{dV}{dp} \tag{9.2.9}$$

The derivative dV/dp is obtained by applying the linear momentum equation to the transformed steady flow of Fig. 9.1, yielding

$$-A\,dp = \rho(V + a)A\,dV \tag{9.2.10}$$

which, for low Mach numbers reduces to the Joukowsky relationship

$$\Delta p = -\rho a \Delta V \tag{9.2.11}$$

Inserting dV/dp from Eq. (9.2.10) into Eq. (9.2.9) yields the wave speed for a circular pipe of diameter D

$$a = \left[\frac{K/\rho}{1 + \frac{D}{e}\frac{K}{E - 2\sigma}}\right]^{1/2} \tag{9.2.12}$$

Equation (9.2.12) is valid for both elastic deformation, in which for metal pipes the tangent modulus becomes Young's modulus and $E >> 2\sigma$, and plastic deformation, in which σ can be of the same order of magnitude as the tangent modulus E. For elastic behavior of metal pipes

$$a = \left[\frac{K/\rho}{1 + \frac{D}{e}\frac{K}{E}}\right]^{1/2} \tag{9.2.13}$$

The effect of pipe wall elasticity relative to fluid compressibility is represented by the second term of the denominator of Eq. (9.2.13). For a rigid pipe, or for wave propagation in an infinite fluid medium, Eq. (9.2.13) reduces to the familiar expression for the acoustic velocity

$$a_o = \sqrt{K/\rho} \tag{9.2.14}$$

Using $K = 0.3 \times 10^6$ psi, the speed of sound in water without any pipe wall effects is 4720 ft/sec from Eq. (9.2.14). Because of elasticity of the pipe wall, waterhammer pressures are quite dependent upon the size, thickness, and nature of the pipe itself. For steel pipe, in which $E = 30 \times 10^6$ psi, the wave speed a from Eq. (9.2.13) varies from 4450 ft/sec for D/e = 10 to 2400 ft/sec for D/e = 300.

In conduits with a noncondensible gas present, either by entrainment or by dissolution from the liquid as a result of depressurization or temperature rise, the acoustic wave speed can be greatly reduced. From Wallis [5] the pulse propagation speed in a two-phase homogeneous mixture is given by

$$a^2 = \frac{1}{[\alpha\rho_g + (1-\alpha)\rho_\ell]\left[\dfrac{\alpha}{\rho_g a_g^2} + \dfrac{1-\alpha}{\rho_\ell a_\ell^2}\right]} \tag{9.2.15}$$

in which ρ_g and ρ_ℓ are mass densities of the gas and liquid phases, respectively; a_g and a_ℓ are single-phase wave speeds; and α is the void fraction.

Waterhammer as a Result of Instantaneous Valve Closure

In order that the concepts of waterhammer may be explained, a discussion of the sequence of events subsequent to the instantaneous closure of a downstream valve in a reservoir-pipeline-valve system is given. For simplification the valve closure is assumed to be instantaneous, the initial flow is assumed to be so small that the hydraulic grade line (HGL) in the pipe is essentially reservoir level, and there is no friction loss in the flow before or during the transient. The initial condition of steady flow is shown in Fig. 9.3a. Following valve closure at t = 0 the sequence of events is as follows.

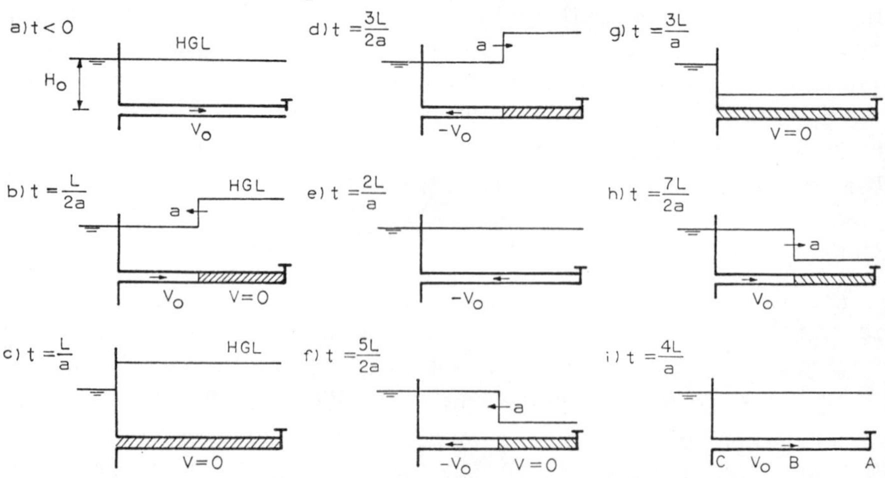

Fig. 9.3. Sequence of events subsequent to instantaneous valve closure

Initially only a few layers of liquid near the valve are brought to rest. The deceleration necessary to instantaneously stop layer by layer of the liquid leads to a pressure wave propagating toward the reservoir at the speed of sound a. If L is the length of the pipe, all liquid in the downstream half of the pipe is compressed and at rest at t = L/2a, as shown in Fig. 9.3b. The rise in the HGL is given by Eq. (9.2.11) in terms of head, aV_0/g, since the velocity change is $-V_0$ and since $\Delta H = \Delta p/\rho g$. As more and more liquid is brought to rest, the wave finally reaches the reservoir at t = L/a, as shown in Fig. 9.3c. At this instant an unstable condition exists as the pressure or head in the pipe is greater than that in the reservoir. Since according to Eq. (9.2.11), a pressure change must be accompanied by a velocity change, the liquid begins to flow back in and toward the reservoir layer by layer, resulting in a negative wave (or pressure drop) propagating downstream.

At t = 3L/2a, Fig. 9.3d, the negative wave has reached the midpoint. Since the round trip time of travel for a wave to leave the valve and return is 2L/a, the pressure at the valve remains at its elevated level until t = 2L/a. At this time, as shown in Fig. 9.3e, the pressure in the pipe is at the original level but all the liquid is flowing toward the reservoir, resulting in another unstable or temporary condition. Since the liquid at the valve must be arrested, a negative wave is propagated upstream, decelerating the column of water layer by layer. The resulting pressure necessary to decelerate the liquid has led to a pressure less than the original static pressure, as shown in Figs. 9.3f and 9.3g, which respectively correspond to the wave being at the midpoint at t = 5L/2a and at the reservoir at t = 3L/a. The overall reduced pressure in the pipe allows a positive wave to propagate downstream, accelerating each layer of the otherwise static water. The effect of the wave, shown for t = 7L/2a in Fig. 9.3h and for t = 4L/a in Fig. 9.3i, brings the liquid back to its original pressure and velocity. In fact, at t = 4L/a the conditions are indentical to those for $t < 0$.

Because of the assumption of no friction or boundary resistance in the pipe during the transient and the assumption of no loss of kinetic energy as the liquid is flowing into the reservoir, the phenomenon continues forever, having a period of 4L/a. The pressure head at the valve and at the midpoint is shown in Fig. 9.4. The velocity at the reservoir is also shown in Fig. 9.4. During the transient the velocity at the valve is zero and the head at the reservoir end of the pipe remains constant at the reservoir level. If the reduced pressure at the valve at t = 2L/a is less than the liquid vapor pressure at that temperature, a vapor pocket will form, greatly altering the results presented in Figs. 9.3 and 9.4. In reality the results are altered because no valve closure is exactly instantaneous and friction always tends to resist and dampen motion.

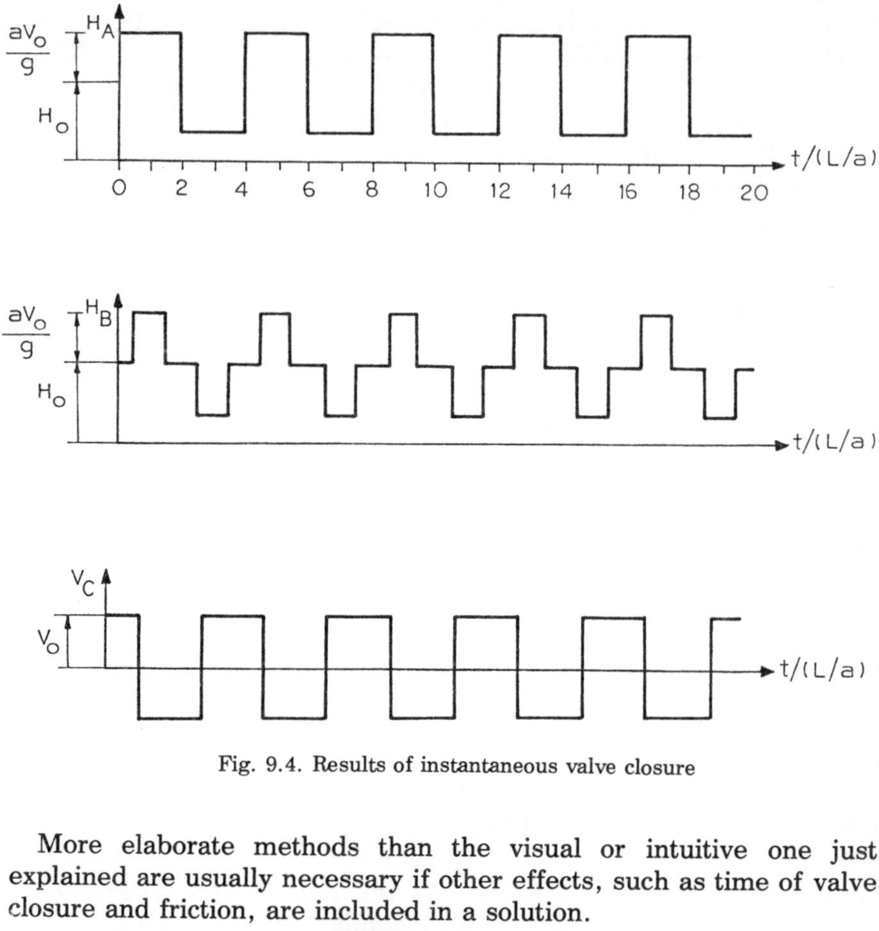

Fig. 9.4. Results of instantaneous valve closure

More elaborate methods than the visual or intuitive one just explained are usually necessary if other effects, such as time of valve closure and friction, are included in a solution.

Wave Reflections and Attenuation

If boundary friction is neglected in an elastic, rigidly held pipe, a pressure wave will propagate undiminished in a single-phase liquid in a uniform conduit. Reflection and transmission of pressure waves at pipe junctions are usually adequately represented by one-dimensional equations with proper boundary conditions if minor losses in head are not too great. Pressure waves can also be attenuated, however, by plastic deformation, by piping motion, mechanical or structural damping, or by fittings and bends. Indeed, for the catastrophic type of waterhammer witnessed in some feedwater piping systems, structural damping in the above modes may be much greater than fluid friction, which is frequently the only type included in transient analyses.

The reflection and transmission of pressure waves through pipe junctions can be reasonably predicted by applying simple theory if waves are assumed to remain plane. For the multiple pipe junction shown in Fig. 9.5, the cross-sectional areas are A_i and A_j where A_i corresponds to the pipe with the incident wave and A_j corresponds to attached pipes. Similar notation is applied to the wave speeds a_i and

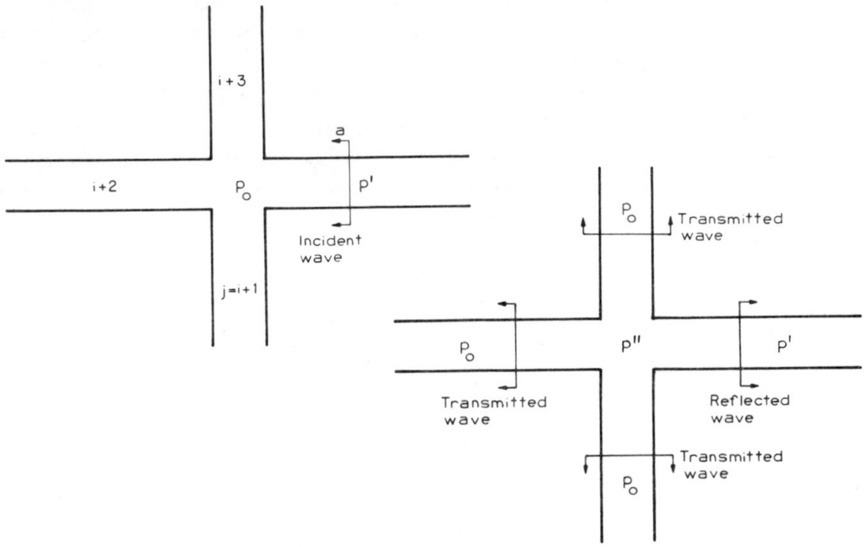

Fig. 9.5. Reflection and transmission of waves at junction

a_j . Satisfying the Joukowsky relationship, Eq. (9.2.11), for the incident, reflected, and transmitted waves shown in Fig. 9.5 yields a transmission coefficient

$$s = \frac{p'' - p_o}{p' - p_o} = \frac{2A_i/a_i}{\dfrac{A_i}{a_i} + \displaystyle\sum_{j=i+1}^{i+n} \frac{A_j}{a_j}} \tag{9.2.16}$$

in which p_o is the initial junction pressure, p' is the pressure behind the incident wave, p'' is the final junction pressure behind the reflected wave and transmitted waves, and n is the number of pipes attached to pipe i. The reflection coefficient r for pipe i is simply

$$r = s - 1 \tag{9.2.17}$$

For a two-pipe junction for which $a_1 = a_2$, $i = 1$, $n = 1$, and $A_2/A_1 = 2$ (pipe expansion), s = 2/3 and r = -1/3, meaning that 2/3 of the pressure change of the incident wave is transmitted with like sign, while the reflected wave is of different sign but 1/3 the amplitude of the incident wave. The incident pressure pulse can in fact be intensified if the wave approaches a junction for which $A_2 < A_1$ (pipe contraction) and/or $a_2 > a_1$. For example, for $a_1 = a_2$ and $A_2 = A_1/2$, $s = 4/3$ and $r = 1/3$. It should be emphasized that these reflection coefficients are implicitly included in the subsequent 1-D analysis using the method of characteristics by proper application of the junction boundary conditions. The plane wave assumption usually yields reasonable results for engineering purposes. The two-dimensional character of waves at a junction and the complex interaction problem has been analyzed by Shin and Daley [6] and Shin and Valentin [7]. Figure 9.6 shows a comparison of results of a two-dimensional method of characteristics with the plane wave approach with fluid friction neglected in both instances.

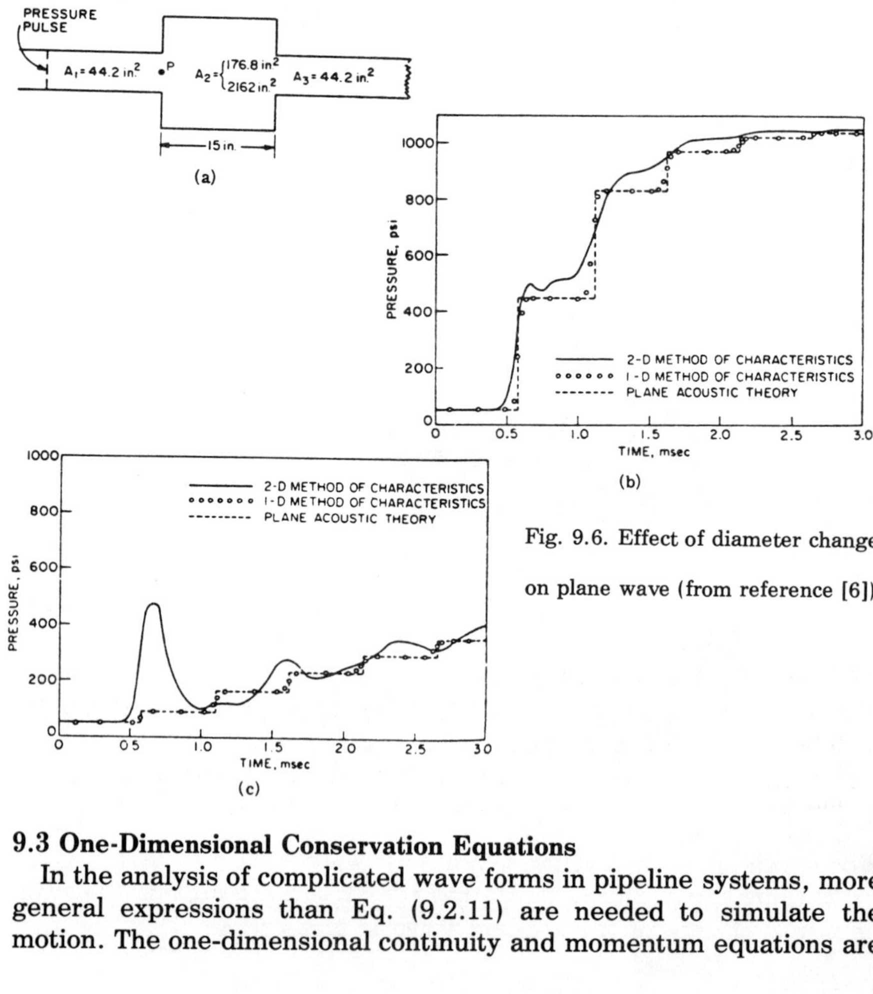

Fig. 9.6. Effect of diameter change on plane wave (from reference [6])

9.3 One-Dimensional Conservation Equations

In the analysis of complicated wave forms in pipeline systems, more general expressions than Eq. (9.2.11) are needed to simulate the motion. The one-dimensional continuity and momentum equations are derived by considering a mass balance and force balance on a fluid volume element in the conduit. For details or a thorough derivation the reader is referred to Wylie and Streeter [2]. The fluid is considered to be single phase and the conduit uniform in area and elastic in wall behavior. The equations are one-dimensional in that only variations in the flow or wave passage direction are explicitly considered.

Conservation of Mass

In the absence of any phase change the mass balance equation applied to a control volume of fluid in an elastic conduit yields the 1-D continuity equation

$$\frac{\partial}{\partial t}(\rho A) + \frac{\partial}{\partial x}(\rho A V) = 0 \tag{9.3.1}$$

in which V, as before, is the cross-sectional average velocity. If the material differential operator is defined as

$$\frac{d}{dt} \equiv \frac{\partial}{\partial t} + V\frac{\partial}{\partial x} \tag{9.3.2}$$

Eq. (9.3.1) is rewritten as

$$\frac{1}{\rho}\frac{d\rho}{dt} + \frac{1}{A}\frac{dA}{dt} + \frac{\partial V}{\partial x} = 0 \tag{9.3.3}$$

Using Eqs. (9.2.4) through (9.2.8) and the wave speed, Eq. (9.2.13), Eq. (9.3.3) becomes

$$\frac{\partial H}{\partial t} + V\frac{\partial H}{\partial x} + \frac{a^2}{g}\frac{\partial V}{\partial x} = 0 \tag{9.3.4}$$

in which H is the piezometric head, defined as

$$H \equiv \frac{p}{\rho g} + z \tag{9.3.5}$$

where z is the elevation head.

Linear Momentum

The application of Newton's Second Law to the same fluid control volume yields the conservation equation for linear momentum

$$-\frac{\partial H}{\partial x} - \frac{4\tau_W}{\rho g D} = \frac{1}{g}\frac{dV}{dt} \tag{9.3.6}$$

for a circular pipe, in which the boundary shear stress is τ_W . It is common to represent the unsteady flow boundary shear stress τ_W in terms of the steady flow Darcy-Weisbach factor

$$\tau_W = \frac{f_W}{8} \rho V|V| \tag{9.3.7}$$

for which the absolute value sign allows for boundary shear stress reversal with flow reversal. It is implicitly assumed that the fluid resistance is in phase with the mean velocity, a condition that is not exactly satisfied in reality. Employing Eq. (9.3.7) and the definition for the total acceleration

$$\frac{dV}{dt} \equiv \frac{\partial V}{\partial t} + V \frac{\partial V}{\partial x} \tag{9.3.8}$$

the momentum equation becomes

$$\frac{\partial V}{\partial t} + V \frac{\partial V}{\partial x} + g \frac{\partial H}{\partial x} + \frac{f_W}{2D} V|V| = 0 \tag{9.3.9}$$

9.4 Method of Characteristics

Eqs. (9.3.4) and (9.3.9) represent the one-dimensional continuity and momentum relationships in partial differential form, respectively. In Eq. (9.3.4) the ratio of the second term $V\,\partial H/\partial x$ to the first term $\partial H/\partial t$ is equivalent to the Mach number relative to unity, as is the ratio of the second to the first term in Eq. (9.3.9). For most single-phase pressure transient problems with liquids $M << 1$, allowing the second term in the two equations to be neglected, resulting in

$$\frac{\partial H}{\partial t} + \frac{a^2}{g} \frac{\partial V}{\partial x} = 0 \tag{9.4.1}$$

and

$$\frac{\partial V}{\partial t} + g \frac{\partial H}{\partial x} + \frac{f_W}{2D} V|V| = 0 \tag{9.4.2}$$

Wave Equations

For the frictionless or ideal fluid case, for which $f_W = 0$, Eqs. (9.4.1) and (9.4.2) can be operated on by $\partial/\partial t$ and $\partial/\partial x$ to yield the classical one-dimensional wave equations of mathematical physics

$$\frac{\partial^2 H}{\partial t^2} = a^2 \frac{\partial^2 H}{\partial x^2} \tag{9.4.3}$$

and

$$\frac{\partial^2 V}{\partial t^2} = a^2 \frac{\partial^2 V}{\partial x^2} \tag{9.4.4}$$

The general solution to Eqs. (9.4.3) and (9.4.4) is

$$V = V_o - \frac{g}{a} [F_1(t + x/a) - F_2(t - x/a)] \tag{9.4.5}$$

and

$$H = H_o + F_1(t + x/a) + F_2(t - x/a) \tag{9.4.6}$$

in which the functions F_1 and F_2 correspond to waves or wavelets propagating in the negative and positive x-directions, respectively. Although a few so-called closed form mathematical solutions to Eqs. (9.4.3) and (9.4.4) exist, the most common use of Eqs. (9.4.5) and (9.4.6) is by application of the graphical technique, Parmakian [8]. In that case initial and boundary conditions are introduced graphically so that the functions F_1 and F_2 are evaluated. Even friction can be added artificially.

With the advent of high-speed digital computers, it is more practical to solve Eqs. (9.4.1) and (9.4.2) numerically, especially for complex pipe networks and complicated boundary conditions. In the following section, the method of characteristics is developed, as well as finite-difference techniques, which can include the nonlinear frictional characteristic of turbulent flow.

Transformation of Partial Differential Equations

By a suitable choice of multipliers Eqs. (9.4.1) and (9.4.2) are combined to form what are known as the **characteristic equations**. The procedure consists of multiplying Eq. (9.4.1) by $\pm g/a$, and adding the result to Eq. (9.4.2) yielding

$$\pm \frac{g}{a}\frac{\partial H}{\partial t} + g\frac{\partial H}{\partial x} + \frac{\partial V}{\partial t} \pm a\frac{\partial V}{\partial x} + \frac{f_W}{2D} V|V| = 0 \tag{9.4.7}$$

The first four terms of Eq. (9.4.7) can be expressed as two total derivatives if the operator

$$\frac{d}{dt} \equiv \frac{\partial}{\partial t} + \frac{dx}{dt}\frac{\partial}{\partial x} \tag{9.4.8}$$

is defined such that

$$\frac{dx}{dt} \equiv \pm a \tag{9.4.9}$$

in contrast to Eq. (9.3.2), for which dx/dt = V. The two total derivatives

$$\frac{dH}{dt} = \frac{\partial H}{\partial t} \pm a \frac{\partial H}{\partial x} \tag{9.4.10}$$

and

$$\frac{dV}{dt} = \frac{\partial V}{\partial t} \pm a \frac{\partial V}{\partial x} \tag{9.4.11}$$

are defined along the so-called characteristic lines C^+ and C^-, for which Eq. (9.4.9) is valid. Using Eqs. (9.4.10) and (9.4.11), the combined continuity and momentum relationship, Eq. (9.4.7), becomes

$$\pm \frac{g}{a}\frac{dH}{dt} + \frac{dV}{dt} + \frac{f_W}{2D} V|V| = 0 \tag{9.4.12}$$

The positive, or forward characteristic is C^+, along which

$$\frac{dx}{dt} = a \tag{9.4.13}$$

and

$$\frac{g}{a}\frac{dH}{dt} + \frac{dV}{dt} + \frac{f_W}{2D} V|V| = 0 \tag{9.4.14}$$

The equation of the negative or backward characteristic C^- is

$$\frac{dx}{dt} = -a \tag{9.4.15}$$

along which

$$-\frac{g}{a}\frac{dH}{dt} + \frac{dV}{dt} + \frac{f_W}{2D} V|V| = 0 \tag{9.4.16}$$

If a is constant, then Eqs. (9.4.13) and (9.4.15) are straight lines on the x-t plane, along which disturbances propagate, as shown in Fig. 9.7. It should be stressed that Eq. (9.4.14) must be satisfied along the C^+ characteristic, while Eq. (9.4.16) applies only on the C^- characteristic. Actually pressure waves travel at a velocity $V \pm a$ rather than $\pm a$, but in most cases of waterhammer $V << a$, allowing for the simplification made earlier. In the more general case the characteristic lines are curved, but in the approximation assumed by the use of Eq. (9.4.13) and the assumption a is constant, they are straight. Solution to these four equations with appropriate initial and boundary conditions constitutes a solution to the given boundary-value problem. As the boundary friction is usually nonlinear, the only practical manner to solve the equations in this form is numerically.

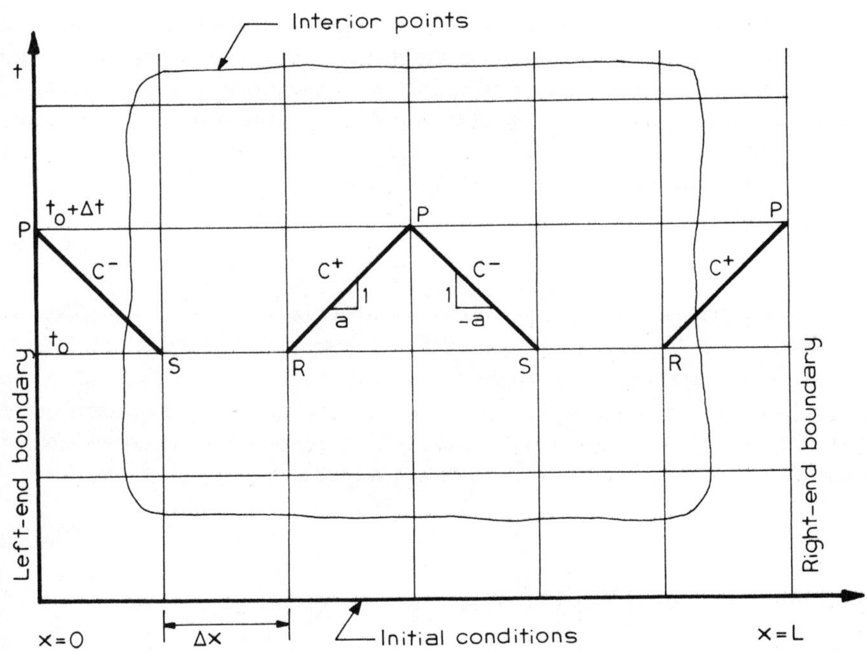

Fig. 9.7. Characteristic lines on x-t plane

Riemann Invariants

For cases in which friction is assumed negligible ($f_W = 0$), Eq. (9.4.12) becomes

$$\frac{d}{dt}\left[\pm\frac{gH}{a} + V\right] = 0 \qquad (9.4.17)$$

or

$$H + \frac{aV}{g} = H_1 + \frac{aV_1}{g} \qquad (9.4.18)$$

on the C^+ characteristic line and

$$H - \frac{aV}{g} = H_1 - \frac{aV_1}{g} \qquad (9.4.19)$$

on the C^- characteristic line, for which H_1 and V_1 are constants. Eqs. (9.4.18) and (9.4.19) are known as the Riemann invariants.

9.5 Numerical Integration

The numerical integration procedure employed herein is very simple, allowing for an explicit non-iterative solution in most cases. As shown in Fig. 9.7, the x-t plane is divided into a rectangular grid comprised of constant Δx and Δt increments. It can be proved that the time increment Δt is related to the distance increment by the Courant criterion in order to insure numerical stability

$$\Delta t \leqslant \frac{\Delta x}{a} \qquad (9.5.1)$$

For a single pipe analysis for which a is constant, equality is recommended for Eq. (9.5.1) in order to improve numerical accuracy.

The boundary-value problem to be solved for V(x,t) and H(x,t) is explained by reference to Fig. 9.7. The initial conditions are specified at t = 0. In most cases the initial conditions correspond to a known steady flow at t = 0. The velocity is constant along x for the uniform pipe

$$V(x,0) = V_o \qquad (9.5.2)$$

and the variation in head is related to the rate of head loss

$$H(x,0) = H_o - \frac{f_W x}{D}\frac{V_o^2}{2g} \qquad (9.5.3)$$

in which H_o is the head at x = 0.

Boundary conditions are specified at the left end (x = 0) and right end (x = L) of the pipe either as functions of time or interrelationships between head and velocity. Typical boundary conditions are a reservoir, an operating valve, pump or turbine, junction of pipes, an air chamber, and surge tank, to name a few. Once a relationship is established between the head and discharge (or pipe velocity) for the boundary condition, this relationship must be solved in conjunction with the characteristic equation that applies from the characteristic line that joins the adjacent grid point in the pipe or pipes. A reservoir boundary condition at the left end of the pipe could be simply

$$H(0,t) = H_o \tag{9.5.4}$$

in which H_o is either a constant or a function of time $H_o(t)$. A closed valve or dead end boundary condition at the right end is

$$V(L,t) = 0 \tag{9.5.5}$$

It should be mentioned that for subsonic flow only one of the two dependent variables, V or H, can be **a priori** specified as a boundary condition. The other quantity, which is an unknown, is solved by coupling the solution along the particular characteristic line, C^+ or C^-, with the boundary condition. Frequently the boundary condition is not a time-dependent function but an interrelationship between H and V. A good example is an operating valve, as is shown later.

The x-t plane shown in Fig. 9.7 is divided into initial (t = 0) and boundary (x = 0 and x = L) condition lines, along which there are boundary points, and interior grid points or nodes. It is assumed that, for any time t_o, values of H and V are known for all points along x. The problem is to determine the unknown values of H and V at points P at the next time $t_o + \Delta t$. The point P may be an interior point, for which the solution is determined by simultaneously integrating along the C^+ characteristic line from R to P and the C^- characteristic line from S to P. The boundary value at x = 0 is determined by coupling the integration along C^- from S to P with the boundary condition. The solution at the right end is similarly obtained by incorporating the integration along C^+ from R to P with the specified boundary condition.

Integration Along Characteristic Lines

The equations of the straight lines from R to P and S to P are given simply by Eqs. (9.4.13) and (9.4.15), respectively. Along R-P, however, the characteristic Eq. (9.4.14) must be satisfied, while along S-P Eq.

(9.4.16) governs. If a and f_w are assumed constant the integrations may be expressed

$$C^+ : \quad \frac{g}{a}\int_{H_R}^{H_P} dH + \int_{V_R}^{V_P} dV + \frac{f_w}{2D}\int_{t_o}^{t_o+\Delta t} V|V|\,dt = 0 \qquad (9.5.7)$$

$$C^- : \quad -\frac{g}{a}\int_{H_S}^{H_P} dH + \int_{V_S}^{V_P} dV + \frac{f_w}{2D}\int_{t_o}^{t_o+\Delta t} V|V|\,dt = 0 \qquad (9.5.8)$$

The complication arises with the third term of Eqs. (9.5.7) and (9.5.8) because of the unknown variation of V along R to P and S to P, respectively. The first two terms of each equation would produce the Riemann invariants if $f_W = 0$. Typically the friction term is either evaluated using a simple rectangular integration, which along the C^+ is approximated by

$$\int_{t_o}^{t_o+\Delta t} V|V|\,dt = V_R|V_R|\,\Delta t \qquad (9.5.9)$$

or the trapezoidal rule, which along the C^- is

$$\int_{t_o}^{t_o+\Delta t} V|V|\,dt = \frac{1}{2}\left[V_S|V_S| + V_P|V_P|\right]\Delta t \qquad (9.5.10)$$

It is apparent that Eqs. (9.5.7) and (9.5.8) become algebraically linear if the rectangular rule is applied, or nonlinear in terms of V_P^2 in the case of the trapezoidal rule. It has been demonstrated that, in conduits for which friction is not the predominant force, the rectangular rule usually suffices. If Eq. (9.5.9) is used for integration along C^+ and its counterpart is applied along C^-, then Eqs. (9.5.7) and (9.5.8) can be expressed as linear relationships between the unknowns H_P and V_P

C^+ :

$$V_P = V_R + \frac{g}{a}H_R - \frac{f_W}{2D}V_R|V_R|\,\Delta t - \frac{g}{a}H_P \qquad (9.5.11)$$

or

$$V_P = C_3 - C_2 H_P \qquad (9.5.12)$$

$$C^{-}: \quad V_P = V_S - \frac{g}{a} H_S - \frac{f_W}{2D} V_S \left| V_S \right| \Delta t + \frac{g}{a} H_P \tag{9.5.13}$$

or

$$V_P = C_1 + C_2 H_P \tag{9.5.14}$$

The quantities C_3 and C_1 are constants for the respective points R and S for the time t_o and $C_2 = g/a$. The linear relationships, Eqs. (9.5.12) and (9.5.14) facilitate use of the method of characteristics. For interior points labeled P the unknowns H_P and V_P are determined by simultaneous solution of Eqs. (9.5.11) and (9.5.13). Equation (9.5.12) would be applied along the C^{+} characteristic from R to P for solution at the right boundary, while Eq. (9.5.14) would be incorporated with the left end boundary condition.

Interior Points

For any interior point P the simultaneous solution of Eqs. (9.5.11) and (9.5.13) yields

$$H_P = \frac{1}{2} [H_R + H_S + \frac{a}{g} (V_R - V_S) - \frac{a}{g} \frac{f_W}{2D} \Delta t \, (V_R |V_R| - V_S |V_S|)] \tag{9.5.15}$$

and

$$V_P = \frac{1}{2} [V_R + V_S + \frac{g}{a} (H_R - H_S) - \frac{f_W}{2D} \Delta t \, (V_R |V_R| + V_S |V_S|)] \tag{9.5.16}$$

Equations (9.5.15) and (9.5.16) allow calculation of the unknown head and velocity at all interior points in the x-t plane at time $t_o + \Delta t$ quite irrespective of each other. The solution only depends upon values of H and V at grid points at time t_o . Because the solution for H_P and V_P depends only upon values at R and S, the method is said to be explicit. Other schemes that involve a system of equations for unknowns at time $t_o + \Delta t$ are termed implicit. If the pipe is divided into N reaches of length Δx , then there are N + 1 total grid points and N - 1 interior points. Only the solution for interior points are given by Eqs. (9.5.15) and (9.5.16). The unknowns at x = 0 and x = L result from a simultaneous solution of the appropriate characteristic equation and boundary condition being applied at that particular end.

Left-End Boundary Condition

At x = 0 the C^- characteristic equation from S to P yields.

$$V_P = C_1 + C_2 H_P \tag{9.5.17}$$

For a reservoir boundary condition for which entrance losses and pipe velocity head are assumed negligible, or $H_P = H_o$, the constant head, the solution for V_P is simply

$$V_P = C_1 + C_2 H_o \tag{9.5.18}$$

Another simple case exists for a closed valve or dead end at x = 0, for which $V_P = V_R = 0$, resulting in

$$H_P = -\frac{C_1}{C_2} \tag{9.5.19}$$

For a turbomachine at x = 0, not only must there be included the head-discharge characteristics of the pump, but also the unbalanced torque on the shaft, especially following power failure of a pump motor. The details of transient calculations associated with hydraulic machinery are included in a subsequent section.

Right-End Boundary Condition

For a right-end boundary Eq. (9.5.12) applies along the C^+ characteristic from R to P

$$V_P = C_3 - C_2 H_P$$

A very common right-end boundary condition is an operating valve. The head-discharge relationship for the valve must be solved simultaneously with the above linear equation. If, as shown in Fig. 9.8, the head loss across the valve is ΔH, the flow is given by

$$Q = C_d A_v \sqrt{2g\Delta H} \tag{9.5.20}$$

in which C_d is a discharge coefficient, and A_v is the cross-sectional area of the valve opening.

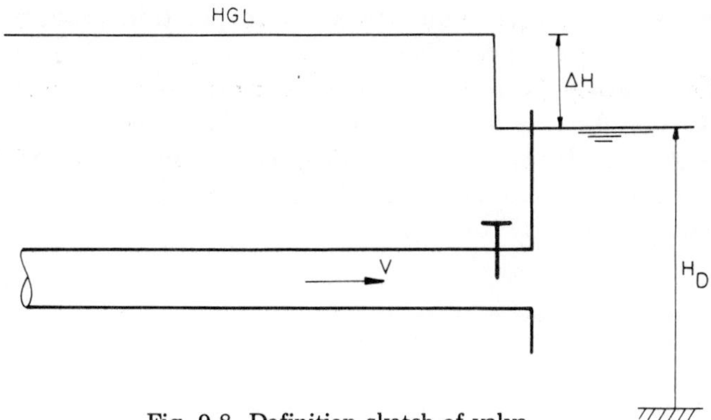

Fig. 9.8. Definition sketch of valve

If V is the mean velocity in the pipe, the head drop can also be expressed as

$$\Delta H = K \frac{V^2}{2g} \tag{9.5.21}$$

in which K is the head loss coefficient. In transient analyses it is very common to relate the flow at any opening to that at fully open position, namely

$$Q_o = C_{do} A_{vo} \sqrt{2g\Delta H_o} \tag{9.5.22}$$

and

$$\Delta H_o = K_o \frac{V_o^2}{2g} \tag{9.5.23}$$

The ratio of Q to Q_o and V to V_o becomes

$$\frac{Q}{Q_o} = \frac{VA}{V_o A} = \frac{C_d A_v}{C_{do} A_{vo}} \sqrt{\frac{\Delta H}{\Delta H_o}} = \sqrt{\frac{K_o \Delta H}{K \Delta H_o}} \tag{9.5.24}$$

As the ratio of discharge coefficients and areas depends upon valve geometry and a complicated flow pattern, it is more convenient to introduce an effective flow coefficient τ , defined as

$$\tau \equiv \frac{C_d A_v}{C_{do} A_{vo}} = \sqrt{\frac{K_o}{K}} \tag{9.5.25}$$

The value of τ ranges from 0 at a closed position to unity when fully open, as ΔH_o corresponds to the wide open case. Only the shape of th τ curve varies from the open to closed position from one valve design to another, as shown in Fig. 9.9. The stroke may be given in degrees or in distance, depending upon the type of valve. For motor-operated valves for which the motion is linear, time can simply replace the stroke axis on Fig. 9.9.

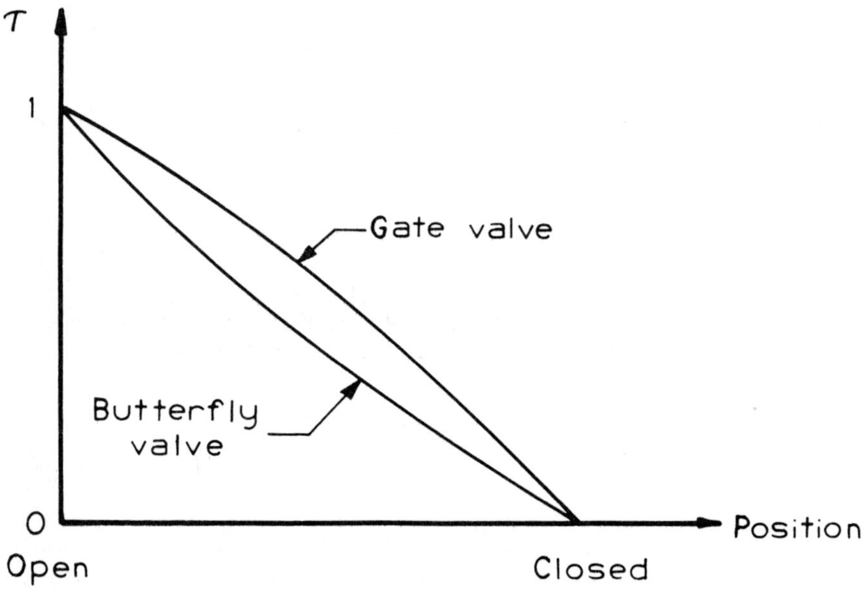

Fig. 9.9. Typical tau curves for valve

The two equations to be solved simultaneously are Eq. (9.5.12) and

$$V_P = \tau V_o \left[\frac{H_P - H_D}{\Delta H_o} \right]^{1/2} \tag{9.5.26}$$

in which H_D is the reservoir level downstream of the valve. Upon combining these two expressions one obtains a quadratic equation in V_P

$$V_P^2 + \frac{\tau^2 V_o^2}{C_2 \Delta H_o} V_P - \frac{\tau^2 V_o^2}{C_2 \Delta H_o}(C_3 - C_2 H_D) = 0 \tag{9.5.27}$$

Defining

$$C_4 = 2\frac{a\tau^2}{K_o} \tag{9.5.28}$$

and

$$C_3' = C_3 - C_2 H_D \tag{9.5.29}$$

then

$$V_P = \frac{1}{2}\left[-C_4 \pm\sqrt{C_4^2 + 4C_3' C_4}\right] \tag{9.5.30}$$

As the flow can be either positive or negative, and since $C_4 > 0$, the plus sign on the radical must be employed. The unknown head upstream of the valve is then given by

$$H_P = \frac{C_3 - V_P}{C_2} \tag{9.5.31}$$

Interpolation

The calculation procedure just discussed will yield quite accurate results from a numerical standpoint inasmuch as the wave speed is a constant, allowing for the intersection of characteristic lines to coincide with the nodes of the grid. In practice the characteristic lines frequently do not span from grid point to grid point because of (1) the usual inequality of $\Delta x/a$ for a multiple pipeline system, and (2) the variability of wave speed due to two-phase flow or plastic deformation.

Even for single-phase flow in a two pipe system, characteristic lines will not intersect as shown on Fig. 9.7 if $\Delta x/a$ differs for the two pipes, as shown in Fig. 9.10. The technique applied earlier will only be valid if H and V are known at points R and S. As explained by Streeter and Wylie [1], interpolation procedures can be employed to approximate the values at R and S from the known values at adjacent grid points. It has been learned since the publication of Ref. 1, however, that unless points R and S are quite close to adjacent grid points, for which the interpolation error is small, numerical errors can be significant.

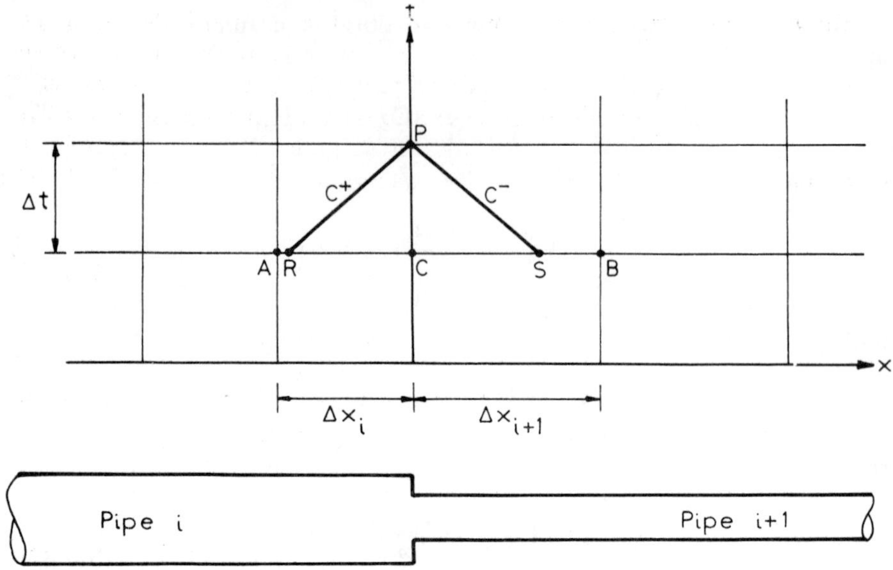

Fig. 9.10. Characteristic lines at junction of two pipes

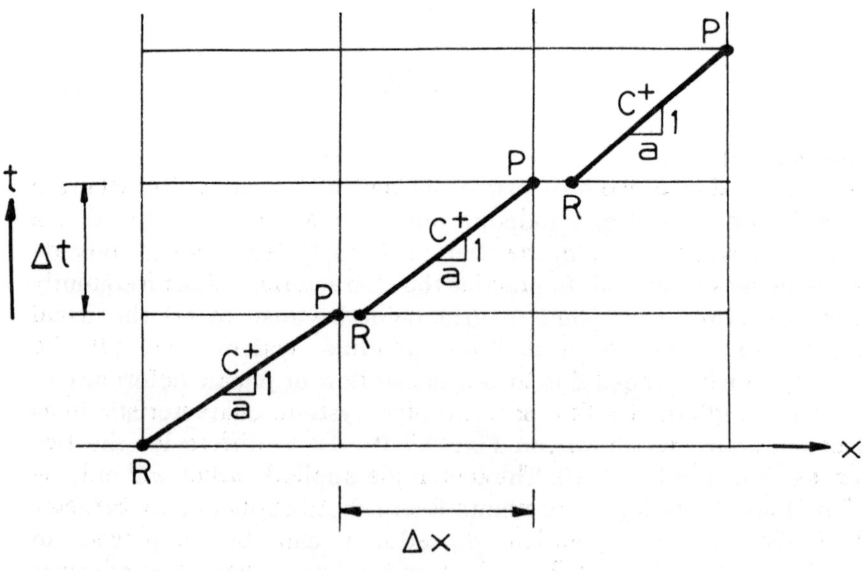

Fig. 9.11. Characteristic lines in pipe with variable wave speed

Another example of non-intersection of characteristic lines at rectangular grid points is the two-phase flow with high variable wave speed shown in Fig. 9.11. In order to minimize numerical error, the extent of interpolation should be minimized by increasing the grid points. As shown by Wiggert and Sundquist [9], the errors of interpolation can be reduced by using a modified method of characteristics.

9.6 Pump Characteristics

Waterhammer in fluid-filled lines is frequently caused by sudden stoppage of centrifugal pumps. Not only are the head and torque characteristics needed for the quadrant of normal pumping, but frequently the machine passes through zones of reverse flow and/or reverse shaft speed, depending upon the transients and pipe system and its controls. In any case a complete set of pump characteristics is usually required for transient analysis. Due to the range of flows and rotational speeds a pump experiences during a transient, the machine characteristics must be represented in terms of dimensionless homologous curves, otherwise the task of handling the data becomes virtually impossible.

Homologous Curves

The dimensional and dimensionless heads, discharges, rotational speeds, and torques are defined in Table 1.

Table 1.
Definition of machine characteristics

Symbol	Definition
H	Head across machine
H_R	Rated head.
$h = H/H_R$	Head ratio
Q	Discharge through machine
Q_R	Rated discharge through machine
$v = Q/Q_R$	Discharge ratio. Positive during pumping and negative during energy dissipation and generation modes.
N	Rotational speed of shaft
N_R	Rated speed of machine
$a = N/N_R$	Speed ratio. Positive during pumping and energy dissipation modes and negative during generating mode.
T	Torque on machine.
T_R	Rated torque.
$\beta = T/T_R$	Torque ratio.
η_R	Efficiency at rated conditions

Pump manufacturers typically represent complete characteristics of their machines in the form of so-called synoptic curves, which form what is known as a Karman-Knapp circle diagram. As shown on Fig. 9.12, lines of constant head h and constant torque β are given in terms of percentage flow v and percentage speed a . For a fixed-vane

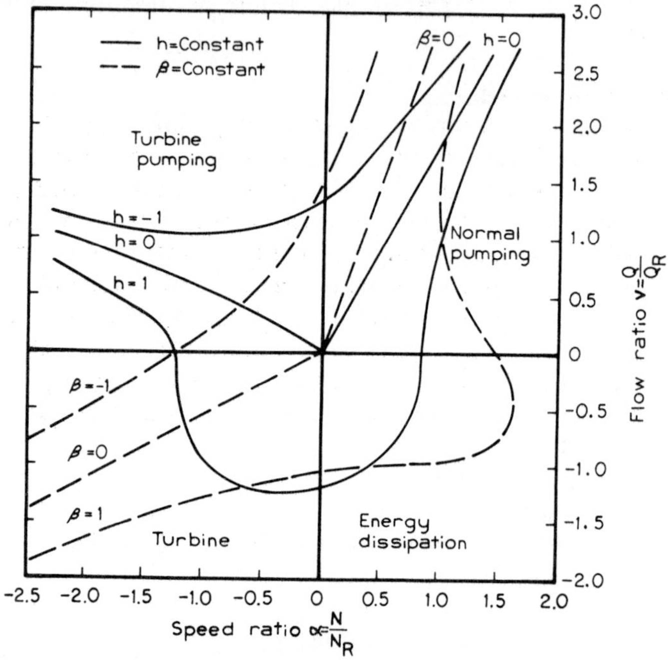

Fig. 9.12. Karman-Knapp circle diagram for centrifugal pump

machine a complete representation of the head and torque curves is accomplished by six single curves; namely, lines of 100 percent head (h = 1) and torque (β = 1), lines of zero head (h = 0) and torque (β = 0), and lines of -100 percent head (h = -1) and torque (β = -1). Although not entirely necessary, curves for intermediate values of head and torque are frequently included as well. The four quadrants are defined as follows:

QUADRANT	DEFINITION	v	a	v/a
I	Normal Pumping	>0	>0	>0
II	Turbine Pumping	>0	<0	<0
III	Normal Turbine	<0	<0	>0
IV	Energy Dissipation	<0	>0	<0

The same data can also be represented in the usual head-discharge and torque-discharge planes in terms of dimensionless ratios

$$\frac{h}{a^2} = F_a(v/a) \tag{9.6.1}$$

and

$$\frac{\beta}{a^2} = G_a(v/a) \tag{9.6.2}$$

for a given pump of fixed size. These relationships are depicted graphically in Figs. 9.13 and 9.14.

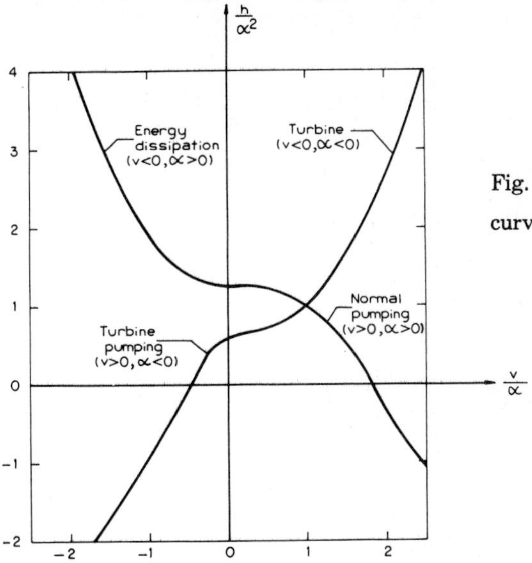

Fig. 9.13. Complete head-discharge curves for centrifugal pump

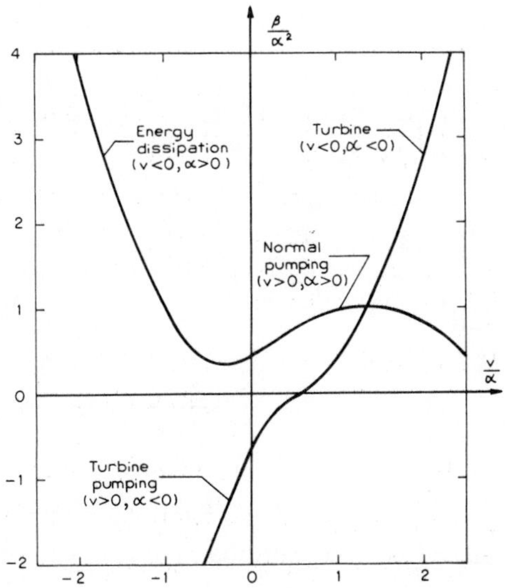

Fig. 9.14. Complete torque-discharge curves for centrifugal pump

Although the curves shown in Figs. 9.13 and 9.14 are useful for clear understanding of pump characteristics, they create difficulties in computer representation for small values of a, for which v/a is large or infinite. Two types of somewhat different representation are currently in use. The first, thoroughly explained by Streeter and Wylie [1], would break each of the four zones of Figs. 9.13 and 9.14 into eight regions by inverting v/a to a/v for $v/a > 1$, allowing for the representation by

$$\frac{h}{a^2} = F_a(v/a) \; ; \left|\frac{v}{a}\right| < 1 \tag{9.6.3}$$

or

$$\frac{h}{v^2} = F_v(a/v) \; ; \left|\frac{v}{a}\right| > 1 \tag{9.6.4}$$

and

$$\frac{\beta}{a^2} = G_a(v/a) \; ; \left|\frac{v}{a}\right| < 1 \tag{9.6.5}$$

or

$$\frac{\beta}{v^2} = G_v(a/v) \; ; \left|\frac{v}{a}\right| > 1 \tag{9.6.6}$$

In the more recent Wylie and Streeter [2], the so-called Suter definition is used; for which

$$\frac{h}{a^2 + v^2} = F_{av}\,(\pi + \tan^{-1} v/a) \tag{9.6.7}$$

and

$$\frac{\beta}{a^2 + v^2} = G_{av}\,(\pi + \tan^{-1} v/a) \tag{9.6.8}$$

The advantage of the representation by Eqs. (9.6.7) and (9.6.8) is that only two continuous curves are needed if principal values of

$\tan^{-1} v/a$ are used. The disadvantage is that the resulting curves are difficult if not impossible to interpret from a physical standpoint.

During a typical transient subsequent to power failure of a pump the machine will pass from quadrant I in Fig. 9.12, through quadrant IV (energy dissipation) of reverse flow and positive rotation, to the zone of turbine operation, quadrant III, for which both the flow and speed are reversed. Quadrants IV and III can be avoided if a discharge valve is closed before flow reversal. In the case of a pump shaft with a non-reversing ratchet, quadrant IV would be avoided subsequent to flow reversal. Centrifugal pumps would rarely, if ever, experience quadrant II during a transient.

Pump Boundary Condition

If a pump is located at the left end of the x-t plane of Fig. 9.7, the C^- characteristic can be written in terms of v and h to give

$$v = \frac{C_1 A}{Q_R} + \frac{C_2 H_R A}{Q_R} h \tag{9.6.9}$$

or

$$v = c_1 + c_2 h \tag{9.6.10}$$

In addition to the machine characteristics, which may be represented by Eqs. (9.6.1) and (9.6.2), by Eqs. (9.6.3) - (9.6.6), or by Eqs. (9.6.7) and (9.6.8), the angular momentum relationship must be satisfied. If T_S is the driving torque on the shaft, and T is the resisting fluid torque, then in general

$$T_S - T = I \frac{d\omega}{dt} \tag{9.6.11}$$

in which I is the rotational moment of inertia of all the rotating parts — pump, motor, shaft, — and the fluid in the impeller. Typically I is defined in terms of WR^2 which is in units of $lb_f\text{-}ft^2$

$$I = \frac{WR^2}{g} \tag{9.6.12}$$

In the case of a pump trip, or loss of power, for which $T_S = 0$, Eq. (9.6.11) becomes

$$-T = \frac{WR^2}{g}\frac{d\omega}{dt} \qquad (9.6.13)$$

In terms of dimensionless ratios α and β

$$\frac{d\alpha}{dt} = -\frac{g^2\rho Q_R H_R}{\omega_R^2 WR^2 \eta_R}\beta = -K_1\beta \qquad (9.6.14)$$

in which K_1 is a constant with units of inverse time.

In summary, the left-end pump boundary condition for the four unknowns h, v, α, and β is solved by simultaneous solution of

$$\frac{d\alpha}{dt} = -K_1\beta$$

$$v = c_1 + c_2 h$$

$$\frac{h}{\alpha^2} = F_\alpha(v/\alpha) \quad \text{or} \quad \frac{h}{v^2} = F_v(\alpha/v)$$

$$\frac{\beta}{\alpha^2} = G_\alpha(v/\alpha) \quad \text{or} \quad \frac{\beta}{v^2} = G_v(\alpha/v)$$

9.7 Applications

In order to demonstrate application of the method of characteristics, several numerical examples are presented. For simplicity each is for a single pipeline with constant wave speed. The first example is the simple reservoir problem of Fig. 9.3 with rapid and gradual valve closure. Transients associated with tripping of a centrifugal pump constitute the second example.

Valve Operation

The reservoir shown in Fig. 9.3 is assumed to have a head H_o = 500 ft. The pipe line has a diameter D = 2.0 ft, a wave speed a = 3000 ft/sec, a length L= 4500 ft, and a Darcy resistance coefficient f_W = 0.020. The right-end boundary condition is a valve with an effective diameter of 0.357 ft and an assumed linear tau curve. The head downstream of the open valve is 100 ft, yielding an initial flow of 15.71 cfs. The valve is allowed to close in (a) 3 seconds and (b) 8 seconds. The results for the head upstream of the valve are shown in Fig. 9.15. Since 2L/a = 3 seconds, the quicker closure is a rapid closure, producing a maximum head at the instant of closure nearly equal to that given by Joukowsky's Eq. (9.2.11), $\Delta H = aV_o/g = 466$ ft. The head rise for the gradual closure of 8 seconds is clearly much less. In both instances the natural period of the reservoir-pipeline-valve system of 4L/ a = 6 seconds is apparent once the valve is completely closed.

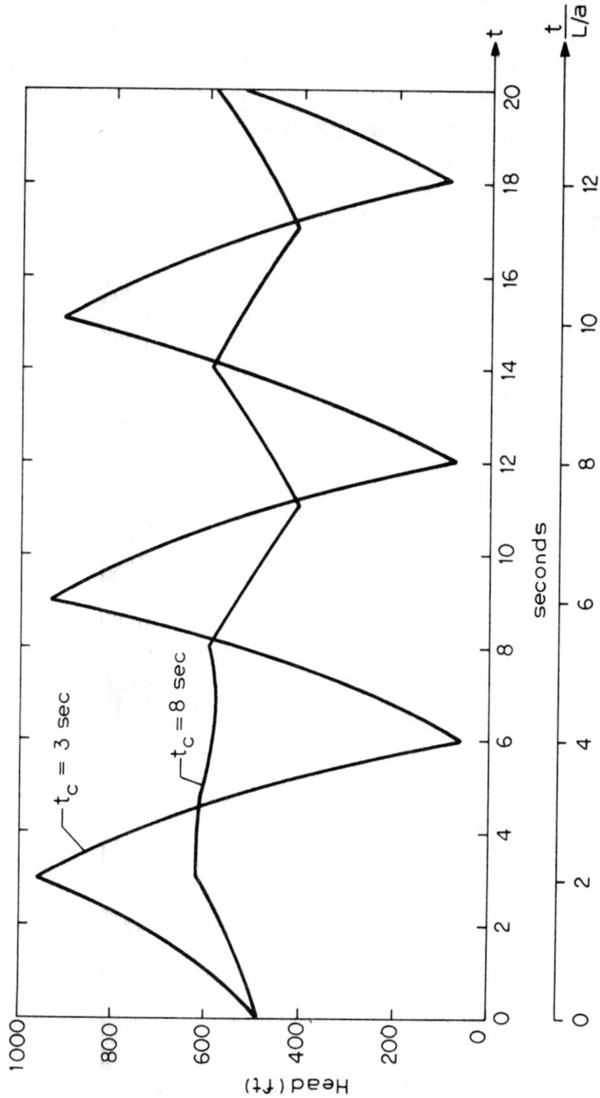

Fig. 9.15. Example of closure of downstream valve

Pump Power Failure

The second example consists of a pump at the left end of a 7000 ft pipe, having a diameter of 5.0 ft, wave speed of 3500 ft/sec, and Darcy f_W of 0.015. The head on the pump is 250 ft at t = 0, at which time the initial flow is 150 cfs. The inertial coefficient in Eq. (9.6.14) $K_1 = 0.1\,\text{sec}^{-1}$. The results are shown in Fig. 9.16 in terms of dimensionless head (h), discharge (v), and shaft speed (a) . The pump is operating in the pumping zone for the first 10 seconds, at which time flow reverses, switching the machine into the zone of energy dissipation. At t = 15.5 seconds the pump shaft reaches zero speed, allowing the machine to reverse and operate as a turbine. Since no valves are operated the machine continues to run in the turbine mode. The change in slope of the discharge-time curve is observed to change every 4 seconds initially, corresponding exactly to 2L/a. The head rise due to reverse flow is only 22 percent above rated — an acceptable value in most cases. Waterhammer is more likely to occur as a consequence of (1) collapse of vapor cavities (liquid column separation) due to too rapid deceleration of the liquid mass, or (2) belated closure of a pump discharge value subsequent to reverse flow through the system.

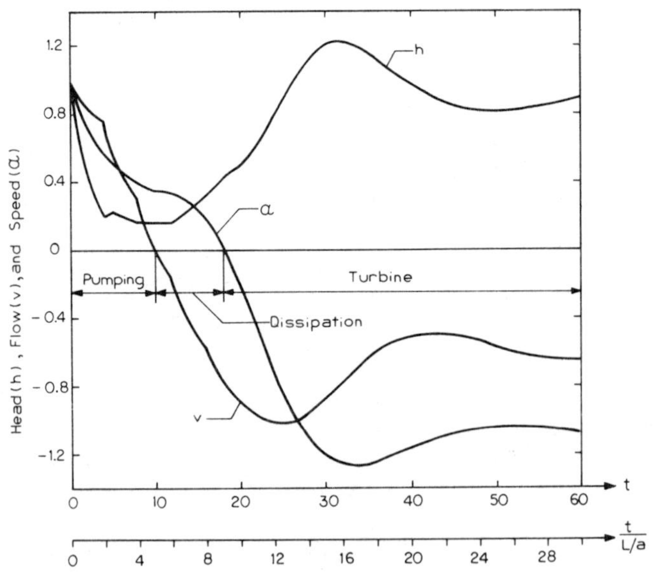

Fig. 9.16. Example of power failure of pump

9.8 Wave Propagation in Two-Phase Flow

There are numerous instances in practice for which a liquid flowing in a conduit contains either vapor or gas, or both, as a mixture. A vapor-liquid mixture of the same chemical substance would be termed two-phase, whereas a gas-liquid combination would, strictly speaking, be called two-component. Typical transient problems associated with two-phase steam-water flows are (1) sudden interaction of steam and water, (2) rapid depressurization of water at high pressure and temperature, and (3) cavitation of cold liquids. In both steady and transient flow, two-component gas-liquid mixtures may occur as a result of free of entrained gas, or because of evolution of dissolved gas from solution due to a drop in pressure or increase in temperature above saturation levels. In this developement the appellation two-phase is applied to both gas-liquid and vapor-liquid combinations.

Gases may be present either in dissolved or entrained state, or both, in cooling-water systems of fossil-fueled and nuclear power stations, in sewage pumping lines, or in crude oil lines. The effect of gas compressibility on wave-propagation speed, and on resulting pressure changes, must be considered in any transient analysis for which even the smallest amount of gas may be present. If pressure changes during a transient lower the pressure to, or near to, the saturation pressure of the liquid, large quantities of gas dissolved in the liquid may evolve from solution and thereby considerably alter the wave propagation speed.

The possible effect of any dissolved gas or free gas in a liquid is frequently ignored in the transient analysis of pressure drop in liquid pipelines subsequent to the loss of power to a pump. In practice free gas probably occurs as a result of evolution of dissolved gas.

Although dissolved gas in a liquid can not affect the acoustic velocity, provided it remains in solution, gas release during a pressure-reducing transient produces the same end result. As shown by Swaffield [10] and Driels [11], the cushioning effect of released gas during a severe pressure drop can significantly reduce the maximum pressure rise during resurge. Kranenburg [12] demonstrated the importance of considering dissolved gas and its evolution on simulation of liquid-column separation and gaseous cavitation in liquid pipelines.

In large circulating-water systems of nuclear and fossil-fueled power plants, the water is usually saturated with gas, especially if the system contains cooling towers in the circuit. Due to the required height of the condenser to accomodate the high number of tubes and the usual low pumping head, the water pressure in the condenser box may even be negative during steady-flow operation, allowing for possible gas release and entrapment during normal operation. The calculation of pressure transients subsequent to pump failure can be complicated by the initial two-phase regime in the condenser as well as further gas release as the pressure reduces toward vapor pressure. Sheer [13] has shown the wide discrepancy between actual experimental transient pressure data in a large cooling-water system and a single-phase theory that completely ignores any presence of gas, dissolved or free. It is very apparent that there is a dire need for (1) good two-phase flow models for transient calculations and (2) a better understanding of the physical chemistry of gas evolution and gas absorption in liquids.

Another example of transient two-phase flow is the postulated loss-of-coolant accident (LOCA) in a nuclear reactor. The development of analytical models to simulate the resulting transient flow through a pipe break is complicated by nonequilibrium thermodynamics, critical flow, heat transfer, two-phase flow through primary coolant pumps, two- and three-dimensional flow in the reactor plenum, etc. The blowdown problem in general has generated considerable interest in two-phase choking flow, however, as evidenced by papers by Moody [14], Henry and Fauske [15], and D'Arcy [16], to cite several sources.

Flow Regimes

In any development of accurate and definitive comprehensive analytical models, whether one-, two- or three-dimensional, one must recognize the fact that two-phase flow in conduits can occur in various regimes. Depending upon a number of factors, but in particular upon the mass flow rates in the respective phases, the flow pattern may range from completely dispersed gas (bubbly flow) to completely dispersed liquid (mist or annular mist). Gravity can produce a significant effect on the flow pattern development as the conduit inclination is changed. For steady flow some of the flow regimes that have been identified for vertically upward flow and horizontal flow are presented in Fig. 9.17. The number of possible regimes that can be identified in either case depends to some extent on the observer.

The vast difference between some of the flow patterns exhibited in Fig. 9.17 suggests the possibility of developing a universal two-phase analytical model is rather remote. In fact, the effect of relative velocity between phases, phase interaction, and heat, mass and momentum transfer can have a much greater effect for one flow regime than another. Although the most suitable model may vary depending upon the flow regime analyzed, for flow in long conduits the assumption of one-dimensionality is usually not a severe limitation.

As reported by Wallis [5], the most widely used analytical models are the homogeneous model, the separated-flow model, and the drift-flux model. Unless there exists a significant difference in velocities of the respective phases, the homogeneous model can be applied with

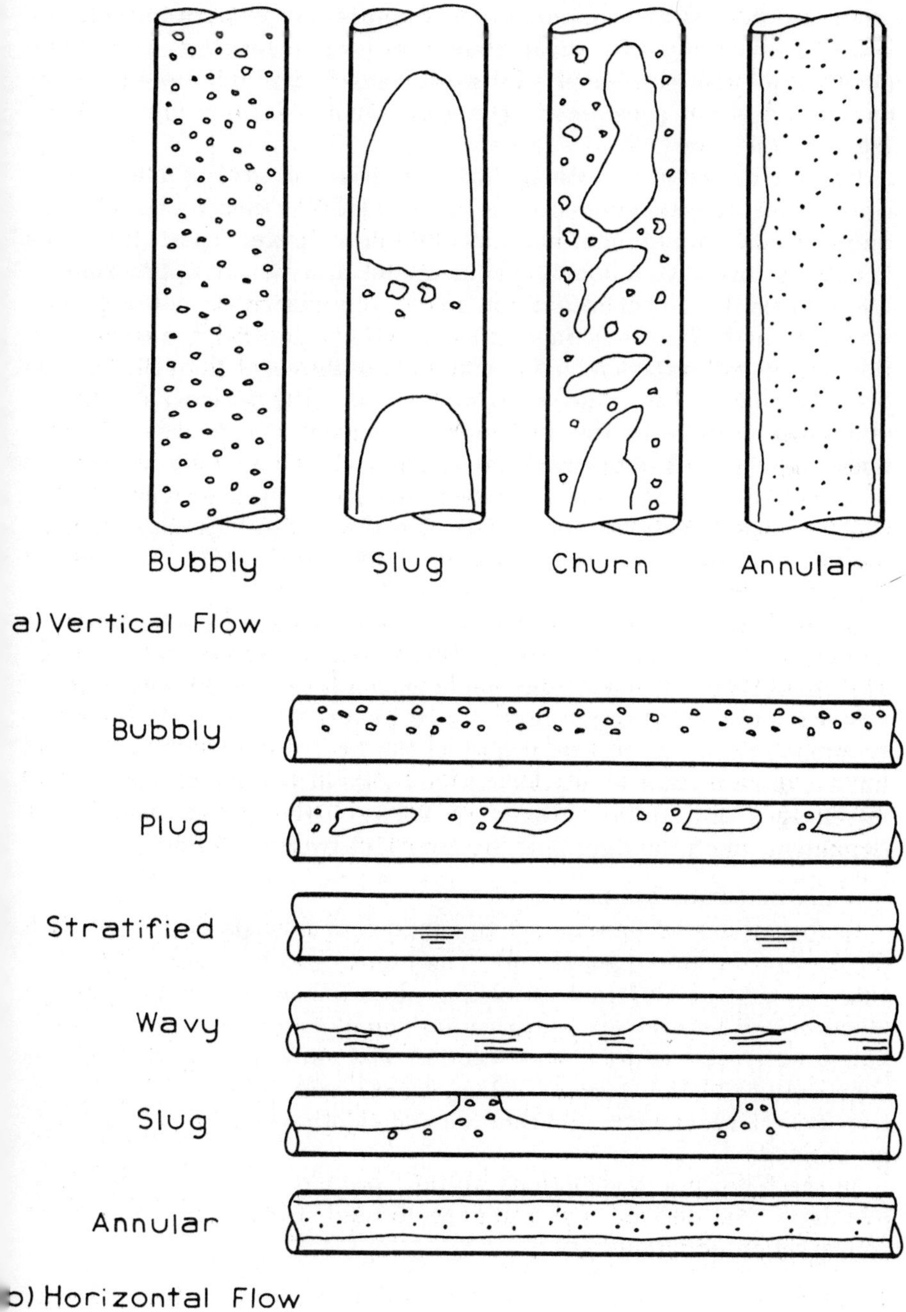

Fig. 9.17. Regimes of two-phase flow

moderate success to dispersed flows, whether bubbly or droplet. In the homogeneous model the components are treated together as a single pseudofluid with average properties. Inasmuch as slip between the gas and liquid is not considered, gas and liquid velocities are assumed equal at every instant.

In some transient and steady flows, inertial and gravitational effects can play an important enough role that the relative velocity between air bubbles and liquid can be considerable. Hence, apart from the simplicity involved, a homogeneous model may not be appropriate in many cases. In the separated-flow model the phases are considered to flow side-by-side, interacting with each other. Generally, a separated-flow model will necessitate six equations to represent the conservation of mass, momentum, and energy of each of the phases. Additional equations describing the interaction between the phases are also needed, severely complicating the problem. Often a simplified version of a separated-flow model is used in which one or more of the conservation equations are written for the mixture rather than for individual phases. In this type of formulation only the velocity difference between the phases is included.

In the drift-flux model the relative motion between the two phases is given attention rather than the motion of the two phases individually. This model is particularly remarkable in that it can include the effect of velocity and concentration profiles. However, since several empirical relationships are an essential part of the drift-flux model, it may not have a general applicability for a wide range of problems. The relative advantages and disadvantages of each of the above models are dependent upon the nature of the transient two-phase flow.

Separated-Flow Model

In the separated-flow model the fluid and flow properties of each phase are considered separately. The phases can be imagined to flow side by side at different velocities. Interphase interaction can be incorporated into the six conservation equations of mass, momentum, and energy. Of course, the solution can only be effected if the constitutive relationships of shear stress, heat transfer, momentum transfer, phase change, and thermal and caloric equations of state can be defined.

In the following development all flow properties are cross-sectional averages. As derived by Yadigaroglu and Lahey [17], the mass conservation equations are

$$\frac{\partial}{\partial t}(\rho_g \alpha A) + \frac{\partial}{\partial x}(\rho_g \alpha V_g A) = \Gamma A \qquad (9.8.1)$$

for the gas or vapor phase and

$$\frac{\partial}{\partial t}\left[\rho_\ell (1-\alpha) A\right] + \frac{\partial}{\partial x}\left[\rho_\ell (1-\alpha) V_\ell A\right] = -\Gamma A \tag{9.8.2}$$

for the liquid phase, in which V_g and V_ℓ are cross-sectional average velocities of the gas and liquid phase, respectively. The quantity Γ represents the mass rate of phase change or gas absorption or production per unit volume.

The conservation equations of momentum for the phases is formulated to include the effect of interfacial shear as well as the effect of mass transfer Γ on the momentum transfer. For the gas or vapor phase

$$\frac{\partial}{\partial t}(\rho_g \alpha V_g A) + \frac{\partial}{\partial x}(\rho_g \alpha V_g^2 A) = -\alpha A \frac{\partial p}{\partial x} - g\rho_g \alpha A \sin\theta$$

$$- \tau_{wg} P_g - \tau_i P_i + \Gamma V_\ell A \tag{9.8.3}$$

The liquid-phase momentum equation is

$$\frac{\partial}{\partial t}\left[\rho_\ell (1-\alpha) V_\ell A\right] + \frac{\partial}{\partial x}\left[\rho_\ell (1-\alpha) V_\ell^2 A\right] = -(1-\alpha) A \frac{\partial p}{\partial x}$$

$$- g\rho_\ell (1-\alpha) A \sin\theta - \tau_{w\ell} P_\ell + \tau_i P_i - \Gamma V_\ell A \tag{9.8.4}$$

The interfacial shear and interfacial perimeter are τ_i and P_i, respectively. The shear stress imparted by the wall to the gas perimeter P_g is τ_{wg}. The corresponding shear stress and portion of the wall wetted by the liquid are $\tau_{w\ell}$ and P_ℓ, respectively. The pipe is inclined to the horizontal at an angle θ. In the derivation of Eqs. (9.8.3) and (9.8.4) it is assumed that the momentum transfer between phases due to mass transfer is related to the liquid velocity V_ℓ.

Although energy conservation equations can be derived separately for each phase, they are combined to form the mixture energy equation

$$\frac{\partial}{\partial t}\left[\rho_g \alpha (e_g - p/\rho_g) A + \rho_\ell (1-\alpha)(e_\ell - p/\rho_\ell) A\right]$$

$$+ \frac{\partial}{\partial x}\left[\rho_g \alpha e_g V_g A + \rho_\ell (1-\alpha) e_\ell V_\ell A\right] = q_w'' P + q''' A \tag{9.8.5}$$

where e is internal energy, q_w'' is the wall heat flux, and q''' the volumetric heat generation rate. The specific energy for each phase is defined by

$$e_k = h_k + \frac{V_k^2}{2} + gz \sin\theta \tag{9.8.6}$$

in which the subscript k refers to either subscript g or ℓ and h_k is the enthalpy for the respective phase. The five conservation Eqs. (9.8.1) - (9.8.5) and Eq. (9.8.6) must be supplemented with constitutive relationships for

$$\rho_g, \rho_\ell, \Gamma, \tau_{wg}, \tau_{w\ell}, \tau_i, h_g, h_\ell, q''_w, \text{and } q'''$$

in order to effect a solution.

Homogeneous Model

For the homogeneous model the two phases or components are treated as a single psuedofluid with average properties. It is assumed that there is no relative motion or slip between phases, or $V_g = V_\ell = V$. The mass, momentum, and energy conservation equations for the psuedofluid are identical to those for a single-phase flow. The conservation of mass is obtained by adding Eqs. (9.8.1) and (9.8.2)

$$\frac{\partial}{\partial t}(\rho_m A) + \frac{\partial}{\partial x}(\rho_m AV) = 0 \qquad (9.8.7)$$

in which the mixture density

$$\rho_m = a\rho_g + (1 - a)\rho_\ell \qquad (9.8.8)$$

The mixture conservation of momentum for Eqs. (9.8.3) and (9.8.4) becomes

$$\frac{\partial}{\partial t}(\rho_m VA) + \frac{\partial}{\partial x}(\rho_m V^2 A) = -A\frac{\partial p}{\partial z} - g\rho_m \sin\theta - \tau_w P \qquad (9.8.9)$$

From Eq. (9.8.5) the conservation of energy for the mixture is

$$\frac{\partial}{\partial t}[\rho_m(e_m - p/\rho_m)A] + \frac{\partial}{\partial x}[\rho_m e_m VA] = q''_w P + q'''A \qquad (9.8.10)$$

The homogeneous model is suitable for flows in which the effect of relative motion is negligible and there is no interfacial mass, momentum, or heat transfer. For example, a separated-flow model is required to assess the effect of vapor or gas production by use of a constitutive relation for Γ .

Numerical Analysis

Numerical solution to the above equations can be accomplished either by using an irregular grid of characteristics or by constructing a grid of specified time and distance intervals in the x-t plane. For rapid pressure increases, shock waves can form in mixtures similar to those in gas dynamics. Like characteristics then begin to intersect, signifying the formation of a shock. A complete simulation of the propagation of a strong shock along a conduit necessitates satisfying Hugoniot shock equations as well. In the case of weak shocks or no shock at all, it is convenient to approximate the characteristic and compatibility equations on a grid of constant Δt and Δx increments. It is possible to allow Δt to vary from time step to time step, however, so that interpolation errors can be minimized.

If the constitutive equations are written in so-called conservation form, the two-step numerical scheme proposed by Lax and Wendroff [18] has the advantage that shock waves and other discontinuities can be treated. Kranenburg [12] has employed this technique for the analysis of gaseous cavitation and liquid-column separation in conduits. Richtmyer and Morton [19] explained in detail the Lax-Wendroff explicit scheme and associated stability conditions. Comparison of numerical results with experiment are reported by Martin, et al. [20] for bubbly two-component flow and by Martin and Padmanabhan [21] for slug flow. Wiggert and Sundquist [22] demonstrate the application of a quasi-homogeneous model with the Γ source term included to simulate the release of dissolved gas during a severe pressure transient.

9.9 Some Examples of Pressure Transients in Nuclear Piping-

In non-nuclear piping the most common sources of waterhammer are (1) valve operation and (2) power failure of pumps. These two sources of pressure-pulse propagation do not normally produce the most severe pressure transients in high pressure nuclear piping because of relatively short pipe lengths. Experience has shown that the most serious waterhammer occurrences have originated from the rapid collapse of steam pockets and the filling of voided pipe lines. The propagation of a high pressure pulse through the tubing of a LMFBR as a result of water-sodium reaction has also been the subject of a number of studies. It has been shown that an initial plastic deformation of the tubing can prevent portions of the remaining tubing from yielding because of the dramatic attentuation associated with the deformation.

Of a lesser extent are the problems of (1) potential waterhammer in circulating water systems due to liquid-column separation and (2) steam hammer in the secondary system due to line break or stop valve closure.

Plastic Deformation

Once the pipe wall material is stressed beyond the elastic range the tangent modulus in Eq. (9.2.12) is no longer equal to Young's modulus, causing the wave speed to be reduced. In fact Fox and Stepnewski [23] show that the wave speed can be decreased as much as an order of magnitude if plastic deformation is severe. An example of the diminished wave speed is shown in Fig. 9.18 by Thorley and Twyman [24].

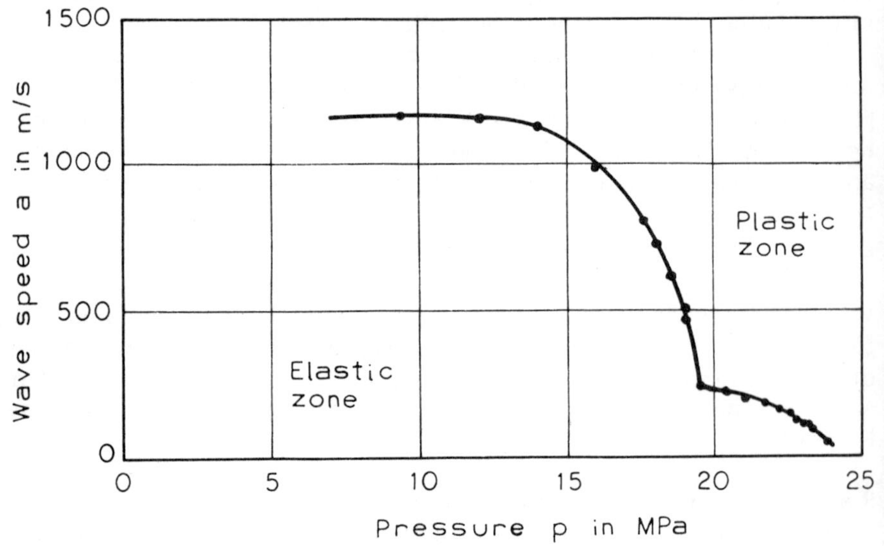

Fig. 9.18. Effect of plastic deformation on wave speed (from Thorley and Twyman [24])

In an effort to understand the potential of pressure wave propagation in tubing of a LMFBR, Romander, et al. [25] conducted a series of tests in straight tubing for which plastic deformation occurred. These well-performed experimental studies have provided the basis for several analytical investigations in an effort to verify computer codes. Fig. 9.19 shows the piping layout and transient pressure traces at six locations for two different experimental runs. The transient was generated at the left end by means of a pulse gun. The dramatic effect of plastic deformation is evident upon comparison of traces P2 in the rigid pipe with trace P4 in the flexible Ni pipe. Lesser plastic deformation further attenuates the pulse as it propagates toward the blind flange. Three different computer codes have been developed and tested using the data in Fig. 9.19 for verification.

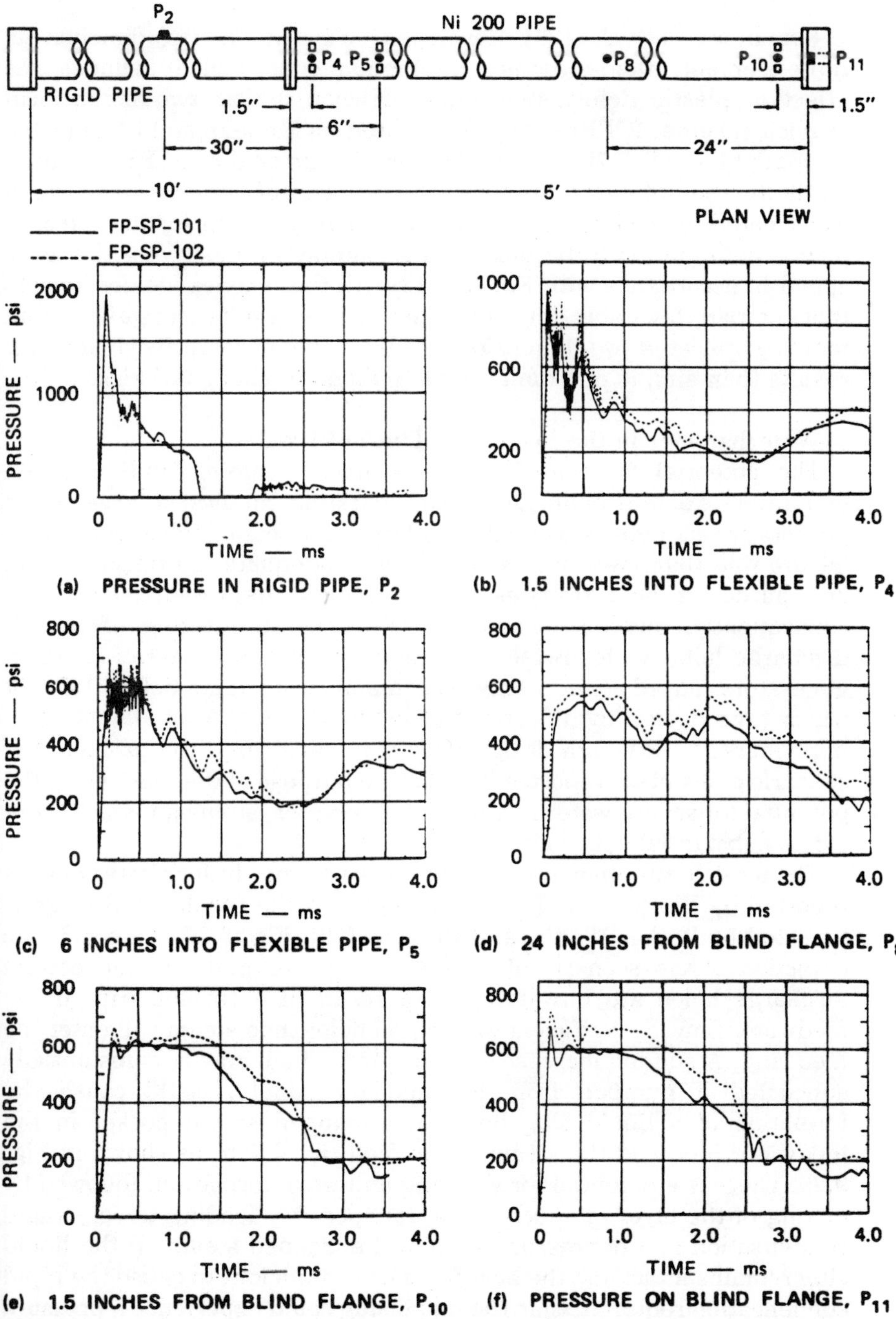

Fig. 9.19. Attenuation of pressure pulse in plastically deforming pipe (from Romander, et al. [25])

Romander, et al. [25] developed WHAM, a two-dimensional finite-element, fluid-structure interaction code which includes the effect of plastic deformation. A somewhat similar two-dimensional implicit scheme, ICEPEL, was formulated by Moneim and Chang [26]. Youngdahl, et al. [27] developed a one-dimensional code PTA-2 based on the method of characteristics. The problem of variable wave speed was treated by means of linear interpolation, allowing for the effect of plastic deformation to be incorporated simply by modifying the wave speed in accordance with Eq. (9.2.12). All three codes yielded results that compare favorably with the experimental results, suggesting that more complicated two-dimensional codes do not necessarily yield better results than simple one-dimensional method of characteristics.

Waterhammer in the Feedwater Lines of PWRs

The potential for waterhammer in nuclear piping and in steam generators exists if under abnormal operating conditions large steam pockets are suddenly surrounded by large amounts of subcooled water. Severe waterhammer can result if there is adequate interfacial mixing and sufficient heat transfer to cause rapid depressurization, and consequently, sudden collapse of void. The unwatering of bottom discharge holes in feedwater sparger rings of first generation steam generators caused some twenty waterhammer incidents before 1976, at which time design modifications to preclude steam entrapment were implemented. Although design fixes may have cured the problem the scenario of events is reported here for the purpose of demonstrating the potential for severe waterhammer in two-phase systems for which there exists substantial subcooling.

The most comprehensive study of waterhammer in feedwater lines is reported by Block, et al. [28]. A summary of the results in Ref. 28 is reported by Rothe, Wallis, and Crowley [29]. Fig. 9.20 is a schematic depiction of the scenario of events. It is postulated that the bottom discharge holes are unwatered as a result of a turbine trip. If the feedwater flow is insufficient to cover all holes then steam will enter the feed ring. A sudden increase of feedwater, which may be substantially subcooled if pumped from an auxiliary supply, could cause the formation of a liquid slug between a trapped steam pocket in the feedwater pipe and the steam in the feed ring itself, as shown in Fig. 9.20. There is a potential for slugging and wave formation, followed by mixing of the layer of subcooled water, possibly leading to very rapid condensation and depressurization of the trapped steam. If the liquid slug remains intact and the heat transfer is sufficient to cause the rapid condensation required, then the liquid slug could experience a pressure difference across it as high as operating steam pressure less the

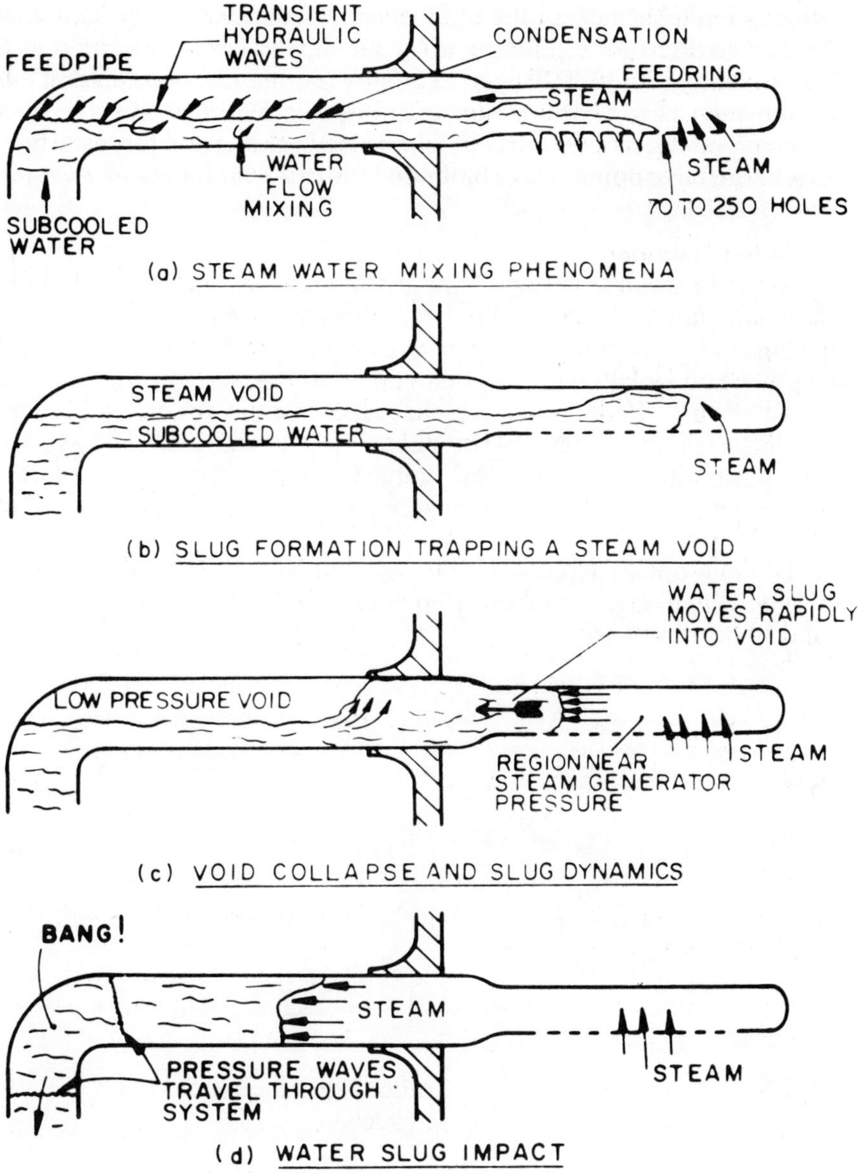

Fig. 9.20. Postulated scenario of waterhammer in feedwater piping of a PWR (from Rothe, et al. [29])

depressurized steam, which could approach nearly absolute zero. Simple analyses yield waterhammer pressure at collapse to exceed 500 atmospheres because of the high velocities attained by the liquid slug. Extensive damage to piping and/or its supports was sustained at four PWR plants prior to 1976 as a result of waterhammer originating either in the feed ring of the steam generator or in the feedwater nozzle. Design modifications which eliminated the troublesome bottom discharge have apparently eliminated the problem for steam generators with feed rings.

Steam Hammer

Steam hammer is defined here as the pressure surges resulting from transient flows in superheated or saturated steam lines, but not the collapse of steam pockets. Typical problems are postulated steam line rupture and isolation of the steam generator or steam turbine by rapid valve closure. Usually the pressure rises are not too large because of the large compressibility of the steam. Typical analyses are based upon the assumption of isentropic behavior and either (1) perfect gas representation or (2) thermodynamic equilibrium and steam table representation.

The one-dimensional equations should now include the convective terms. Frequently boundary friction is neglected, allowing Eqs. (9.3.1) and (9.3.9) to be written

$$\frac{\partial \rho}{\partial t} + V\frac{\partial \rho}{\partial x} + \rho\frac{\partial V}{\partial x} = 0 \tag{9.9.1}$$

and

$$\frac{\partial V}{\partial t} + V\frac{\partial V}{\partial x} = -\frac{1}{\rho}\frac{\partial p}{\partial x} \tag{9.9.2}$$

in which ρ is the mass density of the steam. The assumption of constant entropy

$$a^2 = \left.\frac{\partial p}{\partial \rho}\right|_s = \frac{\gamma p}{\rho} \tag{9.9.3}$$

in which γ is the specific heat ratio. Equations (9.9.1) and (9.9.2) can be put in characteristic form for which

$$\frac{dp}{dV} = \mp \rho a \tag{9.9.4}$$

or

$$\frac{dx}{dt} = V \pm a \tag{9.9.5}$$

Since

$$T\,ds = dh - \frac{dp}{\rho} \tag{9.9.6}$$

in which T is the steam temperature and h the enthalpy, constant entropy yields

$$\mp\, dV = \frac{dh}{a} \tag{9.9.7}$$

Equation (9.9.7) can be integrated along the characteristic lines to yield

$$\frac{V}{a_o} = P\,(p/p_o) + B_1 \tag{9.9.8}$$

and

$$-\frac{V}{a_o} = P\,(p/p_o) + B_2 \tag{9.9.9}$$

in which

$$P\,(p/p_o) = \int_{p/p_o = 1}^{p/p_o} \frac{d\,(h/a_o^2)}{a/a_o} \tag{9.9.10}$$

in which the subscript o represents reference values.

Under the assumption of perfect gas behavior the Riemann invariants are

$$C^+ : \quad \frac{V}{a_o} + \frac{2}{\gamma - 1}\,\frac{a}{a_o} = \text{Constant} \tag{9.9.11}$$

$$C^- : \quad \frac{V}{a_o} - \frac{2}{\gamma - 1}\,\frac{a}{a_o} = \text{Constant} \tag{9.9.12}$$

If the method of characteristics is used, then provision should be made to account for variable wave speed and attendant interpolation errors. Implicit schemes or explicit techniques allowing for passage of weak shocks could be employed. A comparison of several methods is reported by Samra [30].

REFERENCES

1. Streeter, V. L., and Wylie, E. B., **Hydraulic Transients**, McGraw-Hill, New York, 1967.
2. Wylie, E. B. and Streeter, V. L., **Fluid Transients**, McGraw-Hill International, 1978.
3. Fox, J. A., **Hydraulic Analysis of Unsteady Flow in Pipe Networks**, Wiley, New York, 1977.
4. Chaudry, M. H., **Applied Hydraulic Transients**, Van Nostrand Reinhold, 1979.
5. Wallis, G. B., **One-Dimensional Two-Phase Flow**, McGraw-Hill, New York, 1969.
6. Shin, Y. W. and Daley, J. G., Two-dimensional fluid transients by the method of characteristics, **Trans. ANS** 21, 200-201 (1975).
7. Shin, Y. W. and Valentin, R. A., Numerical analysis of the fluid-hammer waves by the method of characteristics, **J. of Computional Physics** 20, 220-237 (1976).
8. Parmakian, J., **Waterhammer Analysis**, Dover, 1963.
9. Wiggert, D. C. and Sundquist, M. J., Fixed-grid characteristics for pipeline transients, **Proc. ASCE, J. Hyd. Div.** 103, 1403-1416 (1977).
10. Swaffield, J. A., A study of the influence of air release on column separation in an aviation kerosene pipeline, **Proc. Inst. Mech. Engrs.** 186, 693-703 (1972).
11. Driels, M. R., An investigation of pressure transients in a system containing a liquid capable of air absorption, **ASME J. of Fluids Engineering** 95, 408-414 (1973).
12. Kranenburg, C., Gas release during transient cavitation in pipes, **Proc. ASCE, J. Hyd. Div.** 100, 1383-1398 (1974).
13. Sheer, T. J., Computer analysis of waterhammer in power station cooling water systems, **Proc. First BHRA Inter. Conf. on Pressure Surges**, Canterbury, Paper D1, 1972.
14. Moody, F. J., Maximum two-phase vessel blowdown from pipes, **ASME J. of Heat Transfer** 88, 285-293 (1966).
15. Henry, R. E. and Fauske, H. K., The two-phase critical flow of one-component mixtures in nozzles, orifices, and short tubes, **ASME J. of Heat Transfer** 93, 179-187 (1971).
16. D'Arcy, D. F., On acoustic propagation and critical mass flux in

two-phase flow, **ASME J. of Heat Transfer** 93, 413-421 (1971).

17. Yadigaroglu, G. and Lahey, R. T., On the various forms of the conservation equations in two-phase flow, **Inter. J. of Multiphase Flow** 2, 477-494 (1976).
18. Lax, P. D. and Wendroff, B., Systems of conservation laws, **Comm. on Pure and Applied Math** 13, 217-237 (1960).
19. Richtmyer, R. D. and Morton, K. W., **Difference Methods for Initial-Value Problems**, Interscience Publications, New York, 1967.
20. Martin, C. S., Padmanabhan, M. and Wiggert, D. C., Pressure wave propagation in two-phase bubbly air-water mixtures, **Proc. Second BHRA Inter. Conf. on Pressure Surges**, London, 1976.
21. Martin, C. S. and Padmanabhan, M., Pressure pulse propagation in two-component slug flow, **ASME J. of Fluids Engineering** 101, 44-52 (1979).
22. Wiggert, D. C. and Sundquist, M. J., The effect of gaseous cavitation on fluid transients, **ASME J. of Fluids Engineering** 101, 79-86 (1979).
23. Fox, G. L. and Stepnewski, D. D., Pressure wave transmission in a fluid contained in a plastically deforming pipe, **ASME J. of Pressure Vessel Technology** 96, 258-262 (1974).
24. Thorley, A. R. D. and Twyman, J. W. R., Propagation of transient-pressure waves in a sodium-cooled fast reactor, **Proc. Second BHRA Inter. Conf. on Pressure Surges**, London, Paper Al, 1976.
25. Romander, C. M., Schwer, L. E. and Cagliostro, D. J., Response of water-filled thin-walled pipes to pressure pulses: Experiments and analysis, ASME Paper 78-PVP-53 (1978).
26. A-Moneim, M. T. and Chang, Y. W., Comparison of ICEPEL code predictions with straight flexible pipe experiments, **Nuclear Eng. and Design** 49, 187-196 (1978).
27. Youngdahl, C. K., Kot, C. A. and Valentin, R. A., Pressure transient analysis in piping systems including the effects of plastic deformation and cavitation, ASME Paper 78-PVP-56 (1978).
28. Block, J. A., Rothe, P. H., Crowley, C. J., Wallis, G. B., and Young, L. R., An evaluation of PWR steam generator waterhammer, Final Report, NUREG-0291, Creare, Inc. (1977).
29. Rothe, P. H., Wallis, G. B. and Crowley, C. J., Waterhammer in the feedwater systems of PWR steam generators, **ASME Symposium on Fluid Transients and Acoustics in the Power Industry**, Winter Annual Meeting, San Francisco, 75-86, 1978.
30. Samra, R. S., Impact energy calculations for a steam check valve following a postulated pipe rupture, ASME Paper 76-WA/FE-8 (1976).

PROBLEMS

1. A metal pipe of 1-inch ID has a wall thickness of 0.05 inch. The tangent modulus of the wall material can be approximated by $E = 15 \times 10^6$ psi for a stress less than 12,000 psi. Between 12,000 psi and 25,000 psi the tangent modulus is given by 4×10^5 psi. Calculate the wave speed as a function of internal fluid pressure for the plastic as well as the elastic range of deformation. The liquid is water for which the bulk modulus is 3×10^5 psi.
2. The head-loss characteristics of a butterfly valve are furnished by the manufacturer in terms of loss coefficients for various angles of inclination. From the tabulated data below complete the table by computing the corresponding tau values.

θ (Degrees)	0	10	20	30	40	50	60	70	80	90
K	1.18	1.49	2.46	5.24	12.5	32.5	122	302	980	∞
τ										

3. The method of characteristics is to be applied to a three pipe system. The pipe lengths and wave speeds are listed below. Complete the table by optimizing the selection of Δx to satisfy the Courant criterion. If necessary modify the wave speeds **slightly** to prevent interpolation.

Pipe	Length [ft]	Wave Speed [ft/sec]	Number of Reaches, N	Δx (ft)
1	6400	3800		
2	4200	3500		
3	1250	4050		

4. An air-water mixture flows in a circular pipe under a pressure of 200 psig. For void fractions of 0.0005 and 0.001, compute the wave speed in the mixture assuming that the single-phase liquid wave speed is 4000 ft/sec and the single-phase gas wave speed is based on adiabatic behavior. The temperature is 80°F.

5. A pressure wave in Pipe i = 1 approaches a junction with two other pipes. All three pipes are of steel for which $E = 30 \times 10^6$ psi. The fluid is water ($K = 3 \times 10^5$ psi). If the incident pressure rise in Pipe 1 is 100 psi, what are the magnitudes of the resulting reflected wave in Pipe 1 and the transmitted waves in Pipes 2 and 3?

Pipe	D [Inches]	e [Inch]
1	8.00	0.25
2	8.00	0.25
3	6.00	0.20

APPENDIX A
CONVERSION FACTORS

Multiply	*By*	*To Obtain*
Area		
Square centimeters	0.15500	square inches
Square feet	0.092903	square meters
Square inches	6.4516	square centimeters
Square meters (m^2)	10.764	square feet
Square meters	1550	square inches
Energy and Heat		
British thermal units (Btu)	251.98	calories
British thermal units	6.584×10^{21}	electron volts
British thermal units	777.98	foot-pounds
British thermal units	1.0756×10^{7}	gram-centimeters
British thermal units	3.9292×10^{-4}	horsepower-hours
British thermal units	1054.8	joules
British thermal units	2.930×10^{-4}	kilowatt-hours
Calories (cal)	3.9685×10^{-3}	Btu
Calories	3.0874	foot-pounds
Electron-volts (ev)	1.5188×10^{-22}	Btu
Ergs (erg)	9.4805×10^{-11}	Btu
Ergs	7.3756×10^{-8}	foot-pounds
Foot-pounds (ft-lb)	0.0012854	Btu
Foot-pounds	5.0505×10^{-7}	horsepower-hours
Foot-pounds	3.7662×10^{-7}	kilowatt-hours
Horsepower-hours (hp-hr)	2545.1	Btu
Horsepower-hours	1.980×10^{4}	foot-pounds
Horsepower-hours	0.7457	kilowatt-hours
Horsepower-hours	3.107×10^{-5}	megawatt-days
Joules (j)	9.478×10^{-4}	Btu
Joules	0.73756	foot-pounds
Joules	3.725×10^{-7}	horsepower-hours
Joules	2.7778×10^{-7}	kilowatt-hours
Kilowatt-hours (kwh, kw-hr)	3413.0	Btu
Kilowatt-hours	2.6552×10^{6}	foot-pounds
Kilowatt-hours	1.3410	horsepower-hours
Kilowatt-hours	2.2472×10^{10}	mev
Megawatt-days (mwd)	8.189×10^{7}	Btu
Megawatt-days (mwd)	24×10^{3}	kilowatt-hours
Megawatt-days (Mwd)	5.3916×10^{23}	Mev
Watt-hours (w-hr)	3.4130	Btu
Force and Mass		
Grams	980.66	dynes
Grams	0.035274	ounces
Grams	0.0022046	pounds
Grams	1×10^{-6}	tons, metric
Kilograms	0.001	tons, metric
Pounds (lb)	4.4482×10^{5}	dynes
Pounds	453.59	grams
Pounds	0.45359	kilograms
Pounds	4.5359×10^{-4}	tons, metric

APPENDIX A (continued)

Multiply	*By*	*To Obtain*
Length		
Centimeters	0.03281	Feet
Centimeters	0.3937	inches
Centimeters	1×10^4	microns
Centimeters	393.7	mils
Feet (ft)	30.480	centimeters
Feet	0.30480	meters
Inches (in)	2.540	centimeters
Inches	1000	mils
Meters	3.2808	feet
Meters	1×10^6	microns
Meters	1.0936	yards
Mils	0.0025400	centimeters
Mils	0.00100	inches
Power		
Btu/hour	0.0003929	horsepower
Btu/hour	0.29293	watts
Foot-pounds/hour (ft-lb/hr)	3.7655×10^{-4}	watts
Foot-pounds/second	1.3558	watts
Horsepower (hp)	2545.1	Btu/hour
Horsepower	33000	foot-pounds/minute
Horsepower	745.7	joules/second
Horsepower	0.7457	kilowatts
Horsepower	550	foot-pounds-second
Kilowatts	3413.0	Btu/hour
Kilowatts	737.56	foot-pounds/second
Kilowatts	1.3410	horsepower
Watts, (w, watt)	3.4137	Btu/hour
Watts	0.73756	foot-pounds/second
Watts	0.0013410	horsepower
Watts	1	joules/second
Pressure		
Atmospheres (atm)	76	centimeters of mercury at 0°C
Atmospheres	33.899	feet of water at 39.2 °F
Atmospheres	1033.3	grams/cm^2
Atmospheres	2116.3	pounds/ft^2
Atmospheres	14.696	pounds/in^2
Bars	0.98692	atmospheres
Bars	14.504	pounds/in^2
Dynes/cm^2	0.0020886	pounds/ft^2
Dynes/cm^2	1.4504×10^{-5}	pounds/in^2
Pounds/in^2	2.3066	feet of water at 39.2 °F
Pounds/in^2	2.0360	inches of mercury at 32°F
Pounds/in^2	144	pounds/ft^2
Pounds/ft^2 (lb/ft^2, psf)	4.7254×10^{-4}	atmospheres
Pounds/ft^2	478.80	dynes/cm^2
Pounds/in^2 (lb/in^2, psi)	0.068046	atmospheres
Pounds/in^2	68947	dynes/cm^2

APPENDIX A (continued)

Multiply	*By*	*To Obtain*
Thermodynamic Properties		
Btu/(hr)(ft^2)	3.1534 x 10^{-4}	watts/cm^2
Btu/(hr)(ft)(°F)	0.01730	watts/cm °C
Btu/(hr)(ft)(°F)	0.01730	joules/(sec)(cm)(°C)
Btu/pound	0.55556	calories/gram
Btu/(lb)(°F)	1.0	calories/(gm)(°C)
Btu/(lb)(°F)	4.186	joules/(gm)(°C)
Btu/second	1.4145	horsepower
Btu/second	1054.8	watts
Btu(thermal)/kwh electrical	2.930 x 10^{-4}	1/efficiency
Kilowatts/liter	9.6644 x 10^4	Btu/ft^3-hr
Watts/cm	104.05	Btu/(hr)(ft)
Watts/cm^2	3171.2	Btu/(hr)(ft^2)
Watts/(cm^2)(°C)	1761.8	Btu/(hr)(ft^2)(°F)
Volume		
Cubic inches (in^3)	5.7870 x 10^{-4}	cubic feet
Cubic inches (in^3)	0.016387	liters
Cubic meters (m^3)	35.314	cubic feet
Cubic meters (m^3)	264.17	gallons
Cubic centimeters (cm^3)	3.5314 x 10^{-5}	cubic feet
Cubic centimeters	0.061023	cubic inches
Cubic centimeters	2.6417 x 10^{-4}	gallons
Cubic feet (ft^3)	28317	cubic centimeters
Cubic feet	1728	cubic inches
Cubic feet	0.028317	cubic meters
Cubic feet	7.4805	gallons
Cubic feet	28.316	liters
Gallons, U.S.	3785.43	cubic centimeters
Gallons, U.S.	0.13368	cubic feet
Gallons, U.S.	3.78533	liters
Liters (1)	0.035316	cubic feet
Liters	61.025	cubic inches
Liters	0.26418	gallons

APPENDIX B
THERMODYNAMIC PROPERTIES
TABLE B.1
Properties of saturated steam and saturated water (temperature)*

Temp F	Press. psia	Volume, ft^3/lb			Enthalpy, Btu/lb			Entropy, Btu/lb x F			Temp F
		Water v_f	Evap v_{fg}	Steam v_g	Water h_f	Evap h_{fg}	Steam h_g	Water s_f	Evap s_{fg}	Steam s_g	
32	0.08859	0.01602	3305	3305	−0.02	1075.5	1075.5	0.0000	2.1873	2.1873	32
35	0.09991	0.01602	2948	2948	3.00	1073.8	1076.8	0.0061	2.1706	2.1767	35
40	0.12163	0.01602	2446	2446	8.03	1071.0	1079.0	0.0162	2.1432	2.1594	40
45	0.14744	0.01602	2037.7	2037.8	13.04	1068.1	1081.2	0.0262	2.1164	2.1426	45
50	0.17796	0.01602	1704.8	1704.8	18.05	1065.3	1083.4	0.0361	2.0901	2.1262	50
60	0.2561	0.01603	1207.6	1207.6	28.06	1059.7	1087.7	0.0555	2.0391	2.0946	60
70	0.3629	0.01605	868.3	868.4	38.05	1054.0	1092.1	0.0745	1.9900	2.0645	70
80	0.5068	0.01607	633.3	633.3	48.04	1048.4	1096.4	0.0932	1.9426	2.0359	80
90	0.6981	0.01610	468.1	468.1	58.02	1042.7	1100.8	0.1115	1.8970	2.0086	90
100	0.9492	0.01613	350.4	350.4	68.00	1037.1	1105.1	0.1295	1.8530	1.9825	100
110	1.2750	0.01617	265.4	265.4	77.98	1031.4	1109.3	0.1472	1.8105	1.9577	110
120	1.6927	0.01620	203.25	203.26	87.97	1025.6	1113.6	0.1646	1.7693	1.9339	120
130	2.2230	0.01625	157.32	157.33	97.96	1019.8	1117.8	0.1817	1.7295	1.9112	130
140	2.8892	0.01629	122.98	123.00	107.95	1014.0	1122.0	0.1985	1.6910	1.8895	140
150	3.718	0.01634	97.05	97.07	117.95	1008.2	1126.1	0.2150	1.6536	1.8686	150
160	4.741	0.01640	77.27	77.29	127.96	1002.2	1130.2	0.2313	1.6174	1.8487	160
170	5.993	0.01645	62.04	62.06	137.97	996.2	1134.2	0.2473	1.5822	1.8295	170
180	7.511	0.01651	50.21	50.22	148.00	990.2	1138.2	0.2631	1.5480	1.8111	180
190	9.340	0.01657	40.94	40.96	158.04	984.1	1142.1	0.2787	1.5148	1.7934	190
200	11.526	0.01664	33.62	33.64	168.09	977.9	1146.0	0.2940	1.4824	1.7764	200
210	14.123	0.01671	27.80	27.82	178.15	971.6	1149.7	0.3091	1.4509	1.7600	210
212	14.696	0.01672	26.78	26.80	180.17	970.3	1150.5	0.3121	1.4447	1.7568	212
220	17.186	0.01678	23.13	23.15	188.23	965.2	1153.4	0.3241	1.4201	1.7442	220
230	20.779	0.01685	19.364	19.381	198.33	958.7	1157.1	0.3388	1.3902	1.7290	230
240	24.968	0.01693	16.304	16.321	208.45	952.1	1160.6	0.3533	1.3609	1.7142	240
250	29.825	0.01701	13.802	13.819	218.59	945.4	1164.0	0.3677	1.3323	1.7000	250
260	35.427	0.01709	11.745	11.762	228.76	938.6	1167.4	0.3819	1.3043	1.6862	260
270	41.856	0.01718	10.042	10.060	238.95	931.7	1170.6	0.3960	1.2769	1.6729	270
280	49.200	0.01726	8.627	8.644	249.17	924.6	1173.8	0.4098	1.2501	1.6599	280
290	57.550	0.01736	7.443	7.460	259.4	917.4	1176.8	0.4236	1.2238	1.6473	290
300	67.005	0.01745	6.448	6.466	269.7	910.0	1179.7	0.4372	1.1979	1.6351	300
310	77.67	0.01755	5.609	5.626	280.0	902.5	1182.5	0.4506	1.1726	1.6232	310
320	89.64	0.01766	4.896	4.914	290.4	894.8	1185.2	0.4640	1.1477	1.6116	320
340	117.99	0.01787	3.770	3.788	311.3	878.8	1190.1	0.4902	1.0990	1.5892	340
360	153.01	0.01811	2.939	2.957	332.3	862.1	1194.4	0.5161	1.0517	1.5678	360
380	195.73	0.01836	2.317	2.335	353.6	844.5	1198.0	0.5416	1.0057	1.5473	380
400	247.26	0.01864	1.8444	1.8630	375.1	825.9	1201.0	0.5667	0.9607	1.5274	400
420	308.78	0.01894	1.4808	1.4997	396.9	806.2	1203.1	0.5915	0.9165	1.5080	420
440	381.54	0.01926	1.1976	1.2169	419.0	785.4	1204.4	0.6161	0.8729	1.4890	440
460	466.9	0.0196	0.9746	0.9942	441.5	763.2	1204.8	0.6405	0.8299	1.4704	460
480	566.2	0.0200	0.7972	0.8172	464.5	739.6	1204.1	0.6648	0.7871	1.4518	480
500	680.9	0.0204	0.6545	0.6749	487.9	714.3	1202.2	0.6890	0.7443	1.4333	500
520	812.5	0.0209	0.5386	0.5596	512.0	687.0	1199.0	0.7133	0.7013	1.4146	520
540	962.8	0.0215	0.4437	0.4651	536.8	657.5	1194.3	0.7378	0.6577	1.3954	540
560	1133.4	0.0221	0.3651	0.3871	562.4	625.3	1187.7	0.7625	0.6132	1.3757	560
580	1326.2	0.0228	0.2994	0.3222	589.1	589.9	1179.0	0.7876	0.5673	1.3550	580
600	1543.2	0.0236	0.2438	0.2675	617.1	550.6	1167.7	0.8134	0.5196	1.3330	600
620	1786.9	0.0247	0.1962	0.2208	646.9	506.3	1153.2	0.8403	0.4689	1.3092	620
640	2059.9	0.0260	0.1543	0.1802	679.1	454.6	1133.7	0.8686	0.4134	1.2821	640
660	2365.7	0.0277	0.1166	0.1443	714.9	392.1	1107.0	0.8995	0.3502	1.2498	660
680	2708.6	0.0304	0.0808	0.1112	758.5	310.1	1068.5	0.9365	0.2720	1.2086	680
700	3094.3	0.0366	0.0386	0.0752	822.4	172.7	995.2	0.9901	0.1490	1.1390	700
705.5	3208.2	0.0508	0	0.0508	906.0	0	906.0	1.0612	0	1.0612	705.5

* Taken from **Steam—its generation and use**, 38th Ed., Babcock & Wilcox Co., New York, 1975 (abstracted from **Thermodynamic and Transport Properties of Steam**, ASME, New York, 1967)

APPENDICES
TABLE B.2
Properties of saturated steam and saturated water (pressure)*

Press. psia	Temp F	Volume, ft³/lb			Enthalpy, Btu/lb			Entropy, Btu/lb x F			Energy, Btu/lb		Press. psia
		Water	Evap	Steam	Water	Evap	Steam	Water	Evap	Steam	Water	Steam	
		v_f	v_{fg}	v_g	h_f	h_{fg}	h_g	s_f	s_{fg}	s_g	u_f	u_g	
0.0886	32.018	0.01602	3302.4	3302.4	0.00	1075.5	1075.5	0	2.1872	2.1872	0	1021.3	**0.0886**
0.10	35.023	0.01602	2945.5	2945.5	3.03	1073.8	1076.8	0.0061	2.1705	2.1766	3.03	1022.3	**0.10**
0.15	45.453	0.01602	2004.7	2004.7	13.50	1067.9	1081.4	0.0271	2.1140	2.1411	13.50	1025.7	**0.15**
0.20	53.160	0.01603	1526.3	1526.3	21.22	1063.5	1084.7	0.0422	2.0738	2.1160	21.22	1028.3	**0.20**
0.30	64.484	0.01604	1039.7	1039.7	32.54	1057.1	1089.7	0.0641	2.0168	2.0809	32.54	1032.0	**0.30**
0.40	72.869	0.01606	792.0	792.1	40.92	1052.4	1093.3	0.0799	1.9762	2.0562	40.92	1034.7	**0.40**
0.5	79.586	0.01607	641.5	641.5	47.62	1048.6	1096.3	0.0925	1.9446	2.0370	47.62	1036.9	**0.5**
0.6	85.218	0.01609	540.0	540.1	53.25	1045.5	1098.7	0.1028	1.9186	2.0215	53.24	1038.7	**0.6**
0.7	90.09	0.01610	466.93	466.94	58.10	1042.7	1100.8	0.3	1.8966	2.0083	58.10	1040.3	**0.7**
0.8	94.38	0.01611	411.67	411.69	62.39	1040.3	1102.6	0.1117	1.8775	1.9970	62.39	1041.7	**0.8**
0.9	98.24	0.01612	368.41	368.43	66.24	1038.1	1104.3	0.1264	1.8606	1.9870	66.24	1042.9	**0.9**
1.0	101.74	0.01614	333.59	333.60	69.73	1036.1	1105.8	0.1326	1.8455	1.9781	69.73	1044.1	**1.0**
2.0	126.07	0.01623	173.74	173.76	94.03	1022.1	1116.2	0.1750	1.7450	1.9200	94.03	1051.8	**2.0**
3.0	141.47	0.01630	118.71	118.73	109.42	1013.2	1122.6	0.2009	1.6854	1.8864	109.41	1056.7	**3.0**
4.0	152.96	0.01636	90.63	90.64	120.92	1006.4	1127.3	0.2199	1.6428	1.8626	120.90	1060.2	**4.0**
5.0	162.24	0.01641	73.515	73.53	130.20	1000.9	1131.1	0.2349	1.6094	1.8443	130.18	1063.1	**5.0**
6.0	170.05	0.01645	61.967	61.98	138.03	996.2	1134.2	0.2474	1.5820	1.8294	138.01	1065.4	**6.0**
7.0	176.84	0.01649	53.634	53.65	144.83	992.1	1136.9	0.2581	1.5587	1.8168	144.81	1067.4	**7.0**
8.0	182.86	0.01653	47.328	47.35	150.87	988.5	1139.3	0.2676	1.5384	1.8060	150.84	1069.2	**8.0**
9.0	188.27	0.01656	42.385	42.40	156.30	985.1	1141.4	0.2760	1.5204	1.7964	156.28	1070.8	**9.0**
10	193.21	0.01659	38.404	38.42	161.26	982.1	1143.3	0.2836	1.5043	1.7879	161.23	1072.3	**10**
14.696	212.00	0.01672	26.782	26.80	180.17	970.3	1150.5	0.3121	1.4447	1.7568	180.12	1077.6	**14.696**
15	213.03	0.01673	26.274	26.29	181.21	969.7	1150.9	0.3137	1.4415	1.7552	181.16	1077.9	**15**
20	227.96	0.01683	20.070	20.087	196.27	960.1	1156.3	0.3358	1.3962	1.7320	196.21	1082.0	**20**
30	250.34	0.01701	13.7266	13.744	218.9	945.2	1164.1	0.3682	1.3313	1.6995	218.8	1087.9	**30**
40	267.25	0.01715	10.4794	10.497	236.1	933.6	1169.8	0.3921	1.2844	1.6765	236.0	1092.1	**40**
50	281.02	0.01727	8.4967	8.514	250.2	923.9	1174.1	0.4112	1.2474	1.6586	250.1	1095.3	**50**
60	292.71	0.01738	7.1562	7.174	262.2	915.4	1177.6	0.4273	1.2167	1.6440	262.0	1098.0	**60**
70	302.93	0.01748	6.1875	6.205	272.7	907.8	1180.6	0.4411	1.1905	1.6316	272.5	1100.2	**70**
80	312.04	0.01757	5.4536	5.471	282.1	900.9	1183.1	0.4534	1.1675	1.6208	281.9	1102.1	**80**
90	320.28	0.01766	4.8777	4.895	290.7	894.6	1185.3	0.4643	1.1470	1.6113	290.4	1103.7	**90**
100	327.82	0.01774	4.4133	4.431	298.5	888.6	1187.2	0.4743	1.1284	1.6027	298.2	1105.2	**100**
120	341.27	0.01789	3.7097	3.728	312.6	877.8	1190.4	0.4919	1.0960	1.5879	312.2	1107.6	**120**
140	353.04	0.01803	3.2010	3.219	325.0	868.0	1193.0	0.5071	1.0681	1.5752	324.5	1109.6	**140**
160	363.55	0.01815	2.8155	2.834	336.1	859.0	1195.1	0.5206	1.0435	1.5641	335.5	1111.2	**160**
180	373.08	0.01827	2.5129	2.531	346.2	850.7	1196.9	0.5328	1.0215	1.5543	345.6	1112.5	**180**
200	381.80	0.01839	2.2689	2.287	355.5	842.8	1198.3	0.5438	1.0016	1.5454	354.8	1113.7	**200**
250	400.97	0.01865	1.8245	1.8432	376.1	825.0	1201.1	0.5679	0.9585	1.5264	375.3	1115.8	**250**
300	417.35	0.01889	1.5238	1.5427	394.0	808.9	1202.9	0.5882	0.9223	1.5105	392.9	1117.2	**300**
350	431.73	0.01913	1.3064	1.3255	409.8	794.2	1204.0	0.6059	0.8909	1.4968	408.6	1118.1	**350**
400	444.60	0.0193	1.14162	1.1610	424.2	780.4	1204.6	0.6217	0.8630	1.4847	422.7	1118.7	**400**
450	456.28	0.0195	1.01224	1.0318	437.3	767.5	1204.8	0.6360	0.8378	1.4738	435.7	1118.9	**450**
500	467.01	0.0198	0.90787	0.9276	449.5	755.1	1204.7	0.6490	0.8148	1.4639	447.7	1118.8	**500**
550	476.94	0.0199	0.82183	0.8418	460.9	743.3	1204.3	0.6611	0.7936	1.4547	458.9	1118.6	**550**
600	486.20	0.0201	0.74962	0.7698	471.7	732.0	1203.7	0.6723	0.7738	1.4461	469.5	1118.2	**600**
700	503.08	0.0205	0.63505	0.6556	491.6	710.2	1201.8	0.6928	0.7377	1.4304	488.9	1116.9	**700**
800	518.21	0.0209	0.54809	0.5690	509.8	689.6	1199.4	0.7111	0.7051	1.4163	506.7	1115.2	**800**
900	531.95	0.0212	0.47968	0.5009	526.7	669.7	1196.4	0.7279	0.6753	1.4032	523.2	1113.0	**900**
1000	544.58	0.0216	0.42436	0.4460	542.6	650.4	1192.9	0.7434	0.6476	1.3910	538.6	1110.4	**1000**
1100	556.28	0.0220	0.37863	0.4006	557.5	631.5	1189.1	0.7578	0.6216	1.3794	553.1	1107.5	**1100**
1200	567.19	0.0223	0.34013	0.3625	571.9	613.0	1184.8	0.7714	0.5969	1.3683	566.9	1104.3	**1200**
1300	577.42	0.0227	0.30722	0.3299	585.6	594.6	1180.2	0.7843	0.5733	1.3577	580.1	1100.9	**1300**
1400	587.07	0.0231	0.27871	0.3018	598.8	576.5	1175.3	0.7966	0.5507	1.3474	592.9	1097.1	**1400**
1500	596.20	0.0235	0.25372	0.2772	611.7	558.4	1170.1	0.8085	0.5288	1.3373	605.2	1093.1	**1500**
2000	635.80	0.0257	0.16266	0.1883	672.1	466.2	1138.3	0.8625	0.4256	1.2881	662.6	1068.6	**2000**
2500	668.11	0.0286	0.10209	0.1307	731.7	361.6	1093.3	0.9139	0.3206	1.2345	718.5	1032.9	**2500**
3000	695.33	0.0343	0.05073	0.0850	801.8	218.4	1020.3	0.9728	0.1891	1.1619	782.8	973.1	**3000**
3208.2	705.47	0.0508	0	0.0508	906.0	0	906.0	1.0612	0	1.0612	875.9	875.9	**3208.2**

THERMODYNAMIC PROPERTIES
TABLE B.3

Properties of superheated steam and compressed water (temp. and press.)*

Abs press. lb/sq in. (sat. temp)		Temperature, F 100	200	300	400	500	600	700	800	900	1000	1100	1200	1300	1400	1500
	v	0.0161	392.5	452.3	511.9	571.5	631.1	690.7								
1	h	68.00	1150.2	1195.7	1241.8	1288.6	1336.1	1384.5								
(101.74)	s	0.1295	2.0509	2.1152	2.1722	2.2237	2.2708	2.3144								
	v	0.0161	78.14	90.24	102.24	114.21	126.15	138.08	150.01	161.94	173.86	185.78	197.70	209.62	221.53	233.45
5	h	68.01	1148.6	1194.8	1241.3	1288.2	1335.9	1384.3	1433.6	1483.7	1534.7	1586.7	1639.6	1693.3	1748.0	1803.5
(162.24)	s	0.1295	1.8716	1.9369	1.9943	2.0460	2.0932	2.1369	2.1776	2.2159	2.2521	2.2866	2.3194	2.3509	2.3811	2.4101
	v	0.0161	38.84	44.98	51.03	57.04	63.03	69.00	74.98	80.94	86.91	92.87	98.84	104.80	110.76	116.72
10	h	68.02	1146.6	1193.7	1240.6	1287.8	1335.5	1384.0	1433.4	1483.5	1534.6	1586.6	1639.5	1693.3	1747.9	1803.4
(193.21)	s	0.1295	1.7928	1.8593	1.9173	1.9692	2.0166	2.0603	2.1011	2.1394	2.1757	2.2101	2.2430	2.2744	2.3046	2.3337
	v	0.0161	0.0166	29.899	33.963	37.985	41.986	45.978	49.964	53.946	57.926	61.905	65.882	69.858	73.833	77.807
15	h	68.04	168.09	1192.5	1239.9	1287.3	1335.2	1383.8	1433.2	1483.4	1534.5	1586.5	1639.4	1693.2	1747.8	1803.4
(213.03)	s	0.1295	0.2940	1.8134	1.8720	1.9242	1.9717	2.0155	2.0563	2.0946	2.1309	2.1653	2.1982	2.2297	2.2599	2.2890
	v	0.0161	0.0166	22.356	25.428	28.457	31.466	34.465	37.458	40.447	43.435	46.420	49.405	52.388	55.370	58.352
20	h	68.05	168.11	1191.4	1239.2	1286.9	1334.9	1383.5	1432.9	1483.2	1534.3	1586.3	1639.3	1693.1	1747.8	1803.3
(227.96)	s	0.1295	0.2940	1.7805	1.8397	1.8921	1.9397	1.9836	2.0244	2.0628	2.0991	2.1336	2.1665	2.1979	2.2282	2.2572
	v	0.0161	0.0166	11.036	12.624	14.165	15.685	17.195	18.699	20.199	21.697	23.194	24.689	26.183	27.676	29.168
40	h	68.10	168.15	1186.6	1236.4	1285.0	1333.6	1382.5	1432.1	1482.5	1533.7	1585.8	1638.8	1992.7	1747.5	1803.0
(267.25)	s	0.1295	0.2940	1.6992	1.7608	1.8143	1.8624	1.9065	1.9476	1.9860	2.0224	2.0569	2.0899	2.1224	2.1516	2.1807
	v	0.0161	0.0166	7.257	8.354	9.400	10.425	11.438	12.446	13.450	14.452	15.452	16.450	17.448	18.445	19.441
60	h	68.15	168.20	1181.6	1233.5	1283.2	1332.3	1381.5	1431.3	1481.8	1533.2	1585.3	1638.4	1692.4	1747.1	1802.8
(292.71)	s	0.1295	0.2939	1.6492	1.7134	1.7681	1.8168	1.8612	1.9024	1.9410	1.9774	2.0120	2.0450	2.0765	2.1068	2.1359
	v	0.0161	0.0166	0.0175	6.218	7.018	7.794	8.560	9.319	10.075	10.829	11.581	12.331	13.081	13.829	14.577
80	h	68.21	168.24	269.74	1230.5	1281.3	1330.9	1380.5	1430.5	1481.1	1532.6	1584.9	1638.0	1692.0	1746.8	1802.5
(312.04)	s	0.1295	0.2939	0.4371	1.6790	1.7349	1.7842	1.8289	1.8702	1.9089	1.9454	1.9800	2.0131	2.0446	2.0750	2.1041
	v	0.0161	0.0166	0.0175	4.935	5.588	6.216	6.833	7.443	8.050	8.655	9.258	9.860	10.460	11.060	11.659
100	h	68.26	168.29	269.77	1227.4	1279.3	1329.6	1379.5	1429.7	1480.4	1532.0	1584.4	1637.6	1691.6	1746.5	1802.2
(327.82)	s	0.1295	0.2939	0.4371	1.6516	1.7088	1.7586	1.8036	1.8451	1.8839	1.9205	1.9552	1.9883	2.0199	2.0502	2.0794
	v	0.0161	0.0166	0.0175	4.0786	4.6341	5.1637	5.6831	6.1928	6.7006	7.2060	7.7096	8.2119	8.7130	9.2134	9.7130
120	h	68.31	168.33	269.81	1224.1	1277.4	1328.1	1378.4	1428.8	1479.8	1531.4	1583.9	1637.1	1691.3	1746.2	1802.0
(341.27)	s	0.1295	0.2939	0.4371	1.6286	1.6872	1.7376	1.7829	1.8246	1.8635	1.9001	1.9349	1.9680	1.9996	2.0300	2.0592
	v	0.0161	0.0166	0.0175	3.4661	3.9526	4.4119	4.8585	5.2995	5.7364	6.1709	6.6036	7.0349	7.4652	7.8946	8.3233
140	h	68.37	168.38	269.85	1220.8	1275.3	1326.8	1377.4	1428.0	1479.1	1530.8	1583.4	1636.7	1690.9	1745.9	1801.7
(353.04)	s	0.1295	0.2939	0.4370	1.6085	1.6686	1.7196	1.7652	1.8071	1.8461	1.8828	1.9176	1.9508	1.9825	2.0129	2.0421
	v	0.0161	0.0166	0.0175	3.0060	3.4413	3.8480	4.2420	4.6295	5.0132	5.3945	5.7741	6.1522	6.5293	6.9055	7.2811
160	h	68.42	168.42	269.89	1217.4	1273.3	1325.4	1376.4	1427.2	1478.4	1530.3	1582.9	1636.3	1690.5	1745.6	1801.4
(363.55)	s	0.1294	0.2938	0.4370	1.5906	1.6522	1.7039	1.7499	1.7919	1.8310	1.8678	1.9027	1.9359	1.9676	1.9980	2.0273
	v	0.0161	0.0166	0.0174	2.6474	3.0433	3.4093	3.7621	4.1084	4.4505	4.7907	5.1289	5.4657	5.8014	6.1363	6.4704
180	h	68.47	168.47	269.92	1213.8	1271.2	1324.0	1375.3	1426.3	1477.7	1529.7	1582.4	1635.9	1690.2	1745.3	1801.2
(373.08)	s	0.1294	0.2938	0.4370	1.5743	1.6376	1.6900	1.7362	1.7784	1.8176	1.8545	1.8894	1.9227	1.9545	1.9849	2.0142
	v	0.0161	0.0166	0.0174	2.3598	2.7247	3.0583	3.3783	3.6915	4.0008	4.3077	4.6128	4.9165	5.2191	5.5209	5.8219
200	h	68.52	168.51	269.96	1210.1	1269.0	1322.6	1374.3	1425.5	1477.0	1529.1	1581.9	1635.4	1689.8	1745.0	1800.9
(381.80)	s	0.1294	0.2938	0.4369	1.5593	1.6242	1.6776	1.7239	1.7663	1.8057	1.8426	1.8776	1.9109	1.9427	1.9732	2.0025
	v	0.0161	0.0166	0.0174	0.0186	2.1504	2.4662	2.6872	2.9410	3.1909	3.4382	3.6837	3.9278	4.1709	4.4131	4.6546
250	h	68.66	168.63	270.05	375.10	1263.5	1319.0	1371.6	1423.4	1475.3	1527.6	1580.6	1634.4	1688.9	1744.2	1800.2
(400.97)	s	0.1294	0.2937	0.4368	0.5667	1.5951	1.6502	1.6976	1.7405	1.7801	1.8173	1.8524	1.8858	1.9177	1.9482	1.9776
	v	0.0161	0.0166	0.0174	0.0186	1.7665	2.0044	2.2263	2.4407	2.6509	2.8585	3.0643	3.2688	3.4721	3.6746	3.8764
300	h	68.79	168.74	270.14	375.15	1257.7	1315.2	1368.9	1421.3	1473.6	1526.2	1579.4	1633.3	1688.0	1743.4	1799.6
(417.35)	s	0.1294	0.2937	0.4307	0.5665	1.5703	1.6274	1.6758	1.7192	1.7591	1.7964	1.8317	1.8652	1.8972	1.9278	1.9572
	v	0.0161	0.0166	0.0174	0.0186	1.4913	1.7028	1.8970	2.0832	2.2652	2.4445	2.6219	2.7980	2.9730	3.1471	3.3205
350	h	68.92	168.85	270.24	375.21	1251.5	1311.4	1366.2	1419.2	1471.8	1524.7	1578.2	1632.3	1687.1	1742.6	1798.9
(431.73)	s	0.1293	0.2936	0.4367	0.5664	1.5483	1.6077	1.6571	1.7009	1.7411	1.7787	1.8141	1.8477	1.8798	1.9105	1.9400
	v	0.0161	0.0166	0.0174	0.0162	1.2841	1.4763	1.6499	1.8151	1.9759	2.1339	2.2901	2.4450	2.5987	2.7515	2.9037
400	h	69.05	168.97	270.33	375.27	1245.1	1307.4	1363.4	1417.0	1470.1	1523.3	1576.9	1631.2	1686.2	1741.9	1798.2
(444.60)	s	0.1293	0.2935	0.4366	0.5663	1.5282	1.5901	1.6406	1.6850	1.7255	1.7632	1.7988	1.8325	1.8647	1.8955	1.9250
	v	0.0161	0.0166	0.0174	0.0186	0.9919	1.1584	1.3037	1.4397	1.5708	1.6992	1.8256	1.9507	2.0746	2.1977	2.3200
500	h	69.32	169.19	270.51	375.38	1231.2	1299.1	1357.7	1412.7	1466.6	1520.3	1574.4	1629.1	1684.4	1740.3	1796.9
(467.01)	s	0.1292	0.2934	0.4364	0.5660	1.4921	1.5595	1.6123	1.6578	1.6990	1.7371	1.7730	1.8069	1.8393	1.8702	1.8998

APPENDICES

TABLE B.3 [cont.]

Properties of superheated steam and compressed water (temp. and press.)

Abs press. lb/sq in. (sat. temp)		Temperature, F														
		100	200	300	400	500	600	700	800	900	1000	1100	1200	1300	1400	1500
600	v	0.0161	0.0166	0.0174	0.0186	0.7944	0.9456	1.0726	1.1892	1.3008	1.4093	1.5160	1.6211	1.7252	1.8284	1.9309
	h	69.58	169.42	270.70	375.49	1215.9	1290.3	1351.8	1408.3	1463.0	1517.4	1571.9	1627.0	1682.6	1738.8	1795.6
(486.20)	s	0.1292	0.2933	0.4362	0.5657	1.4590	1.5329	1.5844	1.6351	1.6769	1.7155	1.7517	1.7859	1.8184	1.8494	1.8792
700	v	0.0161	0.0166	0.0174	0.0186	0.0204	0.7928	0.9072	1.0102	1.1078	1.2023	1.2948	1.3858	1.4757	1.5647	1.6530
	h	69.84	169.65	270.89	375.61	487.93	1281.0	1345.6	1403.7	1459.4	1514.4	1569.4	1624.8	1680.7	1737.2	1794.3
(503.08)	s	0.1291	0.2932	0.4360	0.5655	0.6889	1.5090	1.5673	1.6154	1.6580	1.6970	1.7335	1.7679	1.8006	1.8318	1.8617
800	v	0.0161	0.0166	0.0174	0.0186	0.0204	0.6774	0.7828	0.8759	0.9631	1.0470	1.1289	1.2093	1.2885	1.3669	1.4446
	h	70.11	169.88	271.07	375.73	487.88	1271.1	1339.2	1399.1	1455.8	1511.4	1566.9	1622.7	1678.9	1735.0	1792.9
(518.21)	s	0.1290	0.2930	0.4358	0.5652	0.6885	1.4869	1.5484	1.5980	1.6413	1.6807	1.7175	1.7522	1.7851	1.8164	1.8464
900	v	0.0161	0.0166	0.0174	0.0186	0.0204	0.5869	0.6858	0.7713	0.8504	0.9262	0.9998	1.0720	1.1430	1.2131	1.2825
	h	70.37	170.10	271.26	375.84	487.83	1260.6	1332.7	1394.4	1452.2	1508.5	1564.4	1620.6	1677.1	1734.1	1791.6
(531.95)	s	0.1290	0.2929	0.4357	0.5649	0.6881	1.4659	1.5311	1.5822	1.6263	1.6662	1.7033	1.7382	1.7713	1.8028	1.8329
1000	v	0.0161	0.0166	0.0174	0.0186	0.0204	0.5137	0.6080	0.6875	0.7603	0.8295	0.8966	0.9622	1.0266	1.0901	1.1529
	h	70.63	170.33	271.44	375.96	487.79	1249.3	1325.9	1389.6	1448.5	1504.4	1561.9	1618.4	1675.3	1732.5	1790.3
(544.58)	s	0.1289	0.2928	0.4355	0.5647	0.6876	1.4457	1.5149	1.5677	1.6126	1.6530	1.6905	1.7256	1.7589	1.7905	1.8207
1100	v	0.0161	0.0166	0.0174	0.0185	0.0203	0.4531	0.5440	0.6188	0.6865	0.7505	0.8121	0.8723	0.9313	0.9894	1.0468
	h	70.90	170.56	271.63	376.08	487.75	1237.3	1318.8	1384.7	1444.7	1502.4	1559.4	1616.3	1673.5	1731.0	1789.0
(556.28)	s	0.1289	0.2927	0.4353	0.5644	0.6872	1.4259	1.4996	1.5542	1.6000	1.6410	1.6787	1.7141	1.7475	1.7793	1.8097
1200	v	0.0161	0.0166	0.0174	0.0185	0.0203	0.4016	0.4905	0.5615	0.6250	0.6845	0.7418	0.7974	0.8519	0.9055	0.9584
	h	71.16	170.78	271.82	376.20	487.72	1224.2	1311.5	1379.7	1440.9	1499.4	1556.9	1614.2	1671.6	1729.4	1787.6
(567.19)	s	0.1288	0.2926	0.4351	0.5642	0.6868	1.4061	1.4851	1.5415	1.5883	1.6298	1.6679	1.7035	1.7371	1.7691	1.7996
1400	v	0.0161	0.0166	0.0174	0.0185	0.0203	0.3176	0.4059	0.4712	0.5282	0.5809	0.6311	0.6798	0.7272	0.7737	0.8195
	h	71.68	171.24	272.19	376.44	487.65	1194.1	1296.1	1369.3	1433.2	1493.2	1551.8	1609.9	1668.0	1726.3	1785.0
(587.07)	s	0.1287	0.2923	0.4348	0.5636	0.6859	1.3652	1.4575	1.5182	1.5670	1.6096	1.6484	1.6845	1.7185	1.7508	1.7815
1600	v	0.0161	0.0166	0.0173	0.0185	0.0202	0.0236	0.3415	0.4032	0.4555	0.5031	0.5482	0.5915	0.6336	0.6748	0.7153
	h	72.21	171.69	272.57	376.69	487.60	616.77	1279.4	1358.5	1425.2	1486.9	1546.6	1605.6	1664.3	1723.2	1782.3
(604.87)	s	0.1286	0.2921	0.4344	0.5631	0.6851	0.8129	1.4312	1.4968	1.5478	1.5916	1.6312	1.6678	1.7022	1.7344	1.7657
1800	v	0.0160	0.0165	0.0173	0.0185	0.0202	0.0235	0.2906	0.3500	0.3988	0.4426	0.4836	0.5229	0.5609	0.5980	0.6343
	h	72.73	172.15	272.95	376.93	487.56	615.58	1261.1	1347.2	1417.1	1480.6	1541.1	1601.2	1660.7	1720.1	1779.7
(621.02)	s	0.1284	0.2918	0.4341	0.5626	0.6843	0.8109	1.4054	1.4768	1.5302	1.5753	1.6156	1.6528	1.6876	1.7204	1.7516
2000	v	0.0160	0.0165	0.0173	0.0184	0.0201	0.0233	0.2488	0.3072	0.3534	0.3942	0.4320	0.4680	0.5027	0.5365	0.5695
	h	73.26	172.60	273.32	377.19	487.53	614.48	1240.9	1353.4	1408.7	1474.1	1536.2	1596.9	1657.0	1717.0	1777.1
(635.80)	s	0.1283	0.2916	0.4337	0.5621	0.6834	0.8091	1.3794	1.4578	1.5138	1.5603	1.6014	1.6391	1.6743	1.7075	1.7389
2500	v	0.0160	0.0165	0.0173	0.0184	0.0200	0.0230	0.1681	0.2293	0.2712	0.3068	0.3390	0.3692	0.3980	0.4259	0.4529
	h	74.57	173.74	274.27	377.82	487.50	612.08	1176.7	1303.4	1386.7	1457.5	1522.9	1585.9	1647.8	1709.2	1770.4
(668.11)	s	0.1280	0.2910	0.4329	0.5609	0.6815	0.8048	1.3076	1.4129	1.4766	1.5269	1.5703	1.6094	1.6456	1.6796	1.7116
3000	v	0.0160	0.0165	0.0172	0.0183	0.0200	0.0228	0.0982	0.1759	0.2161	0.2484	0.2770	0.3033	0.3282	0.3522	0.3753
	h	75.88	174.88	275.22	378.47	487.52	610.08	1060.5	1267.0	1363.2	1440.2	1509.4	1574.8	1638.5	1701.4	1761.8
(695.33)	s	0.1277	0.2904	0.4320	0.5597	0.6796	0.8009	1.1966	1.3692	1.4429	1.4976	1.5434	1.5841	1.6214	1.6561	1.6888
3200	v	0.0160	0.0165	0.0172	0.0183	0.0199	0.0227	0.0335	0.1588	0.1987	0.2301	0.2576	0.2827	0.3065	0.3291	0.3510
	h	76.4	175.3	275.6	378.7	487.5	609.4	800.8	1250.9	1353.4	1433.1	1503.8	1570.3	1634.8	1698.3	1761.2
(705.08)	s	0.1276	0.2902	0.4317	0.5592	0.6788	0.7994	0.9708	1.3515	1.4300	1.4866	1.5335	1.5749	1.6126	1.6477	1.6806
3500	v	0.0160	0.0164	0.0172	0.0183	0.0199	0.0225	0.0307	0.1364	0.1764	0.2066	0.2326	0.2563	0.2784	0.2995	0.3198
	h	77.2	176.0	276.2	379.1	487.6	608.4	779.4	1224.6	1338.2	1422.2	1495.5	1563.3	1629.2	1693.6	1757.2
	s	0.1274	0.2899	0.4312	0.5585	0.6777	0.7973	0.9508	1.3242	1.4112	1.4709	1.5194	1.5618	1.6002	1.6358	1.6691
4000	v	0.0159	0.0164	0.0172	0.0182	0.0198	0.0223	0.0287	0.1052	0.1463	0.1752	0.1994	0.2210	0.2411	0.2601	0.2783
	h	78.5	177.2	277.1	379.8	487.7	606.9	763.0	1174.3	1311.6	1403.6	1481.3	1552.2	1619.8	1685.7	1750.6
	s	0.1271	0.2893	0.4304	0.5573	0.6760	0.7940	0.9343	1.2754	1.3807	1.4461	1.4976	1.5417	1.5812	1.6177	1.6516
5000	v	0.0159	0.0164	0.0171	0.0181	0.0196	0.0219	0.0268	0.0591	0.1038	0.1312	0.1529	0.1718	0.1890	0.2050	0.2203
	h	81.1	179.5	279.1	381.2	488.1	604.6	746.0	1042.9	1252.9	1364.6	1452.1	1529.1	1600.9	1670.0	1737.4
	s	0.1265	0.2881	0.4287	0.5550	0.6726	0.7880	0.9153	1.1593	1.3207	1.4001	1.4582	1.5061	1.5481	1.5863	1.6216
6000	v	0.0159	0.0163	0.0170	0.0180	0.0195	0.0216	0.0256	0.0397	0.0757	0.1020	0.1221	0.1391	0.1544	0.1684	0.1817
	h	83.7	181.7	281.0	382.7	488.6	602.9	736.1	945.1	1188.8	1323.6	1422.3	1505.9	1582.0	1654.2	1724.2
	s	0.1258	0.2870	0.4271	0.5528	0.6693	0.7826	0.9026	1.0176	1.2615	1.3574	1.4229	1.4748	1.5194	1.5593	1.5962
7000	v	0.0158	0.0163	0.0170	0.0180	0.0193	0.0213	0.0248	0.0334	0.0573	0.0816	0.1004	0.1160	0.1298	0.1424	0.1542
	h	86.2	184.4	283.0	384.2	489.3	601.7	729.3	901.8	1124.9	1281.7	1392.2	1482.6	1563.1	1638.6	1711.1
	s	0.1252	0.2859	0.4256	0.5507	0.6663	0.7777	0.8926	1.0350	1.2055	1.3171	1.3904	1.4466	1.4938	1.5355	1.5735

THERMODYNAMIC PROPERTIES

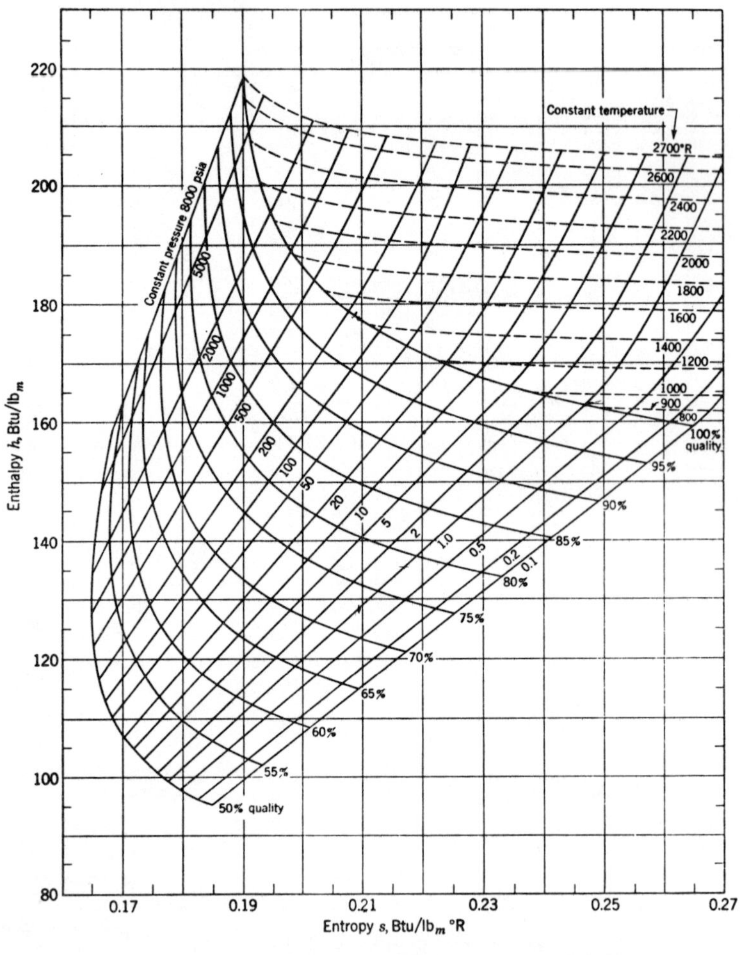

Fig. B.1. Mollier diagram for mercury (taken from El-Wakil, M. M., **Nuclear Power Engineering**, McGraw-Hill, New York, 1962)

APPENDICES

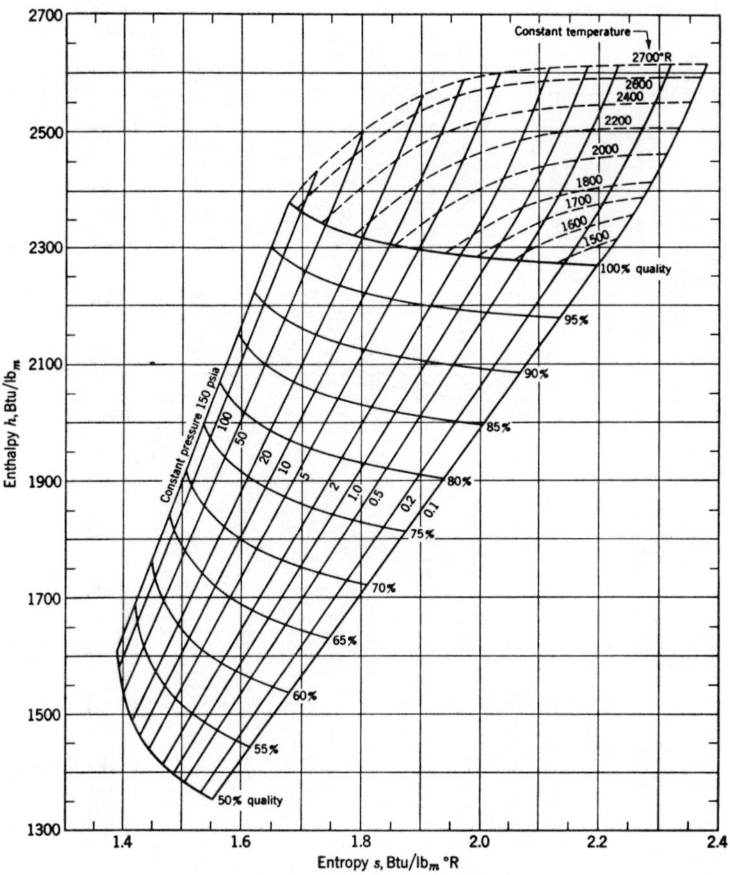

Fig. B.2. Mollier diagram for sodium (taken from El-Wakil, M. M., **Nuclear Power Engineering**, McGraw-Hill, New York, 1962)

APPENDIX C
PROPERTIES OF REACTOR COOLANTS

TABLE C.1
Properties of light liquid water*

Temp. °F	Thermal Conductivity Btu/hr-ft-°F			Specific Heat Btu/lb_m-°F			Dynamic Viscosity lb_m/ft-hr		
	sat. liquid	1000 psia	2000 psia	sat. liquid	1000 psia	2000 psia	sat. liquid	1000 psia	2000 psia
32	0.3185	0.3198	0.3211	1.0083	1.0032	1.0004	4.340	4.309	4.279
50	0.3319	0.3337	0.3356	1.0004	0.9991	0.9963	3.169	3.154	3.140
100	0.3641	0.3659	0.3680	0.9976	0.9932	0.9897	1.650	1.654	1.658
150	0.3836	0.3855	0.3875	0.9996	0.9950	0.9905	1.050	1.059	1.068
200	0.3935	0.3957	0.3980	1.0047	1.0008	0.9958	0.738	0.748	0.757
250	0.3964	0.3992	0.4021	1.0142	1.0095	1.0041	0.568	0.578	0.587
300	0.3952	0.3992	0.4021	1.0289	1.0232	1.0166	0.452	0.460	0.468
350	0.3906	0.3933	0.3965	1.0564	1.0496	1.0411	0.378	0.384	0.390
400	0.3809	0.3840	0.3880	1.0794	1.074	1.062	0.327	0.330	0.335
450	0.3670	0.3697	0.3745	1.1212	1.114	1.100	0.287	0.289	0.293
500	0.3494	0.3510	0.3562	1.1861	1.176	1.154	0.256	0.257	0.260
550	0.3246		0.3318	1.31		1.252	0.230		0.235
600	0.2919		0.2962	1.51		1.448	0.210		0.212
650	0.2454			2.09			0.183		

* Taken from Tipton, C. R. (ed.), **The Reactor Handbook**. Vol. I. **Materials**, Interscience Publishers, Inc., New York, 1960.

Note: The density of water is given in Tables A.1, A.2, and A.3 in Appendix A

TABLE C.2
Properties of heavy liquid water*

Temp. °C	Density g/cm^3	Specific Heat Btu/lb$_m$-°F	Thermal Conductivity $k(D_2O)/k(H_2O)$	Viscosity lb$_m$/ft–hr
10	1.1060	1.009	0.984	4.074
20	1.1054	1.006	0.9806	3.048
30	1.1032	1.005	0.9774	2.352
40	1.1000	1.004	0.974	
60	1.0906		0.9676	1.34
80	1.0782		0.9613	
100	1.0630		0.955	0.781
150	1.0167			0.503
200	0.957			0.365
250	0.881			0.300

* Taken from Tifton, C. R. (ed.), **The Reactor Handbook**. Vol. I. **Materials**, Interscience Publishers, Inc., New York, 1960.

TABLE C.3
Surface tension of liquid water*

Temp. °C	light water lb_f/ft	heavy water lb_f/ft
99.0		4.01×10^{-3}
101.8	4.01×10^{-3}	
110.8	3.87×10^{-3}	3.84×10^{-3}
124.0		3.69×10^{-3}
128.2	3.64×10^{-3}	
145.6		3.34×10^{-3}
148.8	3.35×10^{-3}	
174.5	2.96×10^{-3}	2.91×10^{-3}
193.8		2.61×10^{-3}
201.0	2.57×10^{-3}	
216.0		2.25×10^{-3}
224.4	2.20×10^{-3}	

* Taken from Tifton, C. R. (ed.), **The Reactor Handbook**. Vol. I. **Materials**, Interscience Publishers, Inc., New York, 1960.

TABLE C.4
Properties of sodium*

Temp. °C	Density lb_m/ft^3	Specific Heat Btu/lb_m-°F	Thermal Conductivity Btu/hr-ft-°F	Viscosity lb_m/ft-hr
100	57.9	0.3305	49.7	1.66
168				1.22
200		0.3200	47.1	
250	55.6			0.92
300		0.3116	43.8	
400	53.3	0.3055	41.1	0.65
500		0.3015	38.6	
600		0.2998		
700	48.7	0.3003		0.44

* Melting point is 208°F and the boiling at atmospheric pressure is 1621°F. Property values taken from Lyon, R. N., Liquid-metals handbook, NAVEXOS P-733(Rev.), U.S.A.E.C. (June, 1962) and Tifton, C. R. (ed.), **The Reactor Handbook**. Vol. I. **Materials**, Interscience Publishers, Inc., New York, 1960.

TABLE C.5
Properties of helium (pressure = 50 bar)*

Temp. °C	Density lb_m/ft^3	Specific Heat Btu/lb_m-°F	Thermal Conductivity Btu/hr-ft-°F	Viscosity lb_m/ft-hr
100	0.0802	1.248	0.104	0.0555
200	0.0633	1.248	0.123	0.0650
300	0.0523	1.248	0.141	0.0740
400	0.0446	1.248	0.158	0.0825
500	0.0388	1.248	0.175	0.0906
600	0.0344	1.248	0.190	0.0984
700	0.0308	1.248	0.204	0.106
800	0.0280	1.248	0.218	0.113
900	0.0256	1.248	0.230	0.120
1000	0.0236	1.248	0.242	0.127

* Taken from Tsederberg, N. V., et al., **Thermodynamic and Thermophysical Properties of Helium**, Israel Program for Scientific Translations, Jerusalem, 1971.

APPENDIX D
PROPERTIES OF REACTOR MATERIALS

TABLE D.1
Thermal-physical properties of reactor materials*

Material	Temp.	Density lb_m/ft^3	Specific Heat Btu/lb_m-°F	Thermal Conductivity Btu/hr-ft-°F	Coeff. of Linear Thermal Expansion per °F x 10^6 †	Young's Modulus 10^6 psi	Yield Strength psi	Poisson's Ratio
Aluminum (1100 or 2S)	75°F	169		122(1)	-	10	5000	0.33
(1) is for pure aluminum	212	-	0.225(1)	122(1)	13.1	9.7	4500	-
	300	-		-	-	9.5	3500	-
	400	-		128(1)	13.7	9.0	3000	-
	500	-		-	-	8.5	2000	-
	600	-	0.248(1)	-	14.3	6.5	1500	-
	700	-		131(1)	-	4.0	1000	-
Graphite (approximate values because properties vary upon choice of material and method of forming)	25°C	100–119	0.172	-	-	0.7–1.7	-	0.25
	100	-	0.218	85	-	-	-	-
	250	-	0.304	-	6.5	-	-	-
	300	-	0.324	70	-	-	-	-
	500	-	0.377	51	7.5	-	-	-
	700	-	0.425	41	-	-	-	-
	750	-	0.431	-	8.5	-	-	-
	900	-	0.449	34	-	-	-	-
	1000	-	0.460	-	9	-	-	-
	1100	-	0.469	29	-	-	-	-
	1250	-	0.482	-	9.5	-	-	-

TABLE D.1. Thermal-physical properties of reactor materials (continued)

Material	Temp.	Density lb_m/ft^3	Specific Heat Btu/lb_m-°F	Thermal Conductivity Btu/hr-ft-°F		Coeff. of Linear Thermal Expansion per °F x 10^6 †	Young's Modulus 10^6 psi	Yield Strength psi	Poisson's Ratio
Stainless Steel (Type 347)	68°F	498	0.12	-		-	-	-	-
	212	-	-	9.0		9.2	-	-	-
	300	-	-	-		-	27.5	34000	0.292
	500	-	-	-		-	26.1	32000	0.300
	800	-	-	-		-	24.1	31800	0.312
	900	-	-	12.5		-	23.4	31500	0.316
	1000	-	-	-		-	22.8	-	0.320
	1100	-	-	-		10.0	22.0	28500	0.324
	1200	-	-	-		-	21.4	25500	0.328
Uranium (tensile properties approximate because values depend upon method of manufacture)	80°F	1189	0.028	16.0	Temp.				
	260	1184	0.029	16.4	212°F	8.1	27	30000	0.2
	440	1179	0.032	16.9	392	8.6	26	27000	0.21
	620	1172	0.034	18.1	572	9.1	24	22000	0.23
	800	1166	0.037	19.6	752	9.6	-	17000	-
	980	1157	0.041	21.5	932	10.2	-	12000	-
	1160	1148	0.045	23.7	1132	10.9	-	7000	-

TABLE D.1. Thermal-physical properties of reactor materials (continued)

Material	Temp.	Density lb_m/ft^3	Specific Heat Btu/lb_m-°F	Thermal Conductivity Btu/hr-ft-°F	Coeff. of Linear Thermal Expansion per °F x 10^6 †	Young's Modulus 10^6 psi	Yield Strength psi	Poisson's Ratio
Uranium Dioxide (2) theoretical density	25°C	684(2)	0.056	Consult pages 302 or 305	-	27	-	-
	200	-	0.067		4.0	-	-	-
	400	-	0.072		4.7	-	-	-
	600	-	0.074		5.4	-	-	-
	800	-	0.076		6.2	-	-	-
	1000	-	0.077		6.9	28	-	-
	1500	-	0.081		-	-	-	-
Zircaloy-2	100°F	409	-	6.82	3.3	14.1	66000(20°C)	0.35
	200	-	0.0728	6.89	3.5	13.6	48000(150°C)	-
	400	-	0.0764	7.11	3.9	12.5	40000(250°C)	-
	600	-	0.0789	7.37	4.2	11.5	37000(350°C)	-
	800	-	0.081	7.64	4.6	10.5	27000(500°C)	-
	1000	-	0.0831	-	5.0	9.5	-	-
	1200	-	0.0853	-	5.4	8.5	-	-

† Average value from 68°F to indicated temperatures

* Taken from Tifton, C. R. (ed.), **The Reactor Handbook**. Vol. I. **Materials**, Interscience Publishers, Inc., New York, 1960.

INDEX